W9-CPO-167

To

RUTH JEAN and LOUISE

TABLE OF CONTENTS

7. REGRESSION ANALYSIS

8. CORRELATION ANALYSIS

9. DESIGN OF EXPERIMENTAL INVESTIGATIONS

APPENDIXES

PREFACE

The goals of this revised edition remain the same as they were for the second edition, namely, (1) to provide a book giving statistical methods that have been found useful by workers in many areas of scientific research, (2) to present these methods as integral parts of a complete discipline, and (3) to provide a textbook that will facilitate teaching the science of statistics.

Since this science has its basis in probability, we have expanded Chapter 2 on probability to make the book more self-contained. Thus this edition is usable not only as a text on statistical methods but also as a text for a combined course on probability and statistics. However, the expanded material on probability does not lessen the book's usefulness as a teaching and reference text on statistical methods. Readers who have had prior exposure to probability or who do not feel the need to study probability in depth before going into statistical methods may use Chapter 2 as reference material and move directly to Chapter 3 on sampling and descriptive statistics.

This edition also recognizes more fully the availability of large-scale computers and packaged computer programs that can be used to perform many of the computations associated with statistical analyses. This is particularly true in the chapters on regression analysis (Chapter 7) and the analysis of variance (Chapters 10–13). Although we are cognizant of the usefulness of computers, we also recognize the value of the computing formulas for calculating the needed sums of squares. Thus we have retained the classical analysis of variance tables, while indicating how the same analysis can be achieved using standard regression techniques. We feel that this approach will help the reader recognize that analysis of variance is a meaningful specialized regression analysis.

As in the previous editions, considerable attention is given to the assumptions underlying the techniques presented. Without a thorough understanding of the limitations of various techniques, one might apply them in situations where they should not be used. The learning of methods is easy; learning when and where to use them is not so simple. We have attempted to achieve a reasonable balance between these two ends.

This edition should continue to prove useful as a text for a course in statistical methods, a text for an integrated course in probability and statistics for students in engineering and the physical sciences, and a reference book for research workers and other users of statistical methods.

We are indebted to Sir Ronald A. Fisher, to Dr. Frank Yates, and to Oliver and Boyd Ltd., Edinburgh, for permission to reprint Table III from *Statistical Tables for Biological, Agricultural and Medical Re-*

search. We are also indebted to Dr. O.L. Davies and to Oliver and Boyd Ltd., Edinburgh, for permission to reprint Table 6.G1 from the second edition of *Statistical Methods in Research and Production,* and Tables 7.7, 7.72, E, E.1, G, and H from the second edition of *The Design and Analysis of Industrial Experiments.* Many other persons have also graciously given permission for the reproduction of published material, and acknowledgment has been made at the appropriate places in the text. The authors are deeply appreciative and wish to express thanks to these individuals for their cooperation.

We again acknowledge the encouragement of many of our colleagues during the preparation of this and previous editions. Particular acknowledgment is due Paul G. Homeyer, David V. Huntsberger, Emil H. Jebe, Oscar Kempthorne, George W. Snedecor, and John M. Wiesen for their encouragement during the preparation of the first and second editions. We also wish to express our appreciation to Glen Meeden and Paul N. Somerville, who read some of the material and made several valuable suggestions during the writing of this edition. Thanks are also due Mrs. Nancy Bohlen for her very valuable editorial assistance.

Our greatest personal indebtedness is to our wives, Ruth Jean Ostle and Louise Mensing, not only for their editorial assistance and typing of this and previous editions but, above all, for their unfailing patience and understanding.

STATISTICS IN RESEARCH

CHAPTER 1

THE ROLE OF STATISTICS IN RESEARCH

Every day each of us engages in some observation in which statistics is used. Such common events as noting the weather forecast, weighing oneself, checking the position of a favorite ball team in its league, or testing a new food product are typical. The element of statistics creeps in when you mentally *evaluate* your research. In weighing yourself, you automatically compare your observation with your average weight (deviation from the mean) and conclude the present weight is usual (no significance to the difference) or unusual (a significant difference), basing your judgment upon previous measurements of your weight and your knowledge of the variation generally observed. These common results are easily obtained, are of only local importance, and are soon forgotten. However, the formal research that means so much to improving man's lot is of infinitely greater importance and must be conducted with much greater care. This book is concerned with the latter type of research.

1.1 THE NATURE AND PURPOSE OF RESEARCH

Research, according to Webster, is studious inquiry or examination, i.e., critical and exhaustive investigation or experimentation having for its aim the discovery of new facts and their correct interpretation. It also aims at revising accepted conclusions, theories, or laws in the light of newly discovered facts or the practical applications of such new or revised conclusions. Research therefore means continued search for knowledge and understanding; scientific research is continued research using scientific methods. Scientific research is essentially compounded of two elements: *observation,* by which knowledge of certain facts is obtained through sense perception; and *reasoning,* by which the meaning of these facts, their interrelation, and their relation to the existing body of knowledge are ascertained insofar as the present state of knowledge and the investigator's ability permit.

In any discussion of research, two important facts should be noted. They are: (1) there is an ever-increasing trend toward extreme specialization on the part of individual scientists, and (2) most research problems are such that many disciplines and fields of specialization can contribute in a significant manner to their solutions. Thus it is evident that more and more research will be handled on an interdisciplinary team basis rather than by individual scientists working in "solitary confinement." (**Note:** This is not to say that very little individual research will continue to be performed. Such research always should and always will be performed. The statement was meant to imply only that research is becoming predominantly a team or cooperative effort.)

In summary, research is an inquiry into the *nature of, the reasons for,* and the *consequences of* any particular set of circumstances, whether these circumstances are experimentally controlled or recorded just as they occur. Further, research implies the researcher is interested in more than particular results; he is interested in the repeatability of the results and in their extension to more complicated and general situations.

1.2 RESEARCH AND SCIENTIFIC METHOD

Although the techniques of investigation may vary considerably from one science to another, the philosophy common to all is generally referred to as *scientific method*. There are, perhaps, as many definitions of scientific method as there are workers in research. For our purposes, the following will be used: *since the ideal of science is to achieve a systematic interrelation of facts, scientific method must be a pursuit of this ideal by experimentation, observation, logical arguments from accepted postulates, and a combination of these three in varying proportions.* Therefore, research and scientific method are closely related, if not one and the same thing.

1.3 WHAT IS STATISTICS?

Statistics has often been classified as a method of research along with or in opposition to such methods as case studies, the historical approach, and the experimental method. Since this classification frequently leads to confused and incorrect thinking, it is not wise. It is better to regard statistics as supplying a kit of tools that can be extremely valuable in research. This book will stress gaining an understanding of these tools and learning which tool should be used in various situations arising in scientific research. Only when you know which tool to use, how to use it, and how to interpret your results can you hope to do productive research. To summarize: the science of statistics has much to offer the research worker in planning, analyzing, and interpreting the results of his investigations; and this book is devoted to an exposition of methods and techniques that have proved useful in many fields of inquiry.

As is the case with many words in the English language, the word *statistics* is used in a variety of ways, each correct in its own sphere. In the plural sense it is usually taken to be synonymous with *data*. However, the statistician recognizes another meaning of the word. This meaning is the plural of the word *statistic,* which refers to a quantity calculated from sample observations. (These terms will be defined in considerable detail in later chapters.) In the singular sense statistics is a science, and the word will be employed in this sense most frequently in this book. The *science of statistics* deals with:

1. Collecting and summarizing data.
2. Designing experiments and surveys.
3. Measuring the magnitude of variation in both experimental and survey data.
4. Estimating population parameters and providing various measures of the accuracy and precision of these estimates.
5. Testing hypotheses about populations.
6. Studying relationships among two or more variables.

1.4 STATISTICS AND RESEARCH

As indicated in the preceding section, statistics enters into research and/or scientific method through experimentation and observation. That is, experimental and survey investigations are integral parts of scientific method, and these procedures invariably lead to the use of statistical techniques. Since statistics when properly used makes for more efficient research, it is recommended that all researchers become familiar with the basic concepts and techniques of this useful science.

Because statistics is such a valuable tool for the researcher, it is sometimes overworked. That is, in many cases statistics is used as a crutch for poorly conceived and/or executed research. In addition, in some cases statistics is employed in good faith, but unfortunately insufficient attention is paid to the assumptions required for a valid use of the methods employed. For these and other reasons it is essential that the user of statistics clearly understands the techniques he employs. Consequently, in this book careful attention will be given to both the methods and the underlying assumptions in the hope that such an approach will lead to the proper application and use of statistics in scientific research.

1.5 FURTHER REMARKS ON SCIENCE, SCIENTIFIC METHOD, AND STATISTICS

In the preceding sections, your attention has been called to the close connection that statistics has with experimentation, scientific method, and research. However, in each case, the discussion was quite brief. Because these various topics and their interrelationships are so important to the remainder of this book, a few additional remarks are justified. To expedite the discussion, the following questions and answers have been devised.

What Is Logic? Logic deals with the relation of implication among propositions, i.e., the relation between premises and conclusions. In scientific method, logic aids in formulating our propositions explicitly and accurately so that their possible alternatives become clear. When faced with alternative hypotheses, logic develops their consequences so that when these consequences are compared with observable phenomena, we have a means of testing which hypotheses are to be eliminated and which one is most in harmony with the observed facts.

What Is Science? Science is knowledge that is general and systematic, knowledge from which specific propositions are deduced in accordance with a few general principles. Although all the sciences differ, a universal feature is "scientific method," which consists of searching for general laws that govern behavior and of asking such questions as: Is it so? To what extent is it so? Why is it so? What general conditions or considerations determine it to be so?

What Is Scientific Method? Scientific method is the pursuit of truth as determined by logical considerations. The ideal of science is to achieve a systematic interrelation of facts; scientific method, using the approach of "systematic doubt," attempts to discover what the facts really are.

What Is Experimentation? The function of experimentation is the elimination of untenable theories. Experimentation is used to test hypotheses and to discover new relationships among variables. It must be remembered, however, that no hypothesis that states a general proposition can be demonstrated to be absolutely true; only probable inferences are possible.

What Part Does Experimentation Play in Scientific Method? Experimentation is only a means toward an end. It is a tool of scientific method. Conclusions drawn from experimental data are frequently criticized. Such criticisms are usually based on one or more of the following arguments: (1) the interpretation is faulty, (2) the original assumptions are faulty, or (3) the experiment was poorly designed or badly executed. Obviously, careful attention should be given to the design of the experiment so that the procedures used are both valid and efficient.

What Is Experimental Design? Experimental design is the plan used in experimentation. It involves the assignment of treatments to the experimental units and a thorough understanding of the analysis to be performed when the data become available.

What Is the Relationship between Statistics and Experimental Design? Statistics enters into experimental design because even in the best planned experiments we cannot control all the factors and wish to make inferences based on the observed sample data. To be of any practical use, these uncertain inferences must be accompanied by probability statements expressing the degree of confidence the researcher has in such inferences. To make certain that such probability statements will be possible, the experiments should be designed in accordance with the principles of the science of statistics.

1.6 APPLICATIONS OF STATISTICS IN RESEARCH

Early applications of statistics were mainly concerned with reduction of large amounts of observed data to the point where general trends (if they existed) became apparent. At the same time, emphasis in many sciences turned from the study of individuals to the study of the behavior of aggregates of individuals. Statistical methods were admirably suited to such studies, aggregate data fitting consistently with the concept of a population.

The next major development in statistics arose to meet the need for improved analytical tools in the agricultural and biological sciences. Better analytical tools were needed to improve the processes of interpretation of and generalization from sample data. For example, the farmer is faced with the task of maintaining a high level of productivity of field crops. To aid him, the agronomist conducts an endless number of experiments to determine differences among yields of various crop varieties, effects of various fertilizers, and the best methods of cultivation. On the basis of the results of his experiments he is expected to make accurate and useful recommendations to the farm operator. Clearly then, statistics (being a science of inductive inference using probabilistic methods) should be of great value to the researcher in agronomy.

In early agronomic experimentation, to compare a number of fertilizers, it was thought necessary to devote only a single plot to each treatment and determine yields in order to arrive at valid conclusions concerning relative values of the treatments. However, the agronomists soon found that the yields of a series of plots treated alike differed greatly among themselves even when soil conditions appeared uniform and experimental conditions were carefully designed to reduce errors in harvesting. For this reason it became necessary to find some means for determining whether differences in yields were due to differences in treatments or to uncontrollable factors that also contributed to the variability of plot yields. Statistical methods were applied, and their value in scientific investigation of agronomic practices was soon proved.

Closely related to agronomy is the science of plant breeding. The ultimate objective of any plant-breeding research program is the development of improved varieties or hybrids. A variety may be improved in many possible ways, e.g., in ability to use plant nutrients, disease or insect resistance, cold tolerance, or suitability to the needs or fancies of the grower and/or consumer. Plants are organisms conditioned by genetic factors and by the environment in which they grow. The plant breeder therefore utilizes the principles of genetics in attempting to improve inheritable characteristics of plant varieties, just as the producer attempts to obtain high production by maintaining a favorable environment. However, results of past genetic studies do not provide all the answers relative to the inheritance of plant characteristics. Thus plant breeders continually carry out basic genetic research in each crop, along with practical plant-breeding procedures, to ensure future progress.

Development of a superior new variety by hybridization is seldom a haphazard occurrence. Usually the breeder has in mind the characteristics desired for his particular purpose or area. Growing many plant selections to decide which excel in a quantitatively inherited character requires growing them in a randomized, replicated field design. Choice of design depends on the numbers involved; the uniformity of the soil; the accuracy and precision of the particular estimates deemed necessary to get the desired results; the time, effort, and money available; and perhaps other factors. The data collected are then analyzed in accordance with the plan of the experiment, which was designed to make possible proper comparisons among the strains being tested. The statistical methods employed must have a logical relationship to the biological processes under consideration as well as to the way in which the experiment was conducted if they are to be useful. After the data have been analyzed statistically, the results must be interpreted in view of the assumptions made and the existing knowledge so that some conclusion may be reached with regard to accepting or rejecting the hypotheses being tested. Selection of the strain to be released as a variety or of those to be tested further may then be made with assurance that the decision will in all likelihood be a reasonable one.

Other research areas in which good use of statistical theories and methods is made are poultry breeding, animal breeding, and animal nutrition. Poultry breeding, for example, is concerned with the raising of more efficient and more productive fowl. Increased egg production, egg size, egg color, interior egg quality, more efficient meat production, long life, disease resistance, and high fertility are some of the factors with which the poultry breeder is actively concerned. If a statistically sound research program is adopted, the researcher will be able to reach defensible conclusions and bring about more efficient use of resources.

One of the more important uses of statistics in breeding work is the separation of environmental and hereditary effects. The literature of the field is full of reports dealing with this type of research, both with poultry and domestic animals. For the reader interested in this particular area, we refer to such writers as Hutt (4) and Lush (5).[1]

In the field of animal nutrition many experiments have been devised to discover the significance of various vitamins in the different phases of animal production. In such investigations several groups of animals, as homogeneous as possible, are selected for experimentation. These homogeneous groups are usually formed by considering such criteria as age, weight, sex, heredity, vigor, and previous nutrition. A check group is chosen and fed a standard ration. The other groups are fed different levels of the vitamin in question, one of them on a ration much higher than the standard for the vitamin and another on a ration containing little or none of the vitamin. The remainder of the groups are fed rations somewhere between the extremes. The animals are on the randomly assigned rations for a given period and the researcher records such data as daily gain in weight, economy of gain, livability, etc. If the experiment has been properly designed in accordance with established statistical principles, conclusions of great value to the farmer may then be drawn. Of course, much work of a more complex nature

---

1. Numbers in parentheses designate references listed at the end of the chapter.

than this simple example has also been done in animal nutrition research. Reference to technical journals in this field will reveal many instances where statistics has been of great help.

In the past many persons thought statistics had no place in the so-called "exact sciences" such as chemistry, physics, and the various branches of engineering. These fields are concerned with exact measurement, with quantities that can be measured with a ruler, thermometer, flowmeter, thickness gauge, telescope, or pressure gauge. Therefore, the doubters asked, why use a "pseudoscience" (statistics) that at best merely estimates quantities? As the true meaning of statistics and its application have come to wider attention, these persons have readily admitted there is indeed a place for this important tool in the exact sciences. It has become apparent that all these sciences themselves are based on statistical concepts. For example, it is evident that the pressure exerted by a gas is actually an average pressure, an average effect of forces exerted by individual molecules as they strike the wall of a container. A similar situation is true in regard to temperature.

Since the popularly accepted theory is that all matter is made up of small particles, it does not require much imagination to see that a statistical approach is the logical one to adopt in investigations of the ultimate nature of matter. Such particles are actually part of an almost inconceivably large population, one that is for all practical purposes our closest approach to the infinite population. All these particles exhibit individual behavior characteristics. With the comparatively crude devices of the exact sciences we can generally only note the results of group behavior, an average effect. However, even in these crude applications statistics plays its role. For instance, examine the chart of the elements in any chemistry classroom. The atomic weights shown are actually "weighted averages" of the atomic weights of individual isotopes of the given element, the "weights" being the frequency of occurrence of the element in a normal or naturally occurring mixture.

Statistics has also invaded the fields of meteorology and astronomy. The modern science of meteorology is to a great degree dependent upon statistical methods for its existence. The methods that give weather forecasting the accuracy it has today have been developed using modern sample survey techniques. Thus weather stations throughout the United States are able to give us highly accurate predictions for their individual areas. In addition, by suitable selection of gathering points and proper treatment of the data, an overall picture of the weather for larger areas is pieced together. We may see statistical sampling in action when we turn our attention to snow survey teams, which determine the amount of snow present in a given area and thus the quantity of water to be drained off following a thaw. In the more theoretical aspects of meteorology, statistical inference and analysis are being used to develop new techniques for advancing the field. In astronomy, statistics has long played a major role. One hundred years ago the uncertainty in the measurement of the semimajor axis of the earth's elliptical orbit was 1 part in 20. Today statistical methods have reduced this uncertainty to 1 part in 10,000.

Statistics is now playing an important role in engineering. For example, such topics as the study of heat transfer through insulating materials per unit time, performance guarantee testing programs, production control, inventory control, standardization of fits and tolerances of machine parts, job analyses of technical personnel, studies involving the fatigue of metals (endurance properties), corrosion studies, time and motion studies, operations research and analysis, quality control, reliability analyses, and many other specialized problems in research and development make great use of probabilistic and statistical methods.

Because the above problems are but a small portion of those to which the science of statistics is being applied in industry, the reader can readily appreciate that the application of statistical methods to the field of engineering is not limited to a few areas but is general in nature. As an indication of the wide scope of industrial statistics, P. L. Alger (1) of the General Electric Corporation has listed the following ten major areas of application:

1. Defining the value of observations.
2. Design of experiments.
3. Detection of causes.
4. Production quality control.
5. Getting more out of the inspection dollar.
6. Design specifications.
7. Measurement of human attributes.
8. Operational research.
9. Market research, including opinion polling.
10. Determining trends.

If applied statistics is to play a primary role in the future of engineering or, to be more general, in that of industry, it is quite evident there is a great need for specific training of personnel entering the field. This training is needed for the young engineer as well as for the young businessman, since each must be capable of dealing with combinations of men and machines. S. S. Wilks (7) of Princeton University has made this statement of the problem:

> The statistical problems which the future scientist or engineer will encounter will cut across traditional lines. Therefore, in order that he may be properly equipped to deal with these problems, he should have a fairly broad statistical training. The training should cover not only statistical quality control methods as the term is now understood, but the design of experiments, analysis of variance, and many other topics. It should be built into the training of scientists and engineers, as calculus is now made part of their basic education.

Agricultural engineering, which combines the practices of engineering and agriculture, has also benefited greatly from the use of statistical methods. In this field, statistics has helped the researcher with such

varied projects as the testing of weed-control machinery, certain economic aspects of farm electrification, comparison of various drying
methods for grain, determination of the effects of drying rate on pop-
corn, irrigation research, roofing studies for farm buildings, and methods of cultivation.

Statistics is also proving an important tool in food technology research. Foods exhibit to a marked degree what is widely called "biological variation." Their constitution is heterogeneous, and their complexity is such that duplication is highly improbable. Food properties
are affected not only by the multiplicity of factors influencing their
growth but also by the infinite variety of processing and storage conditions to which they may be subjected. Thus it is impossible to give
a general answer to a question such as "What is the moisture content
of corn?" Before attempting to answer, one would first have to ask
"What variety ... at what stage of its growth or processing cycle ...
where was it grown?" and other such questions. Having obtained the
necessary specifications, the food technologist might be able to quote an
average value. In short, he might specify a frequency distribution of
moisture content of sweet corn under the stated conditions.

This type of problem was encountered by Bard (2) in his investigation of certain palatability factors (tenderness, juiciness, and fiber
cohesiveness) of canned beef as conditions of time and temperature of
processing were varied. In his work a statistical approach dictated the
design of the experiment, and analysis of variance was freely employed
to delineate between variation due to raw material and that caused by
processing treatments. Another example in food technology is provided
by Bernhard (3) in his comparison of several techniques of estimation
of the frequency of occurrence of insect fragments in cream-style corn.
The conventional methods utilized castor oil to separate the insect
fragments by flotation. Bernhard wished to compare the efficiency of
castor oil and lard oil, each at three different temperatures for each
of four different times of mixing oil and food samples. Repeated
samples of any one set of determination conditions could be expected
to yield variable results for the number of insect fragments present.
Thus, statistical methods were required to enable the variation among
mixing times, temperatures, and oils to be analyzed.

One of the most difficult areas of food research is that of evaluating
a food product in terms of consumer reaction. It is well known that
most objective tests of food acceptability (such as laboratory measurements of shear strength, etc.) must be correlated with consumer preference by means of taste-panel observations in order to achieve firm
standing. The problems of the taste panel are many. To what extent
is the taste panel representative of the entire population of tasters?
How is variation from sample to sample of a food product distinguished from variation from taster to taster? How can subjective evaluation of a particular property of food, for example, odor, be separated
from evaluation of another property such as flavor? To what extent can
or should restrictions such as instructions to evaluate a narrow area

be placed upon the taster in view of the fact that the heart of the taste-panel system is the use of the integrated pattern of individual reaction to a complex event?

These and many other such problems of food evaluation are not entirely solved. Even the basic justification for the introduction of statistical analysis is not always clear. For example, a group of tasters may be asked to rank five varieties of corn in order of merit. In searching for a method of evaluating results of this type of problem, many workers have followed the procedure of allotting a number to each rank, e.g., 5 for first, 4 for second, etc. These figures are then treated as numbers and analyzed by analysis of variance to check for significant variation among the five varieties. Such a procedure is not entirely valid because analysis of variance can only be used with numbers, and the ranking figures are not originally set down as quantitative relative estimates of taste reaction. However, in view of the lack of exact methods of analysis, the technique mentioned can provide valuable assistance to the research worker who deals with food products.

In the social sciences, statistical methods also find wide application. Because of their vital interest in public opinion the major political parties have become acquainted with the statistician. In economic research, statistical methods are almost indispensable. Economic laws refer to mass or group phenomena, and the determination of these laws often depends upon the judicious use of statistical techniques.

In marketing research an objective may be increasing consumption of foods shown by nutritional studies to be inadequately supplied in the average diet. The initial role of statistics here is merely one of finding consumption per capita and comparing it with some goal, since the nature of distribution of consumption per capita is as important as the average. Another objective is the analyzing of marketing methods to find the least costly way of doing the job. As a result, a smaller portion of society's efforts need to be expended on product handling.

Measuring demand is another of the many difficult tasks in economics. The research worker must have a knowledge of consumer preferences, supply of money, its distribution, etc. In measuring supply, he must have an intimate acquaintance with marketing functions, services, and costs and be familiar with trends in operational efficiency, both physical and managerial. Data on these particulars can only be digested and made available through statistical procedures.

In production economics probably the most important comparisons are made when two or more characteristics are simultaneously studied or measured. This involves statistical techniques known as regression and correlation. These tools are invaluable to the economist. By their use, one factor can be shown in its relationship with others. For instance, if we made the hypothesis that net income per acre becomes higher as farm size increases, we would want to find the influence of farm size on net income per acre. We might then collect data for ten units of each farm size, ranging from 40 acres to perhaps 480 acres with

40-acre increments. These data, if properly obtained in accordance with the rules of statistical procedure, could then be analyzed to aid the researcher in making a contribution to the theory of production economics.

It is possible to go on almost indefinitely enumerating the fields wherein statistics is being or could be applied. Statistics is utilized for a systematic approach to problems in public health studies, epidemiology, demography, biological assay, psychology, education, sociology, and various areas of home economics. Oddly enough, statistics is not confined to the so-called scientific world, for it also is applied in the arts. It has been used to aid in determining the authorship of certain manuscripts by analyzing the length of sentences. Authenticity of paintings has also been established by analyzing the frequency of brush strokes.

Although statistics is very much in the realm of an applied science, it has its theoretical basis in mathematics. Development of the theoretical branch of the science is as important as that of the applied branch if progress in the field is to continue. Unfortunately, there is a gap between statistical theory and application much the same as exists in other sciences. This gap is steadily being closed, but the job is far from completion. Thus it is not surprising to find statistics in use as both a science and an art. It is a science because its methods are basically systematic and of wide application; it is an art because success in its application depends on the skill, special experience, and knowledge of the person using it. The research worker will appreciate this fact more as he gains a greater understanding of statistical methods and their uses.

1.7 SUMMARY

The scope of statistics might be described as being concerned with the presentation and summarization of data, the estimation of population quantities and the testing of hypotheses, the determination of the accuracy of estimates, the measurement and study of variation, and the design of experiments and surveys. Inherently and inextricably involved in all the above mentioned areas is the process known as methods of reduction of data or the computational aspects of statistics.

The statistical method is one of the devices by which men try to understand the generality of life. Out of a welter of single events they seek endlessly for general trends. Controlled, objective methods by which group trends are abstracted from observations on many separate individuals are called statistical methods. These methods are especially adapted to the elucidation of quantitative data that have been affected by many factors. Statistical methods are fundamentally the same whether employed in the analysis of physical phenomena, the study of educational measurements, the study of data resulting from biological experiments, or the analysis of quantitative material in economics. Agriculturists, biologists, chemists, physicists, and other researchers all attempt to eliminate the many nuisance factors that influence the variables under investigation and to concentrate their attention upon one or two of the most powerful factors affecting the phenomena being studied. Yet many disturbances are always present; thus statistical methods of analysis are vitally necessary. Wherever there is a mass of numerical data that admits of explanation, the statistician should consider its analysis his field of endeavor.

To use statistical methods to advantage, a person should:

1. Be well versed in the subject matter of the field in which the research is to be conducted.
2. Know how to organize masses of data for efficient tabulation and how to lay out economical routines for handling data and computation.
3. Know effective means of presenting data in tabular and graphic form.
4. Have some knowledge of the mathematical theory of statistics in order to have assurance there is a fair correspondence between his data and the assumptions underlying the formulas he uses.
5. Be acquainted with a variety of statistical techniques, the limitations and advantages of each, the assumptions upon which they are based, the place each occupies in a logical analysis of the data, and the interpretations that can be made from them.

Statistics, then, boils down to numerical results, the methods and processes used in obtaining them, the methods and means for estimating their reliability, and the drawing of inferences from them.

During the past half-century the thinking world appears to have awakened to an unusually deep appreciation and respect for numerical facts. There has been a growing tendency to reduce observations and accumulated data to an orderly arrangement, making possible the evaluation of results by means of a systematic method of analysis.

Formerly, many persons believed statistical analysis could be used only in certain highly specialized fields; however, more and more statistical methods are finding their way into scientific workshops in all fields. This is due largely to the fact that some of the enthusiastic supporters of statistical methods have worked faithfully to develop and explain methods useful to and usable by those persons not specifically trained in higher mathematics.

Advancement has been rapid in the field of statistical analysis. Many useful methods are now available for the analysis of data arising from different sources. A clear grasp of simple and standardized statistical procedures will go far to elucidate principles of experimentation. However, we must remember that these procedures are in themselves only a means to a more important end. As fundamental and pervasive as statistical thinking is in the modern world, it must not be considered an end in itself. The statistical method is a tool for organizing facts so they are rendered more available for study. A statistical study can only describe what is; it cannot determine what ought to be, except insofar as it may throw light upon probable concomitants and consequences of certain situations. It is fatuous to suppose the statistical method can provide mechanical substitutes for thinking, although it is often an indispensable aid. Men see increased prevalence of the statistical method in scientific studies; and sometimes failing to grasp underlying reasons for this development, they assume the use of tables, formulas, and numerical summaries is a badge of respectability. As a result some studies, truly subjective in nature, are invested with a false show of objectivity. Thus a vast superstructure of computation is raised upon a foundation inappropriate to such treatment. When such a picture is painted, it is neither good statistics nor good philosophy.

Most statistical studies will not answer all the questions we would like to have answered regarding a given problem. Due to the very nature of statistical work, results are likely to be partial and fragmentary rather than complete and final. Therefore, the researcher must make up his mind that questions must sometimes be left unanswered. He must also on occasion freely admit his study has limitations. Any shortcomings in his work and the danger of attributing more than claimed for his investigation should be pointed out to his readers by the researcher.

It is also imperative that conclusions drawn from observational results should be based on a detailed knowledge of procedures employed in the investigation. The interpretive function in statistical analysis is one of the most important contributions of statistics, and the statistician should plan experiments and investigations that will yield maximum information and valid conclusions from scientific research data. Inference from the particular to the general must be attended with some degree of uncertainty, and research workers in all fields of science must recognize the role statistics plays in this most important aspect of research.

The role of statistics in research is to function as a tool in designing research, analyzing its data, and drawing conclusions therefrom. A greater and more important role can scarcely be envisioned. In utility to research, statistics is second only to the mathematics and common sense from which it is derived. Clearly the science of statistics cannot be ignored by any research worker even though he may not have occasion to use applied statistics in all its detail and ramifications.

PROBLEMS

1.1 Discuss the following terms or phrases: (a) observation and description; (b) cause and effect; (c) analysis and synthesis; (d) assumption, postulate, and hypothesis; (e) testing of hypotheses; (f) deduction and induction.

1.2 What do you believe operations researchers mean by the phrase "measure of effectiveness?"

1.3 Saaty (6), in a chapter entitled "Some Remarks on Scientific Method in Operations Research," refers to: (a) the judgment phase, (b) the research phase, and (c) the action phase. Give your interpretations of these three phases. Then compare your views with those of Saaty.

1.4   Read Chapter 12, "Some Thoughts on Creativity," in Saaty (6). Then prepare a brief report on your reactions to his ideas.
1.5   Prepare a report on the pros and cons of (a) individual research and (b) interdisciplinary team research.
1.6   Discuss the similarities and dissimilarities of pure and applied research.
1.7   Prepare a report on the subject of "scientific method."
1.8   Prepare a report on your interpretation of "the role of statistics in research."
1.9   By consulting the technical journals in your area of specialization, prepare and submit a list of references (properly documented) that illustrate the use of statistical methods.
1.10   Submit a list of publications (books, monographs, papers, etc.) that you believe would be worthwhile additions to the references presented with this chapter.

REFERENCES

1. Alger, P. L. 1948. The growing importance of statistical methods in industry. *General Electric Rev.* Vol. 51, No. 12.
2. Bard, J. C. 1950. Changes in tenderness, fiber cohesiveness and moisture content of canned beef due to thermal processing. M.S. thesis, Iowa State University, Ames.
3. Bernhard, F. L. 1951. Recovery and identification of insect fragments from cream style corn. M.S. thesis, Iowa State University, Ames.
4. Hutt, F. B. 1949. *Genetics of the Fowl.* McGraw-Hill, New York.
5. Lush, J. L. 1945. *Animal Breeding Plans.* Iowa State University Press, Ames.
6. Saaty, T. L. 1959. *Mathematical Methods of Operations Research.* McGraw-Hill, New York.
7. Wilks, S. S. 1947. Statistical training for industry. *Anal. Chem.* Vol. 19.

FURTHER READING

Ackoff, R. L. 1962. *Scientific Method.* Wiley, New York.
Anderson, J. A. 1945. The role of statistics in technical papers. *Trans. Am. Assoc. Cereal Chemists,* Vol. 3.
Baird, D. C. 1962. *Experimentation: An Introduction to Measurement Theory and Experiment Design.* Prentice-Hall, Englewood Cliffs, N.J.
Bartee, E. M. 1968. *Engineering Experimental Design Fundamentals.* Prentice-Hall, Englewood Cliffs, N.J.
Beveridge, W. I. B. 1957. *The Art of Scientific Investigation,* rev. ed. Random House, New York.
Bush, G. P.; and Hattery, L. H. (eds.). 1953. *Teamwork in Research.* American University Press, Washington, D.C.
Chapanis, A. R. E. 1957. *Research Techniques in Human Engineering.* Johns Hopkins Press, Baltimore.
Churchman, C. W. 1948. *Theory of Experimental Inference.* Macmillan, New York.
Churchman, C. W.; Ackoff, R. L.; and Arnoff, E. F. 1957. *Introduction to Operations Research.* Wiley, New York.
Fiebleman, J. K. 1972. *Scientific Method.* Martinus Nijhoff, The Hague, Netherlands.
Freedman, P. 1960. *The Principles of Scientific Research,* 2nd ed. Pergamon Press, New York.
Hillway, T. 1964. *Introduction to Research,* 2nd ed. Houghton Mifflin, Boston.
Jeffreys, H. 1957. *Scientific Inference,* 2nd ed. Cambridge University Press, London.
Johnson, P. O. 1951. Modern statistical science and its function in educational and psychological research. *Sci. Monthly* Vol. 72.
Luszki, M. E. B. 1958. *Interdisciplinary Team Research: Methods and Problems.* New York University Press, New York.
Mandel, J. 1964. *The Statistical Analysis of Experimental Data.* Wiley, New York.

Ostle, B. 1957. Statistics in engineering. *J. Eng. Educ.* Vol. 47 no. 5, Jan.

Ostle, B.; and Tischer, R. G. 1954. Statistical methods in food research. *Adv. Food Res.* Vol. 5.

Popper, K. R. 1968. *The Logic of Scientific Discovery,* 3rd ed. Hutcheson and Co., Ltd., London.

Schenck, H., Jr. 1968. *Theories of Engineering Experimentation,* 2nd ed. McGraw-Hill, New York.

Snedecor, G. W. 1950. The statistical part of the scientific method. *Ann. N.Y. Acad. Sci.,* Vol. 52.

Taton, R. (trans., A. J. Pomerans). 1957. *Reason and Chance in Scientific Discovery.* Philosophical Library, New York.

Whitney, F. L. 1950. *The Elements of Research,* 3rd ed. Prentice-Hall, New York.

Wilson, E. B. 1952. *An Introduction to Scientific Research.* McGraw-Hill, New York.

Worthing, A. G.; and Geffner, J. 1943. *Treatment of Experimental Data.* Wiley, New York.

CHAPTER 2
PROBABILITY

It is difficult to achieve a clear understanding of statistical methods without discussing the underlying theory to some extent. Since statistical theory is intimately associated with probability, it is desirable to begin the study of statistical methodology with a discussion of probability. This we do in this chapter. The basic concepts of probability, probability laws, random variables, distributions, and expectations are discussed. Some of the more commonly encountered probability distributions are presented.

2.1 INTRODUCTION

In general, statistics enters into scientific method through experimentation or observation. Any investigation is only a means to an end. It is a device for testing a hypothesis or for acquiring an amount of knowledge, however small, from which a conclusion can be drawn. Most statements resulting from scientific investigation are only inferences. They are uncertain in character. If, for example, we are interested in measuring the effect of changing the input parameters (e.g., temperature, acid concentration, mixing speeds, etc.) on the yield from a chemical process, we could set up a laboratory experiment in which the parameters are varied and the response variable(s) measured. From such an experiment we would than make inferences about the effects of changing these parameters. Clearly, any output of such an experiment is subject to uncertainties due to the many uncontrollable or chance variations that can enter into an experimental procedure; hence, any conclusions are uncertain in nature. *The measurement of this uncertainty by use of the theory of probability is one of the most important contributions of statistics.*

To understand how this uncertainty is measured, we begin by introducing the basic ideas of probability and its role in experiments in which the outcomes are uncertain. The application of probability to experimental situations also involves the notions of random variables and probability distributions. These are introduced in Sections 2.6 and 2.7. The remaining sections of this chapter are background topics useful for understanding the statistical methods covered in the remainder of the book.

2.2 PROBABILITY AND PROBABILITY MODELS

An experiment in which there is uncertainty as to the ultimate outcome can be referred to as a *random experiment,* which is characterized by the property that, prior to conducting the experiment, the outcome is nondeterministic. A simple random experiment is the "random" toss of a coin. Which outcome will occur, a head or a tail, is unknown and cannot be determined prior to the actual toss. Another example is the measurement of physical characteristics (length, diameter, power, etc.) when such measurements are subject to uncontrollable or "chance" errors. Again, what the value of the physical measurement will be cannot be determined prior to taking it.

To analyze the outcomes of such experiments, it is desirable to construct a model, i.e., a mathematical abstraction of the physical experiment that describes the random experiment and can be used in the analysis. Models for random experiments are called *probability models*.

2.2.1 Probability Models

For any random experiment a probability model consists of three components. First, the model includes a description of the *sample space* or set of all possible outcomes of the experiment. We denote the sample space by Ω.

Example 2.1

If the experiment consists of "randomly" rolling a die and observing the number of dots on the top side at the end of the roll, the sample space consists of the outcomes 1, 2, 3, 4, 5, and 6. That is, $\Omega = \{1, 2, 3, 4, 5, 6\}$.

Example 2.2

Consider a card with a spinner and a circle such that the circle is continuously calibrated between zero and one. The random experiment consists of spinning the spinner and observing the point at which the dial stops. Since it can conceptually stop at any point on the circle, the sample space is $\Omega = [0, 1)$. (**Note:** When we describe intervals of real numbers, we will use the symbols [and (or combinations of the two. The former symbol means the endpoint is included in the interval while the latter means the endpoint is excluded. Thus $\Omega = [0, 1)$ means that the sample space is all the real numbers in the unit interval except the value 1. This is because when calibrating the circle, the points 0 and 1 are coincident.)

In any experiment we are generally interested in the occurrence of an "event" rather than of a single outcome. For example, when rolling a die, we may be interested in the "event" that there is an odd number of dots visible. This occurs whenever the outcome is an element of the subset $E = \{1, 3, 5\}$. Thus the subset E is the set of outcomes for which the "event" (an odd number of dots occur) is true. Similarly, in Example 2.2 we may be interested in the spinner stopping among the points on the circle having values greater than 0.5. This "event" occurs whenever the outcome is in (0.5, 1). For purposes of the probability model this subset in Ω is referred to as the *event*.

Definition 2.1   An *event* is a subset of the sample space.

If E_1 and E_2 are events (subsets) in a sample space Ω then the *union,* denoted $E_1 \cup E_2$, is the subset of Ω consisting of points that are in either E_1 or E_2. The *intersection,* denoted $E_1 \cap E_2$, is the set of points that are in both E_1 and E_2. (**Note:** When we have several events $E_1, \ldots, E_n$, the union is denoted $\cup_{i=1}^{n} E_i$ and the intersection $\cap_{i=1}^{n} E_i$ Of course, $\cup_{i=1}^{n} E_i$ is the subset of points that are in at least one of the E_i's, and $\cap_{i=1}^{n} E_i$ is the set of points that are in all of the E_i's.) The complement of an event E, denoted E', is the set of outcomes in Ω that are not in E. We say two events E_1 and E_2 are *mutually exclusive* if they contain no common outcomes, i.e., if $E_1 \cap E_2 = \phi$ where ϕ is the null or empty set.

The second component of a probability model is recognition of the class of events. In general, it will be possible to associate the occurrence of some "event" with any event (subset) of Ω. Thus, in general, the class of events relevant to any experiment is the class of all possible subsets of the sample space. This is true when the sample space contains a finite number of outcomes, as in Example 2.1, and when Ω is countably infinite. A problem arises only when Ω is a continuum as in Example 2.2. We refer the reader to some of the more advanced texts in probability listed in the references for further discussion of this situation since it is not a serious problem in statistical methods. It is sufficient for us to recognize that the events are subsets in the sample space.

The third element of a probability model consists of the assignment of a number $P(E)$ to each event E such that

1. $0 \le P(E) \le 1$
2. $P(E_1 \cup E_2) = P(E_1) + P(E_2)$ if E_1 and E_2 are *mutually exclusive*

This number $P(E)$ is called the *probability of E.*

In summary, a probability model consists of

1. A description of the sample space Ω
2. Recognition of all possible events E of Ω
3. An assignment of probability $P(E)$ to each event such that
   (a) $0 \le P(E) \le 1$ for all E
   (b) $P(E_1 \cup E_2) = P(E_1) + P(E_2)$ if $E_1 \cap E_2 = \phi$

Example 2.3

Suppose we want to construct a model for the roll of a die. If we are interested in the number of dots visible on top of the die at the end of its roll, as we indicated earlier, $\Omega = \{1, 2, 3, 4, 5, 6\}$.

The events associated with this experiment are all subsets of Ω. There are $2^6 = 64$ such subsets. Some examples are:

$$E_1 = \{2\} \qquad E_3 = \{2, 4, 6\} \qquad E_5 = \{4, 6\}$$
$$E_2 = \{1, 3\} \qquad E_4 = \{4, 5, 6\} \qquad E_6 = \{2, 4, 5, 6\}$$

where E_1 is the event a 2 occurs, E_2 is the event a 1 or 3 occurs, and E_3 is the event an even number occurs. Note that $E_5 = E_3 \cap E_4$ and $E_6 = E_3 \cup E_4$. One possible assignment of probabilities to the events of this experiment is based on the assumption that each side is equally likely to be on top. Then,

$$P(E_1) = 1/6 \qquad P(E_3) = 3/6 \qquad P(E_5) = 2/6$$
$$P(E_2) = 2/6 \qquad P(E_4) = 3/6 \qquad P(E_6) = 4/6$$

Having outlined the elements of a probability model, we must now deal with the fundamental problem of how to construct a model for a physical situation. Generally, it is not hard to specify the sample space for a random experiment. Sometimes the sample space will consist of numerical elements, as in the case for the roll of a die; namely, $\Omega = \{1, 2, 3, 4, 5, 6\}$. On the other hand, Ω can also consist of nonnumerical outcomes, as would be the case if the experiment is a random toss of a coin and we are interested in which side is up when the coin lands. In that case, $\Omega = \{head, tail\}$.

The most difficult part of the specification of a probability model is the assignment of probabilities to the events of the experiment. The question is, Just what do we mean by the probability of E and how does one generate the value $P(E)$ for any event E? So far we have defined $P(E)$ to be a real number having certain properties without considering what the actual values should be in any given situation. Actually, in order to study probability models, all we have to do is to consider that the probabilities are known. We do not need to know where the values came from or how they were generated to study the model. We only require that the values used for $P(E)$ satisfy the conditions (1) and (2) stated earlier.

However, if we are going to apply a particular probability model to a physical situation, the values assigned should be consistent with what we observe. For example, suppose we assign probabilities $1/2$ to each of the outcomes H and T in the toss of a particular coin. That is, the model we select is,

$$\Omega = \{H, T\} \qquad P(H) = 1/2 \qquad P(T) = 1/2$$

Now, suppose we conduct the experiment of tossing the coin 1000 times or more and find that a head occurs 70 percent of the times and a tail 30 percent of the times. The experimental results seem to contradict the model. Perhaps the coin used was biased rather than "fair" as the model proposes. This suggests that a different model would better describe the random toss of this coin.

There are several ways one can interpret probability when applying probability models to physical situations. One such concept that seems to be meaningful and useful in most physical situations is the relative frequency concept.

2.2.2. Relative Frequency Concept of Probability

To introduce the *relative frequency* concept of probability, consider an experiment $\mathcal{E}$ that may be repeated, at least conceptually, many times under identical conditions. Each time the experiment is performed, observe whether an event E does or does not occur. If f is the number of times E occurs in the first n repetitions of the experiment, then the ratio f/n is the relative frequency of E in the first n repetitions of $\mathcal{E}$. In general, the ratio f/n will tend to become more or less constant for large n. This phenomenon is sometimes referred to as *statistical regularity*. It is now conjectured that for a given $\mathcal{E}$ and E we should be able to find a number p such that, as n gets large, the ratio f/n should be approximately equal to p.

Definition 2.2   If an experiment $\mathcal{E}$ can be conceptually repeated under uniform conditions, then by assigning the value p to probability of event E (i.e., letting $P(E) = p$), we mean that in a long series of repetitions of $\mathcal{E}$ it is practically certain that the relative frequency of E will be approximately equal to p.

Example 2.4

Return to the roll of a die in Example 2.3. Suppose we are interested in evaluating $P(E_1)$ where $E_1 = \{2\}$. Imagine rolling the same die many times and looking at the proportion (relative frequency) of rolls in which a 2 is the outcome. Experience has shown that this proportion will eventually settle down to a number p after many rolls. If the die is unbiased, we expect $p = 1/6$. Thus we assign $P(E_1) = 1/6$ in the case of an unbiased die. Likewise, if $E_2 = \{1, 3\}$ and the die is unbiased, we expect E_2 to occur over the long run in about 2 out of 6 repetitions. Thus $P(E_2) = 2/6$.

Example 2.5

Consider the spinner experiment in Example 2.2. If the spinner is spun "at random," experience has shown that over the long run the proportion of times the spinner stops in the interval $(0.5, 1)$ is about 50 percent of the time. Thus if $E_1 = (0.5, 1)$, we assign $P(E_1) = 0.5$. Similarly, if $E_2 = [0.25, 0.5)$, $P(E_2) = 0.25$. In general, if $E = (a, b)$ for $0 \leq a < b \leq 1$, $P(E) = b - a$.

2.3 CONDITIONAL PROBABILITY

For some random experiments it is sometimes convenient to describe the probability of an event E_1 knowing that a second event E_2 has occurred. This latter probability is called the *conditional probability* of E_1 given E_2, i.e., the probability of E_1 conditional on E_2 occurring.

Definition 2.3   For two events E_1 and E_2 in Ω, the *conditional probability* of E_1 given E_2, denoted $P(E_1 \mid E_2)$, is

$$P(E_1 \mid E_2) = P(E_1 \cap E_2)/P(E_2) \qquad (2.1)$$

if $P(E_2) > 0$. If $P(E_2) = 0$, then $P(E_1 \mid E_2)$ is undefined.

Example 2.6

Suppose in the roll of a die we are interested in the event, call it E_1, that the side with 6 dots occurs. That is, $E_1 = \{6\}$. If the die is unbiased, then the appropriate value for $P(E_1)$ is 1/6. On the other hand, suppose it is known that the number of dots visible is at least 4. With this additional bit of information we are more interested in a conditional probability of E_1, namely, the conditional probability of E_1 given that the outcome is 4 or more. Let E_2 be that event, $E_2 = \{4, 5, 6\}$. The conditional probability of E_1 given E_2 is

$$P(E_1 \mid E_2) = P(E_1 \cap E_2)/P(E_2) = (1/6)/(3/6) = 1/3$$

since $E_1 \cap E_2 = \{6\}$ and $P(\{6\}) = 1/6$.

Conceptually, we can view conditional probability, just as we did probability, in terms of relative frequency. Now in a long series of repetitions of $\mathcal{E}$ the relevant relative frequency is the proportion of times E_1 occurs when E_2 has occurred. The proportion of times E_2 occurs is about $P(E_2)$, and the proportion of times both E_1 and E_2 occur is about $P(E_1 \cap E_2)$. Thus of the total number of occurrences of E_2, the event E_1 occurs approximately $P(E_1 \cap E_2)/P(E_2)$ of the times. Thus, the long-run relative frequency of E_1 given E_2 is approximately

$$P(E_1 \mid E_2) = P(E_1 \cap E_2)/P(E_2) \qquad (2.2)$$

which agrees with the definition of the conditional probability of E_1 given E_2.

2.4 INDEPENDENCE

A concept in probability theory that has very important applications in statistical methodology is the idea of *statistical (stochastic) independence*.

Definition 2.4 Two events E_1 and E_2 in the same sample space are said to be *statistically (stochastically) independent* if

$$P(E_1 \cap E_2) = P(E_1)P(E_2) \qquad (2.3)$$

This is equivalent to saying that E_1 and E_2 are (statistically) independent if $P(E_1 \mid E_2) = P(E_1)$ and $P(E_2 \mid E_1) = P(E_2)$.

To extend the idea of independence to more than two events, consider the case of three events E_1, E_2, and E_3 all defined on the same sample space. We say E_1, E_2, and E_3 are *(mutually) independent* if

1. E_1, E_2, E_3 are pairwise independent, i.e., $P(E_1 \cap E_2) = P(E_1)P(E_2)$, $P(E_1 \cap E_3) = P(E_1)P(E_3)$, and $P(E_2 \cap E_3) = P(E_2)P(E_3)$
2. $P(E_1 \cap E_2 \cap E_3) = P(E_1)P(E_2)P(E_3)$ (2.4)

Note that the pairwise independence of three events is not enough to assure mutual independence of the events (see Example 2.7).

Example 2.7

Let Ω be the set of points, $\Omega = \{1, 2, 3, 4\}$, such that each outcome is equally likely; i.e., $P(\{1\}) = P(\{2\}) = P(\{3\}) = P(\{4\}) = 1/4$. Define the events $E_1 = \{1, 4\}$, $E_2 = \{2, 4\}$, and $E_3 = \{3, 4\}$. Then E_1, E_2, and E_3 are pairwise independent since $P(E_1 \cap E_2) = P(E_1)P(E_2) = (1/2)$

$(1/2) = 1/4$, $P(E_1 \cap E_3) = P(E_1)P(E_3) = (1/2)(1/2) = 1/4$,   and $P(E_2 \cap E_3) = P(E_2)P(E_3) = (1/2)(1/2) = 1/4$. However, we note that $E_1 \cap E_2 \cap E_3 = \{4\}$; therefore, $P(E_1 \cap E_2 \cap E_3) = P(\{4\}) = 1/4$. On the other hand, $P(E_1)P(E_2)P(E_3) = (1/2)^3 = 1/8$. Since $P(E_1 \cap E_2 \cap E_3) \neq P(E_1)P(E_2)P(E_3)$, we conclude that E_1, E_2, and E_3 are *not* mutually independent.

2.5 ALGEBRA OF PROBABILITY

Now that we have introduced the basic ideas of probability, we are in a position to indicate some of the properties that allow the use of probability models in more than the simplest experiments. These properties of probability are given without proof; we leave it to the reader to supply his own if he so desires.

Let $E_1, E_2, \ldots, E_n$ denote events in the same sample space. Some properties are:

1. *Addition rule:* If E_1 and E_2 are any two events, then

$$P(E_1 \cup E_2) = P(E_1) + P(E_2) - P(E_1 \cap E_2) \qquad (2.5)$$

If E_1 and E_2 are mutually exclusive, i.e., $E_1 \cap E_2 = \phi$, then

$$P(E_1 \cup E_2) = P(E_1) + P(E_2) \qquad (2.6)$$

The extension of the addition rule to three events E_1, E_2, and E_3 is

$$P(E_1 \cup E_2 \cup E_3) = P(E_1) + P(E_2) + P(E_3) - P(E_1 \cap E_2)$$
$$- P(E_1 \cap E_3) - P(E_2 \cap E_3) + P(E_1 \cap E_2 \cap E_3) \qquad (2.7)$$

2. If E' is the event "not E," then

$$P(E') = 1 - P(E) \qquad (2.8)$$

3. *Multiplication rule:* For any two events E_1 and E_2 such that $P(E_2) > 0$,

$$P(E_1 \cap E_2) = P(E_1 \mid E_2)P(E_2) \qquad (2.9)$$

If E_1 and E_2 are independent, then

$$P(E_1 \cap E_2) = P(E_1)P(E_2) \qquad (2.10)$$

4. For events E_1 and E_2

$$P(E_1 \cup E_2) = 1 - P(E_1')P(E_2' \mid E_1') \qquad (2.11)$$

5. If $E_1, \ldots, E_n$ are mutually exclusive events such that $\bigcup_{i=1}^{n} E_i = \Omega$, and if F is an arbitrary event in Ω such that $P(F) \neq 0$, then

$$P(F) = \sum_{i=1}^{n} P(F \cap E_i) = \sum_{i=1}^{n} P(F \mid E_i)P(E_i) \qquad (2.12)$$

6. *Bayes rule:* Combining (3) and (5),

$$P(E_1 \mid F) = P(F \mid E_1)P(E_1) \Big/ \sum_{i=1}^{n} P(F \mid E_i)P(E_i) \quad (2.13)$$

and similarly for $E_2, \ldots, E_n$.

Many more identities could be stated. Individuals interested in additional properties are referred to more advanced references on probability.

2.6 RANDOM VARIABLES

In most instances the outcomes of random experiments are numerical values. If not, one is usually interested in some function of the outcomes rather than the outcomes themselves thus resulting in numerical observations. Such observations can be thought of as values of a *random variable;* i.e., for any random experiment a random variable is a function (rule) that assigns a real number to each outcome in the sample space.

Example 2.8

Suppose the random experiment is the toss of a coin three times. The sample space consists of the eight outcomes

$$\Omega = \{(HHH)(HHT)(HTH)(HTT)(THH)(THT)(TTH)(TTT)\}$$

Generally, rather than being interested in these outcomes, one is more concerned with the number of heads (or tails) in the three tosses. Thus one is interested in the number of heads being *one*, not whether the outcome was (HTT), (THT), or (TTH). That is, we are interested in the values of the random variable X = number of heads in the 3 tosses, where X can take on the value 0, 1, 2, or 3; i.e., the set of outcomes of interest is the set $\{0, 1, 2, 3\}$.

We shall denote a random variable by a capital letter, e.g., X, Y, W, etc., and the values of the random variable will be denoted x, y, w, etc. The set of all possible values of a random variable defines a new sample space. As before, an event is a subset of the sample space. For the random variable in Example 2.8, the sample space is $\Omega_X = \{0, 1, 2, 3\}$. The event that the value of X is either 0 or 1 is the subset $E_1 = \{0, 1\}$. We denote this by $X \in E_1$.

Example 2.9

Consider the spinner experiment in Example 2.2. A random variable for this experiment could be: X = location on the calibrated circle at which the spinner stops. This random variable can take on any value in the interval $[0, 1)$. The event that the spinner stops between 0.25 and 0.75, in terms of the values of X, is the interval $[0.25, 0.75]$. We denote this event by $(0.25 \leq X \leq 0.75)$.

Just as probabilities were associated with the events in the original sample space, we desire to assign probabilities to the events of the sample space of a random variable. To be sure this can be done, we assume that a random variable X is a function such that any event in Ω_X corresponds to a specific event in Ω. Then the probability assigned to any event in Ω_X is the probability associated with the corresponding event in Ω.

Definition 2.5   A random variable X is a real valued function defined on a sample space Ω such that an event in Ω_X corresponds to an event in Ω.

In general, any real valued function defined on a sample space can be assumed to be a random variable. Examples 2.8 and 2.9 illustrate two different types. The sample space for the random variable in Example 2.8 is a set of discrete numbers. When a random variable can take on only discrete values, it is called a *discrete* random variable. A second type, called a *continuous* random variable, can take on any value in a continuous range. Many measurement type variables, such as length, weight, voltage, etc., that can, at least conceptually, take on any value in a continuous range are examples of continuous random variables. The random variable in Example 2.9 for which $\Omega_X = [0, 1)$ is also an example of a continuous random variable. The distinction between discrete and continuous random variables is important when describing the probability model for a random variable. This distinction is made more precise in Section 2.8.

2.7 PROBABILITY DISTRIBUTIONS

In Section 2.2 we introduced the probability model as an abstraction of a random experiment. When the outcomes of an experiment are values of a random variable, the probability model is called the *probability distribution* of the random variable. Thus for a specified random variable its probability distribution is an assignment of probability to all events in the sample space of the random variable. For example, if $\Omega_X = [0, 1]$ for a random variable X, the probability distribution is a statement of the values $P(a < X < b)$ for any subinterval (a, b) in Ω_X.

Rather than present the probabilities $P(a < X < b)$ for all events in Ω_X, one customarily describes the probability distribution of a random variable using either the distribution function $F(x)$ or the probability or density function $f(x)$.

2.8 DISTRIBUTION, PROBABILITY, AND DENSITY FUNCTIONS

As mentioned in the previous section, the probability distribution of a random variable is usually expressed in terms of the distribution, probability, or density function. Given either of these, i.e., either the distribution function or the probability function for a discrete random variable (distribution function or density function for a continuous random variable), one can easily evaluate the probability of any event of the random variable. We now define these functions.

Definition 2.6   For any random variable X the *distribution function* (cumulative distribution function) $F(x)$ is a function defined for all real values x such that

$$F(x) = P(X \leq x) \qquad -\infty < x < \infty \qquad (2.14)$$

Some properties of the distribution function are:

1. $F(-\infty) = 0$
2. $F(\infty) = 1$
3. If $x_1 < x_2$, then $F(x_1) \leq F(x_2)$; i.e., $F(x)$ is a nondecreasing function of x
4. If $(a < X \leq b)$ is an event in Ω_X, $P(a < X \leq b) = F(b) - F(a)$
5. $P(X = x) = F(x^+) - F(x^-)$, where $F(x^+)$ is the value of $F(x^*)$ as x^* approaches x from above and $F(x^-)$ is the value of $F(x^*)$ as x^* approaches x from below.

Writing any event as a union of events of the types $(a < X \leq b)$ and $(X = c)$ and using properties (4) and (5), the distribution function can be used to evaluate the probability of any event.

To define the probability and density functions, we must distinguish between *discrete* and *continuous* random variables.

Definition 2.7   A random variable that can take on only a finite or countable number of discrete values is a *discrete random* variable.

When dealing with a discrete random variable, the probability function can be used to describe the probability distribution.

Definition 2.8   For a discrete random variable X, the *probability function* $f(x)$ is a function defined for all x such that

$$f(x) = P(X = x) \qquad (2.15)$$

Although $f(x)$ is defined for all x, it will have a nonzero value only for those x's in Ω_X. Some properties of $f(x)$ are:

1. $f(x) \geq 0$ for all x
2. $\sum_{\text{allx}} f(x) = 1$
3. If E is an event in Ω_X, i.e., E is a subset of Ω_X, then $P(X \in E) = \sum_{\text{all } x \in E} f(x)$
4. $F(t) = \sum_{x \leq t} f(x)$

As was true for the distribution function, we can use the probability function to evaluate the probability of any event by applying property (3).

Example 2.10

Suppose the random experiment is the toss of a coin three times and the random variable of interest is: X = number of heads in the three tosses. The sample space is $\Omega_X = \{0, 1, 2, 3\}$. Assuming a fair coin and independent tosses, the probability function for X is

$$
\begin{aligned}
f(x) &= 1/8 && x = 0, 3 \\
&= 3/8 && x = 1, 2 \\
&= 0 && \text{otherwise}
\end{aligned}
$$

Thus the probability that there will be two or more heads is

$$P(X \geq 2) = P(X = 2) + P(X = 3) = f(2) + f(3) = 4/8$$

The second type of random variable is the *continuous random variable*.

Definition 2.9   A random variable X is a *continuous random variable* if there exists a nonnegative function $f(x)$ such that

$$P(X \in E) = \int_E f(x)\,dx \qquad (2.16)$$

for any event E in Ω_X.

Thus a random variable is continuous if we have available a nonnegative function that can be integrated to evaluate the probability of any event in Ω_X. Certainly, if X is a continuous random variable, X can take on a continuum of values.

Definition 2.10   For X, a continuous random variable, the function $f(x)$ is called the *density function* of X.

The relationship between the distribution and density functions for a continuous random variable is

$$F(x) = \int_{-\infty}^{x} f(t)\,dt \qquad (2.17)$$

$$f(x) = \frac{dF(x)}{dx} \qquad (2.18)$$

Example 2.11

Let X be a continuous random variable such that $\Omega_X = [0, 1]$ and

$$
\begin{aligned}
f(x) &= 1 && 0 \leq x \leq 1 \\
&= 0 && \text{otherwise}
\end{aligned}
$$

Then if E is the event $(0.2 < X < 0.8)$,

$$P(0.2 < X < 0.8) = \int_{0.2}^{0.8} dx = 0.8 - 0.2 = 0.6$$

Also,

$$F(x) = 0 \quad x < 0$$
$$= x \quad 0 \le x \le 1$$
$$= 1 \quad x > 1$$

2.9 SOME SPECIAL PROBABILITY DISTRIBUTIONS

Certain distributions occur so often in the application of probability models and statistical methods that they merit special attention. Some of these are tabulated in Tables 2.1 and 2.2. Examples of some of these distributions as probability models are given below:

Distribution	Application
Binomial (n, p)	Let $\mathcal{E}$ be an experiment for which the outcome is either of type E or not of type E and such that the probability that the outcome is of type E is p. If $\mathcal{E}$ is repeated independently n times and X = the number of trials in which E occurs, then X has a binomial (n, p) distribution. Selecting defective or nondefective units, selecting people who use or do not use a product, tossing a coin, and observing if a unit operates or does not operate at n times are examples of the application of the binomial model.
Poisson (λ)	Let E be a "rare" event that occurs over time (space) such that E occurs, on the average, δ times per unit time (space). If X = the number of occurrences of E after t units, then X has a Poisson $(\lambda = \delta t)$ distribution. Examples of the application of the Poisson model are: gamma rays emitted from a radioactive source, cars passing a point on an interstate highway, a "flaw" in a roll of wire, and a "weak" spot in the aluminum body of an aircraft.
Exponential (θ)	Let T be the length of time between occurrences of a "rare" event that occurs, on the average, θ times per unit of time. Then T is an exponential (θ) random variable. Examples of the application of the exponential model are gap lengths between cars, lifetimes of electronic and mechanical units, and arrivals of customers at a service counter.
Uniform (a, b)	Let X be a variable with the property that if I_1 and I_2 are any two nonoverlapping subintervals of $[a, b]$ of equal length, then $P(X \in I_1) = P(X \in I_2)$. Then X has a uniform (a, b) distribution. The random variable in Example 2.9 is an illustration of a uniform $(0, 1)$ random variable.

TABLE 2.1–Special Discrete Probability Distributions

Name	Functional Form of Probability Function	Restriction	Mean	Variance
Hypergeometric	$f(x) = \dfrac{C(D,x)C(N-D,n-x)}{C(N,n)}$	$x = a, a+1, \ldots, b-1, b$ $a = \max(0, n-N+D)$ $b = \min(D,n)$ $n = 1, 2, \ldots, N$ $N = 1, 2, \ldots$	nD/N	$\dfrac{nD(N-D)(N-n)}{N^2(N-1)}$
Binomial	$f(x) = C(n,x)p^x(1-p)^{n-x}$	$x = 0, 1, \ldots, n$ $0 < p < 1$ $n = 1, 2, \ldots$	np	$np(1-p)$
Poisson	$f(x) = e^{-\lambda}\lambda^x/x!$	$x = 0, 1, \ldots$ $\lambda > 0$	λ	λ
Negative binomial	$f(x) = C(x+r-1, r-1)p^r q^x$	$x = 0, 1, \ldots$ $0 < p < 1$ $q = 1 - p$	rq/p	rq/p^2
Geometric	$f(x) = pq^x$	$x = 0, 1, \ldots$ $0 < p < 1$ $q = 1 - p$	q/p	q/p^2

TABLE 2.2–Special Continuous Probability Distributions

Name	Functional Form of Density Function	Restriction	Mean	Variance
Normal	$f(x) = (2\pi b^2)^{-1/2} \exp[-(x-a)^2/2b^2]$	$-\infty < x < \infty$ $b > 0$	a	b^2
Standard normal	$g(z) = (2\pi)^{-1/2} \exp[-z^2/2]$	$-\infty < z < \infty$	0	1
Exponential	$f(x) = (1/\theta)e^{-x/\theta}$	$x > 0$ $\theta > 0$	θ	θ^2
Gamma	$f(x) = \dfrac{x^a e^{-x/b}}{\Gamma(a+1)b^{a+1}}$	$x > 0$ $a > -1$ $b > 0$	$b(a+1)$	$b^2(a+1)$
Beta	$f(x) = \dfrac{\Gamma(a+b+2)x^a(1-x)^b}{\Gamma(a+1)\Gamma(b+1)}$	$0 < x < 1$ $a > -1$ $b > -1$	$\dfrac{a+1}{a+b+2}$	$\dfrac{(a+1)(b+1)}{(a+b+2)^2(a+b+3)}$
Weibull	$f(x) = abx^{b-1}\exp[-ax^b]$	$x > 0$ $a > 0$ $b > 0$	$\alpha^\beta \Gamma(\beta+1)$ where $\alpha = 1/a$ and $\beta = 1/b$	$\alpha^{2\beta}[\Gamma(2\beta+1) - \Gamma^2(\beta+1)]$ where $\alpha = 1/a$ where $\beta = 1/b$
Chi-square	$f(x) = \dfrac{x^{(\nu/2)-1}e^{-x/2}}{2^{\nu/2}\Gamma(\nu/2)}$	$x > 0$ $\nu = 1,2,\ldots$	ν	2ν
"Student's" t	$f(t) = \dfrac{\Gamma[(\nu+1)/2]}{\sqrt{\pi}\Gamma(\nu/2)(1+t^2/\nu)^{(\nu+1)/2}}$	$-\infty < t < \infty$ $\nu = 1,2,\ldots$	0	$\nu/(\nu-2)$ for $\nu > 2$
F	$g(F) = \dfrac{KF^{(\nu_1-2)/2}}{(1+\nu_1 F/\nu_2)^{(\nu_1+\nu_2)/2}}$ where $K = \dfrac{\Gamma[(\nu_1+\nu_2)/2]}{\Gamma(\nu_1/2)\,\Gamma(\nu_2/2)}\left(\dfrac{\nu_1}{\nu_2}\right)^{\nu_1/2}$	$F > 0$ $\nu_1 = 1,2,\ldots$ $\nu_2 = 1,2,\ldots$	$\dfrac{\nu_1}{\nu_2-2}$	$\dfrac{\nu_2^2(\nu_1+2)}{\nu_1(\nu_2-2)(\nu_2-4)}$

Normal (μ, σ^2) If X can be expressed as $X = \mu + \epsilon$ where μ is the "true" value of some physical measurement and ϵ is the cumulation of "random" errors that occur in taking such measurements experience has shown that X is approximately distributed as a normal random variable with parameters μ and σ^2, the variance of the random error. Also, for n large, a binomial random variable is approximately normal with $\mu = np$ and $\sigma^2 = np(1 - p)$.

Some of the distributions tabulated in Tables 2.1 and 2.2 occur most prominently as sampling distributions, i.e., distributions of sample statistics. These are the "Student's" t, chi-square, and Snedecor's F. Discussion of these distributions is delayed until Chapter 4.

Since most applications involving these distributions require the use of probabilities associated with the distributions, it is convenient to have available adequate tables of such probabilities. Accordingly, tables for the Poisson, standard normal, chi-square, "Student's" t, and F distributions are presented in Appendixes 2 through 6. Each of these tables is given in cumulative form, i.e., in terms of the cumulative distribution function, so that the reader will have to learn only one method of reading the tables.

2.10 EXPECTED VALUES AND OTHER DESCRIPTIVE MEASURES

To aid in the description of probability distributions, it is helpful to know something about their properties. For example, it may be helpful to know where the "center" of the distribution is or to know something about the dispersion. Of special importance are those properties associated with the concept of the *expected value*.

The expected value of any function $g(X)$ of the random variable X is the weighted average of the function (weighted by the probability of its occurrence) over all possible values of the variables.

Definition 2.11   For any function $g(X)$ of the random variable X, the *expected value of $g(X)$*, denoted $E[g(X)]$, can be evaluated by

$$E[g(X)] = \sum_{\text{all } x} g(x)f(x) \qquad (2.19)$$

if X is discrete with probability function $f(x)$, and

$$E[g(X)] = \int_{-\infty}^{\infty} g(x)f(x)\,dx \qquad (2.20)$$

if X is continuous with density function $f(x)$.

The symbol $E[\cdot]$ will be used to denote the expected value of whatever appears within the brackets. Thus if $g(X)$ is the function $g(X) = X^2$, the symbol $E[X^2]$ denotes the expected value of the square of X.

From the definition it can be shown that the following properties hold for expected values:

1. If $g(X) = c$, a constant, then

$$E[c] = c \qquad (2.21)$$

2. If $g(X) = aX$,

$$E[aX] = aE[X] \qquad (2.22)$$

3. If $g(X) = aX + b$,

$$E[aX + b] = aE[X] + b \qquad (2.23)$$

4. If $g_1(X)$ and $g_2(X)$ are two functions of X,

$$E[g_1(X) + g_2(X)] = E[g_1(X)] + E[g_2(X)] \qquad (2.24)$$

5. If $g_1(X) \leq g_2(X)$ for all values of X,

$$E[g_1(X)] \leq E[g_2(X)] \qquad (2.25)$$

6.

$$|E[g(X)]| \leq E[|g(X)|] \qquad (2.26)$$

The expectations of certain types of functions play an important role in describing a probability distribution. One of these is called the kth *moment* of X.

Definition 2.12   The *kth moment with respect to the origin of X is the expectation $E[X^k]$, denoted by μ_k'.*

If X is continuous,

$$\mu_k' = E[X^k] = \int_{-\infty}^{\infty} x^k f(x)\, dx \qquad (2.27)$$

or if X is discrete,

$$\mu_k' = \sum_{\text{all } x} x^k f(x) \qquad (2.28)$$

The first moment $\mu_1' = E[X]$, commonly denoted μ, is called the *mean* or *expected value* of X and is a measure of the "center" of the probability distribution. It is often used as a measure of location of a distribution.

Another class of expectations of importance are the moments about the mean.

Definition 2.13   The *kth moment about the mean of X is the expectation $E[(X - \mu)^k]$, denoted by μ_k.*

If X is continuous,

$$\mu_k = \int_{-\infty}^{\infty} (x - \mu)^k f(x)\, dx \qquad (2.29)$$

or if X is discrete,

$$\mu_k = \sum_{\text{all } x} (x - \mu)^k f(x) \qquad (2.30)$$

Using Eq. (2.29) or (2.30), we see that $\mu_1 = 0$. The second moment about the mean μ_2 is called the *variance* and is often denoted by σ^2. That is,

$$\sigma^2 = E[(X - \mu)^2] \qquad (2.31)$$

This expectation is used as a measure of the variability or dispersion of the probability distribution. Using the properties of expected values,

$$\sigma^2 = E[(X - \mu)^2] = E[X^2] - \mu^2 = \mu_2' - \mu^2 \qquad (2.32)$$

The positive square root of the variance, denoted by σ, is called the *standard deviation* and is often used instead of σ^2 to measure variability.

Another function frequently used in statistical theory is the *moment generating function*.

Definition 2.14   The *moment generating function* of X, $\Phi_X(t)$, defined for all values t, is

$$\Phi_X(t) = E[e^{tX}] \qquad (2.33)$$

We call $\Phi_X(t)$ the moment generating function because it can be used to obtain the moments of the random variable. In particular, $E[X^n]$ is the nth derivative of $\Phi_X(t)$, evaluated at $t = 0$. Also, $\Phi_X(t)$ is useful for evaluating the distribution of a function of X.

2.11 MULTIVARIATE PROBABILITY DISTRIBUTIONS

We often encounter situations involving two or more random variables defined on the same sample space. To make probability statements about the simultaneous behavior of several random variables, we introduce the idea of a multivariate probability model or *multivariate probability distribution*.

To develop the ideas of a multivariate (joint) distribution, we look at the special case of two random variables. The results given here are easily extended to the general case of n random variables.

If X and Y are two random variables, outcomes consist of points (x, y) in the two-dimensional plane. Thus the sample space $\Omega_{X,Y}$ is a subset of the two-dimensional plane, and an event E is a subset in $\Omega_{X,Y}$. The joint probability distribution is the assignment of probabilities, $P_{X,Y}(E)$, to all two-dimensional events E. Just as in one dimension, $P_{X,Y}(E)$ is the probability that the outcome (x, y) is in the event E.

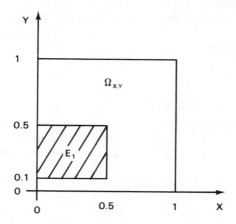

Fig. 2.1—Sample space and an event for X and Y.

Example 2.12

Suppose X and Y are two random variables such that $\Omega_{X,Y} = \{(x,y): 0 \le x \le 1, 0 \le y \le 1\}$. One event in the sample space is $E_1 = \{(x,y): 0 \le x \le 0.5, 0.1 \le y \le 0.5\}$. These are shown in Figure 2.1.

We denote the event E_1 by $(0 \le X \le 0.5, 0.1 \le Y \le 0.5)$. Also, the probability $P_{X,Y}(E_1)$ of E_1 is denoted by $P(0 \le X \le 0.5, 0.1 \le Y \le 0.5)$. Likewise, the event $E_2 = \{(x,y): 0.5 \le x < 1, 0 \le y < 1\}$ is denoted by $(0.5 \le X < 1, 0 \le Y < 1)$ and the probability of E_2 is $P(0.5 \le X < 1, 0 \le Y < 1)$.

As we did for one-dimensional probability distributions, we now indicate some functions useful for describing joint probability distributions.

2.11.1 Joint (cumulative) Distribution Function

The joint distribution function $F(x,y)$ is useful for evaluating the probability distribution of any pair of random variables X and Y.

Definition 2.15 For any two random variables X and Y the *joint distribution function* $F(x,y)$ is a function defined for all real values x and y such that

$$F(x,y) = P(X \le x, Y \le y)$$
$$-\infty < x < \infty, -\infty < y < \infty \qquad (2.34)$$

Given $F(x,y)$ one can evaluate joint probabilities of the type $P(a < X \le b, c < Y \le d)$ for $a < b$ and $c < d$ by using the identity

$$P(a < X \le b, c < Y \le d) = F(b,d) - F(a,d) - F(b,c) + F(a,c)$$
$$(2.35)$$

In turn, by writing any event as the union of such rectangles, the probability of the event can be evaluated using the distribution function.

Some additional properties of the distribution function are:

1. $F(-\infty, y) = F(x, -\infty) = F(-\infty, -\infty) = 0$ for any x, y
2. $F(\infty, \infty) = 1$
3. $F(\infty, y) = P(Y \leq y)$
4. $F(x, \infty) = P(X \leq x)$

2.11.2 Joint Probability Function

If the two random variables X and Y are each discrete and hence $\Omega_{X,Y}$ contains a finite or countable number of points, the *joint probability function* can also be used to describe the probability distribution.

Definition 2.16   For any two discrete random variables X and Y the *joint probability function* $f(x, y)$ is a function defined for all real x and y such that

$$f(x, y) = P(X = x, Y = y) \qquad (2.36)$$

That is, $f(x, y)$ is the probability that simultaneously $X = x$ and $Y = y$.

Properties of the joint probability function are:

1. $f(x, y) \geq 0$ for all (x, y)
2. $\sum_{\text{all } x} \sum_{\text{all } y} f(x, y) = 1$
3. $P(a < X \leq b, c < Y \leq d) = \sum_{a<x\leq b} \sum_{c<y\leq d} f(x, y)$ for $a < b, c < d$

Example 2.13

Suppose a taxicab company has two cars such that at any time each has probability 1/2 of being in service. If a cab is in service, the number of passengers in the cab is equally likely to be 1, 2, 3, or 4. Let X be the random variable denoting the number of cabs in service and Y the random variable denoting the total number of passengers being carried by the cabs. Both X and Y are discrete random variables. The joint probability function is

$$
\begin{aligned}
f(x, y) &= 1/4 && \text{if } x = 0, y = 0 \\
&= 1/8 && \text{if } x = 1, y = 1, 2, 3, 4 \\
&= 1/64 && \text{if } x = 2, y = 2, 8 \\
&= 2/64 && \text{if } x = 2, y = 3, 7 \\
&= 3/64 && \text{if } x = 2, y = 4, 6 \\
&= 4/64 && \text{if } x = 2, y = 5
\end{aligned}
$$

2.11.3 Joint Density Function

We say the random variables X and Y are jointly continuous if there exists a nonnegative function $f(x, y)$ such that

$$P(a < X \leq b, c < Y \leq d) = \int_a^b \int_c^d f(x, y)\, dy\, dx$$

$$\text{for } a < b, c < d \qquad (2.37)$$

Thus the random variables X and Y are jointly continuous if we have available a nonnegative function that can be integrated to evaluate the probability of any event in the sample space.

Definition 2.17   For X and Y jointly continuous random variables, the function $f(x,y)$ is called the *joint density function* for X and Y.

If $f(x,y)$ exists for the random variables X and Y, then

$$f(x,y) = \frac{\partial^2 F(x,y)}{\partial x \, \partial y} \tag{2.38}$$

whenever this derivative exists. Hence

$$F(x,y) = \int_{-\infty}^{x} \int_{-\infty}^{y} f(s,t)\, dt\, ds \quad \text{for all } (x,y) \tag{2.39}$$

Example 2.14

Let X and Y be two random variables defined on $[0, 1]$ such that

$$f(x,y) = 1 \quad 0 \leq x \leq 1, 0 \leq y \leq 1$$
$$= 0 \quad \text{otherwise}$$

This distribution is the joint uniform distribution on the unit square.

2.12 MARGINAL PROBABILITY DISTRIBUTIONS

If we have a pair of random variables X, Y with a joint probability distribution, we can also talk about the probability distribution of one of the random variables. Such probability distributions are called *marginal probability distributions* and refer to the probabilities $P(X \in E)$ where E is an event in the sample space of X.

The marginal probability distribution of X (or Y) can be obtained from knowledge of the joint probability distribution. For example, if X and Y are discrete random variables and the joint probability function $f(x,y)$ is known, the probability function for X is

$$f_X(x) = P(X = x) = \sum_{\text{all } y} f(x,y) \tag{2.40}$$

Similarly, the probability function for Y is

$$f_Y(y) = P(Y = y) = \sum_{\text{all } x} f(x,y) \tag{2.41}$$

If X, Y are jointly continuous with joint density function $f(x,y)$, the *marginal density functions* for X and Y are respectively

$$f_X(x) = \int_{-\infty}^{\infty} f(x,y)\, dy \tag{2.42}$$

and

$$f_Y(y) = \int_{-\infty}^{\infty} f(x,y)\, dx \tag{2.43}$$

On the other hand, if the joint distribution function $F(x, y)$ is known, the marginal distribution functions are respectively

$$F_X(x) = P(X \leq x) = F(x, \infty) \tag{2.44}$$

and

$$F_Y(y) = P(Y \leq y) = F(\infty, y) \tag{2.45}$$

2.13 CONDITIONAL PROBABILITY DISTRIBUTIONS

In Section 2.3 we introduced the notion of conditional probability of an event A, given the occurrence of a second event B, denoted $P(A \mid B)$. The concept of conditional probability also plays an important role in the case of jointly distributed random variables.

Example 2.15

Suppose X and Y are joint *discrete* random variables with probability function $f(x, y)$. A conditional probability of possible interest might be $P(X + Y = a \mid X = b)$ for $b < a$ and $P(X = b) > 0$. Using the definition of conditional probability in Eq. (2.1),

$$P(X + Y = a \mid X = b) = \frac{P(X = b, X + Y = a)}{P(X = b)}$$

$$= \frac{P(X = b, Y = a - b)}{P(X = b)}$$

$$= f(b, a - b)/f_X(b) \tag{2.46}$$

where $f(\cdot)$ and $f_X(\cdot)$ are the joint probability function of X and Y and the marginal probability function of X respectively.

If X and Y are discrete, a conditional probability of particular importance is of the type $P(Y \in E \mid X = b)$, where E is some event in Ω_Y and b is an outcome in Ω_X such that $P(X = b) > 0$. Such conditional probabilities constitute the *conditional probability distribution* of Y given $X(= b)$. Similarly, we can talk about the conditional probability of X given Y.

As with other probability distributions, we have functions that can be used to describe the conditional probability distribution. These are the *conditional distribution function* of Y given X, denoted $F_{Y \mid X}(y \mid x)$, and the *conditional probability function* $f_{Y \mid X}(y \mid x)$.

If the joint probability function $f(x, y)$ for X and Y is known, the conditional distribution and probability function for Y given X are respectively

$$F_{Y \mid X}(y \mid x) = P(Y \leq y \mid X = x) = \sum_{t \leq y} f(x, t)/f_X(x) \tag{2.47}$$

and

$$f_{Y \mid X}(y \mid x) = P(Y = y \mid X = x) = f(x, y)/f_X(x) \tag{2.48}$$

Also, for X given Y

$$F_{X|Y}(x \mid y) = P(X \leq x \mid Y = y) = \sum_{s \leq x} f(s,y)/f_Y(y) \qquad (2.49)$$

and

$$f_{X|Y}(x \mid y) = P(X = x \mid Y = y) = f(x,y)/f_Y(y) \qquad (2.50)$$

Similarly, if X and Y are jointly continuous, the *conditional distribution functions* and the *conditional density functions* can be evaluated from knowledge of the joint density function as follows:

$$F_{X|Y}(x \mid y) = \int_{-\infty}^{x} [f(s,y)/f_Y(y)]\,ds \qquad (2.51)$$

$$F_{Y|X}(y \mid x) = \int_{-\infty}^{y} [f(x,t)/f_X(x)]\,dt \qquad (2.52)$$

$$f_{X|Y}(x \mid y) = f(x,y)/f_Y(y) \qquad (2.53)$$

$$f_{Y|X}(y \mid x) = f(x,y)/f_X(x) \qquad (2.54)$$

Example 2.16

Consider the pair of discrete random variables X and Y such that the joint probability function for X, Y is given in Table 2.3. As can be seen, the marginal probability function for X is

$$
\begin{aligned}
f_X(x) &= 0.7 & \text{if } x = 1 \\
&= 0.3 & \text{if } x = 2 \\
&= 0 & \text{otherwise}
\end{aligned}
$$

This comes directly from Eq. (2.40) since

$$f_X(1) = \sum_{y=1}^{2} f(1,y) = 0.6 + 0.1 = 0.7$$

$$f_X(2) = \sum_{y=1}^{2} f(2,y) = 0.2 + 0.1 = 0.3$$

To evaluate the conditional probability function of X given $Y = 1$, we see from Table 2.3 that $f_Y(1) = 0.8$. Thus, using Eq. (2.50),

$$
\begin{aligned}
f_{X|Y}(x \mid 1) &= 0.6/0.8 = 0.75 & \text{if } x = 1 \\
&= 0.2/0.8 = 0.25 & \text{if } x = 2 \\
&= 0 & \text{otherwise}
\end{aligned}
$$

TABLE 2.3–Joint Probability Function $f(x,y)$
for the Two Random Variables X and Y

x \ y	1	2	$f_X(x)$
1	0.6	0.1	0.7
2	0.2	0.1	0.3
$f_Y(y)$	0.8	0.2	. . .

2.14 INDEPENDENT RANDOM VARIABLES

For jointly distributed random variables a very important concept is the idea of *independent random variables*.

Definition 2.18   Two random variables X and Y are *independent* if and only if

$$F_{X,Y}(x,y) = F_X(x) F_Y(y) \text{ for all } (x,y) \qquad (2.55)$$

That is, a pair of random variables are independent if and only if

$$P(X \leq x, Y \leq y) = P(X \leq x) P(Y \leq y) \qquad (2.56)$$

for all values (x,y).

If X and Y are jointly continuous and independent, it follows from Eq. (2.55) that the joint density function is a product of the marginal density functions; i.e.

$$f(x,y) = f_X(x) f_Y(y) \text{ for all } (x,y) \qquad (2.57)$$

The same identity holds for the probability functions in the case of two discrete random variables.

2.15 EXPECTED VALUES OF MULTIVARIATE RANDOM VARIABLES

If $g(X, Y)$ is a function of the random variables X and Y, the *expected value* of $g(X, Y)$ is evaluated by the identity

$$E[g(X, Y)] = \sum_{\text{all } x} \sum_{\text{all } y} g(x,y) f(x,y) \qquad (2.58)$$

if X, Y are jointly discrete random variables and by

$$E[g(X, Y)] = \int_{-\infty}^{\infty} \int_{-\infty}^{\infty} g(x,y) f(x,y) \, dx \, dy \qquad (2.59)$$

when X, Y are jointly continuous.

The expectations of certain functions play an important role in characterizing a joint probability distribution. We list some of these (assuming jointly continuous random variables):

1. *Product moments* of X and Y:

$$\mu'_{rs} = E[X^r Y^s] = \int_{-\infty}^{\infty} \int_{-\infty}^{\infty} x^r y^s f(x,y) \, dy \, dx \qquad (2.60)$$

2. *Mean* of X:

$$\mu_X = \mu'_{10} = E[X] = \int_{-\infty}^{\infty} \int_{-\infty}^{\infty} x f(x,y) \, dy \, dx = \int_{-\infty}^{\infty} x f_X(x) \, dx$$

Mean of Y:

$$\mu_Y = \mu'_{01} = E[Y] = \int_{-\infty}^{\infty} \int_{-\infty}^{\infty} yf(x,y)\,dx\,dy = \int_{-\infty}^{\infty} yf_Y(y)\,dy$$

(2.61)

3. *Central product moments* of X and Y:

$$\mu_{rs} = E[(X - \mu_X)^r (Y - \mu_Y)^s]$$

(2.62)

4. *Variance* of X:

$$\sigma_X^2 = \mu_{20} = E[(X - \mu_X)^2] = \int_{-\infty}^{\infty} \int_{-\infty}^{\infty} (x - \mu_X)^2 f(x,y)\,dy\,dx$$

$$= \int_{-\infty}^{\infty} (x - \mu_X)^2 f_X(x)\,dx$$

Variance of Y:

$$\sigma_Y^2 = \mu_{02} = E[(Y - \mu_Y)^2] = \int_{-\infty}^{\infty} \int_{-\infty}^{\infty} (y - \mu_Y)^2 f(x,y)\,dx\,dy$$

$$= \int_{-\infty}^{\infty} (y - \mu_Y)^2 f_Y(y)\,dy$$

(2.63)

5. *Covariance* of X and Y:

$$\sigma_{XY} = \mu_{11} = E[(X - \mu_X)(Y - \mu_Y)]$$

(2.64)

The covariance is used as a measure of the systematic association or linear correlation between X and Y. If $\sigma_{XY} = 0$, the variables are said to be linearly *uncorrelated*. (For a further discussion of this concept see Chapter 8.)

The same identity holds for the variance in the multivariate case as for the univariate random variable; namely,

$$\sigma_X^2 = E[X^2] - \mu_X^2; \qquad \sigma_Y^2 = E[Y^2] - \mu_Y^2$$

(2.65)

An identity, similar to Eq. (2.65) for the variances, relating the covariance to the product moments of X and Y is

$$\sigma_{XY} = E[XY] - \mu_X \mu_Y = \mu_{11} - \mu'_{10}\mu'_{01}$$

(2.66)

This latter identity is often useful when it is necessary to evaluate the covariance of two random variables.

If the two random variables X and Y are independent, then for any functions g and h,

$$E[g(X)h(Y)] = E[g(X)]E[h(Y)]$$

(2.67)

In particular,

$$E[XY] = E[X]E[Y] = \mu_X \mu_Y$$

(2.68)

Thus if X and Y are independent, X and Y are uncorrelated. The converse, however, is not necessarily true. Generally, if X and Y are uncorrelated, they need not be independent. There is a special case, though, for which this is true. If X and Y are joint normal random variables and $\sigma_{XY} = 0$ (i.e., X and Y are uncorrelated), then X and Y are also independent.

Another parameter closely related to the covariance of X and Y is the *correlation* between X and Y defined by

$$\rho_{XY} = \sigma_{XY}/\sigma_X\sigma_Y \tag{2.69}$$

where σ_X and σ_Y are the *standard deviations* of X and Y respectively. It can be shown that $-1 \le \rho_{XY} \le 1$.

2.16 EXPECTATIONS OF SUMS OF RANDOM VARIABLES

In the previous section we introduced the idea of the expected value of a function of several random variables. Of special interest in statistical analysis are functions that arise as the sum of several random variables.

If $X_1, \ldots, X_n$ are n jointly distributed random variables, consider a linear combination of the X_i's such as

$$U = \sum_{i=1}^{n} a_i X_i \tag{2.70}$$

where the a_i are known constants. The expected value of the linear combination is

$$E[U] = \sum_{i=1}^{n} a_i E[X_i] = \sum_{i=1}^{n} a_i \mu_i \tag{2.71}$$

where μ_i is the mean of X_i. Also, the variance of U, denoted here by $V(U)$, is

$$V(U) = \sum_{i=1}^{n} a_i^2 V(X_i) + 2\sum_{i<j} a_i a_j \operatorname{Cov}(X_i, X_j)$$

$$= \sum_{i=1}^{n} a_i^2 \sigma_i^2 + 2\sum_{i<j} a_i a_j \sigma_{ij} \tag{2.72}$$

where $V(X_i) = \sigma_{X_i}^2 = \sigma_i^2$ and $\operatorname{Cov}(X_i, X_j) = \sigma_{X_i X_j} = \sigma_{ij}$.

If the random variables $X_1, \ldots, X_n$ are uncorrelated (i.e., $\sigma_{ij} = 0$ for $i \ne j$), Eq. (2.72) for $V(U)$ simplifies to

$$V(U) = \sum_{i=1}^{n} a_i^2 \sigma_i^2 \tag{2.73}$$

Recalling that $X_1, \ldots, X_n$ being mutually independent assures that they are uncorrelated, we see that Eq. (2.73) gives the variance of a linear combination of independent random variables.

Example 2.17

Suppose $X_1, \ldots, X_n$ are independent identically distributed normal (μ, σ^2) random variables, and let $U = \sum_{i=1}^{n} a_i X_i$ be a linear combination of the X_i's. Then,

$$E[U] = \mu \sum_{i=1}^{n} a_i; \quad V(U) = \sigma^2 \sum_{i=1}^{n} a_i^2$$

Of particular interest in statistics is the linear combination, $U = \sum_{i=1}^{n} X_i/n$. For this combination in which each $a_i = 1/n$, $E[U] = \mu$, and $V(U) = \sigma^2/n$.

2.17 DISTRIBUTIONS OF FUNCTIONS OF RANDOM VARIABLES

Frequently in statistics one encounters the need to describe the probability distribution of a function of one or more random variables. Let X be a random variable with known probability distribution $P_X(x)$, and let $Y = g(X)$ be a function of X. A general approach to evaluating the probability distribution is to find the distribution function $F_Y(y)$ for Y. From $F_Y(y)$ one can then obtain the density or probability function $f_Y(y)$. The appropriate relationship for evaluating $F_Y(y)$ is

$$F_Y(y) = P(Y \leq y) = P_X(\{x : g(x) \leq y\}) \tag{2.74}$$

i.e., $F_Y(y)$ is the probability of that event in Ω_X described by the set of x for which $g(x) \leq y$.

Similarly, if $Y = g(X_1, \ldots, X_n)$ is a function of n jointly distributed random variables, the distribution function for Y is evaluated using the identity

$$F_Y(y) = P_{X_1, \ldots, X_n}[\{(x_1, \ldots, x_n) : g(x_1, \ldots, x_n) \leq y\}] \tag{2.75}$$

Example 2.18

Suppose X is a uniform random variable on $[0, 1]$. Then,

$$F_X(x) = 0 \quad x < 0$$
$$= x \quad 0 \leq x \leq 1$$
$$= 1 \quad x > 1$$

Let $Y = g(X) = X + 3$. Using Eq. (2.74),

$$F_Y(y) = P(Y \leq y) = P(\{x : x + 3 \leq y\}) = P(\{x : x \leq y - 3\})$$
$$= F_X(y - 3) = 0 \quad y < 3$$
$$= y - 3 \quad 3 \leq y \leq 4$$
$$= 1 \quad y > 4$$

We now consider two special cases in which the density or probability function for Y can be obtained directly from the corresponding function for X.

Case 1: X is discrete, g is 1-1

By a 1-1 function we mean that each value of x is related to one and only one value $y = g(x)$ and each value y is related to one and only one value $x = h(y)$, where $h(y)$ is obtained by solving $y = g(x)$ for x in terms of y. If X is discrete and g is 1-1, then Y is discrete and

$$f_Y(y) = P_X[X = h(y)] = f_X[h(y)] \qquad (2.76)$$

Example 2.19

Suppose X is a discrete uniform random variable with $\Omega_X = \{0, 1, 2\}$ and

$$\begin{aligned} f_X(x) &= 1/3 \qquad x = 0, 1, 2 \\ &= 0 \qquad \text{otherwise} \end{aligned}$$

Let $Y = g(X) = 2X + 3$, so $\Omega_Y = \{3, 5, 7\}$. Then $h(Y) = (Y - 3)/2$ and

$$\begin{aligned} f_Y(y) = f_X[h(y)] &= 1/3 \qquad y = 3, 5, 7 \\ &= 0 \qquad \text{otherwise} \end{aligned}$$

Case 2: X is continuous, g is 1-1 and differentiable

If X is continuous and g is 1-1 and differentiable, we can solve for x in terms of y, say $x = h(y)$. The density function for Y is given by

$$f_Y(y) = f_X[h(y)] \left| \frac{dh}{dy} \right| \qquad (2.77)$$

where $|\dots|$ refers to the absolute value. The derivative dh/dy is often called the Jacobian of the transformation.

Example 2.20

Suppose X is a continuous random variable on $[0, 1]$ such that

$$\begin{aligned} f_X(x) &= 2x \qquad 0 \le x \le 1 \\ &= 0 \qquad \text{otherwise} \end{aligned}$$

Let $Y = g(X) = X^2$. Since X is a positive random variable, g is 1-1 and differentiable with $X = h(Y) = Y^{1/2}$. Thus

$$\frac{dh}{dy} = y^{-1/2}/2$$

and

$$\begin{aligned} f_Y(y) &= 1 \qquad 0 < y < 1 \\ &= 0 \qquad \text{otherwise} \end{aligned}$$

For additional special cases the reader is referred to one of the more advanced texts on probability theory given in the references at the end of the chapter. Of particular interest may be the cases involving several functions of several random variables.

An alternative procedure sometimes helpful in evaluating the distributions of certain kinds of functions is to use the *moment generating function*, $\Phi(t)$. The basis for the use of $\Phi(t)$ in evaluating the distribution of a function of X is its uniqueness. That is, if X and Y are two random variables such that $\Phi_X(t) = \Phi_Y(t)$ for all values of t, then X and Y have the same probability distribution. Three special cases (of relationships involving moment generating functions) will be given to indicate the usefulness of this approach.

Case 1

If $Y = a + X$, then

$$\Phi_Y(t) = e^{at}\Phi_X(t) \tag{2.78}$$

Case 2

If $Y = aX$, then

$$\Phi_Y(t) = \Phi_X(at) \tag{2.79}$$

Case 3

If $X_1, \ldots, X_n$ are independent random variables and $Y = \sum_{i=1}^{n} a_i X_i$, then

$$\Phi_Y(t) = \prod_{i=1}^{n} \Phi_{X_i}(a_i t) \tag{2.80}$$

Example 2.21

Suppose X_1 and X_2 are independent identically distributed exponential (θ) random variables (see Table 2.2). Then

$$\Phi_{X_i}(t) = E[e^{tx}] = \frac{1}{\theta} \int_0^\infty e^{-x/\theta} e^{tx}\, dx$$

$$= \frac{1}{\theta} \int_0^\infty e^{-x(1-\theta t)/\theta}\, dx = (1 - \theta t)^{-1}$$

Let $Y = X_1 + X_2$. Then, using Eq. (2.80) with $a_1 = a_2 = 1$, $\Phi_Y(t) = [\Phi_X(t)]^2 = (1 - \theta t)^{-2}$. It can be shown that $(1 - \theta t)^{-2}$ is the moment generating function of a gamma random variable with $a = 1, b = \theta$, thus Y has a gamma distribution.

2.18 DISTRIBUTION OF A SUM OF NORMAL RANDOM VARIABLES

In Section 2.17 we discussed some ideas relating to describing the probability distribution of a function of several random variables. In this section we discuss a special case of importance in statistics, namely, the distribution of a sum of normal random variables.

Case 1

Let $X_1, \ldots, X_n$ be n *independent* normal random variables such that $E[X_i] = \mu_i$ and $V(X_i) = \sigma_i^2$. If

$$U = \sum_{i=1}^{n} a_i X_i \tag{2.81}$$

is a linear combination of the X_i's, then U is a *normal* random variable with $E[U] = \sum_{i=1}^{n} a_i \mu_i$ and $V(U) = \sum_{i=1}^{n} a_i^2 \sigma_i^2$. The fact that the mean and variance are as given comes directly from Eqs. (2.71) and (2.73) in Section 2.16. The former fact that U is a normal random variable can be verified using Eq. (2.80) in Section 2.17 and the uniqueness property of the moment generating function.

Case 2

Let $X_1, \ldots, X_n$ be n *nonindependent* (i.e., correlated) normal random variables and

$$U = \sum_{i=1}^{n} a_i X_i \tag{2.82}$$

Again it can be shown that U is a *normal* random variable. Using Eqs. (2.71) and (2.72), we see that $E[U] = \sum_{i=1}^{n} a_i \mu_i$ and $V(U) = \sum_{i=1}^{n} a_i^2 \sigma_i^2 + 2 \sum_{i<j} a_i a_j \, \text{Cov}(X_i, X_j)$.

2.19 TCHEBYCHEFF'S INEQUALITY

An inequality useful in probability and statistics is due to Tchebycheff. If X is a random variable such that its mean and variance exist, i.e., $E[X] = \mu < \infty$ and $V(X) = \sigma^2 < \infty$, then Tchebycheff's inequality for any $k > 0$ is given by

$$P(\,|X - \mu| \geq k\sigma) \leq 1/k^2 \tag{2.83}$$

This inequality is often expressed in the following alternative form:

$$P(\,|X - \mu| < k\sigma) > 1 - 1/k^2 \tag{2.84}$$

or

$$P(\mu - k\sigma < X < \mu + k\sigma) > 1 - 1/k^2 \tag{2.85}$$

Tchebycheff's inequality shows how σ may be used as a measure of variation. It can be applied in a wide variety of cases for it assumes only the existence of μ and σ^2. That is, no assumption is made concerning the form of the population but only that the mean and variance exist.

If we restrict our attention to unimodal distributions, the inequality may be sharpened. Under such a restriction we obtain

$$P(\,|X - MO| \geq kB) \leq 4/9k^2 \tag{2.86}$$

where MO is the mode and $B^2 = \sigma^2 + (MO - \mu)^2$. An alternative form is

$$P(\,|\,X - \mu\,|\, \geq k\sigma) \leq \frac{4}{9} \cdot \frac{1 + A^2}{(k - |A|)^2} \qquad (2.87)$$

where $A = (\mu - MO)/\sigma$. It should be noted that if the distribution is not only unimodal but also symmetric (i.e., $\mu = MO$) Eqs. (2.86) and (2.87) reduce to

$$P(\,|\,X - \mu\,|\, \geq k\sigma) \leq 4/9k^2 \qquad (2.88)$$

[**Note:** The mode MO is the value of $x \in \Omega_X$ for which $f(x)$ is maximum.]

PROBLEMS

2.1   A sample of 3 TV sets is selected from a lot of 30. If there are 5 defective sets in the lot, what is the probability the sample will contain no defectives? 3 defectives? 1 defective and 2 nondefectives?

2.2   A buyer will accept a lot of 10 TV sets if a sample of 3, selected at random, contains no defective sets. What is the probability of accepting a lot of 10 that contains 5 defectives?

2.3   An electrical circuit consists of 4 switches in series. Assume that the operations of the 4 switches are statistically independent. If for each switch the probability of failure (i.e., remaining open) is 0.02, what is the probability of circuit failure?

2.4   Rework the preceding problem for the case where the circuit consists of 4 switches in parallel.

2.5   Defects are classified as type A, B, or C, and the following probabilities have been determined from available production data: $P(A) = 0.20$, $P(B) = 0.16$, $P(C) = 0.14$, $P(AB) = 0.08$, $P(AC) = 0.05$, $P(BC) = 0.04$, and $P(ABC) = 0.02$. What is the probability that a randomly selected item of product will exhibit at least one type of defect? If an item exhibits at least one type of defect, what is the probability that it exhibits both A and B defects?

2.6   An electrical assembly consists of 2 parts connected in series in the order: A followed by B. The probability that part A is defective is 0.025 and the probability that part B is defective is 0.011. What is the probability of having a defective assembly? A nondefective assembly? An assembly that fails only because part B is defective?

2.7   Suppose the probability that a certain piece of airborne electronic equipment will not be in working order after its first flight is 0.40, and the probability of failure drops to one-half its previous value after each succeeding flight. (Assume no repair and replacement.) What is the probability the equipment will be in working order after 3 flights? After 4 flights, given it has survived 2 flights?

2.8   Consider a 4-engine aircraft (2 on each wing) where the probability of an engine failure is 0.05. Assume that the probability of 1 engine failing is independent of the behavior of the others. What is the probability of a crash if the plane can fly on any 2 engines? If the plane requires at least 1 engine operating on each side in order to remain in the air?

2.9   Suppose 3 defective dry cells are mixed in with 7 nondefectives, and you start testing them one at a time. What is the probability that you will find the last defective on the sixth test?

2.10   Three operators (A, B, and C) alternate in operating a certain machine. The number of parts produced by A, B, and C are in the ratio 3:4:3 and, of the parts produced, 1 percent of A's, 2 percent of B's, and 5 percent of C's are defective. If a part is drawn at random from the output of their machine, what is the probability it will be defective?

2.11   Referring to Problem 2.10, what is the probability that if a defective part is selected, it was produced by A? by B? by C?

2.12   If $f(x, y) = \exp[-(x + y)]$ for $x > 0$, $y > 0$, find: (a) $f_X(x)$, (b) $f_Y(y)$, (c) $F_X(x)$, (d) $F_Y(y)$, (e) $f_{Y|X}(y \mid x)$, (f) $f_{X|Y}(x \mid y)$, (g) $F(x, y)$, (h) $F_{Y|X}(y \mid x)$, (i) $F_{X|Y}(x \mid y)$.

2.13   If $f(x, y) = 3x$ for $0 < y < x$, $0 < x < 1$, find the same functions as asked for in the preceding problem.

2.14   If $f(x, y) = 24y(1 - x - y)$ over the triangle bounded by the axes and the line $x + y = 1$, find the same functions as asked for in Problem 2.12.

2.15   During the course of a day a machine turns out either 0, 1, or 2 defective items with probabilities $1/5$, $3/5$, and $1/5$ respectively. Calculate the mean and variance.

2.16   Given that the number of accidents occurring at a particular intersection between 10:00 P.M. and midnight on Saturday is 0, 1, 2, 3, or 4 with probabilities 0.90, 0.04, 0.03, 0.02, 0.01 respectively, determine the expected number of accidents.

2.17   Suppose that the life in hours of a certain type of tube has the density function $f(x) = a/x^2$, $x \geq 500$, and $f(x) = 0$, $x < 500$. Find the distribution function. Determine the mean and variance. What is the probability a tube will last at least 1000 hours?

2.18   A submarine carries 3 missiles. Assuming the only error is in one direction (e.g., a range error but no sideways error) and that a hit within 40 miles of the target is considered a success, compute the probability of a successful operation (i.e., an operation in which at least one hit is a success) if all 3 missiles are launched and the error density function is:

$$
\begin{aligned}
f(x) &= (100 + x)/10{,}000 & -100 < x < 0 \\
&= (100 - x)/10{,}000 & 0 < x < 100 \\
&= 0 & \text{elsewhere}
\end{aligned}
$$

2.19   Referring to the previous problem, the submarine can carry 8 missiles of a smaller size. However, in this case a hit must be within 15 miles to be successful. Assuming the same density function, should the light or heavy missiles be used?

2.20   A service station will be supplied with gasoline once a week. Its weekly volume of sales in thousands of gallons is predicted by the density function $f(x) = 5(1 - x)^4$ for $0 < x < 1$. Determine what the capacity of its underground tank should be if the probability that its supply will be exhausted in a given week is to be 0.01.

2.21   Show that the correlation between two random variables is 0 if they are statistically independent.

2.22   Let X have the marginal density $f_X(x) = 1$ for $-1/2 < x < 1/2$, and let the conditional density of Y given X be

$$
\begin{aligned}
f_{Y|X}(y \mid x) &= 1 & x < y < x + 1, \; -1/2 < x < 0 \\
&= 1 & -x < y < 1 - x, \; 0 < x < 1/2 \\
&= 0 & \text{elsewhere}
\end{aligned}
$$

Find the correlation between X and Y. Discuss the relationship between correlation and statistical independence.

2.23 A process is producing parts that are on the average 1 percent defective. Ten parts are selected at random from the process, and the process is stopped if one or more of the 10 are defective. What is the probability that the process will be stopped?

2.24 In inspecting 1000 welded joints performed by a certain welder using a specific process, 150 defective joints were discovered. If the welder is about to weld 5 joints, what is the probability of getting no defective joints? of 1? of 2? of 2 or more? Discuss any assumptions you make in solving this problem.

2.25 A large number of rivets is used in assembling an airplane. It has been determined that the probability distribution for the number of defective rivets is Poisson with $\lambda = 2$. Find the probability that the number of defective rivets in a plane will be no more than 2.

2.26 Suppose there is an average of 1 typographical error per 10 pages in a certain book. What is the probability that a 30-page chapter will contain no errors?

2.27 A telephone switchboard handles on the average 600 calls during the rush hour. The board can make a maximum of 20 connections per minute. What is the probability the board will be overtaxed in any given minute during the rush hour?

2.28 Assuming a normal distribution, find:
(a) $P(-3 < Y \leq -1)$; given $\mu = 0, \sigma = 1$
(b) $P(-3 \leq Y < 0.5)$; given $\mu = 0, \sigma = 1$
(c) $P(-8 < Y < 0)$; given $\mu = 2, \sigma^2 = 4$
(d) $P(4 < Y \leq 50)$; given $\mu = -0.1, \sigma^2 = 4$
(e) $P(Y \geq 3)$; given $\mu = 0, \sigma^2 = 1$
(f) $P(Y < -3)$; given $\mu = 2, \sigma^2 = 4$

2.29 Assuming a chi-square distribution, find:
(a) $P(\chi^2 \geq 26.1)$ for $\nu = 14$
(b) $P(\chi^2 < 26.1)$ for $\nu = 15$
(c) $P(23.3 < \chi^2 \leq 36.4)$ for $\nu = 24$
(d) $P(\chi^2 \leq 0.70)$ for $\nu = 6$

2.30 Assuming a t distribution, find:
(a) $P(|t| > 2.015)$ for $\nu = 5$
(b) $P(t > 2.015)$ for $\nu = 5$
(c) $P(-1.341 < t < 2.131)$ for $\nu = 15$
(d) $P(t \leq 1.5)$ for $\nu = 20$

2.31 Assuming an F distribution, find:
(a) $P(F > 7.79)$ for $\nu_1 = 11, \nu_2 = 6$
(b) $P(F > 4.03)$ for $\nu_1 = 11, \nu_2 = 6$
(c) $P(F > 7.79)$ for $\nu_1 = 6, \nu_2 = 11$
(d) $P(0.221 < F \leq 2.62)$ for $\nu_1 = 5, \nu_2 = 24$

2.32 The finished diameter on armored electric cable is normally distributed with mean 0.77 in. and standard deviation 0.01 in. What is the probability the diameter will exceed 0.795 in.? If the engineering specifications are 0.78 ± 0.02 in., what is the probability of a defective piece of cable?

2.33 If the density function for the life of a certain type of component is $f(x) = (1/100) \exp(-x/100)$ for $x > 0$, what is the probability that a randomly selected component will last 400 hours? That it will last 400 hours given that it has already survived 200 hours? If an assembly uses three of these components in series, what is the probability that an assembly incorporating three randomly selected components will not fail because of component failure?

2.34   The *hazard rate* is defined as $f(x)/[1 - F(x)]$. If $f(x) = (1/\theta) \exp(-x/\theta)$ for $x > 0$, what is the hazard rate?

FURTHER READING

Drake, A. W. 1967. *Fundamentals of Applied Probability Theory.* McGraw-Hill, New York.

Feller, W. *An Introduction to Probability Theory and Its Applications,* Vol. 1, 3rd ed. (1968) and Vol. 2, 2nd ed. (1971). Wiley, New York.

Goldberg, S. 1960. *Probability: An Introduction.* Prentice-Hall, Englewood Cliffs, N.J.

Hays, W. L.; and Winkler, R. L. 1971. *Statistics: Probability, Inference, and Decision.* Holt, Rinehart and Winston, New York.

Johnson, N. L.; and Kolz, S. 1972. *Distributions in Statistics: Continuous Multivariate Distributions.* Wiley, New York.

————. *Distributions in Statistics: Discrete Distributions* (1969); *Continuous Univariate Distributions,* Vol. I and II (1970). Houghton Mifflin, Boston.

Kempthorne, O.; and Folks, L. 1971. *Probability, Statistics, and Data Analysis.* Iowa State University Press, Ames.

Larson, H. J. 1974. *Introduction to Probability Theory and Statistical Inference,* 2nd ed. Wiley, New York.

Lippman, S. A. 1971. *Elements of Probability and Statistics.* Holt, Rinehart and Winston, New York.

Meyer, P. L. 1970. *Introductory Probability and Statistical Applications,* 2nd ed. Addison-Wesley, Reading, Mass.

Mood, A. M.; Graybill, F. A.; and Boes, D. C. 1974. *Introduction to the Theory of Statistics,* 3rd ed. McGraw-Hill, New York.

Mosteller, F.; Rourke, R. E. K.; and Thomas, G. B. 1970. *Probability with Statistical Applications,* 2nd ed. Addison-Wesley, Reading, Mass.

Parzen, E. 1960. *Modern Probability Theory and Its Applications.* Wiley, New York.

Ross, S. M. 1972. *Introduction to Probability Models.* Academic Press, New York.

Wadsworth, G. P.; and Bryan, J. G. 1960. *Introduction to Probability and Random Variables.* McGraw-Hill, New York.

CHAPTER 3

ELEMENTS OF SAMPLING AND DESCRIPTIVE STATISTICS

In this chapter we shall discuss the basic ideas of sampling and the presentation of sample data. Certain useful statistics of a summarizing nature will be defined and efficient methods of calculation outlined.

3.1 INTRODUCTION

In many experimental situations one of the primary objectives of the investigation is the determination of a model that describes the experimental results. We may know the form of the model (e.g., the variable observed has a normal distribution or an expression for the "true" value of an observation taken at time t is $\alpha + \beta t$), but all necessary information about the population model is not known. In the case of a normal population we may not know the value of μ and/or σ^2. Ideally, if we could observe the entire population, we would have all the necessary information; but in most cases we cannot. Rather, we must rely on the results obtained from part of the population (a sample). There are many reasons for this situation. Three causes are: (1) it is impossible to observe the entire population due to limitations of time, money, and/or personnel; (2) the population as defined does not physically exist; and (3) the observations involve the destruction of the unit being observed. We expand further on the distinction between a population and a sample and then discuss some different types of samples.

3.2 THE POPULATION AND THE SAMPLE

In statistical analysis it is important that we distinguish between the relevant population and the sample. We must know if the data we have is a complete population or if it is a sample selected from some population. Also, if we have a sample, we must be cognizant of the population from which the sample was taken. To distinguish between a population and a sample, we define each and then give some examples.

Definition 3.1 A *population* is the totality of all possible values (measurements, counts, etc.) of a particular characteristic for a specific group of objects.

Such a specified group of objects is called a *universe*. Obviously, a universe can have several populations associated with it. Some examples of universes and populations are:

1. Universe: the employees of Arizona State University as of 5:00 P.M. on December 4, 1962. Populations: the population of blood types, the population of weights, the population of heights, etc.
2. Universe: all single-family dwellings in Ames, Iowa, on September 18, 1972. Populations: the number of rooms per unit, the number of residents in each unit, etc.
3. Universe: a single piece of pipe. Populations: all possible measurements of its inside diameter, of its length, etc.
4. Universe: all transistors of a specific type manufactured by a given manufacturer under similar conditions. Populations: lengths of life, function on test, etc.

These examples should suffice to impress upon the reader the importance of clearly defining the population under investigation. The concept of a sample, as opposed to a population, is very important.

Definition 3.2   A *sample* is part of a population selected according to some rule or plan.

To illustrate the concept of a sample, consider the universe (4) and the population of lengths of life of the transitors. Some samples from that population are:

1. Lengths of life of all transistors produced during an eight-hour day shift at the manufacturing plant.
2. Lengths of life of 50 "randomly" selected transistors of the specified type.
3. Lengths of life of every 100th transistor produced.

The set of observations (life lengths) comprise a sample from the population mentioned. Since we are most often confronted with a sample, it is important that we recognize (1) we are dealing with a sample and (2) which population has been sampled.

If we are dealing with the entire population, our statistical work will be primarily *descriptive*. On the other hand, if we are dealing with a sample, the statistical work will not only describe the sample but also *provide information about the sampled population.*

3.3 TYPES OF SAMPLES

Several types or classes of samples are encountered in practice. The characteristics that distinguish one type from another are (1) the manner in which the sample was obtained, (2) the number of variables recorded, and (3) the purpose for which the sample was drawn. The last two characteristics are easily understood in any practical situation, although (3) is frequently not clearly stated and perhaps even forgotten. The manner of obtaining the sample is very important and will be discussed further.

Samples may be grouped into two broad classes when their method of selection is considered, namely, those selected by judgment and those selected according to some chance mechanism. Samples selected according to some chance mechanism are known as *probability samples* if every item in the population has a known probability of being in the sample. A particular probability sample of importance in statistical analysis is known as a (simple) *random sample*. If the sample is chosen so that each value in the population has an equal and independent chance of being included in the sample, the sample is a (simple) random sample. The observations in such a sample are independent and have the same distribution. Thus we can define a random sample as follows.

Definition 3.3 If $X_1, \ldots, X_n$ denote the n observations in a sample, we say the sample is (simple) random if $X_1, \ldots, X_n$ are *independent identically distributed random variables.*

We shall assume throughout the rest of the book that whenever a sample is taken from a population it is a (simple) random sample. Why are random samples preferred to subjectively selected samples? An answer to this question may be formulated as follows: A good sample is one from which generalizations to the population can be made; a bad sample is one from which they cannot be made. To generalize from a sample to a population, we need to be able to deduce from any assumptions about the population whether the observed sample is within the range of sampling variation that might occur for that population under the given method of sampling. Such deductions can be made if and only if the laws of mathematical probability apply. The purpose of randomness is to ensure that these laws do apply. If we had equally well-established and stable laws of personal bias, subjective sampling could be used.

We can select a (simple) random sample from different populations in various ways:

1. Consider a population for which the relative frequency of values in the population can be approximately described by a density function. Thus the variable under consideration can be thought of as a continuous random variable. To ensure that a random sample is drawn, the only concern is that the observations are made independently and at "random." Observations can be taken from the population with or without replacement.
2. Suppose a random sample is to be selected from an "infinite" population associated with a discrete variable; i.e., the population is described by a probability function. Again, the only concern is that the observations are taken independently and at "random." The sample can be selected with or without replacement.
3. Suppose the population is finite (specified by a probability function). A random sample, as defined above, can be drawn only if the sampling is performed with replacement. Sampling with replacement effectively makes the population infinite.

4. If a finite population is sampled without replacement, we no longer have a random sample as defined earlier. Sometimes, a "random" sample for this situation is defined as one in which each set of n objects has an equal chance of being the sample of size n.

Other types of random samples of a specialized type are sometimes encountered. Two of these are:

1. *Stratified random sample.* The population is first subdivided into subpopulations or strata. Then a simple random sample is drawn from each stratum.
2. *Systematic random sample.* Consider the N units in the population to be arranged in some order. If a sample of size n is required, take a unit at random from the first $k = N/n$ units and then take every kth unit thereafter.

Having defined various types of sampling frequently encountered, the following caution is noted: *The methods of analysis will not be the same for each type of sampling.* Great care must be exercised to use the proper method of analysis; failure to do so can lead to serious errors in judgment when the decision-making stage is reached.

3.4 SAMPLING FROM A SPECIFIED POPULATION

How do we go about selecting a random sample from a specified population? Some examples will serve as explanation:

1. Suppose the population consists of only 2 values. A random sample of 1 can be selected by tossing an unbiased coin.
2. Consider a population consisting of 100 values and a desired sample of 5. One hundred numbered tickets (corresponding to our population of values) can be placed in a bowl and 5 tickets selected in a "random" manner.
3. In the previous example the sample values could have been selected using a table of random numbers.

To illustrate the use of a table of random numbers, consider the problem of sampling the lives of $n = 5$ batteries from a lot of $N = 25$ batteries. That is, we desire to take a random sample of size 5 from the population of 25 lives. First number the batteries 01, 02, ..., 25. Second, refer to a table of random numbers such as given in Appendix 7 and proceed through the following steps:

1. Select by any method one of the four pages of tabled values.
2. Without direction, bring a pencil point down on the printed page so as to hit a random digit.
3. Read this digit and the next three to the right, e.g., 2167.
4. Let the first two of these specify the row and the last two the column.
5. Go to this point in the table of random numbers and read the specified digit and the next one to the right. This reads 73. However,

the only possible numbers of use in the specified problem are 01, 02, ... , 25. Thus, it is necessary to run down the column until five suitable numbers are observed. In order the numbers observed are 73, 48, 54, <u>01</u>, <u>18</u>, 38, 60, 70, 44, 30, 41, 86, <u>23</u>, 64, 31, 71, 68, 64, <u>13</u>, <u>12</u>. The numbers specifying the five batteries to be included in the sample have been underscored.

6. Appropriate changes should be made in step (5) to handle different problems.

3.5 PRESENTATION OF DATA

Having obtained a random sample from a specified population, some way of reducing it to an understandable form is required. To illustrate

TABLE 3.1–Function Times of 201 Explosive Actuators, Measured in Milliseconds (hypothetical data)

64.0	61.5	69.0	65.25	69.0	66.0
63.5	65.25	66.25	67.25	67.25	62.5
61.75	63.5	63.75	66.5	66.0	65.5
65.25	66.5	64.5	67.75	64.5	68.0
63.75	68.0	70.5	68.0	65.0	62.0
62.75	61.5	60.0	65.75	66.0	62.0
65.75	60.75	63.75	62.0	70.25	64.75
68.5	65.0	66.5	64.0	67.0	67.0
63.0	64.0	67.0	63.25	65.25	67.5
65.0	67.5	64.5	68.0	63.5	68.75
63.0	66.25	67.0	65.25	64.0	65.25
63.0	67.0	65,5	62.0	64.5	66.25
65.0	63.75	67.5	65.5	64.75	67.0
68.0	59.0	64.5	67.0	67.75	63.25
63.25	65.5	64.0	67.0	64.5	67.5
65.0	61.0	64.5	63.0	66.5	66.0
65.0	61.25	69.5	64.0	68.0	64.5
66.5	64.25	65.0	62.25	63.5	63.0
67.0	65.25	65.0	65.0	65.25	65.25
63.0	65.5	65.0	62.0	64.0	62.5
64.75	61.5	62.75	68.5	63.5	63.0
64.5	67.0	61.75	66.25	64.75	65.5
62.75	68.5	61.5	63.0	65.5	65.5
63.0	65.5	66.75	69.5	65.25	63.5
66.0	62.25	62.5	61.5	68.0	63.75
66.0	64.0	67.0	67.75	65.25	67.75
68.0	63.5	63.25	63.0	61.75	69.0
65.0	62.5	62.0	64.75	64.0	66.75
66.0	64.5	64.25	62.5	66.5	66.75
64.5	60.0	65.0	66.0	64.5	66.25
65.75	65.5	64.5	62.0	65.25	64.25
63.0	64.0	66.75	65.25	63.75	67.0
61.0	70.0	70.0	65.5	65.25	64.5
67.5	65.75	70.0	. . .	. . .	. . .

TABLE 3.2–Tally Sheet for Data of Table 3.1

Function Time (msec)	Tally	Frequency	Function Time (msec)	Tally	Frequency
59.0	1	1	65.0	卌 卌 11	12
59.25			65.25	卌 卌 1111	14
59.5			65.5	卌 卌 1	11
59.75			65.75	1111	4
60.0	11	2	66.0	卌 111	8
60.25			66.25	卌	5
60.5			66.5	卌 1	6
60.75	1	1	66.75	1111	4
61.0	11	2	67.0	卌 卌 11	12
61.25	1	1	67.25	11	2
61.5	卌	5	67.5	卌	5
61.75	111	3	67.75	1111	4
62.0	卌 11	7	68.0	卌 111	8
62.25	11	2	68.25		
62.5	卌	5	68.5	111	3
62.75	111	3	68.75	1	1
63.0	卌 卌 1	11	69.0	111	3
63.25	1111	4	69.25		
63.5	卌 11	7	69.5	11	2
63.75	卌 1	6	69.75		
64.0	卌 卌	10	70.0	111	3
64.25	111	3	70.25	1	1
64.5	卌 卌 1111	14	70.5	1	1
64.75	卌	5			

the usual techniques for presenting such data, consider the data in Table 3.1.

In this form the data are, to say the least, confusing. It is not easy to visualize any pattern in the observed values, nor is it easy to estimate the average function time. We find it convenient, therefore, to arrange the values in a *frequency distribution* as in Table 3.3. To accomplish this, we first make use of a *tally sheet* as shown in Table 3.2. Incidentally, Table 3.3 provides us with an *array,* i.e., the values arranged in order of magnitude.

Upon examination of Table 3.3 we note that all the observations are greater than or equal to 59 msec and less than or equal to 70.5 msec. That is, we have established the *range* of our data. Further, we can roughly estimate the *average* function time to be 65 msec.

However, since it takes too long to scan all the values in Table 3.3, the data are still in rather cumbersome form. To remedy this, it is customary to condense the data even more by tabulating only the frequencies associated with certain intervals, usually referred to as *class intervals*. To set up class intervals, a good working rule is to have no fewer than 5 and no more than 15. Also, the limits of the class intervals

TABLE 3.3–Frequency Distribution for Data of Table 3.1

Function Time (msec)	Number of Actuators Exhibiting Given Function Time = Frequency	Relative Frequency
59.0	1	0.005
60.0	2	0.010
60.75	1	0.005
61.0	2	0.010
61.25	1	0.005
61.5	5	0.025
61.75	3	0.015
62.0	7	0.035
62.25	2	0.010
62.5	5	0.025
62.75	3	0.015
63.0	11	0.055
63.25	4	0.020
63.5	7	0.035
63.75	6	0.030
64.0	10	0.050
64.25	3	0.015
64.5	14	0.070
64.75	5	0.025
65.0	12	0.060
65.25	14	0.070
65.5	11	0.055
65.75	4	0.020
66.0	8	0.040
66.25	5	0.025
66.5	6	0.030
66.75	4	0.020
67.0	12	0.060
67.25	2	0.010
67.5	5	0.025
67.75	4	0.020
68.0	8	0.040
68.5	3	0.015
68.75	1	0.005
69.0	3	0.015
69.5	2	0.010
70.0	3	0.015
70.25	1	0.005
70.5	1	0.005
Total	201	1.005*

*Total exceeds 1.000 because of errors of rounding.

TABLE 3.4–Frequency Distribution (using class intervals)
for Data of Table 3.1

Function Time (msec)	Number of Actuators with Function Time in Specified Class Interval = Frequency	Relative Frequency
$58 < X \leq 59$	1	0.005
$59 < X \leq 60$	2	0.010
$60 < X \leq 61$	3	0.015
$61 < X \leq 62$	16	0.080
$62 < X \leq 63$	21	0.104
$63 < X \leq 64$	27	0.134
$64 < X \leq 65$	34	0.169
$65 < X \leq 66$	37	0.184
$66 < X \leq 67$	27	0.134
$67 < X \leq 68$	19	0.095
$68 < X \leq 69$	7	0.035
$69 < X \leq 70$	5	0.025
$70 < X \leq 71$	2	0.010
Total	201	1.000

should be chosen so that there is no ambiguity in assigning observed values to the classes. This latter requirement is most easily satisfied by (1) selection of class limits that carry one more decimal place than the original data or (2) proper use of inequality and equality signs. We shall adopt the second of these procedures for use in this text. Using class intervals of 1 msec width, we get the data in the form of Table 3.4. In this table we have used the letter X to represent the various function times in milliseconds. To interpret the values and frequencies, we proceed as follows: one actuator had a function time of more than 58 msec but less than or equal to 59 msec, two actuators had a function time of more than 59 msec but less than or equal to 60 msec, and so on. Note that we have less information available in Table 3.4 than in Table 3.3 because we no longer know the individual values but know only in which class interval they fall, but the loss in accuracy is balanced to some extent by the gain in conciseness. The column headed "relative frequency" tells us what proportion of the total observations fall in each class. The values are found by dividing each class frequency by the total frequency.

Conforming to the adage that a "picture is worth ten thousand words," we often represent our distribution by a chart or *frequency histogram*. This is illustrated in Figure 3.1. The dotted line pictures a *frequency polygon*. Note that the frequency histogram is formed by erecting rectangles over the class intervals; the height of each rectangle agrees with the class frequency if the left-hand scale is read and with the class relative frequency if the right-hand scale is read. The fre-

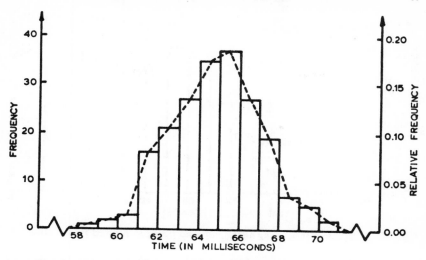

Fig. 3.1–Frequency histogram and polygon plotted from Table 3.4.

quency polygon is formed by joining the midpoints at the tops of the rectangles.

Also note that the frequency histogram and polygon as well as the frequency distribution give us not only an estimate of the average value but also an idea of the amount of *variability* present in the data.

Another convenient way of tabulating data is to prepare a *cumulative*

TABLE 3.5–Cumulative Frequency Distribution Formed from Table 3.4

Function Time (X)	Number of Actuators with Function Time Less Than or Equal to the Specified Value = Cumulative Frequency	Relative Cumulative Frequency
58	0	0.000
59	1	0.005
60	3	0.015
61	6	0.030
62	22	0.109
63	43	0.214
64	70	0.348
65	104	0.517
66	141	0.701
67	168	0.836
68	187	0.930
69	194	0.965
70	199	0.990
71	201	1.000

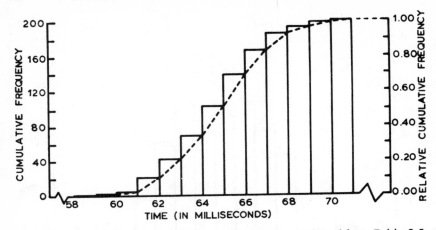

Fig. 3.2–Cumulative frequency histogram and polygon plotted from Table 3.5.

frequency distribution showing the number of observations *less than or equal to* a specified value. The figures are obtained by adding, in cumulative fashion, the frequencies recorded in Table 3.4. This is illustrated in Table 3.5. The graph that arises from this table is shown in Figure 3.2 and is quite helpful in interpreting the observed data. Note that the cumulative (ogive) curve is plotted by joining the right-hand endpoints at the tops of the rectangles. This curve (see dotted line) is formed as just mentioned because it represents the cumulative frequency up to and including the upper class limit.

3.6 SOME DESCRIPTIVE STATISTICS

The following statistics are used to calculate representative values that summarize much of the information available in a set of data. Not all the statistics are of equal importance; however, we present several so that the reader will be aware of their existence, uses, advantages, and disadvantages.

We assume throughout that we have a sample from some population. We let $X_1, \ldots, X_n$ be the n measurements in the sample.

3.6.1 Arithmetic Mean

Perhaps the most common of these representative values is the arithmetic mean, usually called the *sample mean* when the data is a sample. The arithmetic mean, denoted by $\overline{X}$, is defined as the arithmetic average of all the values. The formula for computing the sample mean is

$$\overline{X} = (X_1 + \cdots + X_n)/n = \sum_{i=1}^{n} X_i/n \qquad (3.1)$$

where there are n observations in the sample.

Example 3.1

Given the sample values 3, 4, −2, 1, and 4, the sample mean is $\overline{X} = [3 + 4 + (-2) + 1 + 4]/5 = 10/5 = 2$.

The above example illustrates the method of computing the arithmetic mean. It is to be noted that the arithmetic mean is affected by every item in the sample and greatly so by extreme values. Two interesting properties of the arithmetic mean are: (1) the sum of the deviations from the arithmetic mean is zero and (2) the sum of the squares of the deviations from the arithmetic mean is less than the sum of the squares of the deviations from any other value.

As might be expected, the arithmetic mean has both advantages and disadvantages. Its advantages are: (1) it is the most commonly used average, (2) it is easy to compute, (3) it is easily understood, and (4) it lends itself to algebraic manipulation. The one major disadvantage is that it is unduly affected by extreme values and may therefore be far from representative of the sample.

Before proceeding to a second representative measure for describing samples, it will pay us to look at methods of calculating the arithmetic mean when our data are in the form of a frequency distribution. If for each different value of X we have a frequency f, the sample mean is given by

$$\overline{X} = \frac{f_1 X_1 + f_2 X_2 + \cdots + f_d X_d}{f_1 + f_2 + \cdots + f_d} = \sum_{i=1}^{d} f_i X_i / n \qquad (3.2)$$

where there are d different values of X. Note that we have made use of the fact that $\sum_{i=1}^{d} f_i = n$.

Example 3.2

The data in Table 3.3 are of the type just described. Then

$$\overline{X} = [(1)(59.0) + (2)(60.0) + \cdots + (1)(70.5)]/201$$
$$= 13{,}071.75/201$$
$$= 65.034 \text{ msec}$$

Many times our data appear in frequency tables where we no longer know the actual values of the observations but only to which class interval they belong. In these instances the best we can do is to *approximate* the sample mean. To obtain this approximation, we assume that the values in a particular class interval are uniformly distributed over the interval. (Some writers assume that all the values in an interval are concentrated at the midpoint.) This permits us to use the midpoint for each observation in the interval when calculating the mean. Thus if we denote the midpoint of the ith interval by ξ_i and there are k intervals, the sample mean is approximately

$$\overline{X} \cong \frac{f_1\xi_1 + \cdots + f_k\xi_k}{f_1 + \cdots + f_k} = \frac{\sum_{i=1}^{k} f_i\xi_i}{n} \tag{3.3}$$

Example 3.3

Considering the data of Table 3.4, it is seen that

$\overline{X} \cong [(1)(58.5) + (2)(59.5) + \cdots + (2)(70.5)]/201 = 13,032.5/201$

$= 64.838$ msec

3.6.2 The Midrange

Another representative value of importance, especially when a quick average is needed, is the midrange. The midrange is defined as

$$MR = (X_{min} + X_{max})/2 \tag{3.4}$$

where X_{min} is the smallest (minimum) sample value and X_{max} is the largest (maximum) sample value. It must be realized that even though the midrange is quick and easy to compute, it is often inefficient because all information contained in the intermediate values has been ignored. Also, it can be quite unrepresentative if either the smallest or largest value is decidedly atypical of all the data.

3.6.3 The Mode

Another value used to describe a sample is the mode. The *mode* is defined as the value that occurs most frequently in the sample. The mode of the sample will be denoted by MO. It should be obvious that the mode will not always be a central value; in fact it may often be an extreme value. Then too, a sample may have more than one mode. We should at this time distinguish between an *absolute mode* and a *relative mode*. An absolute mode is what we defined above (there may, of course, be more than one absolute mode); a relative mode is a value that occurs more frequently than neighboring values even if it is not an absolute mode.

Example 3.4

Given a sample consisting of the values 6, 7, −1, 4, and 3, we may say there is no mode or there are five modes since each value occurs only once.

Example 3.5

Considering the data of Table 3.3, we see there are two absolute modes, 64.5 and 65.25 msec, since each of these values occurs 14 times and no other value occurs that frequently.

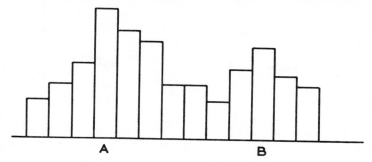

Fig. 3.3—Example of a bimodal frequency histogram.

Example 3.6

Given a frequency histogram like that shown in Figure 3.3, we would say there are two relative modes: one in Class A and one in Class B. However, the mode in Class A is the only absolute mode.

If our data are grouped in class intervals, it will be impossible to locate the mode exactly. Under such circumstances the best we can do is to approximate the value of the mode. The first step is to locate the modal class. This is accomplished quickly by picking out the class interval that shows the highest frequency. The sample mode is then approximated by

$$MO \cong L_{MO} + [d_1/(d_1 + d_2)] (w) \qquad (3.5)$$

where L_{MO} = lower limit of the modal class
d_1 = the difference (sign neglected) between the frequency of the modal class and the frequency of the preceding class
d_2 = the difference (sign neglected) between the frequency of the modal class and the frequency of the following class
w = width of the modal class

Example 3.7

Consider the sample given in Table 3.4. The modal class is from 65 to 66 msec with a frequency of 37. Therefore,

$$MO = 65 + [3/(3 + 10)] (1) = 65.23 \text{ msec}$$

Since the mode is by definition the most typical value, it is often considered the most descriptive of the representative values discussed so far. However, its importance diminishes as the number of observations becomes limited.

3.6.4 The Median

A representative value frequently employed as an aid in describing a set of data is the median. The sample median, denoted by M, is the

$[(n + 1)/2]$th observation when the values are arrayed in order of magnitude. Theoretically, one-half the observations should have a value less than the median and one-half should have a value greater than the median. However, in practice it does not always work out quite this way due to clustering of the observations (see Example 3.8). Regardless, the median is important as a measure of position or location.

Example 3.8

If we consider the data of Table 3.3 where $n = 201$, the median is the $[(201 + 1)/2 = 101]$st item in the array. Counting down the frequencies in Table 3.3 we find the 101st item to be 65. Thus $M = 65$ msec.

Example 3.9

Given the sample (2, 3, 4, 6, 6, 7), the median is the $[(6 + 1)/2 = 3.5]$th observation in the array. To avoid ambiguity, it is agreed that the median will be halfway between the third and fourth observations in the array. Thus $M = 5$.

When data are grouped in class intervals as in Table 3.4, the median cannot be located exactly. However, if we assume that the observations in each class interval are uniformly distributed over the interval, a close approximation to the median may be obtained. The first step is to locate the class in which the median belongs. This is done by adding up the class frequencies until we find the class that contains the $[(n + 1)/2]$th observation. If a cumulative frequency distribution has been formed as in Table 3.5, the median class is easily located. Then the sample median may be approximated using the equation

$$M \cong L_M + \left\{ \frac{[(n + 1)/2] - S}{f_M} \right\} (w) \tag{3.6}$$

where L_M = lower limit of the median class
n = number of observations in the sample
S = sum of the frequencies in all classes preceding the median class
f_M = frequency in the median class
w = width of the median class

Example 3.10

Considering the data of Table 3.4, we see that

$$M \cong 64 + [(101 - 70)/34] (1) = 64.91 \text{ msec}$$

It is possible to approximate the median graphically from a cumulative frequency (ogive) curve using the relative cumulative frequency scale. This will be illustrated in Section 3.6.5.

The median, a measure of position, is affected by the number of items but not by the magnitude of extreme values. Two characteristics of the median that are of interest are: (1) the sum of the absolute values of the deviations from the median is less than the sum of the absolute values of the deviations from any other point of reference and (2) theo-retically the probability is 1/2 that an observation selected at random from a set of data will be less than (greater than) the median.

Some advantages and disadvantages of the median, with which one should be familiar if he wants to make proper use of this statistic, will now be mentioned. The advantages are: (1) it is easy to calculate and (2) it is often more typical of all the observations than is the arithmetic mean, since it is not affected by extreme values. The disadvantages are: (1) the items must be arrayed before the median can be obtained and (2) it does not lend itself to algebraic manipulation.

3.6.5 Percentile, Decile, and Quartile Limits

In this section we shall consider locating various values that divide the sample into groups according to the magnitude of the observations. The median (see Section 3.6.4) was obviously one such value, since it divided the array into two groups, each containing 50 percent of the observations. We now wish to determine other such values.

Let us consider the most general case first. If we want to locate a value (say P_p) such that p percent $(0 < p < 100)$ of the observations are less than or equal to P_p and $100 - p$ percent of the observations are greater than P_p, we call P_p *the upper limit of the pth percentile,* and ap-proximate P_p by the $[p(n + 1)/100]$th observation in the sample array if we start counting from the smallest value. For example, P_{67} is the upper limit of the 67th percentile and is approximated by the $[67(n + 1)/100]$th observation in the sample array. Similarly, P_{66} is the upper limit of the 66th percentile and is approximated by the $[66(n + 1)/100]$th observation in the sample array. If we refer to the 67th per-centile, we mean the interval from P_{66} to P_{67}; in general, the pth per-centile is the interval from P_{p-1} to P_p.

Example 3.11

If we consider the data of Table 3.3, what is the upper limit of the 80th percentile? P_{80} is approximately the $[80(201 + 1)/100 = 161.6]$th ob-servation in the array, which is 67 msec.

Example 3.12

What is the upper limit of the 35th percentile in the sample given in Example 3.9? P_{35} is approximated by the $[35(6 + 1)/100 = 2.45]$th ob-servation in the array. To avoid ambiguity, we agree to set P_{35} forty-five one hundredths of the way from the second to the third observation in the array when we count from the smallest value. Thus P_{35} is 0.45 of the way between 3 and 4; i.e., $P_{35} = 3.45$.

The reason for the word percentile should now be clear. If we locate $P_1, P_2, \ldots, P_{99}$, we have (theoretically) split our array into 100 parts (percentiles), each containing 1 percent of the observations.

The meaning of such terms as decile limits and quartile limits is now almost obvious. The *decile limits* $D_1, D_2, \ldots, D_9$ theoretically split our array into ten parts (deciles), each containing 10 percent of the observa-

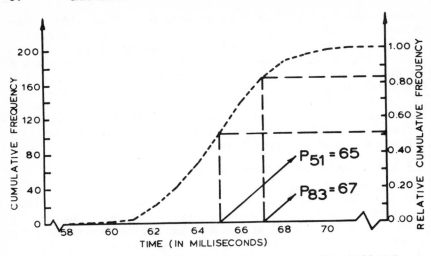

Fig. 3.4–Cumulative frequency (ogive) "curve" plotted from Table 3.5.

tions. The *quartile limits* Q_1, Q_2, and Q_3 theoretically divide our array into four parts (quartiles), each containing 25 percent of the observations. No particular methods of calculation will be presented for decile or quartile limits since they are only special cases of percentile limits. This is clear once we observe that

$$P_{10} = D_1 \quad P_{40} = D_4 \qquad\qquad P_{75} = Q_3$$
$$P_{20} = D_2 \quad P_{50} = D_5 = Q_2 = M \quad P_{80} = D_8$$
$$P_{25} = Q_1 \quad P_{60} = D_6 \qquad\qquad P_{90} = D_9$$
$$P_{30} = D_3 \quad P_{70} = D_7$$

In Section 3.6.4 we mentioned the possibility of estimating the median from the graph of the relative cumulative frequency distribution. To illustrate this technique, we shall undertake the location of percentile limits in general. Consider the cumulative frequency curve of Figure 3.2, which we reproduce here as Figure 3.4. The procedure is as follows: Since P_p is the upper limit of the pth percentile, which says that p percent of the observations are less than or equal to P_p, all we have to do is locate $p/100$ on the relative cumulative frequency scale, draw a horizontal line from this point to the ogive curve, and from here drop a vertical line down to the horizontal axis, thus locating P_p. This is illustrated in Figure 3.4 for P_{51} and P_{83}.

3.6.6 The Range

All the representative values discussed in the preceding sections have been some sort of average or measure of position. It must be clear, though, that they are not sufficient by themselves to describe most samples adequately. This statement may be verified easily if we consider two sets of data that have the same mean, the same median, and the same mode but differ greatly in the amount of variation present in each set of data. It would seem then that some measure of the variation or dispersion among the individual values is also needed. Several such measures have been devised, and we shall mention four of these in this and following sections.

A measure of dispersion, to be suitable, should be large when the values vary over a wide range (and there are quite a few extreme values) and should be small when the range of variation is not too great.

The simplest measure of variation is one that has been mentioned before, i.e., the range. If we denote the smallest (minimum) sample value by X_{min}, and the largest (maximum) sample value by X_{max}, the *sample range* is given by

$$R = X_{max} - X_{min} \tag{3.7}$$

The sample range, though easy to obtain, is often termed inefficient because it ignores all the information available from the intermediate sample values. However, for small samples ($n < 10$) the efficiency (relative to other measures of variation yet to be defined) is quite high. For a more explicit discussion of the efficiency of the range relative to the standard deviation the reader is referred to Section 3.6.7. We find the sample range enjoying a favorable reception and wide use because of ease in computation in such applications as statistical quality control where small samples are the rule rather than the exception.

Example 3.13

For the sample given in Example 3.9 we obtain $R = 7 - 2 = 5$.

3.6.7 The Standard Deviation and Variance

Perhaps the best known and most widely used measure of variability is the standard deviation. Of almost equal importance is the square of the standard deviation, this quantity being known as the variance. We shall explain both of these measures by defining the variance.

The *sample variance*, denoted by s^2, is defined by

$$s^2 = [(X_1 - \bar{X})^2 + \cdots + (X_n - \bar{X})^2]/(n - 1)$$

$$= \sum_{i=1}^{n} (X_i - \bar{X})^2/(n - 1) \tag{3.8}$$

If we let $x_i = X_i - \bar{X}$; i.e., if we denote deviations about the mean by lowercase letters,

$$s^2 = (x_1^2 + \cdots + x_n^2)/(n - 1) = \sum_{i=1}^{n} x_i^2/(n - 1) \tag{3.9}$$

It is easily seen that Eq. (3.8) can be expressed in the form

$$s^2 = \left[\sum_{i=1}^{n} X_i^2 - \left(\sum_{i=1}^{n} X_i \right)^2 \Big/ n \right] \Big/ (n - 1) \tag{3.10}$$

This is the best form for machine calculation. The *sample standard deviation* is then defined as the positive square root of the variance, namely,

$$s = \sqrt{s^2} \qquad (3.11)$$

The use of $n - 1$ (instead of n) when defining the sample variance may seem peculiar to the reader, since we implicitly used a divisor of N when defining the population variance. Our reason for using $n - 1$ is that, in general, one prefers unbiased estimators to biased estimators, and the use of $n - 1$ gives us an *unbiased estimator* of σ^2. (An *estimator* is a statistic, i.e., a function of the sample values, that will provide us with numerical estimates of a parameter.) If n were used, the resulting function of the sample observations would produce biased estimates of the unknown population variance—biased because *on the average* the estimates would be too small. Thus the student of statistics must resign himself to remembering that while the population variance is defined using a divisor of N, the sample variance requires a divisor of $n - 1$. Incidentally, we refer to $n - 1$ as the *degrees of freedom* associated with the sample variance (and standard deviation).

Example 3.14

For the sample $(13, 5, 8, 5)$ we see that

$$s^2 = [(13 - 7.75)^2 + (5 - 7.75)^2 + (8 - 7.75)^2 + (5 - 7.75)^2]/3$$
$$= [(5.25)^2 + 2(-2.75)^2 + (0.25)^2]/3 = 14.25$$

and thus $s = \sqrt{14.25} = 3.775$.

If we had used Eq. (3.10), the same value of s^2 would have been obtained:

$$s^2 = [13^2 + 5^2 + 8^2 + 5^2 - (13 + 5 + 8 + 5)^2/4]/3$$
$$= \frac{283 - (31)^2/4}{3} = \frac{283 - 240.25}{3} = \frac{42.75}{3} = 14.25$$

If the sample data appear in a frequency distribution, the following forms are appropriate for calculation. When no class intervals are involved (as in Table 3.3),

$$s^2 = \left[\sum_{i=1}^{d} f_i X_i^2 - \left(\sum_{i=1}^{d} f_i X_i \right)^2 \bigg/ n \right] \bigg/ (n - 1) \qquad (3.12)$$

where $n = \sum_{i=1}^{d} f_i$.

Example 3.15

Consider Table 3.6. Using Eq. (3.12), we obtain

$$s^2 = [12{,}700 - (450)^2/18]/17 = 1450/17 = 85.3$$

When class intervals are involved, the appropriate formula is

$$s^2 \cong \left[\sum_{i=1}^{k} f_i \xi_i^2 - \left(\sum_{i=1}^{k} f_i \xi_i \right)^2 \bigg/ n \right] \bigg/ (n - 1) \qquad (3.13)$$

TABLE 3.6–Illustration of the Use of Eq. (3.12)

X	f	fX	fX^2
10	3	30	300
20	5	100	2,000
30	8	240	7,200
40	2	80	3,200
Total	18	450	12,700

Example 3.16

Consider Table 3.7. Using Eq. (3.13),

$$s^2 = [12,700 - (450)^2/18]/17 = 85.3$$

It was mentioned in Section 3.6.6 that for small n the range is reasonably efficient relative to the standard deviation. By this statement was meant that if one wishes to estimate σ, it can be done using either R or s. For sampling from a normal population, the efficiency of the sample range relative to the sample standard deviation as an estimator for the population standard deviation is given in Table 3.8. As an example of the use of this table, if a person desires to use R rather than s, he would estimate σ by calculating

$$\hat{\sigma} = (R)[\text{value of } \sigma/E(R) \text{ for given } n] \qquad (3.14)$$

3.6.8 The Coefficient of Variation

The coefficient of variation has been explained by statisticians in different ways. However, attention usually is called to the rather obvious fact that things with large values tend to vary widely while things with small values exhibit small (numerically small, that is) variation. Thus to afford a valid comparison of the variation among large values and the variation among small values (such as the variation among salaries of industrial executives and the variation among the wages of day laborers), the variation is expressed as a fraction of the mean, and frequently as a percentage. This measure of relative variation is called the *coefficient of variation* and is defined as

$$CV = s/\overline{X} \qquad (3.15)$$

TABLE 3.7–Illustration of the Use of Eq. (3.13)

Class Interval	f	ξ	$f\xi$	$f\xi^2$
$5 < X \le 15$	3	10	30	300
$15 < X \le 25$	5	20	100	2,000
$25 < X \le 35$	8	30	240	7,200
$35 < X \le 45$	2	40	80	3,200
Total	18	...	450	12,700

TABLE 3.8–Efficiency of Range R Relative to Standard Deviation s as an Estimator of σ for a Normal Population

Sample Size (n)	Relative Efficiency	$\sigma/E(R)$
2	1.000	0.886
3	0.992	0.591
4	0.975	0.486
5	0.955	0.430
6	0.933	0.395
7	0.912	0.370
8	0.890	0.351
9	0.869	0.337
10	0.850	0.325
12	0.815	0.307
14	0.783	0.294
16	0.753	0.283
18	0.726	0.275
20	0.700	0.268
30	0.604	0.245
40	0.536	0.231
50	0.490	0.222

In percentage form, this becomes

$$100CV = 100(s/\overline{X}) \text{ percent} \tag{3.16}$$

The coefficient of variation is an ideal device for comparing the variation in two series of data that are measured in two *different units*, e.g., a comparison of variation in height with variation in weight.

Example 3.17

For the sample given in Example 3.14 we see that $CV = 3.775/7.75 = 0.4871$, and in percentage form $100CV = 48.71$ percent.

PROBLEMS

3.1   Plot a frequency histogram and polygon for the following data, which
shows weekly wages of 188 female employees of a shoe manufacturing
company. Make approximate eye estimates of the arithmetic mean, median, and mode.

$20.15	$25.00	$40.39	$25.49	$25.70
24.15	22.54	23.80	29.60	18.74
25.62	23.89	28.37	26.00	16.70
26.00	27.82	24.80	26.52	28.09
27.84	25.80	25.88	25.04	24.98
22.97	23.20	23.24	29.00	24.55
25.48	20.88	21.70	25.76	26.20
28.00	28.92	27.92	25.80	22.45
28.24	25.70	22.75	21.40	27.10
31.37	26.77	26.00	18.64	27.39
24.53	24.25	28.28	30.32	23.00
28.13	26.23	21.55	28.04	25.58
22.78	26.88	26.64	22.83	23.45
25.20	29.29	25.62	23.40	26.12
27.08	24.40	25.49	30.48	27.03
26.11	21.80	20.85	26.79	26.25
22.04	22.54	21.85	25.65	27.50
29.48	25.20	26.00	22.69	25.78
21.77	24.32	26.00	22.52	17.50
26.52	20.48	22.92	23.96	26.00
22.00	22.44	26.00	26.35	25.64
22.48	27.25	24.19	23.75	28.94
21.85	22.99	22.33	24.18	25.65
23.12	22.71	26.48	23.23	23.44
31.00	25.38	25.83	18.60	33.80
30.61	22.00	29.72	23.28	25.65
23.80	26.90	24.55	23.12	29.24
26.00	22.68	24.04	32.60	22.15
25.15	22.53	25.12	23.72	22.99
25.70	27.98	26.34	23.08	24.24
28.00	27.14	23.13	26.38	24.00
26.03	31.60	24.79	24.73	27.48
30.23	22.47	34.99	22.09	19.30
24.55	26.67	24.08	25.78	23.42
30.60	28.32	22.28	24.73	25.65
29.15	27.74	23.69	28.83	25.64
22.48	25.20	23.84	25.68	
28.24	30.72	28.92	25.73	

3.2   Plot a frequency histogram and polygon for the data given below, which gives percent silicon in 236 successive casts of pig iron.

1.13	1.00	0.96	0.67	0.77	0.65	0.83	0.92
0.80	0.94	0.96	0.76	0.34	0.60	0.79	0.73
0.85	0.62	0.60	0.66	0.84	1.00	0.99	0.96
0.60	0.32	0.87	0.89	0.70	0.91	1.20	1.00
0.97	1.00	1.08	0.85	0.71	0.72	0.74	0.96
0.92	1.00	0.67	0.77	0.74	1.32	0.85	0.94
0.94	0.89	0.98	0.87	0.97	0.94	0.60	0.72
0.72	0.65	0.88	1.00	1.09	0.60	0.72	0.88
1.17	1.00	0.75	0.73	0.91	1.11	1.45	1.45
0.87	0.64	0.60	1.00	0.81	1.14	0.68	0.74
0.36	0.85	1.17	1.00	0.82	0.77	0.67	0.70
0.68	0.89	0.93	1.13	1.00	0.80	1.00	0.86
0.73	0.66	0.79	0.51	0.60	0.89	1.00	1.18
0.82	0.60	0.76	1.07	0.84	0.93	0.73	0.60
0.79	0.61	1.14	1.33	1.00	0.80	0.71	0.95
0.87	0.83	0.65	0.64	0.85	0.78	0.86	0.60
0.92	0.87	1.00	0.91	0.72	0.79	0.70	1.00
0.81	0.80	0.81	0.87	0.60	0.86	0.94	1.00
0.97	0.70	0.37	1.00	1.00	0.99	0.84	0.72
0.48	1.50	1.50	1.00	0.99	0.80	0.85	0.84
1.00	0.91	0.60	0.68	0.75	0.47	0.73	0.97
0.92	0.60	0.82	1.14	0.87	0.70	0.80	0.95
0.61	1.02	1.45	0.93	0.57	0.60	0.61	0.69
0.81	1.00	1.25	0.90	0.60	0.82	0.84	0.92
0.71	0.94	0.87	0.84	0.94	0.97	0.90	0.99
0.97	1.06	1.10	0.89	0.69	0.86	0.61	0.38
0.89	0.97	0.87	0.71	0.33	0.80	0.64	0.26
1.16	1.25	0.66	0.56	1.12	0.73	0.62	0.78
0.68	0.61	1.00	1.11	1.00	0.81	0.70	0.85
1.00	1.50	1.18	0.94				

3.3 A random sample of 201 women students, age 18, was obtained and their heights and weights are recorded below. Plot a frequency histogram and polygon for (a) the heights, (b) the weights.

5-4	139	5-5	141¼	5-4½	99	5-2	116½	5-1¾	118½
5-3½	122½	5-4	158	5-4	123	5-8½	151	5-4	110½
5-1¾	108	5-7½	146½	5-4½	152½	5-6¼	118	5-6½	123
5-5¼	127½	5-6¼	117½	5-9½	142½	5-3	119½	5-4½	115
5-3¾	141½	5-7	139½	5-5	119	5-9½	134½	5-5¼	117
5-2¾	127	5-3¾	122	5-5	135¾	5-1½	114	5-3¾	112½
5-5¾	175	4-11	110¾	5-5	130¾	5-7¾	148½	5-5¼	149
5-8½	148	5-5½	124	5-2¾	128	5-3	125¾	5-6	128
5-3	130	5-1	107½	5-1¾	104¼	5-4¾	112½	5-2½	102¼
5-5	129	5-1¼	104	5-1½	107½	5-2½	123¼	5-5½	118½
5-3	110¼	5-4¼	134¾	5-6¾	118½	5-6	142	5-8	152
5-3	132½	5-5¼	137¾	5-2½	120	5-2	96½	5-2	114
5-5	125¼	5-5½	108½	5-7	143	5-5¼	122	5-2	144¾
5-8	146½	5-1½	126¾	5-3¼	126¾	5-5½	149½	5-4¾	133½
5-3¼	126	5-7	130¾	5-2	110¼	5-9	134	5-7	152½
5-5	140	5-8½	130½	5-4¼	115	5-7¼	121½	5-7½	130½
5-5	131	5-5½	125¼	5-5	111	5-6	128	5-8¾	151
5-6½	122½	5-2¼	110	5-4½	128¾	5-4½	127½	5-5¼	125
5-7	162	5-4	117	5-6¾	143½	5-5	129¾	5-6¼	140½
5-3	132½	5-3½	116¼	5-10	151½	5-6	117	5-7	139½
5-4¾	112½	5-2½	97½	5-10	110½	5-10¼	140½	5-3¼	108
5-4½	125	5-4½	126	5-5¼	125¼	5-7	138¾	5-7½	157½
5-2¾	175½	5-0	126	5-7¼	141¼	5-5¼	127½	5-6	118½
5-3	105½	5-5½	121½	5-6½	164½	5-3½	112	5-4½	146¼
5-6	115½	5-4	115½	5-7¾	122	5-4	110	5-3	108
5-6	173	5-10	133½	5-8	128	5-4½	128½	5-5¼	120
5-8	135¼	5-5¾	149½	5-5¾	123	5-4¾	131¼	5-2½	113
5-5	126¼	5-9	138½	5-2	121¾	5-7¾	131½	5-3	128
5-6	115½	5-6¼	139	5-4	124¼	5-4½	130¼	5-5½	120½
5-4½	134½	5-3¾	142	5-3¼	112	5-6½	125½	5-5½	125
5-5¾	126	5-4½	118	5-8	152¾	5-8	130½	5-3½	136
5-3	149½	5-10½	154	5-5¼	127	5-3½	125½	5-3¾	111¼
5-1	147½	5-0	130	5-2	134	5-5¼	130	5-7¾	117½
5-7½	128½	5-3¾	118	5-5½	111½	5-4	122¾	5-9	151¾
5-1½	121½	5-6½	138¾	5-7	122	5-3½	117	5-6¾	134
5-5¼	135½	5-7	145	5-7	138¾	5-4¾	137¾	5-6¾	120¾
5-3½	135	5-4½	118¾	5-3	113¼	5-5½	118¾	5-6¼	141¼
5-6½	140	5-7	113½	5-4	126½	5-5¼	155½	5-4¼	121¼
5-8	147½	5-5½	129	5-2¼	135	5-8	125	5-7	188
5-1½	121¾	5-7½	151	5-5	122	5-5¼	129½	5-4½	125½
5-¾	114½								

Source: University of British Columbia Student Health Service.

3.4   Plot a cumulative frequency curve for the data of Problem 3.1. Estimate the median from this curve.

3.5   Plot a cumulative frequency curve for the data of Problem 3.2 Estimate the median from this curve.

3.6   Plot a cumulative frequency curve for (a) the heights, (b) the weights given in Problem 3.3. From these curves estimate the median height and median weight.

3.7   Given the samples listed below, calculate for each the mean, median, mode, midrange, range, variance, standard deviation, and coefficient of variation:

    (a) $5, 19, -3, 7, 1, 1$ (e) $10, 15, 14, 15, 16$
    (b) $5, -3, 2, 0, 8, 6$ (f) $0, 5, 10, -3$
    (c) $6, 9, 5, 3, 6, 7$ (g) $8, 7, 15, -2, 0$
    (d) $1, 3, 2, -1, 5$

3.8   Suppose that $\overline{Y} = 100$ and $s^2 = 15$. What would the values of $\overline{Y}$ and s^2 become if each original observation were (a) increased by 10 units, (b) multiplied by 10 units?

3.9   Given the observations:

Value (Y)	Frequency (f)
2	10
3	10
4	20
5	50
6	30

Calculate the same statistics as asked for in Problem 3.7.

3.10   Calculate the same statistics as asked for in Problem 3.7 for each of the following sets of data: (a) Problem 3.1, (b) Problem 3.2, (c) Problem 3.3.

3.11   A bag of potatoes was sampled for quality, 5 potatoes being selected at random from the bag. Among the observations recorded were the weights of the potatoes: 17, 15, 10, 12, and 11 oz. Calculate $\overline{Y}$, s^2, s, and R. What property (using the word rather loosely) is common to the sample range and the sample variance (or standard deviation)?

3.12   Given that $n = 25$, $\sum_{i=1}^{25} y_i^2 = 600$, and $\overline{Y} = 204$, calculate the variance, standard deviation, and coefficient of variation.

FURTHER READING

Hays, W. L.; and Winkler, R. L. 1971. *Statistics: Probability, Inference, and Decision.* Holt, Rinehart, and Winston, New York.

Huntsberger, D. V.; and Billingsley, P. 1973. *Elements of Statistical Inference,* 3rd ed. Allyn and Bacon, Boston.

Freund, J. E. 1973. *Modern Elementary Statistics,* 4th ed. Prentice-Hall, Englewood Cliffs, N.J.

Snedecor, G. W.; and Cochran, W. G. 1967. *Statistical Methods,* 6th ed. Iowa State University Press, Ames.

Steel, R. G. D.; and Torrie, J. H. 1960. *Principles and Procedures of Statistics.* McGraw-Hill, New York.

Wine, R. L. 1964. *Statistics for Scientists and Engineers.* Prentice-Hall, Englewood Cliffs, N.J.

CHAPTER 4

SAMPLING DISTRIBUTIONS

Certain sampling distributions pertinent to methods presented in later chapters will be discussed here. The law of large numbers and the central limit theorem will be given. Various approximations to exact sampling distributions will also be considered.

4.1 DISTRIBUTION OF THE SAMPLE; THE LIKELIHOOD FUNCTION

Suppose a random sample of n observations is obtained from a population described by the density (probability) function $f(x)$. Let the random variable $X_i (i = 1, 2, \ldots, n)$ represent the ith measurement observed in the sample. The random variables $X_1, \ldots, X_n$ constitute the random sample from the given population. Since each X_i is a random variable, the sample $X_1, \ldots, X_n$ will have a joint distribution often referred to as the *distribution of the sample*. Let $g(x_1, \ldots, x_n)$ denote the joint density (probability) function for this distribution. Since random sampling implies independent observations, the random variables $X_1, \ldots, X_n$ are jointly independent random variables. Further, for independent random variables the joint density function may be expressed as the product of the marginal density functions. Thus

$$g(x_1, \ldots, x_n) = g_{X_1}(x_1) \ldots g_{X_n}(x_n) \qquad (4.1)$$

where $g_{X_i}(x_i)$ denotes the density function for the ith observation X_i. Since each observation came from the same population, $g_{X_i}(x_i) = f(x_i)$, and

$$g(x_1, \ldots, x_n) = \prod_{i=1}^{n} f(x_i) \qquad (4.2)$$

where $f(x)$ is the density function describing the sampled population. The function $\prod_{i=1}^{n} f(x_i)$ is often referred to as the *likelihood function*. It should be noted that Eq. (4.2) gives the joint density function of the sample *in the order drawn*.

4.2 SAMPLE STATISTICS AND SAMPLING DISTRIBUTIONS

As discussed in Chapter 3, when confronted with a sample, we generally summarize the data available in the sample by looking at certain representative functions or *statistics*.

Definition 4.1   A function $h(X_1, \ldots, X_n)$ of the random sample $X_1, \ldots, X_n$ is called a (sample) *statistic*.

Some common examples of statistics are the *sample mean* $\overline{X}$; the *sample variance* s^2; and the *sample median M*. Since a statistic is a function of the random variables $X_1, \ldots, X_n$, it is also a random variable and hence has an associated population or distribution called the *sampling distribution* of the statistic.

Our interpretation of the sampling distribution follows. Suppose we conceptually consider the possibility of repeatedly sampling from the given population under identical conditions. Not all samples will include the same values $x_1, \ldots, x_n$ of the n observations. In fact, it is unlikely that the values $x_1, \ldots, x_n$ will be the same for any two samples. Likewise, we would expect the value $h(x_1, \ldots, x_n)$ of any statistic to vary from sample to sample. The accumulation of values $h(x_1, \ldots, x_n)$ over all possible samples creates a population or the sampling distribution of the statistic.

In the following sections we indicate some of the sampling distributions pertinent to the statistical methods discussed in later chapters.

4.3 SOME PROPERTIES OF THE SAMPLE MEAN

Before developing some of the pertinent sampling distributions, let us look at some of the properties of an important sample statistic, the sample mean. Suppose $X_1, \ldots, X_n$ is a random sample from some population with mean μ and variance σ^2. Recall that the sample mean is

$$\overline{X} = \sum_{i=1}^{n} X_i/n \tag{4.3}$$

4.3.1 Expected Value (mean) of the Sample Mean

Since $\overline{X}$ is a linear combination of random variables, the expected value of $\overline{X}$ is the same linear combination of the expected values of the X_i's (Section 2.16). That is,

$$\mu_{\overline{X}} = E[\overline{X}] = \sum_{i=1}^{n} E[X_i]/n \tag{4.4}$$

All the X_i's have the same distribution and $E[X_i] = \mu$ for all i; thus

$$\mu_{\overline{X}} = \sum_{i=1}^{n} \mu/n = \mu \tag{4.5}$$

That is, the expected value of the sample mean is the population mean. For this reason we call the sample mean an *unbiased estimator* of the population mean. (This is discussed more fully in Chapter 5.) We interpret the result in Eq. (4.5) as follows: although the value of the sample mean may not be μ for a particular sample, the average of the values of $\overline{X}$ over a large number of samples will tend to the population mean.

4.3.2 Variance of the Sample Mean

We have just stated that the expected value of the sample mean is the population mean; however, it is equally important to know something about the variation among all possible values of $\bar{X}$. That is, we need to know something about the variance of the distribution of $\bar{X}$. Let $\sigma_{\bar{X}}^2$ denote the variance of $\bar{X}$. By the definition of the variance (Sections 2.10 and 2.16),

$$\sigma_{\bar{X}}^2 = V(\bar{X}) = E[(\bar{X} - \mu_{\bar{X}})^2] = E[(\bar{X} - \mu)^2] \tag{4.6}$$

since $\mu_{\bar{X}} = \mu$. After some algebraic manipulation, it is seen that

$$\sigma_{\bar{X}}^2 = E\left[\left\{\sum_{i=1}^{n} (X_i - \mu)\right\}^2\right]/n^2$$

$$= \left\{\sum_{i=1}^{n} E[(X_i - \mu)^2] + n(n - 1)E[(X_i - \mu)(X_j - \mu)]\right\}/n^2$$

$$= \{\sigma^2 + (n - 1)E[(X_i - \mu)(X_j - \mu)]\}/n \tag{4.7}$$

Two cases must now be distinguished: (1) sampling from an "infinite" population or sampling with replacement from a finite population, i.e., (simple) random sampling, and (2) sampling without replacement from a finite population. For these two cases we obtain respectively,

$$\sigma_{\bar{X}}^2 = \sigma^2/n \tag{4.8}$$

and

$$\sigma_{\bar{X}}^2 = (\sigma^2/n)[(N - n)/(N - 1)] \tag{4.9}$$

N being the size of the population.

The preceding result is very important. It says that no matter what the population (as long as it has a finite variance), the distribution of the sample mean becomes more and more concentrated in the neighborhood of the population mean as the sample size is increased. That is, the larger the sample size, the more certain we become that the sample mean will be close to the (unknown) population mean. This result will be expressed more precisely in the following section.

4.3.3 Law of Large Numbers

Since we assume that any population of interest has a finite mean and variance, we can apply Tchebycheff's inequality (Section 2.19) to the sample mean. We have

$$P(\mu - k\sigma_{\bar{X}} < \bar{X} < \mu + k\sigma_{\bar{X}}) > 1 - 1/k^2 \tag{4.10}$$

or

$$P(\mu - k\sigma/\sqrt{n} < \bar{X} < \mu + k\sigma/\sqrt{n}) > 1 - 1/k^2 \tag{4.11}$$

Setting $K = k\sigma/\sqrt{n}$, it is seen that

$$P(\mu - K < \overline{X} < \mu + K) > 1 - \sigma^2/nK^2 \qquad (4.12)$$

Thus when sampling from any population with a finite variance, the sample size may be chosen large enough to make it almost certain that the sample mean will be arbitrarily close to the population mean. This is what is known as the *law of large numbers*.

This result has a great deal of application in reliability and quality control. For example, much attention is given to the number of defectives X in a sample of size n. If we assume the sample is a random sample from a binomial population in which $\mu = p$ and $\sigma^2 = p(1 - p)$ (i.e., we have a random sample from a population of defective and nondefective units in which the proportion of defective units is p), then as n gets large,

$$P[\,|\,(X/n) - p\,| < \epsilon] \to 1 \qquad (4.13)$$

where X is the number of defective items observed in the sample of size n and ϵ is an arbitrarily small positive quantity. That is, as n increases, we become more and more certain that the observed fraction defective will be a good estimate of the true fraction defective in the population.

4.4 DISTRIBUTION OF THE SAMPLE MEAN FOR NORMAL POPULATIONS

Let $\overline{X}$ be the mean of a random sample of size n taken from a normal population with mean μ and variance σ^2. We have indicated in Sections 4.3.1, 4.3.2, and 2.18 respectively that $\mu_{\overline{X}} = \mu$, $\sigma^2_{\overline{X}} = \sigma^2/n$, and $\overline{X}$ is a normal random variable since it is a linear combination of normal random variables.

Thus we see that the sample mean is normally distributed with mean μ and variance σ^2/n; i.e., $\overline{X}$ is $N(\mu, \sigma^2/n)$. It should be clear that the probability density function for $\overline{X}$ is

$$g(\overline{x}) = (\sqrt{n}/\sigma\sqrt{2\pi})e^{-n(\overline{x}-\mu)^2/2\sigma^2} \qquad \begin{array}{l} -\infty < \overline{x} < \infty \\ \sigma > 0; -\infty < \mu < \infty \end{array} \qquad (4.14)$$

and that $Z = \sqrt{n}(\overline{X} - \mu)/\sigma$ is $N(0, 1)$.

4.5 CENTRAL LIMIT THEOREM; DISTRIBUTION OF THE SAMPLE MEAN FOR NONNORMAL POPULATIONS

Without doubt, the most important theorem in statistics is the *central limit theorem*. It is important not only from the theoretical point of view but also because of its impact on statistical methods. Since a proof of this theorem is beyond the scope of this text, it will be stated without proof.

> If a population has a finite variance σ^2 and mean μ, the distribution of the sample mean approaches the normal distribution with variance σ^2/n and mean μ as the sample size n increases.

Note that nothing is said about the form of the sampled population. That is, no matter what the form of the sampled population (provided it has a finite variance), the sample mean (if based on a sufficiently large sample size) will be approximately normally distributed. The size n for which the approximation is close enough depends on the form of the sampled population. For example, suppose the initial population is an exponential (θ) population. The density function for X is

$$f(x) = (1/\theta)e^{-x/\theta} \qquad x > 0; \theta > 0 \qquad (4.15)$$

The density functions for the sample mean for several sample sizes are presented in Figure 4.1 for $\theta = 1$. Notice that although the initial population is nothing like a normal distribution, already for $n = 5$ the normal density is appearing and by $n = 10$ the normal approximation is clearly adequate.

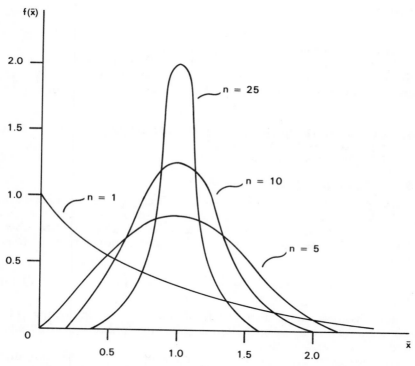

Fig. 4.1—Density functions for $\bar{X}$ from an exponential population.

If the sample size is small, the normal approximation should not be used; rather, one should use the results and techniques of Section 2.17 to get the exact distribution of the sample mean (for any n).

4.6 DISTRIBUTION OF THE DIFFERENCE OF TWO SAMPLE MEANS

If a random sample of n_1 observations is obtained from a population with mean μ_1 and variance σ_1^2 and if a random sample of n_2 observations is obtained from a population with mean μ_2 and variance σ_2^2, what can be said about the distribution of $U = \overline{X}_1 - \overline{X}_2$, where $\overline{X}_1$ is the mean of the first sample and $\overline{X}_2$ is the mean of the second sample?

Regardless of the form of the populations sampled, it is true that

$$\mu_{\overline{X}_1 - \overline{X}_2} = \mu_{\overline{X}_1} - \mu_{\overline{X}_2} = \mu_1 - \mu_2 \qquad (4.16)$$

and

$$\sigma^2_{\overline{X}_1 - \overline{X}_2} = \sigma^2_{\overline{X}_1} + \sigma^2_{\overline{X}_2} = \sigma_1^2/n_1 + \sigma_2^2/n_2 \qquad (4.17)$$

If, however, the sampled populations are both normal, it is also true that $U = \overline{X}_1 - \overline{X}_2$ is normally distributed with mean and variance given by Eqs. (4.16) and (4.17). In this situation,

$$Z = \frac{(\overline{X}_1 - \overline{X}_2) - (\mu_1 - \mu_2)}{\sqrt{\sigma_1^2/n_1 + \sigma_2^2/n_2}} \qquad (4.18)$$

is normally distributed with mean 0 and variance 1.

If the populations are not normal but both sample sizes are sufficiently large, the central limit theorem may be invoked to achieve an approximate normal distribution for the difference of two sample means.

4.7 CHI-SQUARE DISTRIBUTION; DISTRIBUTION OF THE SAMPLE VARIANCE FOR NORMAL POPULATIONS

4.7.1 Chi-Square Distribution

One particular distribution arises quite frequently in applied work and is known as the *chi-square distribution*. When referring to the chi-square distribution, the parameter ν is called the *degrees of freedom*. The probability density function for chi-square with ν degrees of freedom is given by

$$f(u) = \frac{u^{(\nu/2)-1} e^{-u/2}}{2^{\nu/2} \Gamma(\nu/2)} \qquad u > 0 \qquad (4.19)$$

where u is used rather than χ^2 (chi-square) for ease in writing. The cumulative chi-square distribution is tabled in Appendix 4 for all integral values of ν from 1 through 100.

The chi-square distribution has an additive property in the sense that the sum of several independent chi-square random variables also has a chi-square distribution. That is, if $U_1, \ldots, U_k$ are independent chi-square random variables with degrees of freedom $\nu_1, \ldots, \nu_k$ respectively, then

$$U = \sum_{i=1}^{k} U_i \qquad (4.20)$$

is distributed as chi-square with $\nu = \sum_{i=1}^{k} \nu_i$ degrees of freedom.

Some examples of chi-square random variables that arise in statistical analysis are the following:

1. Suppose Z is a standard normal variable; i.e., Z is $N(0, 1)$. Then, Z^2 is chi-square with 1 degree of freedom. Thus it follows that if X is $N(\mu, \sigma^2)$, then

$$Z^2 = [(X - \mu)/\sigma]^2 \qquad (4.21)$$

is chi-square with $\nu = 1$.

2. Let $X_1, \ldots, X_k$ be k independent normal variables with means μ_i and variances σ_i^2 respectively. Then

$$U = \sum_{i=1}^{k} (X_i - \mu_i)^2/\sigma_i^2 \qquad (4.22)$$

is chi-square with k degrees of freedom. In particular, if a random sample $X_1, \ldots, X_n$ of size n is obtained from a normal population with mean μ and variance σ^2,

$$U = \sum_{i=1}^{n} (X_i - \mu)^2/\sigma^2 \qquad (4.23)$$

is distributed as chi-square with n degrees of freedom.

3. Suppose $X_1, \ldots, X_n$ is a random sample from a normal population with mean μ and variance σ^2. It can be shown that the sample mean $\bar{X}$ and the sample variance s^2 are independent. Using this and some of the results given above, it can further be shown that

$$U = (n - 1)s^2/\sigma^2 = \sum_{i=1}^{n} (X_i - \bar{X})^2/\sigma^2 \qquad (4.24)$$

is distributed as chi-square with $\nu = n - 1$ degrees of freedom. This distributional result is used extensively in later sections.

4.7.2 Distribution of the Sample Variance and Related Statistics

Although we do not make extensive use of these, the distributions of $s^2, s, m_2 = \sum_{i=1}^{n} (X_i - \bar{X})^2/n$ and $\sqrt{m_2}$ can be obtained from the preceding results. These are:

$$f(m_2) = \frac{1}{\Gamma\left(\frac{n-1}{2}\right)} \left(\frac{n}{2\sigma^2}\right)^{(n-1)/2} m_2^{(n-3)/2} e^{-nm_2/2\sigma^2} \qquad m_2 > 0 \qquad (4.25)$$

$$f(\sqrt{m_2}) = \frac{2}{\Gamma\left(\frac{n-1}{2}\right)} \left(\frac{n}{2\sigma^2}\right)^{(n-1)/2} (\sqrt{m_2})^{n-2} e^{-nm_2/2\sigma^2} \qquad \sqrt{m_2} > 0 \qquad (4.26)$$

$$f(s^2) = \frac{1}{\Gamma\left(\frac{n-1}{2}\right)} \left(\frac{n-1}{2\sigma^2}\right)^{(n-1)/2} (s^2)^{(n-3)/2} e^{-(n-1)s^2/2\sigma^2} \qquad s^2 > 0 \qquad (4.27)$$

and

$$f(s) = \frac{2}{\Gamma\left(\dfrac{n-1}{2}\right)} \left(\frac{n-1}{2\sigma^2}\right)^{(n-1)/2} s^{n-2} e^{-(n-1)s^2/2\sigma^2} \qquad s > 0 \qquad (4.28)$$

4.8 "STUDENT'S" *t* DISTRIBUTION

Consider two independent random variables Z and U, where Z follows a standard normal distribution and U follows a chi-square distribution with ν degrees of freedom. Let t be the ratio

$$t = Z/\sqrt{U/\nu} \qquad (4.29)$$

Then, the probability density function of t is

$$f(t) = \frac{\Gamma[(\nu+1)/2]}{\sqrt{\pi\nu}\,\Gamma(\nu/2)} \cdot \frac{1}{(1 + t^2/\nu)^{(\nu+1)/2}} \qquad -\infty < t < \infty \qquad (4.30)$$

and it is referred to as the t distribution with ν degrees of freedom. This distribution is extremely useful in many problems of statistical inference. A table of cumulative percentage points of t is given in Appendix 5.

An example of a t statistic encountered in statistical methods arises when a random sample from a normal population is observed and the ratio

$$t = \sqrt{n}(\overline{X} - \mu)/s \qquad (4.31)$$

is formed. This statistic is distributed as t with $\nu = n - 1$ degrees of freedom. We will use this statistic in several later sections.

4.9 F DISTRIBUTION

Given two independently distributed chi-square variates, U with ν_1 degrees of freedom and V with ν_2 degrees of freedom, it may be shown that

$$(U/\nu_1)/(V/\nu_2) = \nu_2 U/\nu_1 V \qquad (4.32)$$

is distributed as F with ν_1 and ν_2 degrees of freedom. The density function is

$$f(F) = \frac{\Gamma\left(\dfrac{\nu_1+\nu_2}{2}\right)}{\Gamma\left(\dfrac{\nu_1}{2}\right)\Gamma\left(\dfrac{\nu_2}{2}\right)} \left(\frac{\nu_1}{\nu_2}\right)^{\nu_1/2} \frac{F^{(\nu_1-2)/2}}{\left(1 + \dfrac{\nu_1 F}{\nu_2}\right)^{(\nu_1+\nu_2)/2}} \qquad F > 0 \qquad (4.33)$$

Of particular interest in applied statistics is the fact that when two random samples are obtained, one from each of two normal populations, the ratio

$$(s_1^2/\sigma_1^2)/(s_2^2/\sigma_2^2) \qquad (4.34)$$

is distributed as F with $\nu_1 = n_1 - 1$ and $\nu_2 = n_2 - 1$ degrees of freedom. This will find application when analyses of variance are discussed later. Appendix 6 gives certain percentage points of the F distribution.

4.10 HYPERGEOMETRIC DISTRIBUTION

Often a universe consists of units that can be classified into one of two classes, A or not A. For example, in industrial applications many manufactured units are often classified either as defective or nondefective. In such cases, sampling usually consists of selecting n units out of the N units in the lot (i.e., the universe) such that the sampling is done *without replacement*. Thus we have a random sample only in the specialized sense that every possible group of n items in the lot has the same chance of comprising the sample. In such a case, if X represents the number of defective items in the sample, the probability function for the distribution of X is

$$f(x) = C(D,x) \cdot C(N - D, n - x)/C(N,n)$$
$$x = a, a + 1, \ldots, b - 1, b$$
$$a = \max(0, n - N + D)$$
$$b = \min(D, n) \tag{4.35}$$

where D represents the number of defective items in the lot and the symbol $C(x, y)$ denotes the number $x!/y!(x - y)!$.

The distribution specified by Eq. (4.35) is known as the *hypergeometric distribution*. It is the distribution underlying practically all acceptance sampling by attributes where an item of product is classified as either defective or nondefective.

Using the theory of earlier chapters, it is seen that

$$\mu = E[X] = nD/N \tag{4.36}$$

and

$$\sigma^2 = E[(X - \mu)^2] = nD(N - D)(N - n)/N^2(N - 1) \tag{4.37}$$

Thus, as expected, the average number of defective items in a sample is equal to the size of the sample multiplied by the fraction of defective items in the lot.

4.11 BINOMIAL DISTRIBUTION

Suppose that a random sample of size n is selected from an "infinite" lot of defective and nondefective items described by

$$g(y) = p^y(1 - p)^{1-y} \qquad y = 0, 1; 0 < p < 1 \tag{4.38}$$

or that a random sample of size n is selected (using sampling with replacement) from a finite population of N items, D of which are defective. In the latter case we can therefore let $p = D/N$.

Both the sampling situations described above lead to the same sampling distribution where X represents the number of defective items in a sample of n items. This distribution is described by the probability function

$$f(x) = C(n,x)p^x(1 - p)^{n-x} \qquad x = 0, 1, \ldots, n; 0 < p < 1 \qquad (4.39)$$

This distribution is known as the *binomial distribution*. Using theory already developed, it can easily be shown that

$$\mu = E[X] = np \qquad (4.40)$$

and

$$\sigma^2 = E[(X - \mu)^2] = np(1 - p) = npq \qquad (4.41)$$

where $q = 1 - p$. These results will prove useful in later work.

Probabilities associated with the binominal probability function of Eq. (4.39) or with its cumulative form have been published by the National Bureau of Standards (3), Robertson (6), and Romig (7). Reference will be made to such tables as the need arises.

4.12 BINOMIAL APPROXIMATION TO THE HYPERGEOMETRIC

Under certain conditions it is permissible to use the binomial distribution as an approximation for the hypergeometric distribution. This approximation is usually invoked to simplify numerical calculations. To see how the approximation is justified, consider the hypergeometric distribution

$$f(x) = C(D,x) \cdot C(N - D, n - x)/C(N,n) \qquad (4.42)$$

Writing this out in detail, we obtain

$$f(x) = C(n,x)\left[\frac{D}{N} \cdot \frac{D - 1}{N - 1} \cdots \frac{D - (x - 1)}{N - (x - 1)} \cdot \frac{N - D}{N - x}\right.$$

$$\left. \cdot \frac{N - D - 1}{N - x - 1} \cdots \frac{N - D - (n - x - 1)}{N - (n - 1)}\right] \qquad (4.43)$$

Setting $D = pN$ and dividing the numerator and denominator of each factor inside the brackets by N, it is seen that

$$f(x) = C(n,x)\left[p \cdot \frac{p - 1/N}{1 - 1/N} \cdots \frac{p - (x - 1)/N}{1 - (x - 1)/N} \cdot \frac{q}{1 - x/N}\right.$$

$$\left. \cdot \frac{q - 1/N}{1 - (x + 1)/N} \cdots \frac{q - (n - x - 1)/N}{1 - (n - 1)/N}\right] \qquad (4.44)$$

where $q = 1 - p$. Letting N get very large, it is clear that

$$f(x) \rightarrow C(n,x)p^x q^{n-x} \qquad (4.45)$$

That is, if N is large, the hypergeometric distribution may be approximated by the binomial distribution. The question of how large N

should be relative to n before using the binomial approximation is one that must be answered. First, since tables of logarithms of factorials are not available for k greater than 2000, calculation of the hypergeometric will be extremely tedious for such cases. Second, and perhaps more to the point, Burr (1) has said that if the lot size N is at least eight times the sample size n, it will be satisfactory to use the binomial as an approximation to the hypergeometric. However, since Burr's statement is only a general comment with no reference to the magnitude of the error involved, it seems only fair to say that each individual case must be considered on its own merits.

4.13 POISSON APPROXIMATION TO THE BINOMIAL

In instances when we do not have access to published tables of the binomial distribution, it becomes necessary to find some way of obtaining the required probabilities without excessive calculation. In such cases we usually seek some form of approximation to the binomial that involves less computation or is associated with more readily available tables. Two such approximations, respectively, involve the Poisson and normal distributions. The first of these will be discussed in this section, while the normal approximation will be examined in Section 4.14.

If p is very small (less than 0.1) and n is quite large (greater than 50), it is sometimes convenient to approximate the binomial probability function by the Poisson probability function in which $\lambda = np$. To see how this approximation is justified, consider the following argument. In

$$
\begin{aligned}
f(x) &= C(n,x)p^x(1-p)^{n-x} \\
&= \frac{n(n-1)\cdots(n-x+1)}{x!}\, p^x(1-p)^{n-x}
\end{aligned}
\tag{4.46}
$$

set $p = \lambda/n$. Then

$$
\begin{aligned}
f(x) &= \frac{n(n-1)\cdots(n-x+1)}{x!}\left(\frac{\lambda}{n}\right)^x\left(1-\frac{\lambda}{n}\right)^{n-x} \\
&= \left(\frac{n}{n}\right)\left(\frac{n-1}{n}\right)\cdots\left(\frac{n-x+1}{n}\right)\frac{\lambda^x}{x!} \\
&\quad\cdot\left(1-\frac{\lambda}{n}\right)^n\left(1-\frac{\lambda}{n}\right)^{-x} \\
&= \left(1-\frac{1}{n}\right)\left(1-\frac{2}{n}\right)\cdots\left(1-\frac{x-1}{n}\right)\frac{\lambda^x}{x!} \\
&\quad\cdot\left(1-\frac{\lambda}{n}\right)^n\left(1-\frac{\lambda}{n}\right)^{-x}
\end{aligned}
\tag{4.47}
$$

If we let $n \to \infty$ and $p \to 0$ such that $np = \lambda$ remains constant,

$$
f(x) \to (1)(1)\cdots(1)(\lambda^x/x!)e^{-\lambda}(1) = e^{-\lambda}\lambda^x/x!
\tag{4.48}
$$

which is the Poisson probability function.

Therefore, if np is large relative to p and n is large relative to np, the Poisson may be used as a reasonable approximation to the binomial. All that is necessary is to set λ in the Poisson distribution equal to np of the binomial distribution we are attempting to approximate. In other words, the means of the two distributions have been equated.

4.14 NORMAL APPROXIMATION TO THE BINOMIAL

The binomial distribution may also be approximated by the normal distribution. As in the preceding section the sample size should be reasonably large before the approximation is employed.

To illustrate the nature of the approximation, consider Figure 4.2. Here, the binomial distribution for $n = 10$ and $p = 1/2$ is pictured by the ordinates at the various values of x. If rectangles of width 1 are erected as shown, the area of the histogram equals 1. This is just an alternative way of expressing the fact that the sum of the ordinates equals 1. Using areas under the normal curve, probabilities associated with various x values may be closely approximated.

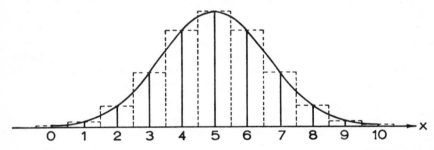

Fig. 4.2–Binomial distribution for $n = 10$ and $p = 1/2$ (solid line ordinates), area representation (dotted line rectangles), and the normal approximation.

To evaluate probabilities associated with a normal distribution, the mean and variance must be known. To specify the mean and variance of the approximating normal, let $\mu = np$ and $\sigma^2 = np(1 - p) = npq$ where np and npq are the mean and variance respectively of the binomial distribution to be approximated. Then for any integers a and $b(a < b)$ in the closed interval $[0, n]$, the approximation takes the form:

$$P(a \leq X \leq b) \cong P\left[\frac{(a - 1/2) - np}{\sqrt{npq}} \leq Z \leq \frac{(b + 1/2) - np}{\sqrt{npq}}\right] \quad (4.49)$$

$$P(a < X \leq b) \cong P\left[\frac{(a + 1/2) - np}{\sqrt{npq}} < Z \leq \frac{(b + 1/2) - np}{\sqrt{npq}}\right] \quad (4.50)$$

$$P(a \leq X < b) \cong P\left[\frac{(a - 1/2) - np}{\sqrt{npq}} \leq Z < \frac{(b - 1/2) - np}{\sqrt{npq}}\right] \quad (4.51)$$

or

$$P(a < X < b) \cong P\left[\frac{(a + 1/2) - np}{\sqrt{npq}} < Z < \frac{(b - 1/2) - np}{\sqrt{npq}}\right] \quad (4.52)$$

Other illustrations could be given, but the foregoing, together with the examples that follow, should be sufficient. The important thing to note is that $1/2$ is added to or subtracted from the limit so as to include or exclude a or b, the proper choice being indicated by the nature of the inequality. This adding or subtracting of $1/2$ is often referred to as a "correction for continuity."

Example 4.1

A random sample of 100 observations is drawn from a binomial population in which $p = 0.2$. Evaluate $P(10 \leq X \leq 25)$. We say that

$$P(10 \leq X \leq 25) \cong P[(9.5 - 20)/4 \leq Z \leq (25.5 - 20)/4]$$
$$= P(-2.62 \leq Z \leq 1.38)$$
$$= G(1.38) - G(-2.62) = 0.91621 - 0.00440$$
$$= 0.91181$$

Example 4.2

Referring to Example 4.1, evaluate $P(10 < X \leq 25)$. We have

$$P(10 < X \leq 25) \cong P[(10.5 - 20)/4 < Z \leq (25.5 - 20)/4]$$
$$= P(-2.37 < Z \leq 1.38) = G(1.38) - G(-2.37)$$
$$= 0.91621 - 0.00889 = 0.90732$$

Example 4.3

Referring to Example 4.1, evaluate $P(X > 26)$. Proceeding as before,

$$P(X > 26) \cong P[Z > (26.5 - 20)/4] = P(Z > 1.62)$$
$$= 1 - G(1.62) = 1 - 0.94738 = 0.05262$$

It is reasonable to ask what error is involved in using the approximation just described. Mood (2) has said that, if $npq > 25$, the error is less than $0.15/\sqrt{npq}$. However, we should realize that for a given n the normal curve gives a better approximation when p is close to $1/2$ than when p is close to 0 or 1. On the other hand, if n is large enough (say 100 or more), the approximation will be satisfactory for most values of p. If p is very close to 0 or 1, the approximation will be less reliable in the tails than near the center of the distribution. Thus, in reliability work, where very small values of p are frequently encountered, the normal approximation may not be too good and one should use either the Poisson approximation or calculate exact probabilities.

4.15 MULTINOMIAL DISTRIBUTION

If a random sample of size n is taken from the multinomial population described by

$$g(y_1, \ldots, y_k) = p_1^{y_1} p_2^{y_2} \cdots p_k^{y_k} \qquad y_i = 0, 1; 0 < p_i < 1$$

$$\textstyle\sum_{i=1}^{k} p_i = \sum_{i=1}^{k} y_i = 1 \qquad (4.53)$$

a multinomial distribution is obtained. This distribution is defined by

$$f(x_1, x_2, \ldots, x_k) = \frac{n!}{x_1! x_2! \ldots x_k!} p_1^{x_1} p_2^{x_2} \cdots p_k^{x_k}$$

$$x_i = 0, 1, \ldots, n; 0 < p_i < 1$$

$$\textstyle\sum_{i=1}^{k} x_i = n; \sum_{i=1}^{k} p_i = 1 \qquad (4.54)$$

where x_i is the number of items occurring in the ith class. The number p_i is the probability of any item being assigned to the ith class, and it is the fraction of the total population belonging to that class. For example, an item of product may be assigned to one of four classes: good, minor defect, major defect, or critical defect. Then the n sample items would be classified into the four groups upon inspection. The number falling in the first group would be denoted by x_1, the number in the second group by x_2, and so on.

4.16 NEGATIVE BINOMIAL DISTRIBUTION AND GEOMETRIC DISTRIBUTION

A sampling distribution encountered fairly often in industrial applications is the *negative binomial distribution*. Suppose p is the probability of a defective item and $q = 1 - p$ is the probability of a nondefective item. If random sampling is being carried out, it is frequently of importance to know the probability that the rth defective unit will occur on the $(x + r)$th unit sampled.

To obtain the probability just described, it is noted that (1) the last unit must be defective and (2) in the preceding $x + r - 1$ units sampled there must be exactly $r - 1$ defective units. Then

$$f(x) = [C(x + r - 1, r - 1)p^{r-1}q^x](p)$$
$$= C(x + r - 1, r - 1)p^r q^x \qquad x = 0, 1, \ldots$$
$$0 < p < 1 \qquad (4.55)$$

Another way of saying this is that the probability of the rth defective unit occurring on the mth unit sampled is

$$g(m) = C(m - 1, r - 1)p^r q^{m-r} \qquad m = r, r + 1, \ldots$$
$$0 < p < 1 \qquad (4.56)$$

It is sometimes of interest to know the probability of the rth defective unit occurring on the rth or $(r + 1)$st or $\ldots$ or nth unit sampled. This is given by

$$\sum_{m=r}^{n} C(m-1, r-1)p^r q^{m-r} = \sum_{m=r}^{n} C(n,m)p^m q^{n-m} \quad (4.57)$$

and the last expression may be found by consulting tables of the cumulative binomial distribution.

If in Eq. (4.55) we let $r = 1$, the negative binomial distribution simplifies to the *geometric distribution*.

4.17 ORDER STATISTICS

Observations on a chance variable usually occur in random order. However, in certain cases, observations ordered according to magnitude are encountered. This can happen in two ways: (1) the observations were obtained in random order but were subsequently reordered according to magnitude and (2) the observations naturally became available in order of magnitude. As an example of the latter consider the life testing of a group of vacuum tubes. The first observation to arise is that associated with the weakest tube (i.e., the tube with the shortest life), the second observation is associated with the next weakest tube, etc. Since such data occur fairly often in industrial applications, some sampling distributions associated with *order statistics* will now be discussed.

Consider a population specified by $f(x)$, $a \leq x \leq b$. Denote the smallest and largest values in a random sample of n observations from this population by u and v respectively. Then it may be shown that

$$g(u, v) = n(n-1)f(u)f(v)[F(v) - F(u)]^{n-2} \quad a \leq u \leq v \leq b \quad (4.58)$$

where $F(\cdot)$ is the distribution function of the population.

The marginal density functions of u and v are

$$g_1(u) = nf(u)[1 - F(u)]^{n-1} \quad a \leq u \leq b \quad (4.59)$$

and

$$g_2(v) = nf(v)[F(v)]^{n-1} \quad a \leq v \leq b \quad (4.60)$$

These distributions are very useful when dealing with problems involving *extreme values*.

Order statistics are also valuable when dealing with the sample range, $R = v - u = X_{max} - X_{min}$. If in Eq. (4.58) we let $v = u + R$, we obtain

$$g(u, R) = n(n-1)f(u)f(u+R)[F(u+R) - F(u)]^{n-2} \quad (4.61)$$

Then

$$h(R) = \int_a^{b-R} g(u, R)\, du \quad 0 \leq R \leq b - a \quad (4.62)$$

It should be noted that if instead of dealing with the joint distribution of the range and the smallest sample value we deal with the joint distribution of the range and the largest sample value, namely,

$$g(v, R) = n(n - 1) f(v - R) f(v)[F(v) - F(v - R)]^{n-2} \quad (4.63)$$

then

$$h(R) = \int_{a+R}^{b} g(v, R) \, dv \qquad 0 \le R \le b - a \qquad (4.64)$$

Eqs. (4.62) and (4.64) will produce the same result.

Example 4.4

If $f(x) = 1$, $0 < x < 1$, then $g(R) = n(n - 1)R^{n-2}(1 - R)$ where $0 < R < 1$.

Example 4.5

There is no simple expression for the distribution of the range when sampling from a normal population. Pearson (4) gave the values of the mean and standard deviation for ranges from a standardized normal distribution. Pearson and Hartley (5) evaluated the probability integral of the range for sample sizes of 2 to 20. Incidentally, the mean and standard deviation of the range when a standard normal population has been sampled are denoted by d_2 and d_3 respectively. That is, when sampling from any normal population, $\mu_R = d_2\sigma$ and $\sigma_R = d_3\sigma$. Selected values of d_2 and d_3 are given in Appendix 8.

PROBLEMS

4.1 How large a sample should be taken if we want to be 95 percent sure that $\bar{X}$ will not fall farther than $\sigma/2$ from μ?

4.2 A book of 400 pages contains 400 misprints. Estimate the probability that a page contains at least three misprints.

4.3 A lot contains 1400 items. A sample of 400 items is selected. If no more than two defective items appear in the sample, the lot will be accepted. Evaluate the probability that the lot will be accepted, assuming that the lot is 1 percent defective.

4.4 The width of a slot on a forging is normally distributed with mean 0.900 in. and standard deviation 0.003 in. The specifications are 0.900 ± 0.005 in. What percentage of forgings will be defective?

4.5 Refer to Problem 4.4. Samples of size 5 are obtained daily and their means computed. What percentage of these sample averages will be outside specifications?

4.6 The diameters of some shafts and some bearings are each normally distributed with standard deviation equal to 0.001 in. If the shaft has a mean diameter of 0.500 in. and the bearing has a mean diameter of 0.503 in., what is the probability of interference?

4.7 Three resistors are connected in series. Their nominal ratings are 10, 15, and 20 ohms respectively. If it is known that the resistors are normally distributed about the nominal ratings, each having a standard deviation of 0.5 ohm, what is the probability that an assembly will have a resistance in excess of 46.5 ohms?

4.8 Rework Problem 4.7 assuming that the standard deviation is 5 percent of nominal in each case.

4.9 A "1-pound" box of candy is machine packed to contain 32 pieces. If the weights of the pieces of candy are normally distributed with a mean of

0.5 oz and a standard deviation of 0.05 oz, what are the probabilities that a customer receives (a) less than 1 lb, (b) less than 15 oz, (c) more than 1 lb, (d) more than 16.2 oz, (e) exactly 1 lb?

4.10   Referring to Problem 4.9 and assuming the standard deviation remains unchanged, how should you change the mean of the process so that only 1 customer in 100 will receive less than the advertised weight?

4.11   A factory assembles stoves at the rate of 500 per week. On the average, 5 percent of the stoves are found to be defective when inspected following final assembly. What is the probability that next week's production will contain less than 20 defective stoves?

4.12   Review all parts of the book pertaining to the Poisson, normal, chi-square, t, and F distributions and be certain you know how to use the tables in Appendixes 2 through 6.

REFERENCES

1. Burr, I. W. 1953. *Engineering Statistics and Quality Control.* McGraw-Hill, New York.
2. Mood, A. M. 1950. *Introduction to the Theory of Statistics.* McGraw-Hill, New York.
3. National Bureau of Standards. 1949. Tables of the binomial probability distribution. Applied Mathematics Ser. 6. USGPO, Washington, D.C.
4. Pearson, E. S., 1932. The percentage limits for the distribution of range in samples from a normal population. *Biometrika* 24 (Nov.):404–17.
5. Pearson, E. S.; and Hartley, H. O. 1942. The probability integral of the range in samples of N observations from a normal population. *Biometrika* 32(Apr.):301–10.
6. Robertson, W. H. 1960. Tables of the binomial distribution function for small values of p. Sandia Corp. Monogr. SCR-143, Albuquerque, N. Mex.
7. Romig, H. G. 1953. *50–100 Binomial Tables.* Wiley, New York.

FURTHER READING

Bennett, C. A.; and Franklin, N. L. 1954. *Statistical Analysis in Chemistry and the Chemical Industry.* Wiley, New York.

Bowker, A. H.; and Lieberman, G. J. 1972. *Engineering Statistics,* 2nd ed. Prentice-Hall, Englewood Cliffs, N.J.

Brownlee, K. A. 1965. *Statistical Theory and Methodology in Science and Engineering,* 2nd ed. Wiley, New York.

Dixon, W. J.; and Massey, F. J., Jr. 1969. *Introduction to Statistical Analysis,* 3rd ed. McGraw-Hill, New York.

Freund, J. E. 1973. *Modern Elementary Statistics,* 4th ed. Prentice-Hall, Englewood Cliffs, N.J.

Hays, W. L.; and Winkler, R. L. 1971. *Statistics: Probability, Inference and Decision.* Holt, Rinehart and Winston, New York.

Huntsberger, D. V.; and Billingsley, P. 1973. *Elements of Statistical Inference,* 3rd ed. Allyn and Bacon, Boston.

Miller, I.; and Freund, J. E. 1965. *Probability and Statistics for Engineers.* Prentice-Hall, Englewood Cliffs, N.J.

Snedecor, G. W.; and Cochran, W. G. 1967. *Statistical Methods,* 6th ed. Iowa State University Press, Ames.

Steel, R. G. D.; and Torrie, J. H. 1960. *Principles and Procedures of Statistics.* McGraw-Hill, New York.

Walpole, R. E.; and Myers, R. H. 1972. *Probability and Statistics for Engineers and Scientists.* Macmillan, New York.

Wine, R. L. 1964. *Statistics for Scientists and Engineers.* Prentice-Hall, Englewood Cliffs, N.J.

CHAPTER 5

STATISTICAL INFERENCE: ESTIMATION

In this chapter, general concepts associated with that part of statistical inference referred to as "estimation and prediction" will be examined. Examples dealing with particular populations frequently encountered in applied work will also be given.

5.1 INTRODUCTION

Suppose we have a population of interest described by some distribution of *known* form (normal, exponential, binomial, etc.). In general, we do not know the values of the parameters of the distribution nor the values of the population mean and variance. To estimate the values of these parameters, we commonly obtain a random sample from the specified population. The sample $X_1, \ldots, X_n$ is then used to derive approximate values of the population parameters. This procedure is called *estimation,* and we are interested in developing methods for estimating the values of the parameters for some frequently encountered distributions. We first define what we mean by an *estimator.*

Definition 5.1   An *estimator* is a statistic—a function $h(X_1, \ldots, X_n)$ of the sample—that is used to estimate the value of a parameter of a distribution.

Thus an estimator is a function $h(X_1, \ldots, X_n)$ of the sample, which provides us with an "approximate" value of a parameter based on the sample observed. The value $h(x_1, \ldots, x_n)$ of an estimator for a specific sample is called an *estimate.*

If the estimator provides a single value as an estimate of a parameter, it is called a *point estimator,* and its value for a specific sample is called a *point estimate* or just an *estimate.* On the other hand, if the estimator defines an interval of values as the estimate, the estimator is called an *interval estimator.* The interval defined for a specific sample is called a *confidence interval.* These latter estimates, i.e., confidence interval estimates, are very important in statistical methodology and, if at all possible, we obtain an estimate of this type. We illustrate the difference between point and interval estimates by a short example.

Example 5.1

If we wish to estimate the average weight of the people in a classroom, we could take a random sample of five people, record their weights, and average them. The resulting average (suppose it turned out to be 160 lb) would be our point estimate. However, this may not be sufficient for our purposes. If we say that the true average weight of all the people in the room (they are the population) is between 0 and 300 lb, we are very confident of ourselves (in fact, we are almost certain of our statement). If we make our interval much smaller (and in practice the interval should be as small as possible) our degree of confidence in our interval estimate will become less. For example, if we say we think that the average weight of all people in the room is between 158 and 162 lb, our degree of confidence may be quite small.

If we wish to be able to evaluate our degree of confidence for any interval estimate, it is customary to make certain assumptions concerning the distribution of the observations being obtained. Several examples of such confidence intervals will be studied later in this chapter.

5.2 POINT ESTIMATORS

A point estimator provides a single value as an estimate of a population parameter. Before looking at specific point estimators, we should state some criteria useful for judging or evaluating different ones. We do this by specifying some properties of a good estimator; these may then be used as criteria to distinguish between good and bad estimators. Other criteria may be found in the literature, but the three given here are perhaps the most important from a practical point of view.

If $X_1, \ldots, X_n$ is a random sample from a population with parameter θ, let $\hat{\theta} = h(X_1, \ldots, X_n)$ be a point estimator of θ.

1. *Unbiased.* An estimator $\hat{\theta}$ is *unbiased* if $E[\hat{\theta}] = \theta$. That is, $\hat{\theta}$ is unbiased if the average of all possible values of $\hat{\theta}$ is θ.
2. *Consistent.* If, as the sample size n gets very large (i.e., approaches N where N is the number of items in the population), the probability that $\hat{\theta}$ will be very close to θ approaches 1, or certainty, then $\hat{\theta}$ is called a *consistent estimator* of θ. That is, $\hat{\theta}$ is a consistent estimator of θ if $P(|\hat{\theta} - \theta| < \epsilon) \to 1$ as $n \to \infty$. In other words, if we take a larger and larger sample, we expect to get an estimate very close to the true value, and the probability that we will do so is close to 1.
3. *Minimum variance unbiased.* $\hat{\theta}$ is a *minimum variance unbiased estimator* if $V(\hat{\theta}) \leq V(\theta^*)$ for any other unbiased estimator θ^*. Thus if θ_1 and θ_2 are both unbiased estimators of θ and if $\sigma^2_{\theta_1} < \sigma^2_{\theta_2}$, then we prefer θ_1 to θ_2. In general, we prefer the estimator (in the class of unbiased estimators) that has the minimum variance.

Next we look at how a point estimator can be derived. Several principles of estimation, leading to routine mathematical procedures, have been proposed for obtaining good estimators. These include:

1. The principle of moments.
2. Minimum chi-square.
3. The method of least squares.
4. The principle of maximum likelihood.

The application of these principles in particular cases will lead to estimators that may differ and hence possess different attributes of "goodness." A principle much in use, yielding estimators with many desirable attributes of goodness and obtained by easily applied routine mathematical procedures, is that of *maximum likelihood* devised by Fisher (5, 6). This important principle of estimation will be used in the remainder of the chapter.

The procedure for determining the maximum likelihood estimate of a population parameter θ is as follows:

1. Assuming that the random sample $x_1, \ldots, x_n$ is obtained from the population $f(x; \theta)$, determine the density (probability) function of the sample,

$$g(x_1, \ldots, x_n; \theta) = \prod_{i=1}^{n} f(x_i; \theta)$$

Recall that in Section 4.1 $g(x_1, \ldots, x_n; \theta)$ was referred to as the *likelihood function*. [**Note:** we use $g(x_1, \ldots, x_n; \theta)$ instead of $g(x_1, \ldots, x_n)$ to show that $g(\cdot)$ is a function of the parameter θ.]

2. The value of θ that maximizes $g(x_1, \ldots, x_n; \theta)$ for the given sample $x_1, \ldots, x_n$ is called the *maximum likelihood estimate*. Frequently, the procedure for deriving this value is: (a) determine

$$L(x_1, \ldots, x_n; \theta) = \log \prod_{i=1}^{n} f(x_i; \theta)$$

and (b) determine the value of θ that maximizes L by solving the equation $\partial L / \partial \theta = 0$.

Rather than detailing the derivation of *maximum likelihood estimators*, the results for several of the more common distributions are presented in Table 5.1.

TABLE 5.1–Maximum Likelihood Estimators Associated
with Certain Distributions

Distribution	Parameter	Maximum Likelihood Estimator
Binomial	p	$\hat{p} = X/n = $ observed relative frequency
Poisson	$\lambda(=\mu)$	$\hat{\lambda} = \hat{\mu} = \bar{X} = \sum X/n$
Normal	$a(=\mu)$	$\hat{a} = \hat{\mu} = \bar{X} = \sum X/n$
	$b^2(=\sigma^2)$	$\hat{b}^2 = \hat{\sigma}^2 = m_2 = \sum (X - \bar{X})^2/n$
Exponential	$\theta(=\mu)$	$\hat{\theta} = \hat{\mu} = \bar{X} = \sum X/n$

5.3 CONFIDENCE INTERVALS

A point estimate of a parameter is usually not very meaningful without some measure of the possible error in the estimate. An estimate $\hat{\theta}$ of a parameter θ should be accompanied by some interval about $\hat{\theta}$, possibly of the form $\hat{\theta} - d$ to $\hat{\theta} + d$, together with some measure of assurance that the true parameter θ does lie within the interval. Estimates are often given in such form. For example, the activated life of a thermal battery may be estimated to be 300 ∓ 20 sec with the idea that the life is unlikely to be less than 280 sec or greater than 320 sec. The development engineer engaged in research on capacitors may estimate the mean life of a certain type of capacitor under stated conditions to be 300 ∓ 50 hr with the implication that the correct average life very probably lies between 250 and 350 hr. The failure rate for a specific component might be estimated as being less than 0.02 with the feeling that the true failure rate is most likely no greater than the stated limit. In this last case the point estimate might have been anywhere between 0 and 0.02.

Confidence intervals enable us to obtain a useful type of information about population parameters without the necessity of treating such parameters as statistical variables. It should be clearly understood that we are merely betting on the correctness of the rule of procedure when applying confidence interval techniques to a given experiment. It will be observed in the following sections that this technique may be applied to various familiar population parameters such as the mean and variance.

An examination of the following sections will reveal that the method for finding confidence intervals consists in first finding a random variable, call it Z, that involves the desired parameter θ but the distribution of which does not depend upon any other unknown parameters. Next, two numbers Z_1 and Z_2 are chosen such that

$$P(Z_1 < Z < Z_2) = \gamma \qquad (5.1)$$

where γ is the desired confidence coefficient, such as 0.95. Then the two inequalities are manipulated so that the probability statement assumes the form

$$P(L < \theta < U) = \gamma \qquad (5.2)$$

where L and U are random variables depending on Z but not involving θ. Finally, we substitute the sample values in L and U to obtain a numerical interval that is the desired confidence interval. It is clear that any number of confidence intervals can be constructed for a parameter by choosing Z_1 and Z_2 differently each time or by choosing different random variables of the Z type.

The above has been concerned with what is called a two-sided confidence interval. However, we sometimes do not care how much our estimate may err in one direction provided it is not too far off in the other. For example, we may be estimating a standard deviation that we hope will be small. We would be concerned only about an upper limit and hence would want an interval of the form

$$P(\theta < U) = \gamma \qquad (5.3)$$

The theory of one-sided intervals is basically the same as for two-sided intervals.

5.4 CONFIDENCE INTERVALS FOR THE MEAN OF A POPULATION

It has already been said that the sample mean $\overline{X}$ is an unbiased estimator of the population mean μ. We now make use of some of the sampling distributions in Chapter 4 and the procedure discussed in the previous section to construct a confidence interval for μ.

5.4.1 Confidence Interval for the Mean of a Normal Population

Case 1: σ^2 unknown

If the population is normal with unknown variance, the statistic[1]

$$t = (\overline{X} - \mu)/s_{\overline{X}} = \sqrt{n}(\overline{X} - \mu)/s \tag{5.4}$$

has a t distribution with $n - 1$ degrees of freedom. Thus we can make the statement,

$$P[-t_{0.975(n-1)} < (\overline{X} - \mu)/s_{\overline{X}} < t_{0.975(n-1)}] = 0.95 \tag{5.5}$$

where $t_{\delta(n-1)}$ is that value of a t random variable with $n - 1$ degrees of freedom such that $P[t < t_{\delta(n-1)}] = \delta$. After some algebraic manipulation we can rewrite the statement as

$$P[\overline{X} - t_{0.975(n-1)}s_{\overline{X}} < \mu < \overline{X} + t_{0.975(n-1)}s_{\overline{X}}] = 0.95 \tag{5.6}$$

or

$$P[\overline{X} - t_{0.975(n-1)}s/\sqrt{n} < \mu < \overline{X} + t_{0.975(n-1)}s/\sqrt{n}] = 0.95 \tag{5.7}$$

Our interpretation of this statement is: the probability (before the sample is drawn) that the random interval (L, U), where L and U are defined by Eqs. (5.8) and (5.9) respectively, will cover or include the true population mean μ is equal to 0.95. Thus if a random sample is obtained from a normal population with mean μ and variance σ^2, and the two quantities

$$L = \overline{X} - t_{0.975(n-1)}s_{\overline{X}} \tag{5.8}$$

and

$$U = \overline{X} + t_{0.975(n-1)}s_{\overline{X}} \tag{5.9}$$

are computed, we can say we are 95 percent confident that the true mean μ will be in the interval (L, U). We do *not* say that the probability is 0.95 that μ lies between L and U but only that we are 95 percent confident that μ does lie between L and U. This distinction is made because μ either does or does not fall between L and U; the probability is either 0 or 1, for μ is a constant and does not possess a probability distribution. The distinction made above is a subtle one and the concept may not be fully appreciated at this time. However, it is a distinction that must be made.

---

1. The symbol $s_{\overline{X}}$ is known as the *standard error of the mean*, and it is clearly an estimator of $\sigma_{\overline{X}}$ as defined by Eq. (4.6).

Example 5.2

Consider the estimation of the mean breaking strength of some particular material. We take at random a number of samples (for this example, 6) and subject them to test, recording the pressure at which they break. These values might be as follows:

2206 lb	2205 lb	2206 lb
2209 lb	2203 lb	2207 lb

Averaging these values, we obtain a point estimate, i.e., one value, of 2206 lb. This means that from our sample a reasonable estimate of the true (population) average breaking strength of the material is 2206 lb. However, we do not have any measure of our degree of confidence in this estimate. If we are willing to assume a normal distribution, we can find:

$$L = \overline{X} - t_{0.975(n-1)}s_{\overline{X}} = 2206 - 2.571(0.8165) = 2203.9 \text{ lb}$$
$$U = \overline{X} + t_{0.975(n-1)}s_{\overline{X}} = 2206 + 2.571(0.8165) = 2208.1 \text{ lb}$$

where $t_{0.975(5)} = 2.571$ was obtained from the table in Appendix 5 and all other values were calculated from our sample. We can now say that we are 95 percent confident that the true population mean breaking strength lies between 2203.9 and 2208.1 lb. A 99 percent confidence interval can be found in a similar manner using $t_{0.995(5)} = 4.032$.

In general, a 100γ percent confidence interval $0 < \gamma < 1$ is given by the random interval (L, U) where

$$\left.\begin{array}{c} L \\ U \end{array}\right\} = \overline{X} \mp t_{[(1+\gamma)/2](n-1)}s_{\overline{X}} \qquad (5.10)$$

Rather than to proceed as in Example 5.2, we might have wanted only a lower confidence limit. That is, we might have no interest in an upper limit on breaking strength, since ordinarily no harm can result from the material being too strong. The statement needed (assuming a 0.95 confidence coefficient) is then

$$P[\overline{X} - t_{0.95(n-1)}s_{\overline{X}} < \mu] = 0.95 \qquad (5.11)$$

Note that here $t_{0.95(n-1)}$ is used instead of $t_{0.975(n-1)}$ since we want the entire 0.05 error risk to be on one side of the limit rather than to be split equally beyond two limits. Thus we would obtain

$$L = \overline{X} - t_{0.95(n-1)}s_{\overline{X}} = 2206 - 2.015(0.8165) = 2204.4 \qquad (5.12)$$

and could then state that we are 95 percent confident that the true population mean breaking strength is above 2204.4 lb. In general, the lower limit would be $L = \overline{X} - t_{\gamma(n-1)}s_{\overline{X}}$.

It should be clear that if only a 100γ percent upper confidence limit is desired, the procedure would be to calculate

$$U = \overline{X} + t_{\gamma(n-1)}s_{\overline{X}} \qquad (5.13)$$

Case 2: σ^2 known

If the variance is known, we use the fact that

$$Z = (\overline{X} - \mu)/\sigma_{\overline{X}} \qquad (5.14)$$

is a standard normal random variable. A 100γ percent confidence interval for μ is given by (L, U) where

$$\left.\begin{array}{r}L \\ U\end{array}\right\} = \overline{X} \mp z_{(1+\gamma)/2}\sigma_{\overline{X}} \qquad (5.15)$$

and $z_{(1+\gamma)/2}$ is the $100(1 + \gamma)/2$ fractile of the standard normal distribution. In a similar way, for an upper or lower confidence interval the appropriate value of z replaces t and $\sigma_{\overline{X}}$ replaces $s_{\overline{X}}$.

5.4.2 Confidence Interval for the Mean of a Nonnormal Population

A question that might logically arise is "What can we do if we want a confidence interval estimate of the mean of a nonnormal population?" The central limit theorem discussed in Chapter 4 provides us with an answer that is often satisfactory. That is, unless the distribution is much different from normal and the sample size is extremely small, the distribution of sample means will be nearly normal so that the normal theory may be applied with only a small error.

However, if the error introduced by the approximate procedure suggested in the preceding paragraph cannot be tolerated, we always have recourse to exact methods associated with the particular population distribution involved. No attempt will be made to list all the different situations. Rather, we shall state only that the basic approach is always the same as outlined in Section 5.3. If the need arises for an exact answer for a nonnormal distribution, the reader is referred to many such examples in the literature. If the particular case in question cannot be located in this manner, a mathematical statistician should be consulted.

5.5 CONFIDENCE INTERVALS FOR THE DIFFERENCE BETWEEN THE MEANS OF TWO POPULATIONS

Many practical problems in statistics involve the comparison of two populations. We now consider computing confidence intervals for the difference of the means of two normal populations. Let $X_{11}, \ldots, X_{1n_1}$ and $X_{21}, \ldots, X_{2n_2}$ denote two random samples from normal populations $N(\mu_1, \sigma_1^2)$ and $N(\mu_2, \sigma_2^2)$ respectively. We are interested in computing confidence intervals for the difference $\mu_1 - \mu_2$. There are several cases that we must consider.

Case 1: Samples are independent

We assume the variables X_{1j} and X_{2j} are independent.

Case 1a: $\sigma_1^2 = \sigma_2^2$

If it can be assumed that the two normal populations have equal variances; i.e., if we can assume a common variance σ^2, then the statistic

$$[(\bar{X}_1 - \bar{X}_2) - (\mu_1 - \mu_2)]/s_{\bar{X}_1 - \bar{X}_2}$$
$$= [(\bar{X}_1 - \bar{X}_2) - (\mu_1 - \mu_2)]/[s^2(1/n_1 + 1/n_2)]^{1/2} \qquad (5.16)$$

is distributed as "Student's" t with $n_1 + n_2 - 2$ degrees of freedom if s^2 is calculated by means of the formula

$$s^2 = \frac{(n_1 - 1)s_1^2 + (n_2 - 1)s_2^2}{n_1 + n_2 - 2} = \frac{\sum_{j=1}^{n_1}(X_{1j} - \bar{X}_1)^2 + \sum_{j=1}^{n_2}(X_{2j} - \bar{X}_2)^2}{n_1 + n_2 - 2}$$

$$(5.17)$$

where s_1^2 and s_2^2 are the usual variance estimates derived from the two samples. It should be noted that s^2 is often referred to as the *pooled estimate of variance*.

Under the assumptions stated above, 100γ percent confidence limits for $\mu_1 - \mu_2$ may be found by calculating

$$\left.\begin{array}{c}L\\U\end{array}\right\} = (\bar{X}_1 - \bar{X}_2) \mp t_{[(1+\gamma)/2](n_1 + n_2 - 2)} s_{\bar{X}_1 - \bar{X}_2} \qquad (5.18)$$

Example 5.3

There are two methods of measuring the moisture content of heat-processed beef. For Method 1 we obtain $\bar{X}_1 = 88.6$, $s_1^2 = 109.63$, and $n_1 = 41$. For Method 2 the comparable results are $\bar{X}_2 = 85.1$, $s_2^2 = 65.99$, and $n_2 = 31$. Thus

$$s^2 = [40(109.63) + 30(65.99)]/70 = 90.93$$

and

$$s_{\bar{X}_1 - \bar{X}_2} = [(90.93/41) + (90.93/31)]^{1/2} = 2.27$$

Finally, assuming an 80 percent confidence interval is desired, we obtain $L = 3.5 - (1.294)(2.27) \cong 0.6$ and $U = 3.5 + (1.294)(2.27) \cong 6.4$.

Case 1b: $\sigma_1^2 \neq \sigma_2^2$

If there is reason to believe that the two populations have different variances, the procedure just discussed is not appropriate. What then can be done? If we are willing to assume that $s_1^2 = \sigma_1^2$ and $s_2^2 = \sigma_2^2$, an approximate 100γ percent confidence interval may be found by calculating

$$\left.\begin{array}{c}L\\U\end{array}\right\} = (\bar{X}_1 - \bar{X}_2) \mp z_{(1+\gamma)/2}(s_1^2/n_1 + s_2^2/n_2)^{1/2} \qquad (5.19)$$

where $z_{(1+\gamma)/2}$ is the $100(1 + \gamma)/2$ fractile of the standard normal distribution. However, because of the doubtful validity of the assumption that the sample variances equal the population variances, this procedure provides only a very crude estimate of the true mean difference. Consequently, the procedure should be used with extreme caution. Other approximations using the t distribution are discussed in Snedecor (13).

Case 2:   Paired observations; $n_1 = n_2 = n$

If the two samples are of equal size and the observations in one sample can logically be paired with the observations in the other, a modified procedure applies. Thus we assume that X_{11} is related to X_{21}, X_{12} is related to X_{22}, etc. In the language of a later chapter, the variables X_1 and X_2 are said to be *correlated*. For example, if the variables X_1 and X_2 are the expenditures for recreational activities in 1973 and 1974 and if the *same* households were sampled in both years, X_1 and X_2 are correlated.

When such a pairing of observations exists, an appropriate procedure is to calculate the differences $D_i = X_{1i} - X_{2i}$ for $i = 1, \ldots, n$. The 100γ percent confidence limits for $\mu_D = \mu_1 - \mu_2$ are given by

$$\left.\begin{array}{r} L \\ U \end{array}\right\} = \bar{D} \mp t_{[(1+\gamma)/2](n-1)}s_{\bar{D}} \qquad (5.20)$$

where $s_{\bar{D}}^2 = s_D^2/n$ and $s_D^2 = \sum_{i=1}^{n}(D_i - \bar{D})^2/(n - 1)$.

Example 5.4

It is desired to compare the prices of Delicious and McIntosh apples. On a certain day, prices (per box) were obtained from a random selection of 11 markets. Assuming (1) prices to be normally distributed and (2) the price of one variety in a market would be influenced by the price of the other variety in the same market, the method of paired observations will be used. The data are given in Table 5.2. Calculation yields $\bar{D} = 0.14$, $s_D^2 =$

TABLE 5.2–Price per Box of Delicious and McIntosh Apples

Market	Delicious	McIntosh	Difference
1	$2.15	$2.32	$0.17
2	2.16	2.34	0.18
3	2.13	2.30	0.17
4	2.25	2.40	0.15
5	2.20	2.34	0.14
6	2.18	2.16	−0.02
7	2.27	2.42	0.15
8	2.21	2.36	0.15
9	2.23	2.40	0.17
10	2.16	2.30	0.14
11	2.20	2.34	0.14

0.0030, and $s_{\bar{D}}^2 = 0.0030/11$. Therefore, a 95 percent confidence interval for μ_D is specified by

$$\left.\begin{matrix} L \\ U \end{matrix}\right\} = 0.14 \mp (2.228)(0.0030/11)^{1/2} \cong (\$0.10, \$0.18)$$

As with the estimate of a single mean, one-sided confidence limits are also possible. All that is necessary is a change in the value of t (or z) in the equation for L or U.

5.6 CONFIDENCE INTERVAL FOR THE VARIANCE OF A NORMAL POPULATION

Using a technique similar to that outlined for the mean, a confidence interval for estimating the variance of a normal population can be found. This time, however, we use the fact that the statistic

$$(n - 1)s^2/\sigma^2 \tag{5.21}$$

is a chi-square random variable with $n - 1$ degrees of freedom. Thus we can make the probability statements

$$P\{\chi^2_{[(1-\gamma)/2](n-1)} < (n - 1)s^2/\sigma^2 < \chi^2_{[(1+\gamma)/2](n-1)}\} = \gamma$$

or

$$P\left\{\frac{(n - 1)s^2}{\chi^2_{[(1+\gamma)/2](n-1)}} < \sigma^2 < \frac{(n - 1)s^2}{\chi^2_{[(1-\gamma)/2](n-1)}}\right\} = \gamma \tag{5.22}$$

which we read: the probability (before the sample is drawn) that the random interval (L, U) where

$$L = (n - 1)s^2/\chi^2_{[(1+\gamma)/2](n-1)} = \sum_{i=1}^{n} (X_i - \bar{X})^2/\chi^2_{[(1+\gamma)/2](n-1)} \tag{5.23}$$

and

$$U = (n - 1)s^2/\chi^2_{[(1-\gamma)/2](n-1)} = \sum_{i=1}^{n} (X_i - \bar{X})^2/\chi^2_{[(1-\gamma)/2](n-1)} \tag{5.24}$$

will include the true population variance σ^2 is equal to γ. Or, as it is more often phrased, we are 100γ percent confident that the true population variance σ^2 will be in the interval (L, U).

Example 5.5

For the data in Example 5.2 we find the 90 percent confidence interval for σ^2 to be $(1.8, 17.5)$.

As with means, we can determine a one-sided confidence interval. This would be defined by

$$P\left[\sigma^2 < \sum_{i=1}^{n} (X_i - \bar{X})^2/\chi^2_{(1-\gamma)(n-1)}\right] = \gamma \tag{5.25}$$

if an upper limit is desired. Although a lower limit is conceivable, it would seldom be of interest.

If we are interested in a confidence interval for estimating σ rather than σ^2, the confidence limits $L' = \sqrt{L}$ and $U' = \sqrt{U}$, where L and U are the confidence limits for σ^2, may be computed. It should be noted that this is not the exact solution. However, it is sufficiently accurate for most purposes.

5.7 CONFIDENCE INTERVAL FOR THE RATIO OF THE VARIANCES OF TWO NORMAL POPULATIONS

The problem of estimating the ratio of two population variances (or standard deviations) is also frequently encountered. If the two populations are normal and independent, we use the fact that the statistic

$$(s_1^2/\sigma_1^2)/(s_2^2/\sigma_2^2) \tag{5.26}$$

has an F distribution with $n_1 - 1$ and $n_2 - 1$ degrees of freedom. The procedure is to calculate

$$L = (s_2^2/s_1^2) F_{[(1-\gamma)/2](n_1-1, n_2-1)} \tag{5.27}$$

and

$$U = (s_2^2/s_1^2) F_{[(1+\gamma)/2](n_1-1, n_2-1)} \tag{5.28}$$

and these limits define a 100γ percent confidence interval for σ_2^2/σ_1^2. If only an upper (or lower) limit is desired, it can easily be found by using F_γ in Eq. (5.28) or $F_{1-\gamma}$ in Eq. (5.27). One other useful result is the following: if only an abbreviated F table is available (e.g., one that contains only the upper percentage points), the identity

$$F_{p(\nu_1, \nu_2)} = 1/F_{(1-p)(\nu_2, \nu_1)} \tag{5.29}$$

permits the calculation of F values at the left-hand tail of the distribution.

Example 5.6

Using the data given in Example 5.3, 99 percent confidence limits for σ_2^2/σ_1^2 are found to be:

$$L = (65.99/109.63)(0.416) = 0.25$$
$$U = (65.99/109.63)(2.52) = 1.52$$

If a confidence interval for the ratio of the standard deviations of two normal populations is desired (i.e., if we wish to estimate σ_2/σ_1), it is appropriate to calculate $L' = \sqrt{L}$ and $U' = \sqrt{U}$ where L and U are defined by Eqs. (5.27) and (5.28).

5.8 CONFIDENCE INTERVAL FOR p, THE PARAMETER OF A BINOMIAL POPULATION

It has already been suggested in Section 5.2 that the best point estimator of p is

$$\hat{p} = X/n = \text{observed relative frequency} \tag{5.30}$$

If a two-sided 100γ percent confidence interval for p is desired, the following two equations must be solved for p:

$$\sum_{y=X}^{n} C(n,y)p^y(1 - p)^{n-y} = (1 - \gamma)/2 \tag{5.31}$$

$$\sum_{y=0}^{X} C(n,y)p^y(1 - p)^{n-y} = (1 - \gamma)/2 \tag{5.32}$$

The solution of Eq. (5.31) is L, while the solution of Eq. (5.32) is U. If $X = 0$, L is taken to be 0; if $X = n$, U is taken to be 1. We may then state that we are 100γ percent confident that $L \leq p \leq U$.

Because the computation involved in solving Eqs. (5.31) and (5.32) is tedious, convenient tables and graphs useful for finding L and U are available. See Hald (7), Snedecor and Cochran (13), Clopper and Pearson (3), Calvert (2) or Muench (8) for such tables and graphs.

An alternative method for establishing a confidence interval for p makes use of the F distribution. Using this distribution, 100γ percent confidence limits for p are:

$$L = \frac{XF_{[(1-\gamma)/2],[2X,2(n-X+1)]}}{(n - X + 1) + XF_{[(1-\gamma)/2],[2X,2(n-X+1)]}} \tag{5.33}$$

$$U = \frac{(X + 1)F_{[(1+\gamma)/2],[2(X+1),2(n-X)]}}{(n - X) + (X + 1)F_{[(1+\gamma)/2],[2(X+1),2(n-X)]}} \tag{5.34}$$

Example 5.7

Consider an industrial process producing parts that are classified either as defective or nondefective. In a random sample of 200 units 6 are found to be defective. Thus an estimate of the proportion defective produced by this process is $\hat{p} = 6/200 = 0.03$. To obtain a 95 percent confidence interval for p, we use Eqs. (5.33) and (5.34) and have

$$L = (6)(0.367)/[195 + (6)(0.367)] \cong 0.011$$
$$U = (7)(1.83)/[194 + (7)(1.83)] \cong 0.061$$

Thus, we are 95 percent confident that p is between 0.011 and 0.061.

We can also make use of the normal approximation to the binomial for large n to get an approximate confidence interval. Approximate 100γ percent confidence limits are given by:

$$\left.\begin{matrix} L \\ U \end{matrix}\right\} = \hat{p} \mp z_{(1+\gamma)/2} \sqrt{\hat{p}(1 - \hat{p})/n}$$

5.9 TOLERANCE LIMITS

One common method used by engineers to specify the quality of manufactured product is the method of tolerance limits. When such limits are quoted, it is expected that a certain percentage of the product will have a quality between the stated limits. For example, suppose electrical gaps are judged by the characteristic, "transfer time." It is then desirable to be able to quote two limits A and B such that we are fairly certain that, say, 98 percent of all gaps produced will exhibit transfer times between A and B. Such limits clearly provide us with a measure of the quality of the product under consideration. For certain weapon applications, it is convenient to be able to set a one-sided tolerance limit. An example of such a case is the following: 99 percent of all Type XYZ batteries will yield an activated life of at least 200 sec. In general, then, tolerance limits are limits within which we are highly confident a certain percentage of the individuals of a statistical population will lie.

To apply tolerance limits in a satisfactory manner, certain conditions must be met. In summary, the conditions upon which tolerance limits are based are the following:

1. All assignable causes of variability must be detected and eliminated so that the remaining variability may be considered random.
2. Certain assumptions must be made concerning the nature of the statistical population under study.

5.10 TOLERANCE LIMITS (TWO-SIDED, ONE-SIDED) FOR NORMAL POPULATIONS

Tolerance limits considered in this section are based on the assumption that the parent population may be described by a normal distribution. If the true mean and standard deviation are known, tolerance limits are formed by adding to and subtracting from the mean some multiple of the standard deviation. That is, if μ and σ are known, tolerance limits take the form $\mu \mp z\sigma$, where z is selected from Appendix 3 and depends only on the proportion of the population to be included within the calculated limits. For example, the limits $\mu \mp 1.645\sigma$ include 90 percent of a normal population with mean μ and standard deviation σ. One-sided tolerance limits may be obtained by considering $\mu + z\sigma$ or $\mu - z\sigma$ as the problem requires.

In a practical situation, μ and σ are unknown. Only estimates $\overline{X}$ and s are available. While it was true that the limits $\mu \mp 1.645\sigma$ will include 90 percent of the population, the same statement cannot be made concerning $\overline{X} \mp 1.645s$. Just what proportion of the population will lie between $\overline{X} \mp Ks$ depends on how closely $\overline{X}$ and s estimate μ and σ. Note that K is used here to represent the constant used with $\overline{X}$ and s in contrast with the z used with μ and σ.

Since $\overline{X}$ and s (hence $\overline{X} \mp Ks$) are random variables, it is impossible to state with certainty that $\overline{X} \mp Ks$ will always contain a specified proportion P of the population. That is, it is impossible to choose K so that the calculated limits will always contain a specified proportion P of the population. However, it is possible to determine K so that in many random samples from a normal population a certain fraction γ of the intervals $(\overline{X} \mp Ks)$ will contain $100P$ percent or more of the population. When this notation is used, P is referred to as the coverage and γ as the confidence coefficient. This terminology is used since we are 100γ percent confident that the tolerance range specified by $\overline{X} \mp Ks$ will include at least $100P$ percent of the normal population sampled.

Intuitively, it is reasonable to expect that values of K used with $\overline{X}$ and s will be larger than values of z used with μ and σ. It is also clear that if K is taken large enough, the probability that $\overline{X} \mp Ks$ will contain at least $100P$ percent of the population may be made very close to 1. However, the smaller K is taken, the more meaningful and useful the tolerance range becomes. The engineer is thus faced with a decision: make broad statements with little risk of error or make precise statements (i.e., a narrow tolerance range) with greater risk of error. The problem, statistically speaking, becomes that of finding the smallest value of K consistent with a specified confidence coefficient γ, proportion P, and sample size n.

We must not forget that one-sided tolerance limits are frequently more appropriate than two-sided ones. That is, it is often desirable to specify a single limit such that a given percentage of the population will be less than (or greater than) this limit. Such a limit is known as a one-sided tolerance limit and is usually of the form $\overline{X} + Ks$ (or $\overline{X} - Ks$). Both one-sided and two-sided tolerance limits for normal populations will be discussed in the following paragraphs.

TABLE 5.3–Two-Sided Tolerance Factors

	P			
n	0.7500	0.9000	0.9500	0.9900
5	3.002	4.275	5.079	6.634
6	2.604	3.712	4.414	5.775
7	2.361	3.369	4.007	5.248
8	2.197	3.136	3.732	4.891
9	2.078	2.967	3.532	4.631
10	1.987	2.836	3.379	4.433
17	1.679	2.400	2.858	3.754
37	1.450	2.073	2.470	3.246
145	1.280	1.829	2.179	2.864
∞	1.150	1.645	1.960	2.576

Note: Factors K such that the probability is 0.95 that at least a proportion P of the distribution will be included between $\overline{X} \mp Ks$ where $\overline{X}$ and s are computed from a sample of size n.

Table 5.3 is an abbreviated table of K factors for two-sided tolerance intervals. Values of K taken from this table give a 95 percent confidence that at least a fraction P will be included in the interval $(\overline{X} \mp Ks)$. Table 5.4 is an abbreviated table of K factors for one-sided tolerance intervals. Values of K taken from this table give a 95 percent confidence that at least a fraction P will be above (below) $\overline{X} - Ks(\overline{X} + Ks)$. Much more extensive tables can be found in Bowker and Lieberman (1), Eisenhart et al. (4), Owen (12), and Weissberg and Beatty (14).

Example 5.8

Using the data of Example 5.2, find tolerance limits such that you are 95 percent confident of including at least 99 percent of the sampled population. These limits are given by

$$\overline{X} \mp Ks = 2206 \mp 5.775(2) = (2194.45, 2217.55)$$

The foregoing discussion of tolerance limits and the K factors given in Tables 5.3 and 5.4 depend squarely on the assumption of a random sample from a normal population. If tolerance limits are calculated using these tables when the sampled population is definitely non-normal, considerable error is possible.

TABLE 5.4–One-Sided Tolerance Factors

	P			
n	0.7500	0.9000	0.9500	0.9900
5	2.150	3.412	4.212	5.751
6	1.895	3.008	3.711	5.065
7	1.733	2.756	3.400	4.644
8	1.618	2.582	3.188	4.356
9	1.532	2.454	3.032	4.144
10	1.465	2.355	2.911	3.981
17	1.220	2.002	2.486	3.414
37	1.014	1.717	2.149	2.972
145	0.834	1.481	1.874	2.617
∞	0.674	1.282	1.645	2.326

Note: Factors K such that the probability is 0.95 that at least a proportion P of the distribution will lie above (below) $\overline{X} - Ks(\overline{X} + Ks)$ where $\overline{X}$ and s are computed from a sample of size n.

5.11 DISTRIBUTION-FREE TOLERANCE LIMITS

Sometimes it is desirable to set tolerance limits that do not depend on the assumption of normality. That is, we recognize that it is not always possible to justify the assumption of a normal distribution. If we are dealing with a statistical variable that can be described by a continuous distribution, one very simple set of *distribution-free tolerance limits* is specified by X_{min} and X_{max}, the smallest and largest values in a random sample of size n. Clearly, the confidence in such limits will depend on n. Persons interested in reading further on this topic are referred to Murphy (9), Ostle (10), Owen (11), and Wilks (15).

PROBLEMS

5.1   As a physicist or chemist, you would soon become acquainted with such "constants" as Planck's constant and Euler's constant. To consider a specific case, Planck's constant is defined as "the quantum of energy radiated from black bodies ÷ frequency of radiation." Suppose you were attempting to find the value of this constant by experimental methods. You ran 6 experiments and obtained the following estimates of h (Planck's constant):

$$6.53 \times 10^{-27} \qquad 6.58 \times 10^{-27} \qquad 6.55 \times 10^{-27}$$
$$6.54 \times 10^{-27} \qquad 6.56 \times 10^{-27} \qquad 6.55 \times 10^{-27}$$

What inferences can you make about the true value of h? Be careful to state explicitly any assumptions you make.

5.2   Given that $n = 9$, $\overline{Y} = 20$, and $\sum_{i=1}^{n} (Y_i - \overline{Y})^2 = 288$, calculate a 95 percent confidence interval for μ on the assumption that you have a random sample from a normal population. Interpret this confidence interval.

5.3   From a random sample of 100 aptitude test scores drawn from a normal population, the 95 percent confidence interval for μ is calculated to be (45, 55). Fifty other random samples, each of size 100, are drawn from the same population, but only 10 of their means fall within the above limits. Is it correct to expect 95 percent of such sample means to be between 45 and 55? Explain your answer.

5.4   A random sample of 25 observations from a normal population had a mean of 20 and a sum of squares of the deviations from the mean of 2400. Compute and interpret the 90 percent confidence interval for the population mean.

5.5   It has been reasonably well established that a particular machine produces nails whose length is a random variable with a normal distribution. A random sample of 5 nails yields the following results: 1.14 in., 1.15 in., 1.14 in., 1.12 in., 1.10 in. Calculate 99 percent confidence limits for μ.

5.6   The density of each of 27 explosive primers was determined, with the sample average being 1.53 and the sample standard deviation being 0.04. Determine a 90 percent upper confidence limit for μ.

5.7   The firing of 101 rockets yielded an average range (i.e., distance flown) of 3000 yd and a standard deviation of 40 yd. Determine an 85 percent lower confidence limit for μ.

5.8   Using the data of Problem 5.1, compute a 95 percent confidence interval for σ^2.

5.9   Using the data of Problem 3.11, compute a 90 percent confidence interval for σ.

5.10 If in a sample of 14 bolts the estimate of the population standard deviation of their lengths was $s = 0.021$, what are the 98 percent confidence limits for the standard deviation of the population σ? What assumptions must be made to determine these limits?

5.11 Using the data of Problem 5.10, determine a 90 percent upper confidence limit for σ.

5.12 Using the data of Problem 5.5, determine a 97.5 percent upper confidence limit for σ.

5.13 Using the data of Problem 5.6, determine a 90 percent upper confidence limit for σ.

5.14 In 1954 the mean earnings of 63 physicians in communities from 10,000 to 25,000 was $13,944, with $s = 4616. Find the 90 percent confidence limits for the population standard deviation. State your assumptions.

5.15 In a random sample of 400 farm operators, 65 percent were owners and 35 percent were nonowners. Determine 95 percent confidence limits for the true percentage of farm owners in the population of operators sampled.

5.16   In a random sample of 600 light bulbs, 12 were defective. Determine a 95 percent upper confidence limit for the true fraction defective.

5.17   Using the data of Problem 3.1 and the results found in Problem 3.10, determine (a) 95 percent confidence limits for μ and (b) 80 percent confidence limits for σ^2. State all assumptions.

5.18   Using the data of Problem 3.2 and the results found in Problem 3.10, determine (a) a 99 percent upper confidence limit for μ and (b) a 95 percent upper confidence limit for σ. State all assumptions.

5.19   Using the data of Problem 3.3 and the results found in Problem 3.10, determine 50 percent confidence limits for each of the means and 50 percent upper confidence limits for each of the standard deviations. State all assumptions.

5.20   You are engaged as a testing engineer in an electrical manufacturing plant. One of the products being produced is an electric fuse, and the most important characteristic of this fuse is the length of time before it "blows" when subjected to a specified load. A testing program was undertaken and the following sample data (in seconds) were obtained.

Day 1	Day 2
42	69
45	109
68	113
72	118
90	153

Place a 90 percent confidence interval on the true difference between the means of the two different days' productions. Assume that each day's production may be represented by a normal population. State all other assumptions that you make and interpret your numerical answer.

5.21   Given that $\overline{Y}_1 = 75, n_1 = 9, \sum_{i=1}^{9}(Y_{1i} - \overline{Y}_1)^2 = 1482, \overline{Y}_2 = 60, n_2 = 16, \sum_{j=1}^{16}(Y_{2j} - \overline{Y}_2)^2 = 1830$, and assuming that the 2 samples were randomly selected from 2 normal populations in which $\sigma_1^2 = \sigma_2^2$, calculate an 80 percent confidence interval for $\mu_1 - \mu_2$.

5.22   Two barley varieties have been grown at a number of locations over several years in an area and their general adaptability is under discussion. Which variety would you select for the area on the basis of the following yields in bushels per acre?

        Trebi—41.2, 19.3, 45.5, 63.9, 63.8, 44.2, 42.5, 53.0
        Svanota—39.4, 30.8, 44.5, 51.5, 41.1, 26.5, 35.7

Place confidence limits on the difference between the means.

5.23   Two varieties of tomato were experimented with concerning their fruit-producing abilities, measured in pounds. The study was done in a greenhouse and because of extreme variations (among locations within greenhouses) of temperature, light quality, and light intensity the experimental plants were placed in pairs (one of each variety) at several locations. The following data were obtained. Determine 90 percent confidence limits for the true difference between the expected weights of the two varieties. State all assumptions.

Location	Variety		Difference = $A - B = D$
	A	B	
1	3.03	2.28	0.75
2	3.10	2.68	0.42
3	2.35	2.17	0.18
4	3.86	3.56	0.30
5	3.91	3.73	0.18
6	2.65	1.48	1.17
7	1.72	1.85	−0.13
8	2.30	1.86	0.44
9	2.70	2.76	−0.06
10	3.60	2.68	0.92
Total	29.22	25.05	4.17

5.24 Using the data of Problem 5.20, obtain 95 percent confidence limits for σ_1^2/σ_2^2.

5.25 Using the data of Problem 5.21 and ignoring the assumption used there, namely, that $\sigma_1^2 = \sigma_2^2$, obtain 99 percent confidence limits for σ_1^2/σ_2^2.

5.26 Using the data of Problem 5.22, obtain 95 percent confidence limits for σ_S/σ_T.

5.27 Using the data of Problem 5.2, determine with 95 percent confidence (a) 95 percent tolerance limits and (b) an upper 99 percent tolerance limit.

5.28 Using the data of Problem 5.4, determine with 95 percent confidence (a) 75 percent tolerance limits and (b) a lower 90 percent tolerance limit.

5.29 Using the data of Problem 5.6, determine with 95 percent confidence a 99 percent upper tolerance limit on the densities.

5.30 Using the data of Problem 5.7, determine with 95 percent confidence a 90 percent lower tolerance limit on the ranges.

5.31 Consider the following definitions:

(a) If the expected value of an estimator does not equal the true value being estimated, the difference between the expected value and the true value is known as the *bias* of the estimator.

(b) If an estimator has 0 bias, it is said to be *accurate*.

(c) If an estimator has a small bias, it is said to be *relatively accurate*.

(d) If an estimator has a large bias, it is said to be *inaccurate*.

(e) The *precision* of an estimator is a measure of the *repeatability* of the estimator. Therefore, precision may be expressed in terms of the variance of an estimator, with a large variance signifying lack of precision and a small variance signifying high precision. Obviously, absolute precision implies a 0 variance, an ideal seldom if ever achieved. (**Note:** Sometimes a measure of precision is referred to as a measure of reliability. Because the word "reliability" has another meaning in engineering, this is unfortunate. However, as with many expressions, the phrase is now a part of the language of statistics and will therefore continue to be used.)

It should be observed that an estimator may be (a) both precise and accurate, (b) neither precise nor accurate, (c) precise but not accurate, or (d) accurate but not precise.

(a) Discuss the foregoing concepts and definitions relative to the contents of Section 5.1.

(b) Discuss these ideas, taking cognizance of costs and other economic and physical limitations that continually plague the researcher.

(c) Discuss the accuracy and precision of the various estimators that have been introduced so far in this text.

REFERENCES

1. Bowker, A. H.; and Lieberman, G. J. 1972. *Engineering Statistics,* 2nd ed. Prentice-Hall, Englewood Cliffs, N.J.
2. Calvert, R. L. 1955. The determination of confidence intervals for probabilities of proper, dud, and premature operation. Sandia Corp. Tech. Memo. SCTM 213-55-51, Albuquerque, N. Mex.
3. Clopper, C. J.; and Pearson, E. S. 1934. The use of confidence or fiducial limits illustrated in the case of the binomial. *Biometrika* 26:404–13.
4. Eisenhart, C.; Hastay, M. W.; and Wallis, W. A. (eds.). 1947. *Selected Techniques of Statistical Analysis.* (Statistical Research Group, Columbia University.) McGraw-Hill, New York.
5. Fisher, R. A. 1922. On the mathematical foundations of theoretical statistics. *Phil. Trans. Roy. Soc.* Ser. A, Vol. 222.
6. _____. 1925. Theory of statistical estimation. Proc. Cambridge Phil. Soc. Vol. 22.
7. Hald, A. 1952. *Statistical Tables and Formulas.* Wiley, New York.
8. Muench, J. O. 1960. A confidence limit computer. Sandia Corp. Monogr. SCR-159, Albuquerque, N. Mex.
9. Murphy, R. B. 1948. Non-parametric tolerance limits. *Ann. Math. Stat.* 19(Dec.): 581–89.
10. Ostle, B. 1957. Some remarks on the problem of tolerance limits. *Ind. Qual. Control* 13 (Apr.):11–13.
11. Owen, D. B. 1957. Distribution-free tolerance limits. Sandia Corp. Tech. Memo. SCTM 66A-57-51, Albuquerque, N. Mex.
12. _____. 1958. Tables of factors for one-sided tolerance limits for a normal distribution. Sandia Corp. Monogr. SCR-13, Albuquerque, N. Mex.
13. Snedecor, G. W., and Cochran, W. G. 1967. *Statistical Methods,* 6th ed. Iowa State University Press, Ames.
14. Weissberg, A.; and Beatty, G. H. 1959. *Tables of Tolerance-Limit Factors for Normal Distributions.* Battelle Memorial Institute, Columbus, Ohio.
15. Wilks, S. S. 1942. Statistical prediction with special reference to the problem of tolerance limits. *Ann. Math. Stat.* 13:400–409.

FURTHER READING

See Chapter 4.

CHAPTER 6

STATISTICAL INFERENCE: TESTING HYPOTHESES

The general concepts of hypothesis testing will be explored in this chapter. The tests for several of the frequently encountered distributions are outlined and examples of the test procedures are given.

6.1 INTRODUCTION

A hypothesis is defined by Webster as "a tentative theory or supposition provisionally adopted to explain certain facts and to guide in the investigation of others." A statistical hypothesis is a statement about a statistical population and usually is a statement about the values of one or more parameters of the population. For example, the following could be taken as hypotheses: (1) the probability of a 1 on a toss of a certain die is 1/6, (2) the mean height of American adult males is 5 feet 8.4 inches, (3) the mean length of a certain brand of 6-inch rulers is 5.99 inches and the standard deviation is 0.02 inch.

It is frequently desirable to test the validity of such hypotheses. To do this, an experiment is conducted and the hypothesis is rejected if the results obtained are improbable under this hypothesis. If the results are not improbable, the hypothesis is accepted. For example, we might test hypothesis (1) above by tossing the die 600 times. Intuitively, it is evident that if 600 1's are obtained, the result is improbable under the hypothesized probability of 1/6, and the hypothesis should be rejected. On the other hand, if 100 1's were observed, this result would not be improbable and the hypothesis would undoubtedly be accepted. When results such as these are obtained, intuition (combined with common sense) is sufficient to decide whether to accept the hypothesis. However, in actual practice, experimental results do not usually lead to such obvious conclusions, hence the need for *statistical* tests of hypotheses. It should be pointed out that although we accept or reject a hypothesis, we have not proved or disproved it.

When we make a decision to accept or reject a hypothesis based on the results of an experiment, it is possible that we will make one of two types of wrong decisions. These wrong decisions or errors are called: (1) Type I error, rejection of a hypothesis that is true, or (2) Type II error, acceptance of a hypothesis that is false. The relationship between the decision made, the true situation, and the errors are summarized in Table 6.1.

Since our decisions are based on experimental data, wrong decisions occasionally will be made. Nothing can be done to prevent this. Con-

TABLE 6.1–Types of Errors Associated with Tests of Hypotheses

Decision	True Situation	
	Hypothesis is true	Hypothesis is false
Accept the hypothesis	No error	Type II error
Reject the hypothesis	Type I error	No error

sider the following example:

Example 6.1

Suppose we are concerned about the unbiasedness of a coin. We can evaluate this by testing the hypothesis that the probability of a head in a toss of a coin is $1/2$. To do this, suppose we toss the coin 100 times. Even if the hypothesis is true, it is unlikely that we would get exactly 50 heads. Therefore, let us make our decision to accept or reject the hypothesis in the following way: if X is the number of heads in the 100 tosses, we decide (1) to accept the hypothesis if $48 \leq X \leq 52$ or (2) to reject the hypothesis if $X \leq 47$ or $X \geq 53$. If the experiment is conducted many times and the hypothesis is true, occasionally X will be 47 or less, or 53 or more. Similarly, even if the hypothesis is false, it is possible for X to be between 48 and 52. In either case, the decision-making procedure outlined above results in the wrong decision.

The procedure we will follow in testing a hypothesis includes the following steps, which we illustrate using Example 6.1. Initially, the hypothesis to be tested is stated. This hypothesis, denoted H, is called the *null hypothesis*. One must also specify a hypothesis (even if it is the hypothesis that H is false) that will be accepted if H is rejected. This latter hypothesis is called the *alternative hypothesis,* and we denote it by A. In Example 6.1, if we let p denote the probability of a head occurring, the null hypothesis is $H:p = 1/2$, and an alternative hypothesis could be $A:p \neq 1/2$. Thus we are testing the hypothesis that the coin is unbiased against the alternative hypothesis that the coin is biased. In addition to stating the null and alternative hypotheses, we need to identify a statistic, called the *test statistic,* the value of which will be used to decide whether to accept or reject H. In our example the test statistic considered was X, the number of heads in the 100 tosses of the coin. Note that the test statistic is a function of the number of tosses of the coin, i.e., the sample size. Finally, we specify the decision-making rule. In the example our decision was based on partitioning the set of possible values of X into two regions (1) the set of values ($X \leq 47$ or $X \geq 53$) for which we decide to reject H and accept A and (2) the set of values ($48 \leq X \leq 52$) for which we accept H. The former region is called the *rejection* or *critical region,* and the latter set of values is often called the *acceptance region.*

As pointed out earlier, occasionally we will commit a Type I error (i.e., $X \leq 47$ or $X \geq 53$ even though $p = 1/2$), and occasionally we will make a Type II error (i.e., $48 \leq X \leq 52$ even though $p \neq 1/2$). Let α denote the probability of a Type I error. (**Note:** By α we mean the "long-run" relative frequency of making a Type I error when the experiment is repeated many times under identical conditions and using the same decision rule). That is, α is the probability that we will reject a true null hypothesis using the test selected. [**Note:** 100α (in percent) is commonly referred to as the *significance level*.] In Example 6.1 it is the probability of concluding the coin is biased ($p \neq 1/2$) when in fact it is unbiased (i.e., $p = 1/2$). Similarly, let β denote the probability of a Type II error; i.e., β is the probability of not concluding the coin is biased when in fact it is. The size of α and β will depend on the distribution of the test statistic under the null and alternative hypotheses respectively. Note also, as pointed out earlier, that the distribution of the test statistic, hence the value of α and β, will generally depend upon the sample size.

Example 6.2

Consider a simple hypothesis $H:\mu = \mu_0$ against a single alternative $A:\mu = \mu_1$, where we are dealing with a normal population with known variance σ^2. Let the decision to reject H (accept A) or to accept H (reject A) be based on a single observation obtained *at random* from the population under examination. If the random observation is less than C (see Fig. 6.1), H will be accepted; if the random observation is greater than or equal to C, H will be rejected. That is, $X \geq C$ constitutes the rejection or critical region. The probabilities α and β are represented by the shaded and cross-hatched areas respectively. Besides depending on the choice of C, α depends on the hypothesis under test (the *null hypothesis*) while β depends both on the null hypothesis and on the alternative hypothesis.

An important consideration in discussing the probabilities of Type II errors is the "degree of falseness" of a false null hypothesis. In a given experiment if the null hypothesis is false but is nearly true (such as hypothesizing that a probability is $1/2$ when actually it is $1.0001/2$), β could be quite large. However, if the null hypothesis is grossly false, (such as hypothesizing that a probability is $1/2$ when it is actually 1), β

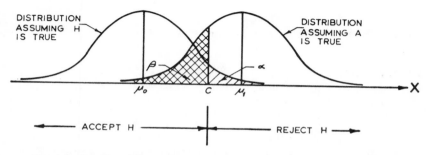

Fig. 6.1–Graphical illustration of the acceptance and rejection regions in Example 6.2.

should be much smaller. For a given experiment testing a specific null hypothesis, the value of $1 - \beta$ is known as the *power of the test*. Since the power depends on the difference between the value of the parameter specified by the null hypothesis and the actual value of the parameter where the latter is unknown, $1 - \beta$ should be expressed as a function of the true parameter. Such a function is known as a *power function* and is expressed as $1 - \beta(\theta)$, where θ represents the true parameter value. The complementary function $\beta(\theta)$ is known as the *operating characteristic* (OC) *function.*

Example 6.3

Modify Example 6.2 to the following extent: consider $H : \mu = \mu_0$ versus the composite alternative $A : \mu > \mu_0$. In this situation α is the same as before but β is now better denoted by $\beta(\mu) = P(\text{accept } H \mid \mu)$. Clearly, $\beta(\mu)$ changes as we think of the "alternative distribution" in Figure 6.1 taking all possible positions for which $\mu_1 > \mu_0$. Thus an OC curve similar to the one shown in Figure 6.2 is generated.

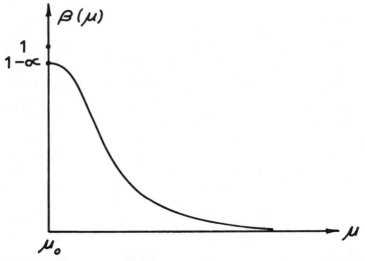

Fig. 6.2–Type of OC curve to be expected in situations similar to Example 6.3.

Example 6.4

Consider a further modification of Example 6.2, namely, $H : \mu = \mu_0$ versus the alternative $A : \mu \neq \mu_0$. The acceptance and rejection regions might be as shown in Figure 6.3, namely, reject if $X \leq C_1 = \mu_0 - k\sigma$ or if $X \geq C_2 = \mu_0 + k\sigma$ and accept if $C_1 < X < C_2$. Only the distribution of the test statistic under H is shown. The distribution under A may be visualized if the reader thinks of sliding the distribution shown to the left and to the right. For this situation, an OC curve similar to the one in Figure 6.4 would result.

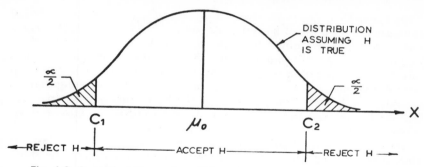

Fig. 6.3–Graphical illustration of the acceptance and rejection regions in Example 6.4.

Once we have stated the hypothesis, selected the test statistic and set up the decision rule, and settled on a certain sample size, the values of α and β (or the OC curve) are determined. On the other hand, it would be preferable to specify desired values of α and β and then, for a given test statistic and decision rule, determine the sample size necessary to meet the values of α and β. It is more common, however, to assure ourselves that neither of these probabilities are too high by specifying an allowable maximum value for at least one of these probabilities (usually α) and either: (1) if the sample size is fixed, selecting the test statistic and decision rule that minimize the second probability β or (2) if the test statistic and decision rule are given, assuring ourselves that n is large enough for the desired value of β. Before indicating how a test procedure is established, we first discuss some ideas on selecting allowable values of α and β.

What constitutes suitably small values of α and β? This is not a question that can be answered unequivocally for all situations. Obviously the values of α and β should depend on the consequences of making

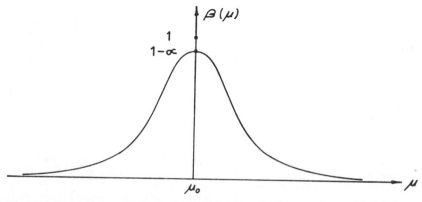

Fig. 6.4–Type of OC curve to be expected in situations similar to Example 6.4.

Type I and II errors respectively. For example, if we are considering the purchase of a lot of batteries (or some other very critical item) for use in weapons, we might hypothesize that the lot is of satisfactory quality. (Actually we should state this hypothesis in more precise terms.) If this hypothesis is true and we reject it, no great harm has been done since we can always wait for the next lot (assuming that we are not in a hurry). Consequently, α can be relatively large (perhaps 0.25 or larger). On the other hand, if the hypothesis is false and we accept it, the result may be a large number of dud weapons. Since this is very undesirable, β should be quite small, (maybe 0.01 or less). It should be pointed out that the supplier might feel differently about these probabilities.

6.2 ESTABLISHMENT OF TEST PROCEDURES

When establishing a test procedure to investigate statistically the credibility of a stated hypothesis, several factors must be considered. Assuming that a clear statement of the problem has been formulated and an associated hypothesis has been stated in mathematical terms, these are:

1. The *nature of the experiment* that will produce the data must be defined.
2. The *test statistic* must be selected. That is, the method of analyzing the data should be specified.
3. The *nature of the critical region* must be established.
4. The *size of the critical region* (i.e., α) must be chosen.
5. A *value should be assigned to $\beta(\theta)$* for at least one value of θ other than the value of θ specified by H. This is equivalent to stating what difference between the hypothesized value of the parameter and the true value of the parameter must be detectable, and with what probability we must be confident of detecting it.
6. The *size of the sample* (i.e., the number of times the experiment will be performed) must be determined.

It should be clear that these steps will not always be taken in the order listed. Not all the steps are independent, and frequently it is necessary to reconsider (several times) the various steps until a reasonable test procedure is formulated. More will be said on this subject later. For now, some explanatory examples will probably be of more value than additional generalizations.

Example 6.5

Reconsider Example 6.1. With respect to a specific coin we have $H:P(\text{head}) = p = 0.5$ and $A:p \neq 0.5$. The experiment will consist of tossing the coin some number of times and counting the times heads occur; rejection of H will take place if either a very small or very large proportion of heads are observed. Let $\alpha = 0.05$. Ignore $\beta(p)$ for the moment. Consider $n = 5$ and the rejection region to consist of either no heads or all heads. Then $P(\text{rejection} \mid p = 0.5) = 1/16 = 0.0625$. Since this is greater than $\alpha = 0.05$, a larger number of tosses is required. Let us try $n = 6$, keeping the same rejection region. Now we have $P(\text{rejection} \mid p = 0.5) = 0.03125$, which is less than $\alpha = 0.05$. Thus an acceptable test procedure has been developed. [**Note:** The probabilities of rejection given were calculated using $f(x) = C(n,x) p^x (1 - p)^{n-x}$.]

Example 6.6

In Example 6.5 we derived the test: "toss the coin six times and reject H if either zero or six heads occurs; otherwise, accept H." Other rejection regions might have been chosen together with different values of n, as long as $P(\text{rejection} \mid p = 0.5) \leq \alpha$. What we found in Example 6.5 was the smallest value of n for the specified rejection region. The reader should investigate some of these other possibilities.

Example 6.7

Now consider $\beta(p)$ for the test derived in Example 6.5. For selected values of p the approximate values of $\beta(p) = 1 - p^6 - (1 - p)^6$ are given in Table 6.2. [**Note:** Only approximate values are given because the exact answers involve an unnecessary number of decimal places. For example, for $p = 0.5$, $\beta(p) = 1 - P(\text{rejection} \mid p = 0.5) = 1 - 0.03125 = 0.96875 \cong 0.97$.] If one did not consider the derived test to be discriminating enough (as evidenced by the OC curve), the discriminatory power could be increased by: (1) changing the sample size and the definition of the critical region or (2) concocting an entirely different test procedure and test statistic. It is clear that we are faced with just this situation in the present case. The test derived in Example 6.5 is good for detecting two-headed or two-tailed coins (nearly as good as looking at both sides of the coin) but is poor for detecting slightly or even moderately biased coins. Thus a modified or new test is required.

In the following sections we outline the test procedures for some of the commonly encountered testing situations. The test statistics and critical regions given are those that minimize β for a given sample

TABLE 6.2–Selected Values of the
OC Function for Example 6.7

p	Approximate Value of $\beta(p)$
0	0
0.1	0.47
0.2	0.74
0.3	0.88
0.4	0.95
0.5	0.97
0.6	0.95
0.7	0.88
0.8	0.74
0.9	0.47
1.0	0

size n. In any case, reference must be made to the OC function to assure that the sample size being considered is adequate.

6.3 NORMAL POPULATION AND TESTS OF THE MEAN μ

Suppose we wish to test a hypothesis about the mean of a normal population. More specifically, suppose we have the random variable X, which can be considered to have a normal distribution, and we want to test the hypothesis $H : \mu = \mu_0$.

If the value of σ^2 is unknown and a random sample is observed from the population, the test statistic to use is

$$t = (\overline{X} - \mu_0)/s_{\overline{X}} = \sqrt{n}(\overline{X} - \mu_0)/s \qquad (6.1)$$

where n is the sample size and $\overline{X}$ and s^2 are the sample mean and variance respectively. Thus the test compares the difference between the sample mean and the hypothesized value μ_0 with the standard deviation of $\overline{X}$. If this difference is large relative to $s_{\overline{X}}$, we would want to reject the hypothesis. If not, we have no reason to reject H.

To set up the critical region we note that if the hypothesis is true the test statistic t will have a "Student's" t distribution with $\nu = n - 1$ degrees of freedom. Let 100α be the level of significance.

Case 1: $H : \mu = \mu_0; A : \mu \neq \mu_0$

For the simple null hypothesis and the two-sided alternative reject H if

$$t \leq -t_{(1-\alpha/2)(n-1)} \quad \text{or} \quad t \geq t_{(1-\alpha/2)(n-1)} \qquad (6.2)$$

where $t_{\delta\nu}$ is the value of a t random variable with ν degrees of freedom such that $P(t \leq t_{\delta\nu}) = \delta$. Otherwise do not reject H.

Example 6.8

A metallurgist makes measurements of the melting point of manganese. He is interested in testing if these measurements are in accord with the hypothesized value of 1260 C. Assuming such measurements are approximately normally distributed, the hypotheses are $H : \mu = 1260C; A : \mu \neq 1260$ C. To make this test, determinations of 1269 C, 1271 C, 1263 C, and 1265 C are observed and the value

$$t = (\overline{X} - \mu_0)/s_{\overline{X}} = (1267 - 1260)/1.862 = 3.83$$

is computed. Assuming $\alpha = 0.05$ and observing that $n = 4$, the critical region is $t \leq -3.182$ or $t \geq 3.182$. Since $3.83 > 3.182$, we reject H and conclude that the hypothesized value is not the "true" average of the measured values. By using a 5 percent significance level, it is recognized that the probability of Type I error will be no greater than 0.05. That is, there is a maximum risk of 5 percent in rejecting the hypothesis that $\mu = 1260$ if the hypothesis is really true.

Case 2: $H: \mu \leq \mu_0; A: \mu > \mu_0$

This is the case of a one-sided alternative. For this case, reject H if

$$t \geq t_{(1-\alpha)(n-1)} \tag{6.3}$$

otherwise do not reject H.

Case 3: $H: \mu \geq \mu_0; A: \mu < \mu_0$

Reject H if

$$t \leq -t_{(1-\alpha)(n-1)} \tag{6.4}$$

otherwise do not reject H. The three cases are summarized in Table 6.3.

TABLE 6.3–Tests for the Mean of a Normal Population

Hypothesis	Test Statistic	Critical Region
1. σ known		
a. $H: \mu = \mu_0; A: \mu \neq \mu_0$	$z = \sqrt{n}(\overline{X} - \mu_0)/\sigma$	$\|z\| \geq z_{(1-\alpha/2)}$
b. $H: \mu \leq \mu_0; A: \mu > \mu_0$		$z \geq z_{(1-\alpha)}$
c. $H: \mu \geq \mu_0; A: \mu < \mu_0$		$z \leq -z_{(1-\alpha)}$
2. σ unknown		
a. $H: \mu = \mu_0; A: \mu \neq \mu_0$	$t = \sqrt{n}(\overline{X} - \mu_0)/s$	$\|t\| \geq t_{(1-\alpha/2)(n-1)}$
b. $H: \mu \leq \mu_0; A: \mu > \mu_0$		$t \geq t_{(1-\alpha)(n-1)}$
c. $H: \mu \geq \mu_0; A: \mu < \mu_0$		$t \leq -t_{(1-\alpha)(n-1)}$

Example 6.9

A manufacturer of television sets purchases tubes from one of the few large suppliers of such specialized material. He will not purchase tubes, however, unless it can be demonstrated that the average length of life will exceed 500 hours. In this case, the hypotheses being considered are, for testing purposes, expressed as $H: \mu \leq 500; A: \mu > 500$. A random sample of 9 tubes is subjected to a "life test," and the following values are obtained: $\overline{X} = 600$ and $s^2 = 2500$. It is assumed that the "lengths of life" (measured in hours) are normally distributed. Shall the hypothesis $H: \mu \leq 500$ be accepted? For this example, $t = (600 - 500)/16.67 = 6.00$ far exceeds $t_{0.95(8)} = 1.860$, and the null hypothesis is rejected. As can be seen, a 5 percent significance level was used. This means that the maximum risk of rejecting $H: \mu \leq 500$ when H is really true is 5 percent. Therefore, the manufacturer of television sets will undoubtedly purchase tubes from this supplier.

If the value of σ^2 were known, the test statistic to use is

$$z = \sqrt{n}(\overline{X} - \mu_0)/\sigma \tag{6.5}$$

When H is true, this test statistic has a standard normal distribution. The critical regions are similar to the previous cases and are given in Table 6.3.

It would also be of interest to examine the OC curves for the tests proposed above. However, it would be necessary to prepare OC curves for many values of α and n. Also, the formula for β associated with "Student's" t test involves the noncentral t distribution, which must be considered beyond the scope of this test. Thus, the reader is referred to examples of such curves given in Bowker and Lieberman (2).

6.4 TWO NORMAL POPULATIONS AND COMPARISON OF THE MEANS

Suppose we have two normal populations with means μ_1 and μ_2 and variances σ_1^2 and σ_2^2 respectively, and we wish to test the null hypothesis that $\mu_1 = \mu_2$. Generally, the variances are unknown so we must consider several cases. To do this, we present the discussion for the hypotheses $H:\mu_1 = \mu_2$; $A:\mu_1 \neq \mu_2$. The tests for the one-sided hypotheses are summarized in Table 6.8. We let $X_{11}, \ldots, X_{1n_1}$ and $X_{21}, \ldots, X_{2n_2}$ be random samples of size n_1 and n_2 respectively from the two populations.

6.4.1 Independent Populations

We begin by considering the cases in which the variables X_{1j} and X_{2j} are independent.

Case 1: $\sigma_1^2 = \sigma_2^2$

If we can assume that the variances of the two populations are equal but unknown, the test statistic for testing the equality of the means is

$$t = (\overline{X}_1 - \overline{X}_2)/[s^2(1/n_1 + 1/n_2)]^{1/2} \qquad (6.6)$$

where $\overline{X}_i$ is the mean of the sample from the ith population and s^2 is the estimate of the common variance,

$$s^2 = [(n_1 - 1)s_1^2 + (n_2 - 1)s_2^2]/(n_1 + n_2 - 2)$$

$$= \left[\sum_{j=1}^{n_1} (X_{1j} - \overline{X}_1)^2 + \sum_{j=1}^{n_2} (X_{2j} - \overline{X}_2)^2 \right] \bigg/ (n_1 + n_2 - 2) \quad (6.7)$$

The critical region for the two-sided alternative hypothesis is

$$t \leq -t_{(1-\alpha/2)(n_1+n_2-2)}; t \geq t_{(1-\alpha/2)(n_1+n_2-2)}$$

since t is distributed as "Student's" t with $(n_1 + n_2 - 2)$ degrees of freedom when H is true.

Example 6.10

Wire cable is being manufactured by two processes. We wish to determine if the processes are having different effects on the mean breaking strength of the cable. Laboratory tests were performed by putting samples of cable under tension and recording the load required to break the cable. Using the data given in Table 6.4 and letting $\alpha = 0.05$, test the hypothesis $H:\mu_1 = \mu_2$ versus $A:\mu_1 \neq \mu_2$. Calculations yield $\overline{X}_1 = 8.17$, $\overline{X}_2 = 11.29$, $s^2 = 5.29$, and $t = -2.44$. Since $t_{0.975(11)} = 2.201$, the hypothesis H is rejected.

TABLE 6.4–Critical Values of the Load (coded data)

Process No. 1	Process No. 2
9	14
4	9
10	13
7	12
9	13
10	8
. . .	10

Example 6.11

Two rations (feeds) are to be compared with respect to their effect on the weight gains of hogs. Ten animals are available, and 5 are fed feed No. 1 while the other 5 are fed feed No. 2. Using $\alpha = 0.10$, test the hypothesis that the two feeds are equally effective in causing hogs to gain weight. The data obtained are given in Table 6.5. Calculations yield $t = -1.58$. Since $t_{0.95(8)} = 1.860$, we are unable to reject $H:\mu_1 = \mu_2$.

TABLE 6.5–Gains in Weight (lb)

Feed No. 1	Feed No. 2
1	4
2	3
4	9
5	10
8	9

The procedure for a one-sided alternative hypothesis is the same as it is for a single population. We illustrate the procedure with an example.

Example 6.12

Two pieces of meat, one a control and the other treated to tenderize the fibers, are to be tested. Tenderness will be measured by the force needed to shear samples of meat. (Lower shear force values indicate tenderer meat.) Given the data in Table 6.6 and letting $\alpha = 0.025$, test the hypothesis $H:\mu_T \geq \mu_C$ versus $A:\mu_T < \mu_C$. Calculations yield $\overline{X}_C = 42.5$, $\overline{X}_T = 41.0$, $s^2 = 101.17$, and $t = (\overline{X}_T - \overline{X}_C)/s_{\overline{X}_T - \overline{X}_C} = -0.28$. Since $t_{0.975(12)} = 2.179$, we cannot reject H. Thus we conclude that the treatment probably does not improve the tenderness of the meat.

The OC curves associated with tests of significance must be examined if we are to be certain that the suggested test procedure is discriminating enough. Once again we shall beg the question and refer the reader to Bowker and Lieberman (2) for samples of such curves.

TABLE 6.6–Shear Force Values for Tenderness Test

Control	Treated
50	46
49	40
39	32
44	23
24	54
50	51
41	. . .
43	. . .

Case 2: $\sigma_1^2 \neq \sigma_2^2$

When we cannot assume that the two populations have equal variances, we do not have an exact test as we did in the previous case. However, a reasonably good procedure is as follows. The test statistic is

$$t' = (\overline{X}_1 - \overline{X}_2)/(s_1^2/n_1 + s_2^2/n_2)^{1/2} \qquad (6.8)$$

and the critical region is the set of values of t' such that

$$t' \leq -(w_1 t_1 + w_2 t_2)/(w_1 + w_2)$$

or

$$t' \geq (w_1 t_1 + w_2 t_2)/(w_1 + w_2) \qquad (6.9)$$

where $w_1 = s_1^2/n_1$, $w_2 = s_2^2/n_2$, and

$$t_1 = t_{(1-\alpha/2)(n_1-1)}; \; t_2 = t_{(1-\alpha/2)(n_2-1)}$$

The critical region for the one-sided alternative uses $t_1' = t_{(1-\alpha)(n_1-1)}$ and $t_2' = t_{(1-\alpha)(n_2-1)}$ instead of t_1 and t_2.

Example 6.13

As an illustration, Example 6.10 will be reworked on the assumption that σ_1^2 does not equal σ_2^2. Thus $\overline{X}_1 = 8.17$, $\overline{X}_2 = 11.29$, $s_1^2 = 5.4$, $s_2^2 = 5.2$, $w_1 = 0.9$, $w_2 = 0.74$, $t' = -2.4$, $t_1 = 2.571$, $t_2 = 2.447$, and the weighted average of t_1 and t_2 is 2.52. Conclusion: accept H.

6.4.2 Paired Populations

We now consider the case where the observations from the two populations cannot be considered to be independent, but corresponding observations from the populations can be paired. This kind of situation arises when the same type of measurement is taken on the same unit at two different times. For example, to measure the absorptive properties of some material, we may weigh it before and after putting it in a water bath. The weight before would be an observation from one population and the weight after would be from a second population. In this case we deal with the differences

$$D_j = X_{1j} - X_{2j} \qquad j = 1, \ldots, n \qquad (6.10)$$

and treat these differences D_j as a random sample from a normal population with mean $\mu_D = \mu_1 - \mu_2$ and some variance σ_D^2. For testing the null hypothesis $H:\mu_D = 0$ (or $\mu_1 = \mu_2$) versus the alternative $A:\mu_D \neq 0$, the test statistic is

$$t = \sqrt{n}\bar{D}/s_D \tag{6.11}$$

where

$$s_D^2 = \sum_{j=1}^{n} (D_j - \bar{D})^2/(n-1) \tag{6.12}$$

We reject the null hypothesis if $t \leq -t_{(1-\alpha/2)(n-1)}$ or $t \geq t_{(1-\alpha/2)(n-1)}$.

Example 6.14

In a Brinell hardness test a hardened steel ball is pressed into the material being tested under a standard load. The diameter of the spherical indentation is then measured. Two steel balls are available (one from each of two manufacturers) and their performance will be compared on 15 pieces of material. Each piece of material will be tested twice, once with each ball. The data obtained are given in Table 6.7. Calculations yield $\bar{D} = 8$, $s_D^2 \cong 121.6$, and $t \cong 2.81$. Using $\alpha = 0.05$, it is seen that $2.81 > t_{0.975(14)} = 2.145$, and thus we reject the hypothesis that the two steel balls give the same average hardness indication.

We summarize these test procedures, as well as the one-sided test procedures, in Table 6.8.

TABLE 6.7–Data Obtained in a Brinell Hardness Test

Sample	Diameter		$D = X - Y$
	X	Y	
1	73	51	22
2	43	41	2
3	47	43	4
4	53	41	12
5	58	47	11
6	47	32	15
7	52	24	28
8	38	43	-5
9	61	53	8
10	56	52	4
11	56	57	-1
12	34	44	-10
13	55	57	-2
14	65	40	25
15	75	68	7

TABLE 6.8 – Tests for Comparing the Means of Two Normal Populations

Hypothesis	Test Statistic	Critical Region
Independent populations		
1. σ_1, σ_2 known		
$H: \mu_1 - \mu_2 = \delta; A: \mu_1 - \mu_2 \neq \delta$	$z = (\bar{X}_1 - \bar{X}_2 - \delta)/(\sigma_1^2/n_1 + \sigma_2^2/n_2)^{1/2}$	$\|z\| \geq z_{1-\alpha/2}$
$H: \mu_1 - \mu_2 \leq \delta; A: \mu_1 - \mu_2 > \delta$		$z \geq z_{1-\alpha}$
$H: \mu_1 - \mu_2 \geq \delta; A: \mu_1 - \mu_2 < \delta$		$z \leq -z_{1-\alpha}$
2. σ_1, σ_2 unknown; $\sigma_1 = \sigma_2$		
$H: \mu_1 - \mu_2 = \delta; A: \mu_1 - \mu_2 \neq \delta$	$t = (\bar{X}_1 - \bar{X}_2 - \delta)/[s^2(1/n_1 + 1/n_2)]^{1/2}$	$\|t\| \geq t_{(1-\alpha/2)(n_1+n_2-2)}$
$H: \mu_1 - \mu_2 \leq \delta; A: \mu_1 - \mu_2 > \delta$		$t \geq t_{(1-\alpha)(n_1+n_2-2)}$
$H: \mu_1 - \mu_2 \geq \delta; A: \mu_1 - \mu_2 < \delta$		$t \leq -t_{(1-\alpha)(n_1+n_2-2)}$
3. σ_1, σ_2 unknown; $\sigma_1 \neq \sigma_2$		
$H: \mu_1 - \mu_2 = \delta; A: \mu_1 - \mu_2 \neq \delta$	$t' = (\bar{X}_1 - \bar{X}_2 - \delta)/(s_1^2/n_1 + s_2^2/n_2)^{1/2}$	$\|t'\| \geq (w_1 t_1 + w_2 t_2)/(w_1 + w_2)$
$H: \mu_1 - \mu_2 \leq \delta; A: \mu_1 - \mu_2 > \delta$		$t' \geq (w_1 t_1' + w_2 t_2')/(w_1 + w_2)$
$H: \mu_1 - \mu_2 \geq \delta; A: \mu_1 - \mu_2 < \delta$		$t' \leq -(w_1 t_1' + w_2 t_2')/(w_1 + w_2)$
Paired observations		
$H: \mu_1 - \mu_2 = \delta; A: \mu_1 - \mu_2 \neq \delta$	$t = \sqrt{n}(\bar{D} - \delta)/s_D$	$\|t\| > t_{(1-\alpha/2)(n-1)}$
$H: \mu_1 - \mu_2 \leq \delta; A: \mu_1 - \mu_2 > \delta$		$t \geq t_{(1-\alpha)(n-1)}$
$H: \mu_1 - \mu_2 \geq \delta; A: \mu_1 - \mu_2 < \delta$		$t \leq -t_{(1-\alpha)(n-1)}$

6.5 SEVERAL NORMAL POPULATIONS; $H:\mu_1 = \mu_2 = \cdots = \mu_k$

In Section 6.4 a t test was proposed for testing $H:\mu_1 = \mu_2$ for two independent normal populations. We now consider a test procedure for handling the comparison of the means of k independent normal populations $k \geq 2$.

Intuitively, it seems reasonable that the validity of the hypothesis $H:\mu_1 = \mu_2 = \cdots = \mu_k$ should be assessed by comparing the sample estimates of $\mu_1, \ldots, \mu_k$. That is, it is to be expected that any suggested test procedure will involve a comparison of the sample means $\overline{X}_1, \ldots, \overline{X}_k$.

Assume we have k *independent* normal populations, $N(\mu_1, \sigma_1^2), \ldots, N(\mu_k, \sigma_k^2)$ such that $\sigma_1^2 = \cdots = \sigma_k^2$, i.e., such that the variances of the populations are homogeneous. Let $X_{i1}, \ldots, X_{in_i}$ be a sample of size n_i taken from the ith population $(i = 1, \ldots, k)$. To test the hypothesis $H:\mu_1 = \cdots = \mu_k$ versus A:at least one inequality, the appropriate test statistic is

$$F = \sum_{i=1}^{k} n_i(\overline{X}_{i.} - \overline{X}_{..})^2/(k - 1) \left/ \sum_{i=1}^{k} \sum_{j=1}^{n_i} (X_{ij} - \overline{X}_{i.})^2 \right/ \sum_{i=1}^{k} (n_i - 1)$$

$$(6.13)$$

where $\overline{X}_{i.} = \sum_{j=1}^{n_i} X_{ij}/n_i$ is the sample mean from the ith population, and $\overline{X}_{..} = \sum_{i=1}^{k} \sum_{j=1}^{n_i} X_{ij}/\sum_{i=1}^{k} n_i$ is the mean of all observations. Under H, F has an F distribution with $\nu_1 = k - 1$ and $\nu_2 = \sum_{i=1}^{k} (n_i - 1)$ degrees of freedom. Thus, if $F \geq F_{(1-\alpha)(\nu_1, \nu_2)}$, the hypothesis H would be rejected. (**Note:** The reader may easily verify that if $k = 2$ the procedure outlined in this section is algebraically equivalent to that of Section 6.4.)

The computations necessary for carrying out this test procedure can be summarized in a convenient tabular form such as Table 6.9. The

TABLE 6.9–Tabular Presentation of the F Test for the Equality of Means of k Normal Populations under the Assumption of Homogeneous Variances

Source of Variation	Degrees of Freedom	Sum of Squares	Mean Square	F Ratio
Mean	1	M_{xx}	$M = M_{xx}/1$	
Among groups	$k - 1$	G_{xx}	$G = G_{xx}/(k - 1)$	G/W
Within groups	$\sum_{i=1}^{k} (n_i - 1)$	W_{xx}	$W = W_{xx}/\sum_{i=1}^{k} (n_i - 1)$	
Total	$\sum_{i=1}^{k} n_i$	$\sum_{i=1}^{k} \sum_{j=1}^{n_i} X_{ij}^2$	$\cdots$	$\cdots$

sums of squares listed in the table are based on the identity

$$\sum_{i=1}^{k} \sum_{j=1}^{n_i} X_{ij}^2 = n(\overline{X}_{..})^2 + \sum_{i=1}^{k} n_i(\overline{X}_{i.} - \overline{X}_{..})^2 + \sum_{i=1}^{k} \sum_{j=1}^{n_i} (X_{ij} - \overline{X}_{i.})^2 \quad (6.14)$$

in which $n = \sum_{i=1}^{k} n_i$ (i.e., the total number of observations in the combined samples). In terms of the notation in Table 6.9 this may be expressed as

$$\sum_{i=1}^{k} \sum_{j=1}^{n_i} X_{ij}^2 = M_{xx} + G_{xx} + W_{xx} \quad (6.15)$$

Such a table is usually referred to as an *analysis of variance* table. The labor involved in calculating the sum of squares can be reduced if we use the following identities:

$$G_{xx} = \sum_{i=1}^{k} X_{i.}^2/n_i - M_{xx} = \sum_{i=1}^{k} G_i^2/n_i - M_{xx} \quad (6.16)$$

where $X_{i.} = G_i = \sum_{j=1}^{n_i} X_{ij}$ is the total of the observations in the ith group or sample, and

$$W_{xx} = \sum_{i=1}^{k} \sum_{j=1}^{n_i} X_{ij}^2 - G_{xx} - M_{xx} \quad (6.17)$$

with $M_{xx} = n(\overline{X}_{..})^2$.

Example 6.15

Consider the data of Table 6.10. Using Eqs. (6.15)–(6.17), the results shown in Table 6.11 were obtained. Since $F = 72 > F_{0.99(3,16)} = 5.29$, the hypothesis $H:\mu_1 = \mu_2 = \mu_3 = \mu_4$ is rejected at the 1 percent significance level.

TABLE 6.10–Sample Data from Four Normal
Populations to Be Used in Example 6.15

		Groups		
	1	2	3	4
	45	35	34	41
	46	33	34	41
	49		35	44
	44		34	43
Observations			33	41
				42
				44
				41
				41
Total	184	68	170	378
Mean	46	34	34	42

TABLE 6.11–Analysis of Variance Using the Data of Table 6.10 to Test the Hypothesis $H: \mu_1 = \mu_2 = \mu_3 = \mu_4$

Source of Variation	Degrees of Freedom	Sum of Squares	Mean Square	F Ratio
Mean	1	32,000	32,000	...
Among groups	3	432	144	72
Within groups	16	32	2	...
Total	20	32,464	...	...

6.6 NORMAL POPULATION; TESTS OF THE VARIANCE σ^2

Suppose we have a random sample of size n from a normal population, which is to be used to test the hypothesis $H: \sigma^2 = \sigma_0^2$; i.e., we want to test if the population variance is σ_0^2. The test statistic to use is

$$\chi^2 = \sum_{i=1}^{n} (X_i - \bar{X})^2 / \sigma_0^2 \tag{6.18}$$

If H is true, χ^2 has a chi-square distribution with $\nu = n - 1$ degrees of freedom. As true for the mean, the critical region will depend on the alternative hypothesis. Let 100α be the level of significance. Suppose $H: \sigma^2 = \sigma_0^2$; $A: \sigma^2 \neq \sigma_0^2$. To test the hypothesis that the variance is σ_0^2 against the alternative that it is not σ_0^2, reject H if

$$\chi^2 \leq \chi^2_{(\alpha/2)(n-1)} \text{ or } \chi^2 \geq \chi^2_{(1-\alpha/2)(n-1)} \tag{6.19}$$

where $\chi^2_{\delta\nu}$ is that value of a chi-square random variable with ν degrees of freedom such that $P(\chi^2 \leq \chi^2_{\delta\nu}) = \delta$. Otherwise H is not rejected.

Example 6.16

Consider the data given in Example 6.8. Do these values support the hypothesis that if repeated measurements are assumed to be normally distributed, the true variance of all such measurements is equal to 2? Here the hypothesis is $H: \sigma^2 = 2$ and the alternative is $A: \sigma^2 \neq 2$. It is determined that $\sum_{i=1}^{n} (X_i - \bar{X})^2 / \sigma_0^2 = 20$. Since $\chi^2_{0.025(3)} = 0.216$ and $\chi^2_{0.975(3)} = 9.35$, we see that H is rejected. You will note that once again a probability of Type I error equal to 0.05 was chosen. That is, we have run a maximum risk of 5 percent of rejecting a true hypothesis.

The critical regions for the tests of σ^2 are summarized in Table 6.12.

Example 6.17

Consider the data of Table 6.13, which were obtained from a random sample of 80 bearings. To test (using $\alpha = 0.05$) $H: \sigma^2 \leq 0.00005$ versus $A: \sigma^2 > 0.00005$, we calculate

$$\chi^2 = \sum_{i=1}^{80} (X_i - \bar{X})^2 / 0.00005 = 0.000474 / 0.00005 = 9.48$$

Since this does not exceed $\chi^2_{0.95(79)} = 100.7$, we are unable to reject H.

TABLE 6.12–Tests for the Variance of a Normal Population

Hypothesis	Test Statistic	Critical Region
$H:\sigma^2 = \sigma_0^2; A:\sigma^2 \neq \sigma_0^2$	$\chi^2 = \sum_{i=1}^{n} (X_i - \bar{X})^2/\sigma_0^2$	$\chi^2 \geq \chi^2_{(1-\alpha/2)(n-1)}$ or $\chi^2 \leq \chi^2_{(\alpha/2)(n-1)}$
$H:\sigma^2 \leq \sigma_0^2; A:\sigma^2 > \sigma_0^2$		$\chi^2 \geq \chi^2_{(1-\alpha)(n-1)}$
$H:\sigma^2 \geq \sigma_0^2; A:\sigma^2 < \sigma_0^2$		$\chi^2 \leq \chi^2_{\alpha(n-1)}$

TABLE 6.13–Number of Bearings
Observed with the
Indicated Diameters

Diameter (in.)	Number
3.573	4
3.574	2
3.575	9
3.576	12
3.577	12
3.578	10
3.579	9
3.580	9
3.581	8
3.582	4
3.583	1
Total	80

The reader is referred to Bowker and Lieberman (2) for examples of OC curves associated with this test.

6.7 TWO NORMAL POPULATIONS; COMPARISON OF THE VARIANCES

If the variables X_{1j} and X_{2j} have independent normal populations with variances σ_1^2 and σ_2^2 respectively, we consider tests of the hypothesis of equal variances $H:\sigma_1^2 = \sigma_2^2$. The test statistic is

$$F = s_1^2/s_2^2 \qquad (6.20)$$

where s_1^2 and s_2^2 are the sample variances of n_1 and n_2 observations from populations 1 and 2 respectively. If H is true, F has an F distribution with $\nu_1 = n_1 - 1$ and $\nu_2 = n_2 - 1$ degrees of freedom so that, against the alternative $A:\sigma_1^2 \neq \sigma_2^2$, we reject H if $F \leq F_{(\alpha/2)(n_1-1, n_2-1)}$ or $F \geq F_{(1-\alpha/2)(n_1-1, n_2-1)}$. The tests are summarized in Table 6.14.

TABLE 6.14–Tests for Comparing the Variances of
Two Normal Populations

Hypothesis	Test Statistic	Critical Region
$H:\sigma_1^2 = \sigma_2^2; A:\sigma_1^2 \neq \sigma_2^2$	$F = s_1^2/s_2^2$	$F \leq F_{(\alpha/2)(n_1-1, n_2-1)}$ or $F \geq F_{(1-\alpha/2)(n_1-1, n_2-1)}$
$H:\sigma_1^2 \leq \sigma_2^2; A:\sigma_1^2 > \sigma_2^2$		$F \geq F_{(1-\alpha)(n_1-1, n_2-1)}$
$H:\sigma_1^2 \geq \sigma_2^2; A:\sigma_1^2 < \sigma_2^2$		$F \leq F_{\alpha(n_1-1, n_2-1)}$

Example 6.18

Using the data of Example 5.3 and letting $\alpha = 0.05$, test the hypothesis $H:\sigma_1^2 = \sigma_2^2$ versus $A:\sigma_1^2 \neq \sigma_2^2$. It is seen that $F = 109.63/65.99 = 1.66$ with $\nu_1 = 40$ and $\nu_2 = 30$ degrees of freedom. Since $F_{0.975(40, 30)} = 2.01$, we are unable to reject H.

6.8 SEVERAL NORMAL POPULATIONS; $H:\sigma_1^2 = \cdots = \sigma_k^2$

In Section 6.7 an F test was proposed for testing $H:\sigma_1^2 = \sigma_2^2$ for two independent normal populations. We now consider a test procedure for comparing the variances of k independent normal populations $k \geq 2$. Several test procedures have been proposed for handling this type of problem, but only the method due to Bartlett (1) will be presented.

Let $X_{ij}(i = 1,\ldots,k; j = 1,\ldots,n_i)$ denote the sample observations. The sample variance from the ith population is

$$s_i^2 = \sum_{j=1}^{n_i} (X_{ij} - \overline{X}_{i.})^2/(n_i - 1) = \sum_{j=1}^{n_i} x_{ij}^2/(n_i - 1) \qquad (6.21)$$

where we have used the notation $x_{ij} = X_{ij} - \overline{X}_{i.}$. A summary of the computations necessary for Bartlett's procedure is given in Table 6.15.

To test the hypothesis $H:\sigma_1^2 = \cdots = \sigma_k^2$ versus A: at least one inequality, the test statistic is

$$\chi^2 = 2.3026\left[(\log_{10}s^2)\sum_{i=1}^{k}(n_i - 1) - \sum_{i=1}^{k}(n_i - 1)\log_{10}s_i^2\right] \qquad (6.22)$$

which under H has approximately a chi-square distribution with $\nu = k - 1$ degrees of freedom. If $\chi^2 \geq \chi^2_{(1-\alpha)(k-1)}$, the hypothesis H would be rejected.

A closer approximation to the chi-square distribution with $\nu = k - 1$ degrees of freedom can be attained by using the corrected χ^2,

$$\text{Corrected } \chi^2 = (1/C)\chi^2 \qquad (6.23)$$

where

$$C = 1 + [1/3(k - 1)]\left\{\sum_{i=1}^{k} 1/(n_i - 1) - \left[1/\sum_{i=1}^{k}(n_i - 1)\right]\right\} \qquad (6.24)$$

TABLE 6.15–Computations for Bartlett's Test for Homogeneity of Variance

Sample	$\sum x_i^2$	Degrees of Freedom	$1/d.f.$	s_i^2	$\log_{10} s_i^2$	$(d.f.) \log_{10} s_i^2$
1	$\sum_{j=1}^{n_1} x_{1j}^2$	$n_1 - 1$	$1/(n_1 - 1)$	s_1^2	$\log_{10} s_1^2$	$(n_1 - 1) \log_{10} s_1^2$
2	$\sum_{j=1}^{n_2} x_{2j}^2$	$n_2 - 1$	$1/(n_2 - 1)$	s_2^2	$\log_{10} s_2^2$	$(n_2 - 1) \log_{10} s_2^2$
$\cdots$		$\cdots$	$\cdots$	$\cdots$	$\cdots$	$\cdots$
k	$\sum_{j=1}^{n_k} x_{kj}^2$	$n_k - 1$	$1/(n_k - 1)$	s_k^2	$\log_{10} s_k^2$	$(n_k - 1) \log_{10} s_k^2$
Total	W_{xx}	$\sum_{i=1}^{k} (n_i - 1)$	$\sum_{i=1}^{k} \frac{1}{n_i - 1}$	$\cdots$	$\cdots$	$\sum_{i=1}^{k} (n_i - 1) \log_{10} s_i^2$

Pooled estimate of variance $= s^2 = W_{xx} / \sum_{i=1}^{k} (n_i - 1)$

We will find it necessary to compute the corrected value of chi-square *only* if the uncorrected chi-square falls close to and above the tabulated value, and then only if we wish to obtain a very accurate evaluation of the exact probability of Type I error.

Example 6.19

Consider the data of Table 6.16. Following the procedure indicated in Table 6.15, the results presented in Table 6.17 are obtained. It is seen that $\chi^2 = 2.81 < \chi^2_{0.95(3)} = 7.81$, and thus the hypothesis of homogeneous variances may not be rejected at the 5 percent significance level. (**Note:** There was no need to calculate the corrected value of chi-square in this example; the computations were carried out only to illustrate the method.)

TABLE 6.16–Four Samples from Normal Populations
$$H: \sigma_1^2 = \sigma_2^2 = \sigma_3^2 = \sigma_4^2$$

1	2	3	4
48	42	33	78
49	39	42	69
67	51	46	60
75	57	47	52
53	75	50	63
33	...	...	45
...	...	...	50
...	...	...	35

6.9 BINOMIAL POPULATION; TESTS OF THE PROPORTION p

We now consider situations in which sample elements can be classified in either of two categories. For example, in industrial quality control, units are classified either as defective or nondefective; or in a public opinion poll, people either vote in favor of or against an issue. In any case, the parameter of interest is the proportion p in one of the categories.

Suppose we wish to consider the hypotheses $H: p = p_0$; $A: p \neq p_0$. That is, we want to test the hypothesis that the proportion in one of the classes (say class 1) is p_0 against the alternative that $p \neq p_0$. For a sample of size n, the test statistic to use is X, the number in the sample in class 1. Under the null hypothesis, X has a binomial distribution with parameters n and $p = p_0$. For a given significance level 100α, the acceptance and rejection regions are determined by solving

$$\sum_{x=0}^{L} C(n,x)p_0^x(1-p_0)^{n-x} = \alpha/2 \tag{6.25}$$

and

$$\sum_{x=U}^{n} C(n,x)p_0^x(1-p_0)^{n-x} = \alpha/2 \tag{6.26}$$

TABLE 6.17–Computations for Barlett's Test: Data from Table 6.16

Sample	$\sum x_i^2$	Degrees of Freedom	$1/d.f.$	s_i^2	$\log_{10} s_i^2$	$(d.f.) \log_{10} s_i^2$
1	1113.0	5	0.2000	222.6	2.34753	11.73765
2	820.8	4	0.2500	205.2	2.31218	9.24872
3	173.2	4	0.2500	43.3	1.63649	6.54596
4	1330.0	7	0.1428	190.0	2.27875	15.95125
Total	3437.0	20	0.8428	...	...	43.48358

Pooled estimate of variance $= s^2 = 3437/20 = 171.85$

$\chi^2 = (2.3026)(44.7030 - 43.48358) = 2.80784$

Correction factor $= C = 1 + [1/3(3)](0.8428 - 1/20) = 1.0881$

Corrected $\chi^2 = 2.80784/1.0881 = 2.5805$

for L and U. The acceptance region defined by these two equations is the set of positive integers between but not including L and U.

Unfortunately, it is usually impossible to find integral values of L and U to satisfy Eqs. (6.25) and (6.26). Therefore, it is customary to choose those values of L and U that make the value of each of the summations as large as possible without exceeding $\alpha/2$. Occasionally, the restriction of being less than or equal to $\alpha/2$ will be relaxed if by so doing the probability of rejecting a true hypothesis will be only slightly larger than the chosen α.

Example 6.20

In a certain cross of two varieties of peas, genetic theory led the investigator to expect one-half of the seeds produced to be wrinkled and the remaining one-half to be smooth. Taking $\alpha = 0.01$ and $n = 40$, determine L and U and thus define the acceptance and rejection regions. Using Eqs. (6.25) and (6.26) with $p_0 = 0.5$, we obtain $L = 11$ and $U = 29$. Therefore, the acceptance region consists of those values of X for which $11 < X < 29$.

Without adequate tables the procedure discussed so far in this section is not very palatable to the researcher. Consequently, some approximate procedures that lend themselves to easy calculation will be investigated. One approximation that can be used is the normal approximation to the binomial distribution. To test $H:p = p_0$ versus $A:p \neq p_0$, the test statistic is

$$Z = [(X + 0.5) - np_0]/[np_0(1 - p_0)]^{1/2} \qquad \text{for } X < np_0$$
$$= [(X - 0.5) - np_0]/[np_0(1 - p_0)]^{1/2} \qquad \text{for } X > np_0 \qquad (6.27)$$

and the null hypothesis is rejected if $|Z| \geq z_{1-\alpha/2}$ where $z_{1-\alpha/2}$ comes from the standard normal distribution found in Appendix 3.

Example 6.21

Consider the situation described in Example 6.20. A random sample of 40 seeds segregated into 30 wrinkled and 10 smooth. Assuming $\alpha = 0.01$ and using Eq. (6.27),

$$Z = [(30 - 0.5) - 40(0.5)]/[40(0.5)(0.5)]^{1/2} \cong 3 \geq z_{0.995} = 2.58$$

Therefore, $H:p = 0.5$ is rejected.

Either of these statistics can be used when the alternative hypothesis is one-sided. The appropriate critical regions are given in Table 6.18.

Example 6.22

From past experience it has been determined that a qualified operator on a certain machine turning out 400 items per day produces 20 or fewer defective items per day. A new operator is hired to run the same machine and the hypothesis is made that he is a qualified operator. Taking $\alpha = 0.03$, determine U and thus define the acceptance and rejection regions. Here the hypothesis is $H:p \leq 0.05$ and $n = 400$. Using an equation similar to Eq. (6.26), we find that $U = 29$. Thus if the new operator produced

TABLE 6.18–Tests for the Parameter p of a Binomial Population

Hypothesis	Test Statistic	Critical Region
$H\!:\!p = p_0;\ A\!:\!p \neq p_0$	$\chi^2 = \displaystyle\sum_{i=1}^{2} (\,\lvert O_i - E_i \rvert\ - 0.5)^2/E_i$ or $Z = \dfrac{(X + 0.5) - np_0}{[np_0(1 - p_0)]^{1/2}} \qquad X < np_0$ $ = \dfrac{(X - 0.5) - np_0}{[np_0(1 - p_0)]^{1/2}} \qquad X > np_0$	$\chi^2 \geq \chi^2_{(1-\alpha)(1)}$ $\lvert Z \rvert \geq z_{1-\alpha/2}$
$H\!:\!p \geq p_0;\ A\!:\!p < p_0$ $H\!:\!p \leq p_0;\ A\!:\!p > p_0$	(Z only) (Z only)	$Z \leq z_{\alpha}$ $Z \geq z_{1-\alpha}$

more than 28 defective items in a run of 400, we would reject the hypothesis that he is a qualified operator.

Example 6.23

Using the normal approximation, test the hypothesis $H:p \leq 0.05$ versus $A:p > 0.05$, given that $\hat{p} = 32/400 = 0.08$. Let $\alpha = 0.03$. Using Eq. (6.27),

$$Z = [(32 - 0.5) - 400(0.05)]/[400(0.05)(0.95)]^{1/2} \cong 2.6 > z_{0.97} = 1.89$$

Thus $H:p \leq 0.05$ is rejected.

Another useful approximation is available when we have the two-sided alternative $A:p \neq p_0$. This approximation is based on the fact that the square of a standard normal variate is distributed as chi-square with one degree of freedom (see Section 4.7). When the chi-square approximation is used, the test statistic is

$$\chi^2 = \sum_{i=1}^{2} (|O_i - E_i| - 0.5)^2/E_i \qquad (6.28)$$

where $O_1 = X$, $O_2 = n - X$, $E_1 = np_0$, and $E_2 = n(1 - p_0)$.

The hypothesis $H:p = p_0$ will be rejected if $\chi^2 \geq \chi^2_{(1-\alpha)(1)}$; otherwise H will be accepted. It should be clear that O stands for *observed* and that E stands for *expected* in Eq. (6.28).

Example 6.24

It will be instructive to rework Example 6.21 using this method. Thus

$$\chi^2 = (|30 - 20| - 0.5)^2/20 + (|10 - 20| - 0.5)^2/20$$
$$= 9.025 > \chi^2_{0.99(1)} = 6.63$$

Therefore, as in the preceding example, $H:p = 0.5$ is rejected. [**Note:** $\chi^2 = 9.025 = Z^2 \cong (3)^2$].

We emphasize again the fact that the chi-square statistic can *only* be used when the alternative hypothesis is $p \neq p_0$. The tests for $H:p = p_0$ are summarized in Table 6.18.

6.10 TWO BINOMIAL POPULATIONS; COMPARISON OF THE PROPORTIONS

Suppose we have two populations in each of which the data is classed in either of two categories [class 1 (or A) or class 2 (or B)]. If p_1 and p_2 denote the proportion in a certain class (say class A) for population 1 and population 2 respectively, a hypothesis that is frequently tested is $H:p_1 = p_2$ versus $A:p_1 \neq p_2$.

The normal approximation to the binomial distribution can again be used to get an approximate test. In this case the test statistic is

$$Z = (\hat{p}_1 - \hat{p}_2)/[\hat{p}(1 - \hat{p})(n_1 + n_2)/n_1 n_2]^{1/2} \qquad (6.29)$$

where $\hat{p}_j = X_j/n_j$, $\hat{p} = (X_1 + X_2)/(n_1 + n_2)$, and X_j is the number in

TABLE 6.19–Sample Data from Two Binomial Populations

Population	Class		Row Total
	1 (or A)	2 (or B)	
1	$O_{11}(= a)$	$O_{12}(= b)$	$R_1 = n_1 = a + b$
2	$O_{21}(= c)$	$O_{22}(= d)$	$R_2 = n_2 = c + d$
Column Total	$C_1 = a + c$	$C_2 = b + d$	$n = n_1 + n_2 = a + b + c + d$

class A in a sample of size n_j taken from the jth population. When H is true, Z has an approximate standard normal distribution. Thus the null hypothesis is rejected if $|Z| \geq z_{1-\alpha/2}$ where $z_{1-\alpha/2}$ comes from the standard normal distribution found in Appendix 3.

A second approximation is also available using the chi-square distribution. Let the observed data be tabulated as in Table 6.19, where O_{ij} represents the number in the jth class in the sample from the ith population. The test statistic is

$$\chi^2 = n(O_{11}O_{22} - O_{12}O_{21})^2/k = n(ad - bc)^2/k \qquad (6.30)$$

where $k = R_1 R_2 C_1 C_2 = (a + b)(c + d)(a + c)(b + d)$. When H is true, χ^2 has an approximate chi-square distribution with $\nu = 1$ degree of freedom. Thus H is rejected at the 100α level of significance if $\chi^2 \geq \chi^2_{(1-\alpha)(1)}$.

As in Section 6.9 a correction for continuity may be used to sharpen the approximation. The corrected test statistic is

$$\chi^2 = n(|O_{11}O_{22} - O_{12}O_{21}| - n/2)^2/k = n(|ad - bc| - n/2)^2/k$$
$$(6.31)$$

Example 6.25

Robertson (4) reported on the analysis of an experiment involving the evaluation of a silicon dip as a protection for vacuum tubes. The data shown in Table 6.20 were obtained. Using a 10 percent significance level and the procedure outlined above, the value of the test statistic

$$\chi^2 = 690[(2)(343) - (7)(338)]^2/(340)(350)(9)(681) = 2.67$$

does not exceed $\chi^2_{(0.9)(1)} = 2.71$, and thus one would conclude there is no difference in the failure rate between protected and unprotected tubes.

TABLE 6.20–Success-Failure Results from an Experiment on 690 Vacuum Tubes

	Failure	Nonfailure
Protected	2	338
Unprotected	7	343

Source: Robertson (4).

When we are comparing two binomial populations, an exact test procedure is not too difficult to apply, especially if a digital computer is available. Thus it seems appropriate to indicate the nature of the exact method.

It can be shown that the exact probability of observing $\hat{p}_1 = a/(a + b)$ and $\hat{p}_2 = c/(c + d)$ as estimates of p_1 and p_2 (the parameters of two binomial populations) when $p_1 = p_2$ is

$$P_1 = \frac{(a + b)!(c + d)!(a + c)!(b + d)!}{a!b!c!d!n!} \tag{6.32}$$

To obtain the final probability to be used in assessing the validity of $H:p_1 = p_2$, it is necessary to add to P_1 the probabilities of more divergent fractions than those observed. Assuming $\hat{p}_1 < \hat{p}_2$ (and the table can always be arranged to make this so), the next more divergent situation would be the one in which a and d are each decreased by unity, and b and c are each increased by unity. For this array, we calculate

$$P_2 = \frac{(a + b)!(c + d)!(a + c)!(b + d)!}{(a - 1)!(b + 1)!(c + 1)!(d - 1)!n!} \tag{6.33}$$

The cell entries are again changed, following the same rule as before, and P_3 is calculated. Continue in this manner until P_{a+1} is calculated. Then if

$$P = \sum_{i=1}^{a+1} P_i \tag{6.34}$$

is less than or equal to α, the hypothesis $H:p_1 = p_2$ should be rejected.

Example 6.26

Using the data as reported in Table 6.20 and the exact procedure outlined above, Robertson (4) found $P = P_1 + P_2 + P_3 = 0.0957$. He then concluded that "the failure rate for protected tubes is just barely significantly less than that for unprotected tubes." Apparently a 10 percent significance level had been decided upon prior to the analysis.

6.11 SEVERAL BINOMIAL POPULATIONS

First we consider the situation when there are several (k) binomial populations and the hypothesis to be tested is that the proportion in Class 1 is the same for all populations. That is, we wish to test the hypothesis $H:p_1 = \ldots = p_k$, where p_i is the proportion in Class 1 in population i. If a sample of size n_i is observed from the ith population, the data can be tabulated as in Table 6.21.

The test statistic to use for testing the hypothesis is

$$\chi^2 = \sum_{i=1}^{k} \sum_{j=1}^{2} (O_{ij} - E_{ij})^2/E_{ij} \tag{6.35}$$

where $E_{ij} = R_i C_j/n$. When H is true, χ^2 has an approximate chi-square

TABLE 6.21–Sample Data from Several Binomial Populations

Population	Class		Row Total
	1	2	
1	O_{11}	O_{12}	$R_1 = n_1$
2	O_{21}	O_{22}	$R_2 = n_2$
.	.	.	.
.	.	.	.
.	.	.	.
i	O_{i1}	O_{i2}	$R_i = n_i$
.	.	.	.
.	.	.	.
k	O_{k1}	O_{k2}	$R_k = n_k$
Column Total	C_1	C_2	$n = \sum\limits_{i=1}^{k} n_i$

distribution with $\nu = k - 1$ degrees of freedom, so H is rejected if $\chi^2 > \chi^2_{(1-\alpha)(k-1)}$.

A situation that occurs frequently in experimental work is the following: a hypothesis is to be tested and several experiments are conducted to produce data that bear on the problem. When this situation prevails, it is natural to think of combining the experimental results. For example, if the hypothesis $H: p = p_0$ is being tested relative to the alternative $A: p \neq p_0$, it is quite common to have available k samples (perhaps of different sizes) as a result of k *replications* or *repetitions* of the basic experiment.

How should the data from the several samples be combined? There are two ways this can be done, and each will be discussed and then illustrated in Example 6.27. It will be noted that the analysis performed involves the chi-square distribution and depends on the previously mentioned additive property of chi-square. Actually, several chi-square values are calculated, and each of these contributes a different item of information relative to the hypothesis under test.

You will note that a chi-square value (with 1 degree of freedom) is found for each sample. Each of these values can be interpreted as in Section 6.9. As the next step in the analysis, we may calculate $\chi^2 = \chi_1^2 + \chi_2^2 + \cdots + \chi_k^2$ with k degrees of freedom. This value will be referred to as the *pooled chi-square,* and it is clearly a pooling or accumulation of the bits of evidence provided by the k independent samples. This value may now be used to assess the validity of the hypothesis under test. An alternative way of pooling the information from several samples is to lump the original data into one large sample and compute the *total chi-square* (with 1 degree of freedom) associated with this super sample. One other statistic should also be obtained, namely, the *heterogeneity chi-square*. This quantity, which has $k - 1$ degrees of freedom, is found by subtracting the total chi-square from the pooled chi-square. It is used to measure the lack of consistency among the several samples.

Example 6.27

Consider again the hypothesis tested in Example 6.20. Now, instead of 1 sample, 8 separate experiments give rise to 8 samples as shown in Table 6.22. Assuming $\alpha = 0.01$, it is seen that: (1) no 1 of the 8 samples leads to rejection; (2) the super sample of 1600 observations yields $\chi^2 = 2.56$, which is not significant; and (3) the pooled chi-square is significant. Why do we get these seemingly contradictory results? The pooled chi-square is significant because we have accumulated enough evidence from each sample to indicate that the hypothesis $H:p = 0.5$ should be rejected. The reason the total chi-square did not give the same answer is that in 3 samples smooth seeds predominated, while in 5 samples wrinkled seeds predominated. This effect was hidden (i.e., the majorities in opposite directions tended to cancel out) when the data were lumped into one large sample. Attention is called to the previously mentioned lack of consistency among the 8 samples by the significant heterogeneity chi-square.

TABLE 6.22–Chi-Square Analysis Combining Data from Several Samples of Smooth and Wrinkled Peas

Sample	Sample Size	Number Wrinkled	Number Smooth	Chi-Square (χ_i^2)	Degrees of Freedom
1	100	60	40	4.00	1
2	200	108	92	1.28	1
3	180	80	100	2.22	1
4	208	118	90	3.77	1
5	300	165	135	3.00	1
6	182	106	76	4.94	1
7	230	105	125	1.73	1
8	200	90	110	2.00	1
Pooled χ^2	...	...	...	$22.94 = \sum\limits_{i=1}^{8} \chi_i^2$	8
Total	1600	832	768	2.56	1
Difference	...	...	...	20.38	7

6.12 MULTINOMIAL DATA

Many times, our sample elements may be assigned to any one of several different classes or categories, rather than simply to one or the other of two classes as in Section 6.9. In such a situation we must work with the multinomial distribution rather than the binomial distribution.

A common problem is to test the hypothesis

$$H:p_i = p_{i0} \qquad i = 1, 2, \ldots, k$$

where there are k classes. Of course,

$$\sum_{i=1}^{k} p_i = \sum_{i=1}^{k} p_{i0} = 1$$

A simple test procedure is available by means of the chi-square approximation. In this case, the degrees of freedom equal $k - 1$, i.e., one less than the number of classes (or parameters). The procedure is to calculate

$$\chi^2 = \sum_{i=1}^{k} (O_i - E_i)^2 / E_i \qquad (6.36)$$

where O_i represents the number observed in the ith class and $E_i = np_{i0}$ represents the number expected in the ith class if H is true. Clearly, $\sum O_i = \sum E_i = n$, the sample size. Then if $\chi^2 \geq \chi^2_{(1-\alpha)(k-1)}$, the hypothesis H is rejected.

Example 6.28

In a particular genetic experiment the observations were classified as follows: Class A—99, Class B—33, Class C—24, and Class D—4; but genetic theory called for a 9:3:3:1 ratio. Using a 5 percent significance level, do the data support the theory? Calculation yields

$$\chi^2 = (99 - 90)^2/90 + (33 - 30)^2/30$$
$$+ (24 - 30)^2/30 + (4 - 10)^2/10 = 6.0$$

This is less than $\chi^2_{0.95(3)} = 7.81$, and thus we are unable to reject the hypothesized theory.

Another problem, similar to the binomial case in Section 6.11, is to compare several multinomial populations. The hypothesis to be tested is

$$H: p_{1j} = p_{2j} = \ldots = p_{rj} \qquad j = 1, \ldots, k$$

where p_{ij} is the proportion in class j in the ith population. If a sample of size n_i is observed from the ith population, the number of elements in each class can be tabulated as in Table 6.23.

An approximate test based on the chi-square approximation is to use the test statistic

$$\chi^2 = \sum_{i=1}^{r} \sum_{j=1}^{k} (O_{ij} - E_{ij})^2 / E_{ij} \qquad (6.37)$$

where O_{ij} = observed number in the jth class in the sample from the ith population and $E_{ij} = R_i C_j / n$ = the expected number in the jth class in the sample from the ith population. When H is true, χ^2 has an approximate chi-square distribution with $\nu = (r - 1)(k - 1)$ degrees of freedom, so H is rejected if $\chi^2 \geq \chi^2_{(1-\alpha)[(r-1)(k-1)]}$.

The discussion relative to the use of the heterogeneity chi-square statistic in Section 6.11 holds as well for multinomial data with the appropriate changes in degrees of freedom.

TABLE 6.23–Sample Data from r Multinomial Populations

Population	Class				Row Total
	1	2	...	k	
1	O_{11}	O_{12}	...	O_{1k}	$R_1 = n_1$
2	O_{21}	O_{22}	...	O_{2k}	$R_2 = n_2$
$\vdots$	$\vdots$	$\vdots$		$\vdots$	$\vdots$
r	O_{r1}	O_{r2}	...	O_{rk}	$R_r = n_r$
Column Total	C_1	C_2	...	C_k	$n = \sum_{i=1}^{r} n_i$

6.13 POISSON DATA

Several processes give rise to observations distributed according to the Poisson probability function

$$f(x) = e^{-\lambda}\lambda^x/x! \qquad x = 0, 1, 2, \ldots; \lambda > 0 \qquad (6.38)$$

Some examples are: (1) radioactive disintegrations, (2) bomb hits on a given area, (3) chromosome interchanges in cells, and (4) flaws in materials.

Obviously, many hypotheses and alternatives could be considered and discussed. However, for purposes of illustrating the methods of analysis, only two will be examined.

To test the hypothesis $H:\lambda \leq \lambda_0$ versus $A:\lambda > \lambda_0$, it would be appropriate to obtain for a sample of 1,

$$P = 1 - F(x - 1) \qquad (6.39)$$

where $F(x)$ is read from Appendix 2 under the assumption $\lambda = \lambda_0$. If $P \leq \alpha$, the hypothesis H would be rejected.

Example 6.29

A random sample of 2 phonograph records shows 1 and 4 defects per record respectively. Assuming $\alpha = 0.01$, test the hypothesis $H:\lambda \leq 0.5$ versus $A:\lambda > 0.5$. (**Note:** This is testing the hypothesis that the *average* number of defects per record is less than or equal to 1/2.) Since we have a total of 5 defects from 2 records, we make use of the fact that $w = x_1 + x_2$ also follows a Poisson distribution with parameter $\lambda' = n\lambda = 2\lambda$. Consulting Appendix 2 for $\lambda' = 2\lambda_0 = 2(1/2) = 1$, we see that $F(w - 1) = F(5 - 1) = F(4) = 0.996$ and thus $P = 1 - F(w - 1) = 0.004$. Since this is less than $\alpha = 0.01$, the hypothesis $H:\lambda \leq 0.5$ is rejected in favor of the alternative $A:\lambda > 0.5$.

The second situation to be examined is of interest from a methodological point of view since it combines the assumption of a Poisson distribution with the chi-square method of analysis. Essentially, it is a comparison of several Poisson distributions to see if the parameters (i.e., the λ's) differ significantly. The procedure is best illustrated by an example.

Example 6.30

Suppose a phonograph record manufacturing company is investigating 5 different production processes. Four records are selected at random from those produced by each process, and the number of defects per record is counted. The data are given in Table 6.24. Chi-square is then computed for each process, using the observed process average as the expected number of defects per record for that process. Each of these chi-squares has 3 degrees of freedom. Using the additive property of chi-square, it is noted that the total, 9.70, has 15 degrees of freedom. It can be verified that none of these 6 values of chi-square is significant at the 1 percent level. Thus there is little question about the uniformity of records produced by the same process. However, if the chi-square representing the variation among processes is calculated, i.e.,

$$\chi^2 = [(64 - 35.2)^2 + (28 - 35.2)^2 + (32 - 35.2)^2 + (32 - 35.2)^2$$
$$+ (20 - 35.2)^2]/35.2 = 32.18$$

we see that $\chi^2 = 32.18 > \chi^2_{0.99(4)} = 13.3$. Therefore, the hypothesis of no differences among processes is rejected. It might be concluded that some processes (probably B, C, D, and E) will allow production of products containing fewer defects.

TABLE 6.24–Number of Defects per Record from XYZ Manufacturing Company

Process	Number of Defects per Record	Process Total	Process Mean	$\chi_i^2 = \sum_{j=1}^{4} \dfrac{(O_{ij} - E_i)^2}{E_i}$
A	11, 16, 17, 20	64	16	42/16 = 2.62
B	5, 7, 5, 11	28	7	24/7 = 3.43
C	11, 9, 7, 5	32	8	20/8 = 2.50
D	8, 10, 7, 7	32	8	6/8 = 0.75
E	5, 6, 5, 4	20	5	2/5 = 0.40
Total	...	176	...	9.70

6.14 CHI-SQUARE TEST OF GOODNESS OF FIT

One thing that is often done, with no justification other than saying it appears reasonable, is to assume that the variate under discussion follows a particular distribution. For example, data are frequently assumed to be samples from a normal population, and you may well question this assumption. At this time, one procedure useful in checking on the validity of such assumptions will be presented.

This procedure makes a comparison between the actual and expected number of observations (expected under the "assumption") for various values of the variate. The expected numbers are usually calculated by using the assumed distribution with the parameters set equal to their sample estimates. The chi-square statistic will be calculated according to Eq. (6.36), and the degrees of freedom will be $k - p - 1$, where p represents the number of parameters estimated by sample statistics. For example, if a normality assumption were under test, μ and σ^2 would be estimated by $\bar{X}$ and s^2 and the degrees of freedom would be $k - 3$, where k represents the number of class intervals used in fitting the distribution. If the assumption of a Poisson distribution were being tested, λ would be estimated by $\bar{X}$, and the degrees of freedom would be $k - 2$.

Rather than continuing the discussion in general terms, an example involving the Poisson distribution will be studied.

Example 6.31

The data given in Table 6.25 show the number of "senders" (a type of automatic equipment used in telephone exchanges) that were in use at a

TABLE 6.25–Number of Busy Senders in a Telephone Exchange

Number Busy	Observed Frequency (O)	Expected Frequency (E)	Deviation $(O - E)$	$\dfrac{(O - E)^2}{E}$
0	0 ⎱	0.11 ⎱	3.74	11.01
1	5 ⎰	1.15 ⎰		
2	14	5.98	8.02	10.76
3	24	20.82	3.18	0.49
4	57	54.33	2.67	0.13
5	111	113.44	−2.44	0.05
6	197	197.38	−0.38	0.00
7	278	294.38	−16.38	0.91
8	378	384.16	−6.16	0.10
9	418	445.63	−27.63	1.71
10	461	465.24	−4.24	0.03
11	433	441.56	−8.56	0.17
12	413	384.15	28.85	2.17
13	358	308.50	49.50	7.94
14	219	230.05	−11.05	0.53
15	145	160.11	−15.11	1.43
16	109	104.47	4.53	0.20
17	57	64.16	−7.16	0.80
18	43	37.21	5.79	0.90
19	16	20.45	−4.45	0.97
20	7	10.67	−3.67	1.26
21	8	5.31	5.69	1.36
22	3	4.51	−1.51	0.51
Total	3754	3753.77	0.23	$\chi^2 = 43.43$

Source: Fry (3), p. 295.

given instant. Observations were made on 3754 different occasions. The expected numbers were calculated from $f(x) = e^{-\lambda}\lambda^x/x!$ where λ was set equal to $\overline{X} = 10.44$. Since $\chi^2 = 43.43 > \chi^2_{0.99(20)} = 37.6$, the hypothesis of a Poisson distribution with $\lambda = 10.44$ is rejected.

One point to be noted in Example 6.31 is the combination of the entries of the top two lines of the table to form a single class. This was done because the expected number on the first line was too small. The reason for avoiding such expected numbers is that they lead to large chi-square values (perhaps even significant values of chi-square) that do not reflect a departure of "observed from expected" but only the smallness of the "expected." In other words, if some expected numbers are too small, the chi-square statistic will be a poor indicator of the validity of the hypothesis under test. Some authors say that "too small" means less than 3; others say less than 5. Since not everyone is agreed on the interpretation of what is too small, you should feel free to use any reasonable definition; we favor the value "3."

6.15 CONTINGENCY TABLES

Suppose n randomly selected items are classified according to two different criteria. The tabulation of the results could be presented as in Table 6.26, where O_{ij} represents the number of items belonging to the ijth cell of the $r \times c$ table. Such data can be used to test the hypothesis that the two classifications, represented by rows and columns, are statistically independent. If this hypothesis is rejected, the two classifications are not independent and we say there is some *interaction* between the two criteria of classification.

TABLE 6.26–An $r \times c$ Table

Row	Column 1	2	...	c	Row Total
1	O_{11}	O_{12}	...	O_{1c}	R_1
2	O_{21}	O_{22}	...	O_{2c}	R_2
.	.	.		.	.
.	.	.		.	.
r	O_{r1}	O_{r2}	...	O_{rc}	R_r
Column Total	C_1	C_2	...	C_c	n

The exact test for independence is difficult to apply. However, if the same size n is sufficiently large, a reasonably good approximate procedure is to calculate

$$\chi^2 = \sum_{i=1}^{r} \sum_{j=1}^{c} (O_{ij} - E_{ij})^2/E_{ij} \tag{6.40}$$

where O_{ij} = observed number in the ijth cell and $E_{ij} = R_i C_j/n$ = expected number in the ijth cell.

The value of chi-square given by Eq. (6.40) has $\nu = (r-1)(c-1)$ degrees of freedom. If $\chi^2 \geq \chi^2_{(1-\alpha)[(r-1)(c-1)]}$, the hypothesis of independence should be rejected.

Example 6.32

A company has to choose among three proposed pension plans. One hypothesis that the company wishes to investigate is: preference for plans is independent of job classification. It asks the opinion of a sample of the employees and obtains the information presented in Table 6.27. The expected numbers for each cell are calculated and appear in Table 6.28. Calculation then yields $\chi^2 = 11 < \chi^2_{0.99(6)} = 16.8$ so the hypothesis cannot be rejected. Thus it is concluded that the employees' choices of pension plans are quite probably independent of their job classifications.

TABLE 6.27–Classification of Employees by Job and Pension Plan Preference

Classification	Number of Employees Favoring			Row Total
	Plan A	Plan B	Plan C	
Factory employees	160	30	10	200
Clerical employees	140	40	20	200
Foremen and supervisors	80	10	10	100
Executives	70	20	10	100
Column Total	450	100	50	600

TABLE 6.28–Expected Number of Observations

Classification	Plan A	Plan B	Plan C
Factory employees	150	100/3	50/3
Clerical employees	150	100/3	50/3
Foremen and supervisors	75	100/6	50/6
Executives	75	100/6	50/6

If the contingency table consists of two rows and two columns, as in Table 6.29, a short-cut method of computing chi-square is available. The appropriate formula is

$$\chi^2 = n(ad - bc)^2/k \qquad (6.41)$$

where $n = a + b + c + d$ and $k = (a+b)(c+d)(a+c)(b+d)$. This will give the same numerical value of chi-square that would be obtained if Eq. (6.40) were used. It should be clear that the chi-square statistic thus obtained will have only 1 degree of freedom.

TABLE 6.29–A 2 × 2 Table

	A_1	A_2	Total
B_1	a	b	$a + b$
B_2	c	d	$c + d$
Total	$a + c$	$b + d$	n

As in Section 6.10, a correction for continuity may be used to sharpen the approximation. This is accomplished by calculating

$$\chi^2 = n(\mid ad - bc \mid - n/2)^2/k$$
$$= \sum_{i=1}^{2} \sum_{j=1}^{2} (\mid O_{ij} - E_{ij} \mid - 0.5)^2/E_{ij} \qquad (6.42)$$

It must be remembered that this correction should not be applied to $r \times c$ tables in which $r > 2$ and $c > 2$.

Example 6.33

A random sample of 250 men and 250 women were polled as to their desires concerning the ownership of television sets. The data in Table 6.30 resulted. Calculation by either method yielded $\chi^2 = 13.33 > \chi^2_{0.99(1)} = 6.63$. Thus the hypothesis that desire to own a television set is independent of sex is rejected.

TABLE 6.30–Results of Sample Poll on Television Ownership

Classification	Men	Women	Total
Want television	80	120	200
Don't want television	170	130	300
Total	250	250	500

6.16 SEQUENTIAL TESTS

In all the test procedures described thus far, the sample size has been decided upon in advance. As has been inferred, the determination of the proper sample size is often difficult. However, given the necessary information (e.g., an estimate of the variability to be encountered and statements concerning the allowable risks associated with incorrect decisions), the required sample size may be specified (see Section 6.18). The reader should realize, though, that a certain "cost" is attached to such an approach. That is, there is an implicit assumption in the fixed (predetermined) sample size approach that a sample of the specified size will be taken and observations recorded for each sample unit regardless of whether all the observations are needed to reach a decision. In view of this and in the hope of achieving economies due to reduced sample sizes, it seems desirable to seek a test procedure in which the sampling may be terminated as soon as it is possible to reach a decision to either accept or reject the hypothesis under test. For certain specific cases, namely, those that involve a simple hypothesis $H:\theta = \theta_0$ and a single alternative $A:\theta = \theta_1$, such a test has been devised. It is known as the *sequential probability ratio test*. In the remainder of this section, the general nature of this procedure will be described and certain specific applications illustrated.

The sequential method of testing proceeds as follows: sample units are randomly selected *one at a time* (i.e., sequentially), and after each observation is obtained, one of the following decisions is made:

1. Accept $H:\theta = \theta_0$ (i.e., reject $A:\theta = \theta_1$).
2. Reject $H:\theta = \theta_0$ (i.e., accept $A:\theta = \theta_1$).
3. Obtain an additional observation.

To determine which of these three decisions is appropriate, the analyst should calculate

$$R_n = \prod_{i=1}^{n} f_1(x_i)/f_0(x_i) \tag{6.43}$$

where $f_0(x)$ is the probability function (or probability density function) under the assumption that $H:\theta = \theta_0$ is true and $f_1(x)$ is the probability function (or probability density function) under the assumption that $A:\theta = \theta_1$ is true. Then, depending on the value of R_n, one of the three decisions previously listed is reached by proceeding according to the following rules:

1. If $R_n \leq \beta/(1 - \alpha)$, accept H(i.e., reject A).
2. If $R_n \geq (1 - \beta)/\alpha$, reject H(i.e., accept A).
3. If $\beta/(1 - \alpha) < R_n < (1 - \beta)/\alpha$, obtain an additional observation.

In the above, α and β are respectively the preassigned risks of: (1) rejecting H when H is true and (2) accepting H when A is true. If a_n and r_n are used to denote the acceptance and rejection values respectively for a test statistic, the decision rule may be restated in the following form:

1. If the value of the test statistic is less than or equal to a_n, accept H (i.e., reject A).
2. If the value of the test statistic is greater than or equal to r_n, reject H (i.e., accept A).
3. If the value of the test statistic is greater than a_n and less than r_n, continue sampling.

It should be clear that the sample size is a variable in a sequential procedure as contrasted to its role as a (predetermined) constant in the classical test procedures. Thus in addition to examining the power of a sequential test procedure by studying its OC function, it is appropriate that its "cost" be assessed by considering the average size of sample required to reach the decision to accept or to reject. This analysis is usually made in terms of the *average sample number* (ASN) function. Rather than going into details concerning the ASN function and the savings due to reduced sample sizes, let us be content with the general statement that the potential savings are considerable, in some cases as much as 50 percent.

Considerable space could be devoted to a detailed discussion of the sequential probability ratio test for each of the commonly encountered situations. However, it is doubtful if such discussions would serve any useful purpose. Accordingly, the tests have been specified in Table 6.31

Example 6.34

Consider a binomial population and the hypothesis $H:p = 0.10$ versus $A:p = 0.20$. Let $\alpha = 0.01$ and $\beta = 0.05$. Then $\log[\beta/(1 - \alpha)] = -2.986$ and $\log[(1 - \beta)/\alpha] = 4.554$. If we represent a sample unit possessing the characteristic associated with p by the symbol d and a unit not possessing this characteristic by g (e.g., defective and nondefective units respectively), then the sequence

$$gggdgdggdgggddgdgddgd$$

would terminate at this point with the decision to reject H and accept A.

Example 6.35

Consider a normal population with known standard deviation, $\sigma = 10$. Test the hypothesis $H:\mu = 50$ versus the alternative $A:\mu = 70$. Let $\alpha = 0.01$ and $\beta = 0.01$. Then $\log[\beta/(1 - \alpha)] = -4.595$ and $\log[(1 - \beta)/\alpha] = 4.595$. If sequential sampling yielded, in the order shown, the following values of $X(60, 75, 65, 70)$, the sampling would terminate at this stage with the decision to reject H and accept A.

6.17 SOME PSEUDO *t* AND *F* STATISTICS

Many times the researcher may feel that the time and effort involved in the calculation of s, the sample standard deviation, is too great for the benefit derived therefrom. Thus it is not surprising that techniques have been devised to use R, the sample range, in place of s. One of these techniques involves the use of a pseudo t statistic defined by

$$\tau_1 = (\overline{X} - \mu)/R \tag{6.44}$$

If we are willing to assume random sampling from a normal population, tests of hypotheses concerning μ or confidence interval estimation of μ may be performed by considering the sampling distribution of τ_1 and utilizing the values recorded in Table 1 of Appendix 17.

Example 6.36

Consider again the data and the hypothesis of Example 6.8. Using Eq. (6.44), we obtain $\tau_1 = (1267 - 1260)/8 = 0.875$. Since this calculated value of τ_1 exceeds the critical value for $n = 4$, namely, $\tau_{1(0.975)} = 0.717$, the hypothesis that $\mu = 1260$ is rejected. (**Note:** This is in agreement with the conclusion reached in Example 6.8.)

TABLE 6.31–Sequential Test Procedures for Certain Commonly Encountered Situations

Hypothesis and Alternative	Assumption	Test Statistic	Acceptance Value (a_n)	Rejection Value (r_n)
$H:p=p_0$ $A:p=p_1$	Binomial; $p_1 > p_0$	X = number of sample observations in the category associated with p	$\dfrac{\log[\beta/(1-\alpha)] + n\log[(1-p_0)/(1-p_1)]}{\log(p_1/p_0) - \log[(1-p_1)/(1-p_0)]}$	$\dfrac{\log[(1-\beta)/\alpha] + n\log[(1-p_0)/(1-p_1)]}{\log(p_1/p_0) - \log[(1-p_1)/(1-p_0)]}$
$H:\sigma=\sigma_0$ $A:\sigma=\sigma_1$	Normal; $\sigma_1 > \sigma_0$ μ known	$\sum (X-\mu)^2$	$\dfrac{2\log[\beta/(1-\alpha)] + n\log(\sigma_1^2/\sigma_0^2)}{1/\sigma_0^2 - 1/\sigma_1^2}$	$\dfrac{2\log[(1-\beta)/\alpha] + n\log(\sigma_1^2/\sigma_0^2)}{1/\sigma_0^2 - 1/\sigma_1^2}$
$H:\sigma=\sigma_0$ $A:\sigma=\sigma_1$	Normal; $\sigma_1 > \sigma_0$ μ unknown	$\sum (X-\overline{X})^2$	$\dfrac{2\log[\beta/(1-\alpha)] + (n-1)\log(\sigma_1^2/\sigma_0^2)}{1/\sigma_0^2 - 1/\sigma_1^2}$	$\dfrac{2\log[(1-\beta)/\alpha] + (n-1)\log(\sigma_1^2/\sigma_0^2)}{1/\sigma_0^2 - 1/\sigma_1^2}$
$H:\mu=\mu_0$ $A:\mu=\mu_1$	Normal; σ known	$\sum X$	$[\sigma^2/(\mu_1 - \mu_0)]\log[\beta/(1-\alpha)] + n(\mu_0 + \mu_1)/2$	$[\sigma^2/(\mu_1 - \mu_0)]\log[(1-\beta)/\alpha] + n(\mu_0 + \mu_1)/2$
$H:\lambda=\lambda_0$ $A:\lambda=\lambda_1$	Poisson; $\lambda_1 > \lambda_0$	$\sum X$	$\dfrac{\log[\beta/(1-\alpha)] + n(\lambda_1 - \lambda_0)}{\log\lambda_1 - \log\lambda_0}$	$\dfrac{\log[(1-\beta)/\alpha] + n(\lambda_1 - \lambda_0)}{\log\lambda_1 - \log\lambda_0}$
$H:\theta=\theta_0$ $A:\theta=\theta_1$	Exponential; $\theta_1 > \theta_0$	$\sum X$	$\dfrac{-\log[\beta/(1-\alpha)] + n\log(\theta_0/\theta_1)}{1/\theta_1 - 1/\theta_0}$	$\dfrac{-\log[(1-\beta)/\alpha] + n\log(\theta_0/\theta_1)}{1/\theta_1 - 1/\theta_0}$

Note: All logarithms used in this table are natural logarithms, i.e., logarithms to the base e.

Example 6.37

Utilizing the same data as in Example 6.36, a 99 percent confidence interval for μ may be obtained as follows:

$$L = \overline{X} - \tau_{1(0.995)} R = 1267 - 1.316(8) = 1256.47$$
$$U = \overline{X} + \tau_{1(0.995)} R = 1267 + 1.316(8) = 1277.53$$

An even simpler statistic that may be used as a substitute for "Student's" t is

$$\tau_2 = [(X_{min} + X_{max})/2 - \mu \ ]/R \qquad (6.45)$$

Critical values of this statistic are given in Table 3 of Appendix 17. Because of the similarity in use of this statistic to the one discussed in the preceding paragraph, no numerical examples will be presented.

When dealing with the difference between two sample means, a third pseudo t statistic may be utilized, namely,

$$\tau_d = (\overline{X}_1 - \overline{X}_2)/(R_1 + R_2) \qquad (6.46)$$

Critical values of τ_d are tabulated for samples of equal size in Table 2 of Appendix 17.

Example 6.38

Consider the data and hypothesis of Example 6.11. Using Eq. (6.46), we obtain $\tau_d = (4 - 7)/(7 + 7) = -3/14 = -0.214$. Since the calculated value to τ_d lies between $-\tau_{d(0.95)} = -0.246$ and $\tau_{d(0.95)} = 0.246$, we are unable to reject the hypothesis that $\mu_1 = \mu_2$. (**Note:** This agrees with the conclusion reached in Example 6.11.)

As might be expected, it is also possible to utilize the range in place of the standard deviation to provide a quick substitute for the familiar F statistic. The suggested statistic is

$$R_1/R_2 = \text{the ratio of the two sample ranges} \qquad (6.47)$$

and critical values are tabulated for certain selected sample sizes in Table 4 of Appendix 17. As with F the critical values for the lower end of the distribution of R_1/R_2 may be found by interchanging n_1 and n_2 (the two sample sizes) and calculating the reciprocals of the tabulated values. Because of the simplicity of this test and its similarity to those discussed in the preceding section, no numerical examples will be given.

6.18 SAMPLE SIZE

A question frequently asked of statisticians is, "How large a sample is needed for this experiment?" The question is deceptively simple, but the answer is hard to find. Before the statistician can provide anything better than an "educated guess," he must retaliate with several questions, the answers to which should enable him to attack the problem with some hope of reaching a valid answer. Frustrating as this may be to the researcher, it frequently serves a very good purpose, for it forces the researcher to give serious thought to several aspects of his problem. To illustrate, some of the questions that might be asked by the statistician are:

1. What is your hypothesis? What are the alternatives?
2. What are you trying to estimate?
3. What significance level are you planning to use? What confidence level?
4. How large a difference do you wish to be reasonably certain of detecting? With what probability?
5. What width confidence interval can you tolerate?
6. What do you expect the variability of your data to be?

When answers to these and other questions are provided by the researcher, the statistician can be of help in determining the needed sample size.

Before you get the impression that all is lost, let us hasten to assure you that the picture is not all black. In some cases, fairly simple formulas are available for *estimating* the required sample size. Also, if OC curves are available for the test procedure to be used, the required sample size may be determined upon examination of these curves. Tables have also been provided for certain procedures, and four of these are reproduced in Appendixes 9 throught 12 for your use. If all these three approaches (i.e., formulas, OC curves, or tables) fail to meet your demands, a professional statistician should be consulted.

Example 6.39

Consider testing the hypothesis $H:\mu = \mu_0$ versus $A:\mu \neq \mu_0$ at the 5 percent significance level. If σ is estimated to be 0.8 and a difference $\delta = |\mu - \mu_0| = 1.2$ is to be detected with probability 0.9 (this is equivalent to setting $\beta = 0.1$ at $\mu = \mu_0 - 1.2$ and at $\mu = \mu_0 + 1.2$), how large a sample is needed? Setting $D = 1.2/0.8 = 1.5$ and consulting Appendix 9, it is found that $n = 7$.

Example 6.40

Consider testing $H:\mu \leq \mu_0$ versus $A:\mu > \mu_0$ at the 1 percent significance level. If σ is estimated to be 1.2 and $\delta = \mu - \mu_0 = 0.9$ is to be detected with probability 0.95, how large a sample is needed? Setting $D = 0.9/1.2 = 0.75$ and consulting Appendix 9, it is found that $n = 31$.

Example 6.41

Consider testing $H:\mu_1 \leq \mu_2$ versus $A:\mu_1 > \mu_2$ at the 2.5 percent significance level. If σ is estimated to be 1.0 and $\delta = \mu_1 - \mu_2 = 1.6$ is to be detected with probability 0.99, how large should the two samples be? Setting $D = 1.6/1.0 = 1.6$ and consulting Appendix 10, it is found that $n_1 = n_2 = 16$.

Example 6.42

Consider testing $H:\mu_1 = \mu_2$ versus $A:\mu_1 \neq \mu_2$ at the 1 percent significance level. If σ is estimated to be 1.5 and $\delta = |\mu_1 - \mu_2| = 1.8$ is to be detected with probability 0.95, how large should the two samples be? Setting $D = 1.8/1.5 = 1.2$ and consulting Appendix 10, it is found that $n_1 = n_2 = 27$.

Example 6.43

Consider testing $H:\sigma^2 \leq \sigma_0^2$ versus $A:\sigma^2 > \sigma_0^2$ at the 5 percent significance level. If a value of $\sigma^2 = 4\sigma_0^2$ is to be detected with probability 0.99, how large a sample is needed? Using $R = 4$ and consulting Appendix 11, it is seen that $15 < \nu < 20$. Crude interpolation suggests $\nu = 19$ or $n = \nu + 1 = 20$.

Example 6.44

Consider testing $H:\sigma^2 \geq \sigma_0^2$ versus $A:\sigma^2 < \sigma_0^2$ at the 5 percent significance level. If a value of $\sigma^2 = 0.33\ \sigma_0^2$ is to be detected with probability 0.99, how large a sample is needed? Since Appendix 11 is constructed for values of $R \geq 1$, a slight change in procedure (from Example 6.43) is required. The table in Appendix 11 is entered with $\alpha' = \beta = 0.01$, $\beta' = \alpha = 0.05$, and $R' = 1/R = 3$. Thus, it is noted that $24 < \nu < 30$. Crude interpolation suggests $\nu = 26$ or $n = \nu + 1 = 27$. (**Note:** Although the roles of α and β were interchanged when Appendix 11 was consulted, the actual test would be carried out at the original value of α which in this example was 0.05.)

Example 6.45

Consider testing $H:\sigma_1^2 \geq \sigma_2^2$ versus $A:\sigma_1^2 < \sigma_2^2$ at the 5 percent significance level. If a value of $\sigma_2^2 = 4\sigma_1^2$ is to be detected with probability 0.99, how large should the two samples be? Using $R = 4$ and consulting Appendix 12, it is noted that $30 < \nu_1 = \nu_2 < 40$. Crude interpolation suggests that $\nu_1 = \nu_2 = 34$ or $n_1 = n_2 = 35$.

PROBLEMS

6.1   A company engaged in the casting of pig iron must be concerned with the percentage of silicon in the pig iron (grams of silicon/100 gm pig iron). The data given below constitute a random sample of the production records. Using $\alpha = 0.02$ and assuming normality, test the hypothesis that the process average is 0.85 grams of silicon per 100 grams of pig iron.

1.13	0.87
0.80	0.92
0.85	0.81
0.60	0.97
0.97	0.48
0.92	1.00
0.94	0.92
0.72	0.61
1.17	0.81
0.87	0.71
0.36	0.97
0.68	0.89
0.73	1.16
0.82	0.68
0.79	1.00

6.2 Consider the following observations to represent the average hourly earnings (cents/hour) during May 1940 of a random selection of 50 male workers in a specified industry.

35	65	68	77	81
52	82	74	73	71
68	79	73	70	67
82	61	77	84	56
29	53	61	83	92
99	80	62	50	64
76	47	59	64	72
55	63	107	48	70
55	70	43	66	85
79	90	39	88	86

(a) What is your best estimate of the average hourly earnings for all male workers in the industry?

(b) How good is your estimate in (a)? What is its standard error?

(c) Establish confidence limits for your estimate in (a). Write out your statement about these confidence limits in words. State your assumptions clearly.

(d) Is your estimate in (a) in agreement with the hypothesized true value of 68 cents/hour for average earnings in May 1940? Explain your answer.

(e) What additional data would you need to estimate the total earnings in the industry for the month of May?

(f) Test the hypothesis that $\mu \leq 80$.

6.3 Test the hypothesis that the mean life (in years) of wooden telephone poles is less than 8 years. State any assumptions you make about the following data:

Life (years)	Number of Poles Replaced
0.5 but under 1.5	4
1.5 but under 2.5	7
2.5 but under 3.5	15
3.5 but under 4.5	32
4.5 but under 5.5	30
5.5 but under 6.5	57
6.5 but under 7.5	61
7.5 but under 8.5	73
8.5 but under 9.5	96
9.5 but under 10.5	104
10.5 but under 11.5	103
11.5 but under 12.5	95
12.5 but under 13.5	91
13.5 but under 14.5	73
14.5 but under 15.5	64
15.5 but under 16.5	38
16.5 but under 17.5	30

(*continued*)

Life (years)	Number of Poles Replaced
17.5 but under 18.5	18
18.5 but under 19.5	5
19.5 but under 20.5	1
20.5 but under 21.5	1
21.5 but under 22.5	2
Total	1000

6.4   A consumer panel report on the economic and geographic distribution of the purchases of a particular product reveals among other things that the nation's families bought on the average 17.5 lb of that product in 1949. This estimate was based on returns from a supposed random sample of 1225 families, and the standard deviation of individual family purchases in this sample was found to be 7.5 lb. From sales and inventory records it is determined that average purchases per family in 1948 must have been at least 18.5 lb, or 1 lb more than the sample estimate for 1949. Could this difference of 1 lb be due to sampling variation, or does it indicate that average consumption of the product by families had decreased in 1949 from the 1948 level of consumption? What assumptions did you make?

6.5   Using the data in Problem 5.1, test the hypothesis $H:\mu = 6.55 \times 10^{-27}$ versus $A:\mu \neq 6.55 \times 10^{-27}$. Let $\alpha = 0.01$.

6.6   Using the data in Problem 5.5 and letting $\alpha = 0.025$, test $H:\mu \leq 1.12$ in. versus $A:\mu > 1.12$ in.

6.7   Using the data of Problem 5.6 and letting $\alpha = 0.05$, test $H:\mu \geq 1.55$ versus $A:\mu < 1.55$.

6.8   Using the data of Problem 5.7 and letting $\alpha = 0.25$, test $H:\mu \leq 2900$ yd versus $A:\mu > 2900$ yd.

6.9   Using the data of Problem 5.1 and letting $\alpha = 0.01$, test $H:\sigma \leq 0.01 \times 10^{-27}$ versus $A:\sigma > 0.01 \times 10^{-27}$.

6.10   Using the data of Problem 5.5 and letting $\alpha = 0.10$, test $H:\sigma^2 \leq 0.0001$ versus $A:\sigma^2 > 0.0001$.

6.11   Using the data of Problem 5.6 and letting $\alpha = 0.005$, test $H:\sigma \geq 0.05$ versus $A:\sigma < 0.05$.

6.12   Using the data of Problem 5.7 and letting $\alpha = 0.01$, test $H:\sigma = 50$ yd versus $A:\sigma \neq 50$ yd.

6.13   In making a certain cross, a geneticist expected a segregation of 15 A's to 1 B. In a random sample of 800 he observed 730 A's and 70 B's. Do the data support the expected ratio? Why?

6.14   In a random sample of 400 farm operators, 65 percent were owners and 35 percent were nonowners. Test the hypothesis that in the population of farm operators 60 percent are owners. Use a probability of Type I error equal to 0.05.

6.15   A manufacturer of light bulbs claims that on the average 1 percent or less of all the light bulbs manufactured by his firm are defective. A random sample of 400 light bulbs contained 12 defectives. On the evidence of this sample, do you believe the manufacturer's claim? Why? Assume that the maximum risk you wish to run of falsely rejecting the manufacturer's claim—the true fraction defective is 0.01—has been set at 2 percent.

6.16 A sampler of public opinion asked 400 randomly chosen persons from some specified population whether they favored candidate A or B; 220 voted for A and 180 for B. Using a probability of Type I error equal to 0.05, do you think that opinion in the population may have been equally divided? Why?

6.17 A supermarket is to be built in a new location. The question arose as to whether provision should be made for individual customer service at the meat counter or whether a self-service counter with all meats ready-cut and packaged would adequately serve customers in the new area. The management decision was that individual customer service would not be supplied unless 40 percent of the prospective customers desired it. A random sample of 160 prospective customers showed only 50 respondents desiring individual service. Does it appear that the proportion of preference in the population of prospective customers equals or exceeds the critical level set by management?

6.18. In a triangular test for selecting judges to compose a taste panel, a prospective judge was successful in selecting the odd sample 11 times in 15 trials. Would you select him for the panel? How many would he have to pick correctly to be chosen? What is the probability of Type I error if we accept the above judge for our panel? Construct the complete table of probabilities, showing them also in cumulative form, for $n = 15$.

6.19 Retail sales data indicate that 1/3 of the families in the WOI-TV area have television sets. A random sample of 900 families from the area is to be taken.
(a) What is the expected number of television families for the sample?
(b) The sample yields 360 families with television sets. Indicate at least two methods by which we may obtain approximate confidence limits for the population proportion of families owning television sets.
(c) Is the observed number, 360, in "reasonable" agreement with the expected number?

6.20 Eighty out of 1000 randomly chosen cases of diphtheria resulted in death. What methods or techniques are available for using these results to test the hypothesis that the true percentage of fatality is 10 percent? State whether the tests are exact or approximate.

6.21 A botanist observed 150 seedlings for the purpose of studying chlorophyll inheritance in corn. The seed came from self-fertilized heterozygous green plants. Hence green and yellow seedlings were expected in proportions of 3 green to 1 yellow. The sample showed 120 green and 30 yellow seedlings. Is this sample in agreement with expectation?

6.22 A metropolitan newspaper was considering a change to tabloid form. A random sample of 900 of its daily readers was polled to secure readership reaction to such a change. Of this sample, 541 persons opposed the change in format for the paper.
(a) Is it likely that more than 50 percent of the readers are in favor of the change?
(b) Describe two or more procedures for obtaining confidence limits for the population proportion opposed to the change.

6.23 From a keg containing 1000 bolts, a random sample of 20 bolts has been presented to you for testing. One hundred percent of the bolts in the sample successfully pass the test. Of all the bolts in the keg, what is your estimate of the percentage that will pass? What limits would you place on the reliability of your estimate; i.e., what confidence statement would you make about the true percentage of all the bolts that will pass the test?

6.24   After a survey of opinion is made, point and interval estimates are calculated. The investigator states that the 95 percent confidence interval is from 60 percent to 75 percent of the population in favor of a law. Describe precisely the meaning of this statement.

6.25   Using the data of Problem 5.20 and letting $\alpha = 0.005$, test $H:\mu_2 \leq \mu_1$ versus $A:\mu_2 > \mu_1$.

6.26   Using the data of Problem 5.21 and letting $\alpha = 0.05$, test $H:\mu_1 = \mu_2$ versus $A:\mu_1 \neq \mu_2$.

6.27   Using the data of Problem 5.22 and letting $\alpha = 0.10$, test $H:\mu_1 = \mu_2$ versus $A:\mu_1 \neq \mu_2$.

6.28   We are told that the mean yields of two corn hybrids were 75 and 85 bushels per acre respectively, and that each had been tried in 16 fields selected at random from some population of fields. Further, assuming that $\sigma_1^2 = \sigma_2^2$, we are told that the standard error of each of the above means was 3. Test the hypothesis that $\mu_1 = \mu_2$.

6.29   The diameter of a cylinder was measured by 16 persons. Each person made 3 determinations using a micrometer caliper and 3 determinations using a vernier caliper. Following are the averages of the 3 determinations (in inches) for each caliper made by the 16 persons. Is there any difference between the means of the populations of measurements represented by the 2 samples? The method to be used is determined by the fact that each person used both calipers. Do you think the difference is attributable to imperfections of the calipers or to the difficulty of setting the vernier caliper?

Micrometer	Vernier	Micrometer	Vernier	Micrometer	Vernier
1.265	1.265	1.270	1.269	1.264	1.267
1.265	1.267	1.267	1.273	1.266	1.272
1.267	1.267	1.268	1.270	1.266	1.273
1.266	1.266	1.267	1.270	1.268	1.267
1.268	1.267	1.267	1.267	1.265	1.268
1.265	1.267				

6.30   The following are the lengths in millimeters of 6-year-old white crappies from East Lake, Lucas County, Iowa, in 1948. Measurements were made by William Lewis and T. S. English. Is there any difference between the lengths of male and female crappies of this age group in East Lake in 1948?

Males		Females		
228	217	219	231	225
219	230	217	222	214
224	220	225	220	221
225	221	228	222	233
239	225	234	222	227
223	223	222	223	234
241	223	225	253	220
233	213	224	235	281
224	212	218	235	231
231	220	224	264	
251	231	223	246	
247	214	241	272	

6.31 In order to test 2 methods of teaching spelling, 40 pupils were randomly assigned to 2 classes and one method was tried on each class. At the end of the trial a test was given. Following are the scores on the tests. Test the hypothesis that the two methods of teaching spelling are equally effective. State all your assumptions.

Method A		Method B	
10	48	20	57
20	50	27	60
25	51	35	63
30	52	40	64
33	54	41	65
37	56	50	67
41	57	50	67
43	65	54	73
46	73	56	83
46	86	57	95

6.32 Using the data of Problem 5.23 and letting $\alpha = 0.01$, test $H:\mu_D = 0$ versus $A:\mu_D \neq 0$.

6.33 A certain stimulus administered to each of 9 patients resulted in the following increases in blood pressure: 5, 1, 8, 0, 3, 3, 5, −2, 4 mm Hg. Can it be concluded that the stimulus will be in general accompanied by an increase in blood pressure?

6.34 Suppose an investigator of group differences in I.Q. finds, for independent random groups A and B of 11 subjects each assumed to be from normal populations of same variance, a difference in sample means of $\overline{Y}_A - \overline{Y}_B = 3.9$ I.Q. points and an estimated standard error of the mean difference of 2.0. He selects 100α as 5 percent.

(a) For the data as given, what hypothesis might he test? Perform the required test and state your conclusions.

(b) Suppose group A had been given special coaching designed to "increase" I.Q., while group B had been maintained as a control. What hypothesis might he test? Perform the required test and give the resulting inferences.

6.35 In examining the resistance to crushing offered by kernels of a single ear of corn, we choose at random two lots of 10 kernels each with the following resistances (in points). Using the method of paired observations, we find the difference between the two means to be significant. We draw 4 more sets of 2 samples, each time with a significant difference. This seems surprising, since all the samples were taken from the kernels of the same ear. Can you explain the results?

Lot I		Lot II	
8	18	8	20
14	20	15	20
16	22	16	24
17	27	18	28
18	30	20	31

6.36 (a) For the data given below, test the hypothesis that the true mean

tensile strength of the screwdriver of the C and C Manufacturing
Company is greater than the corresponding value for its competi-
tor. State all your assumptions.

(b) Ignoring any assumption about variances you may have found it
necessary to make in (a), test the hypothesis that the two population
variances are equal.

C & C Company				Competitor			
Test	Tensile Strength in Pounds Y	Test	Tensile Strength in Pounds Y	Test	Tensile Strength in Pounds Y	Test	Tensile Strength in Pounds Y
1	130.1	19	153.5	1	65.7	20	149.4
2	132.3	20	154.1	2	101.3	21	151.0
3	133.4	21	154.7	3	103.0	22	153.3
4	135.5	22	155.4	4	103.6	23	155.2
5	137.7	23	156.7	5	107.2	24	157.6
6	139.3	24	157.5	6	115.9	25	160.7
7	140.4	25	158.4	7	117.4	26	164.3
8	144.2	26	159.4	8	122.6	27	166.1
9	145.0	27	160.7	9	126.5	28	168.8
10	146.7	28	161.9	10	129.1	29	170.4
11	147.4	29	163.1	11	132.3	30	180.6
12	148.3	30	164.8	12	134.6	31	184.6
13	149.7	31	169.3	13	135.2	32	188.8
14	150.6	32	171.2	14	136.7	33	192.9
15	151.1	33	174.0	15	138.3	34	196.0
16	151.8	34	180.7	16	142.1	35	200.4
17	152.1			17	143.4	36	204.8
18	152.7	Total	5183.7	18	147.2		
				19	148.2	Total	5295.2

6.37   Using the following data for crushing strengths of 248 samples of
Douglas fir, $2'' \times 2'' \times 8''$, test the hypothesis that the true mean
crushing strengths of air-dried and green Douglas fir wood are the same.
(Tested by Forest Products Laboratory, University of British Columbia.)
State all your assumptions and interpret your results.

			Air-dried Douglas Fir											
N 1	4713	N 1	5641	N 9	7145	E 7	6508	S 5	8413	W 3	7446			
2	5516	2	5550	E 3	6200	8	6828	6	7690	5	7941			
3	5956	3	7433	4	7501	9	6098	7	8484	6	8159			
4	5652	4	7097	5	8086	10	6359	8	8139	7	9316			
5	5951	5	7865	6	8055	S 1	5208	9	7595	8	9515			
6	7178	6	8045	7	8042	2	4648	10	7021	9	8171			
7	6630	7	7408	8	8678	3	7153	11	6416	10	9001			
8	6284	8	7344	9	6710	4	6504	W 3	6657	N 3	8161			
9	6246	9	7518	10	7512	5	6562	5	8264	4	7820			
10	4689	10	7280	S 1	6438	6	7105	7	7268	6	8560			

Air-dried Douglas Fir

11	4825	E 4	7174	3	6074	7	7114	8	8101	7	8222
12	4697	5	7234	4	7170	8	6263	9	7066	8	8387
E 3	5757	6	8452	5	7306	W 3	5530	10	7301	9	7500
4	6661	7	8709	6	7760	4	6632	N 1	5961	10	2181
5	6098	8	7710	7	7049	5	6429	2	6254	11	7655
6	5867	9	7609	8	6863	6	6912	3	7247	E 3	7373
7	5573	10	6731	9	6987	7	7053	4	7480	4	7949
8	6282	S 3	6342	10	6511	8	6370	5	8512	5	8199
9	5536	4	6924	W 3	7025	9	7413	6	8911	6	8547
10	4941	5	7712	4	6775	10	6335	7	8988	7	8464
S 2	4003	6	6805	5	7754	N 1	6584	8	9330	8	8594
3	4789	7	7539	6	7495	3	7518	9	9899	9	7092
4	4889	8	7630	7	7990	4	7106	10	9025	10	7433
5	5304	9	7501	8	6149	5	7135	11	8920	S 1	6444
6	5350	10	7531	9	6774	6	7596	E 3	6419	2	6545
7	5601	11	6096	10	7137	7	7573	4	8403	3	7320
8	5932	12	6983	N 1	4858	8	7521	5	8220	4	7886
9	5245	W 3	6212	3	6148	9	7261	6	9501	5	8173
10	5585	4	6530	4	5388	10	6364	7	9250	6	7844
11	4313	5	7800	5	5883	11	6905	8	9479	7	7613
12	4924	6	7713	6	5930	E 3	7608	9	9985	8	8469
W 3	5196	7	7759	7	6252	4	6793	10	9686	9	7675
4	4810	8	7253	8	5920	5	7734	11	8849	10	7371
5	6641	9	6898	9	6260	6	6465	S 3	6693	W 3	7113
6	4625	10	7403	10	6403	7	7499	4	6338	4	7283
7	6704	N 2	6144	11	6644	8	7703	5	5976	5	8337
8	5555	3	6717	12	5841	9	7470	7	8495	6	8509
9	6813	4	7021	E 3	6650	10	7178	8	9184	7	7510
10	6061	5	8096	4	5802	S 1	6201	9	9485	8	8361
11	4959	6	7608	5	7287	3	7878	11	8507	9	7485
12	5618	7	8025	6	6379	4	7155	12	8270	10	8522
14	3958	8	8115								

Green Douglas Fir

N 1	2428	W13	2343	N 7	3639	E 3	3446	S 3	3412	S 7	4088
2	2173	N 2	2603	8	3645	4	2892	4	3904	8	4377
3	2896	3	2911	9	3487	5	3629	5	4030	9	4267
4	2980	4	3158	10	3351	6	3442	6	4212	10	4256
5	3378	5	3553	E 3	3591	7	3412	7	4423	11	4109
6	3167	6	3659	4	2849	8	3477	8	4575	12	3325
7	3208	7	3800	5	3911	9	3474	9	4318	W 3	3297
8	3342	8	3645	6	2591	10	3007	10	3829	4	3606
9	2982	9	3505	7	2769	S 1	2493	11	3933	5	3534
10	3301	10	3834	8	4097	2	2505	12	4608	6	4159
11	2330	E 3	2506	9	3203	3	3449	W 4	3340	7	4393
12	2651	4	2818	10	3179	4	3224	5	3887	8	3992
E 3	2478	5	3775	S 1	2668	5	3485	6	4097	9	4049
4	2665	6	3318	2	2766	6	3667	7	3440	N 1	2813
5	3033	7	3686	3	3280	7	3343	8	4503	2	2574

(*continued*)

Green Douglas Fir

No.	Value	No.	Value	No.	Value	No.	Value	No.	Value	No.	Value
6	3205	8	3705	4	3295	8	3431	9	3806	3	3286
7	3282	9	3543	5	3844	W 3	2643	10	3939	4	3310
8	3229	10	3848	6	4022	4	3039	N 1	2902	5	3610
9	3137	S 3	2778	7	3575	5	3510	2	2869	6	3637
10	2693	4	2743	8	3784	6	3469	3	3610	7	3871
S 1	2128	5	3541	9	3621	7	3635	4	3547	8	3757
2	2200	6	3580	11	3698	8	4016	5	4012	9	3716
3	1977	7	3803	W 3	3032	9	3777	6	3919	E 3	3105
4	2498	8	3787	4	3132	10	3642	7	4585	4	3172
5	2732	9	3623	5	3781	N 3	3257	8	4553	6	3679
6	2920	10	3848	6	4141	4	3426	9	4235	7	3854
7	3102	11	3530	7	3730	5	4001	10	4495	8	3670
8	3050	12	3296	8	4162	6	3993	11	3694	9	3386
9	3230	W 3	2845	9	3559	7	4201	12	3492	10	3368
10	3053	4	3015	10	3532	8	4555	E 3	3173	S 1	2688
11	2993	5	3384	N 1	2296	9	3914	4	3879	3	3089
12	2518	6	3671	2	2458	10	3931	5	3751	4	3212
W 3	2938	7	3794	3	2794	E 3	3769	6	4197	5	3618
4	2272	8	3863	4	3075	4	3622	7	4110	7	3551
5	3144	9	3712	5	3166	5	4168	8	4061	8	3752
6	2904	10	3553	6	3255	6	4246	9	4589	9	3474
7	3314	N 1	2607	7	3233	7	4282	10	3762	10	3556
8	3448	2	2591	8	3600	8	4118	11	2733	W 3	3181
9	3468	3	3042	9	3471	9	3928	S 4	3071	4	3163
10	3289	4	2450	10	3735	S 1	3095	5	3886	9	3733
11	2456	5	3444	11	3329	2	3218	6	3873	10	3823
12	3078	6	3593								

6.38 Two lots of steers, 10 head in each lot, were used in a 90-day feeding trial. Lot 1 received standard ration A. Lot 2 received special ration K. Steers on ration A gained 1.84 lb/head/day, while the animals fed K gained at the rate of 2.36 lb/head/day. Two questions were of interest.

(a) Will daily gains on ration K exceed 2 lb/day? The variance of the mean gain, 2.36, was found to be 0.0144.

(b) Is ration K better than standard ration A in producing gains? The variance of the mean gain, 1.84, for lot 1 was 0.0256; thus we see that the pooled sum of squares for daily gain of the 2 lots is 3.60.

Answer the two questions with the information given above. Why do we use twice the pooled variance in examining the difference in gains between the 2 lots, whereas in answering question (a) we use the variance without such modification?

6.39 A sample of rural and urban families was taken to study differences in coffee purchases by the two groups. The data obtained are listed below in terms of pounds per family purchased annually. Would you attribute the difference in coffee consumption observed in these samples to normal sampling fluctuation, or is there a real difference between rural and urban coffee consumption? Select your own level for control of the Type I error and draw your conclusion accordingly. What is the *specified* population from which these data provide you a sample? State your assumptions.

Family No.	Rural	Urban
1	12.1	8.3
2	6.8	9.3
3	9.1	9.2
4	11.1	11.1
5	11.4	10.7
6	13.3	4.6
7	9.8	9.8
8	11.3	7.9
9	9.4	8.5
10	10.2	9.1
11	...	9.7
12	...	6.2

6.40 It has been suggested that the resistance of wire C is greater than the resistance of wire D. The following data (in ohms) were obtained from tests made on samples of each wire. Assuming that $\sigma_C^2 = \sigma_D^2$, test (using $\alpha = 0.01$) the hypothesis $H:\mu_C \leq \mu_D$ versus $A:\mu_C > \mu_D$. State your conclusion and interpret the results.

C	D
0.140	0.135
0.138	0.140
0.143	0.142
0.142	0.136
0.144	0.137
0.139	...

6.41 If the estimate of the population standard deviation from one sample of 45 is 12 and a corresponding estimate from another sample of 45 is 18, are these samples consistent with the hypothesis that they are from normal populations with the same variance?

6.42 Two methods of determining moisture content of samples of canned corn have been proposed and both have been used to make determinations on portions taken from each of 21 cans. Method I is easier to apply but appears to be more variable than Method II. If the variability of Method I were not more than 25 percent greater than that of Method II, we would prefer Method I. Based on the following sample results, which method would you recommend? $n_1 = n_2 = 21$, $\overline{Y}_1 = 50$, $\overline{Y}_2 = 53$, $\sum (Y_1 - \overline{Y}_1)^2 = 720$, $\sum (Y_2 - \overline{Y}_2)^2 = 340$. (**Hint:** Test $H:\sigma_1^2 = 1.25\sigma_2^2$ against $A:\sigma_1^2 > 1.25\sigma_2^2$. Under this hypothesis $(s_1^2/1.25)/s_2^2$ is distributed as $F_{(\nu_1,\nu_2)}$, where $\nu_1 = \nu_2 = n_1 - 1 = n_2 - 1 = 20$.)

6.43 The amount of surface wax on each side of waxed paper bags is believed to be normally distributed. However, there is reason to believe that there is greater variation in the amount on the inner side of the paper than on the outside. A sample of 25 observations of the amount of wax on each side of these bags was obtained and the following data recorded. Conduct a test (using $\alpha = 0.05$) of the hypothesis $H:\sigma_O^2 \geq \sigma_I^2$ versus $A:\sigma_O^2 < \sigma_I^2$.

Wax in Pounds/Unit Area of Sample	
Outside surface	Inside surface
$\overline{X} = 0.948$	$\overline{Y} = 0.652$
$\sum X^2 = 91$	$\sum Y^2 = 82$

6.44 Using the data of Problem 5.21 and letting $\alpha = 0.05$, test $H:\sigma_1 = \sigma_2$ versus $A:\sigma_1 \neq \sigma_2$.

6.45 Using the data of Problem 5.22 and letting $\alpha = 0.01$, test $H:\sigma_1 = \sigma_2$ versus $A:\sigma_1 \neq \sigma_2$.

6.46 Using the data of Problem 5.20 and letting $\alpha = 0.01$, test $H:\sigma_2 \leq \sigma_1$ versus $A:\sigma_2 > \sigma_1$.

6.47 Using the data of Problem 5.23 and letting $\alpha = 0.05$, test $H:\sigma_A \leq \sigma_B$ versus $A:\sigma_A > \sigma_B$.

6.48 A child psychologist, analyzing personality differences in children by a projective technique, classified the responses of a group of 99 preschool children into three major types: static form of response, 23; outer activity, 51; inner activity, 25. Do these data differ significantly from a chance distribution of responses? Use $\alpha = 0.01$.

6.49 A random sample of 147 women college students were interviewed with regard to their habits concerning the purchase of clothing. The source of each individual's income was also determined. Given the data below and letting $\alpha = 0.10$, test the hypothesis that women purchase clothing without planning in the following proportion: frequently—10 percent, seldom—80 percent, never—10 percent.

Source of Income	Numbers Who Purchased Clothing Items without Planning		
	Frequently	Seldom	Never
Earned all of spending money	2	14	27
Earned part of spending money	8	17	5
Had regular allowance	4	12	7
Money given as needed	15	25	11

6.50 Referring to the data of Problem 6.49 and letting $\alpha = 0.01$, test the hypothesis that frequency of purchasing clothing items without planning is independent of source of income.

6.51 An experimenter testing 3 chemical treatments applied each to 200 randomly selected seeds and then conducted germination tests. The following results were obtained. Test the hypothesis that the percentage of seeds germinating is independent of the chemical used.

Chemical	Number	
	Germinating	Not germinating
A	190	10
B	170	30
C	180	20

6.52 An experimenter fed different rations to 3 groups of chicks. Assume that the chicks were assigned to the rations (groups) at random and that all other management practices for the 3 groups were the same. A record of mortality is given below. Would you attribute the differences among the mortality rates of the 3 groups to rations? Why?

Ration	Number	
	Lived	Died
A	87	13
B	94	6
C	89	11

6.53 A random sample of students at Arizona State University was selected, and they were asked their opinions concerning a proposed radio program. The results are given below. The same number of each sex was included within each class group; i.e., freshmen and sophomores each consisted of 100 men and 100 women, while juniors and seniors each consisted of 50 men and 50 women. Test the hypothesis that opinions are independent of the class groupings.

Class	Number	
	Favoring program	Opposed to program
Freshmen	120	80
Sophomores	130	70
Juniors	70	30
Seniors	80	20

6.54 An agency engaged in market research conducted some of its sampling by mail. For one survey the following results in terms of response to successive mailings were obtained:

Response No.:	1st	2nd	3rd	4th	Original Mailing
Returns:	150	60	40	20	1000

Another agency obtained the following results in a mail sampling of a similar population:

Response No.:	1st	2nd	3rd	4th	Original Mailing
Returns:	200	30	50	25	800

Does it appear that the two mail samplings were homogeneous in eliciting replies from the two populations?

6.55   In a large city the division of the voting strength between two candidates for mayor appeared to be about equal. The campaign manager for candidate A polled a random sample of 2500 voters two weeks before the election. In this sample 1313 of the voters indicated they would vote for A. If the sample is representative of the population of voters in this city, is it likely that A will be elected? Establish 99 percent confidence limits for the proportion of voters favoring A.

6.56   An opinion-polling agency reported the distribution of a sample in the following manner:

Republicans	Democrats	Independents	Total
400	450	150	1000

A newspaper poll in the same area yielded the following distribution in terms of declared political opinion of respondents:

Republicans	Democrats	Independents	Total
300	325	75	700

Are these two samples homogeneous with regard to division of political opinion?

6.57   The following data on number of machine breakdowns were obtained from a random sampling of the records of a specific company. Test the hypothesis that the number of breakdowns on each machine is independent of the shift. Use $\alpha = 0.05$.

	Machine				
	A	B	C	D	Total per Shift
Shift 1	10	6	12	13	41
Shift 2	10	12	19	21	62
Shift 3	13	10	13	18	54
Total per machine	33	28	44	52	157

6.58 Road tests gave the data shown below regarding number of tire failures. Letting $\alpha = 0.05$, test the hypothesis that left-right tire wear is independent of front-rear tire wear.

	Front	Rear	Total
Left	115	65	180
Right	125	95	220
Total	240	160	400

6.59 A car rental firm has a particular car that has experienced 13 breakdowns in the past year. Using a Poisson distribution and letting $\alpha = 0.01$, test $H:\mu \leq 10$ versus $A:\mu > 10$.

6.60 Ignoring the correction for continuity in Eq. (6.28) (i.e., dropping the adjustment of -0.5) show that $\chi^2 = (a - rb)^2/r(a + b)$ where a and b are the observed numbers in the two classes and r equals the hypothesized ratio of type A to type B.

6.61 Work the preceding problem using the correction for continuity and show that $\chi^2 = (|a - rb| - (r + 1)/2)^2/r(a + b)$.

6.62 Rework the problems noted below, using the method described in Section 6.5:

(a) 6.26   (c) 6.28   (e) 6.31   (g) 6.39
(b) 6.27   (d) 6.30   (f) 6.37

6.63 Given the following data (3 random samples from 3 normal populations) and assuming homogeneous variances, test the hypothesis $H:\mu_1 = \mu_2 = \mu_3$. Let $\alpha = 0.10$.

Sample 1	Sample 2	Sample 3
48	72	48
24	24	12
36	48	24
48		

6.64 Assuming homogeneous variances, test the hypothesis that the 4 normal populations, from which the following random samples were obtained, have the same mean. Let $\alpha = 0.025$.

Sample 1	Sample 2	Sample 3	Sample 4
95	45	95	20
50	40	130	55
105	95	15	50
10	65	135	80
60	45	125	

6.65 Using the data of Table 6.16 and assuming homogeneous variances, test the hypothesis $H:\mu_1 = \mu_2 = \mu_3 = \mu_4$. Let $\alpha = 0.01$.

6.66 Using the data of Table 6.10, test the hypothesis $H: \sigma_1^2 = \sigma_2^2 = \sigma_3^2 = \sigma_4^2$. Let $\alpha = 0.05$.

6.67 Letting $\alpha = 0.10$, test the hypothesis of homogeneous variances for each of the following problems: (a) 6.63 and (b) 6.64.

6.68 Using first τ_1 and then τ_2 (as defined in Section 6.17), rework the following problems:

(a) 5.1   (c) 6.5   (e) 6.29   (g) 6.33
(b) 5.5   (d) 6.6   (f) 6.32

6.69 Using τ_d (as defined in Section 6.17), rework the following problems:
(a) 5.20   (c) 6.25   (e) 6.37
(b) 5.23   (d) 6.31

6.70 Using the pseudo F statistic defined in Section 6.17 rework the following problems:
(a) 5.24   (c) 6.45   (e) 6.47
(b) 5.26   (d) 6.46

REFERENCES

1. Bartlett, M. S. 1937. Some examples of statistical methods of research in agriculture and applied biology. *J. Roy. Stat. Soc.* (Suppl.) 4:137.
2. Bowker, A. H.; and Lieberman, G. J. 1972. *Engineering Statistics,* 2nd ed. Prentice-Hall, Englewood Cliffs, N.J.
3. Fry, T. C. 1928. *Probability and Its Engineering Uses.* Van Nostrand, New York.
4. Robertson, W. H. 1960. Programming Fisher's exact method of comparing two percentages. *Technometrics* 2 (Feb.):103–7.

FURTHER READING

See Chapter 4.

CHAPTER 7

REGRESSION ANALYSIS

The methods of analysis studied thus far in this text have been concerned with data on only one characteristic associated with the experimental units. That is, in any given problem we have been working with only one variable; however, as you will realize, many problems involve more than one. Consequently, it is necessary that techniques developed for analyzing multivariate problems be studied. Some of these will be investigated in this chapter.

7.1 INTRODUCTION

In many experimental situations large quantities of information exist in the form of sets of values of several *related* variables. It is natural in such situations to seek ways of expressing the form of the functional relationship. In particular we seek a mathematical function that tells us how the variables are interrelated. In addition it may be of interest to indicate by a quantitative measure the strength of the relationship between the variables. The latter is the topic of *correlation analysis* and is the subject of Chapter 8. The analysis that can be used to examine data and draw conclusions about the functional relationships existing among variables is called *regression analysis*. This is the topic of the present chapter.

We shall be discussing methods useful for obtaining and examining a mathematical equation, the *regression model,* that describes the functional relationship among the variables observed. The methods described here are based on expressing one variable, the *response variable,* as a function of several others, the *independent variables*. Thus we assume the existence of a regression model

$$\eta = \phi(X_1, \ldots, X_p; \theta_1, \ldots, \theta_q) \tag{7.1}$$

where η is the *response* (or *dependent*) variable; $X_1, \ldots, X_p$ are the *independent* variables; $\theta_1, \ldots, \theta_q$ are q unknown parameters; and ϕ is the mathematical function. The function ϕ is commonly called the *regression* or *response function*. An example of a regression model is the linear equation

$$\eta = \theta_0 + \theta_1 X_1 + \theta_2 X_2 \tag{7.2}$$

involving two independent variables X_1 and X_2.

By the independent variables in the regression model we usually mean variables that either can be controlled at fixed values (e.g., operating temperatures, chemical concentrations) or, if they cannot be controlled, are observable with negligible error (e.g., time, outdoor temperature, and wind velocity). With changes in the independent variables there is usually a change in the response variable. How the dependent variable changes as a result of a change in the independent variables is an indication of the effect of the latter variables on the response. This effect is expressed mathematically by the regression function.

In any analysis it is hoped that the postulated regression model represents the exact relationship between the independent variables and the response. However, simply expressed relationships are an exception rather than the rule. Also, our knowledge of the basic mechanism operating among the variables is often not so far advanced as to be able to specify the functional relationship. In any case, it may still be possible to approximate the true relationship by some kind of a regression model. Such a model may not be physically exact, but it may still prove useful as an analytic and predictive tool.

With this in mind a word of warning must be given relative to the interpretation of analyses involving variables in a regression model. This warning is: *just because a particular functional relationship has been assumed and a specific computational procedure followed, do not assume that a causal relationship exists among the variables*. That is, because a function has been found that is a good fit to a set of observed data, we are not necessarily in a position to infer that a change in one variable causes a change in another.

In summary, the only person who can safely say that the basic variables are those used and that the basic mechanism operates in accordance with the selected mathematical function is one well trained in the subject matter field in which the experiment was performed. The regression analysis is only a tool to aid him in the analysis and interpretation of data.

How does one go about choosing a particular regression function as representative of the relationship existing among the variables under investigation? Ideally, an analytical consideration of the phenomenon concerned would lead to a regression function. Often such considerations are impossible or, even if possible, lead to models that are too complex to work with effectively. In such cases, an examination of *scatter diagrams* plotted from the observed data can be employed to suggest some tentative models. While the first method is preferred, the second should not be underrated. If little is known about the basic mechanisms involved, the use of scatter diagrams can be quite helpful.

7.2 METHOD OF LEAST SQUARES

Once we have decided on the type of regression function that represents our concept of the exact relationship existing among the variables, the problem becomes one of choosing a particular model from this family of functions. That is, assuming the regression model is an expression of the true relationship, it is necessary to estimate the parameters of the model. The determination of these estimates and thus the specification of a particular model is called *curve fitting*. This estimation will be based on an observed set of values of the variables in the model.

We are, as in Chapter 5, faced with the question of choosing among several methods of estimating parameters. Our choice should of course provide us with "good" estimates. With this in mind we choose the *method of least squares* which provides us with such results. In fact, if the usual assumption of normality is made, the method of least squares becomes equivalent to the method of maximum likelihood.

To study the method of least squares, assume that we are considering the response variable η, which is related to the variables $X_1, \ldots, X_p$ according to the relationship

$$\eta = \phi(X_1, \ldots, X_p; \theta_1, \ldots, \theta_q) \tag{7.3}$$

Both the form of the function and the values of the parameters must be determined, although in practice the form is usually assumed to be known.

If we could exactly measure the variables in the model, the parameter values could be determined without error. In any experimental situation, even if we can control the independent variables X (or measure them without error), we cannot assume that the response variable is measured without error. Although the true response η for specific inputs $X_1, \ldots, X_p$ is constant, it is not true that the true value η will be observed each time in an experimental situation. In fact, the true value may never actually be observed. There are many uncontrollable factors that can affect the measurements so that the true response η is not observed. These deviations from the true response we call *experimental error*. We also must recognize that the form of the functional relationship ϕ may not be known exactly or that more variables besides $X_1, \ldots, X_p$ may affect the response. We assume, at least initially, that the functional relationship is as stated. Later in the chapter we discuss how the data can be used to test the assumed model.

If we let Y denote the observed variable, our regression model becomes

$$Y = \phi(X_1, \ldots, X_p; \theta_1, \ldots, \theta_q) + \epsilon \tag{7.4}$$

where ϵ denotes the error in the observed value (i.e., ϵ is the error from the true value). Since ϵ is affected by many factors and the value of ϵ will change between observations even if ϕ is constant, it is reasonable to assume that ϵ is a random variable such that its expected value is 0. Further, we assume the variance of ϵ is constant for all values of ϕ and equal to σ_E^2. Thus Y is a random variable with $E[Y] = \phi(X_1, \ldots, X_p; \theta_1, \ldots, \theta_q)$ and $V(Y) = \sigma_E^2$.

Let $\hat{\theta}_1, \ldots, \hat{\theta}_q$ denote the estimators of $\theta_1, \ldots, \theta_q$ and let $\phi(X_1, \ldots, X_p; \hat{\theta}_1, \ldots, \hat{\theta}_q)$ denote the estimator of the regression model. To find these estimators, least squares says to use the values of $\hat{\theta}_j$ that minimize the sum of squared deviations of the observations from the estimated regression model. That is, assuming a sample of n observations, the $\hat{\theta}_j$'s are found by minimizing

$$S(\hat{\theta}_1, \ldots, \hat{\theta}_q) = \sum_{i=1}^{n} [Y_i - \phi(X_{1i}, \ldots, X_{pi}; \hat{\theta}_1, \ldots, \hat{\theta}_q)]^2 = \sum_{i=1}^{n} (Y_i - \hat{Y}_i)^2$$

(7.5)

where $\hat{Y}_i = \phi(X_{1i}, \ldots, X_{pi}; \hat{\theta}_1, \ldots, \hat{\theta}_q)$. This is a familiar calculus problem. To minimize S with respect to the parameters, S is differentiated with respect to each $\hat{\theta}_j$ and each partial derivative is set equal to zero. Thus

$$\frac{\partial S}{\partial \hat{\theta}_j} = 0 \qquad j = 1, \ldots, q$$

(7.6)

is a set of q equations in the q unknowns $\hat{\theta}_j$ ($j = 1, \ldots, q$). Solving this system of equations leads to the estimators $\hat{\theta}_1, \ldots, \hat{\theta}_q$. This assumes that such a solution exists. Further comment on this is delayed until a later section.

Graphically, least squares amounts to minimizing the sum of squares of the vertical distances between the observed values and the value of the estimated regression equation.

Example 7.1

Suppose the model is

$$Y = \theta_1 + \theta_2 X + \epsilon$$

(7.7)

Since the form of ϕ is the equation of a straight line, the estimated regression equation would also be a straight line such as shown in Figure 7.1. The resulting estimated line would be denoted by

$$\hat{Y} = \hat{\theta}_1 + \hat{\theta}_2 X$$

(7.8)

If a more complicated ϕ had been assumed, e.g.,

$$\eta = \theta_1 + \theta_2 X + \theta_3 X^2$$

(7.9)

the problem would be somewhat more complex mathematically, but the principle would be unchanged. The parameters would still be estimated by minimizing the sum of the squares of the vertical deviations about the appropriate curve. This is illustrated in Figure 7.2.

Fig. 7.1–Example of a scatter diagram with a straight line inserted showing the vertical deviations whose sum of squares is to be minimized by the proper choice of straight line.

Fig. 7.2–Example of a scatter diagram with a second-degree polynomial
inserted showing the vertical deviations whose sum of squares
is to be minimized by the proper choice of parabola.

7.3 SIMPLE LINEAR REGRESSION

We first consider in detail the situation involving one independent
variable X_1 in which the postulated model for the response variable Y
is of the form

$$Y = \beta_0 + \beta_1 X_1 + \epsilon \qquad (7.10)$$

The initial problem is to estimate β_0 and β_1 from observed data. To do
this using the method of least squares, we must make certain assumptions about the model.

7.3.1 Assumptions in the Simple Linear Regression Model

Let $(X_{11}, Y_1), \ldots, (X_{1n}, Y_n)$ be the n sample observations. For simple
linear regression we assume:

1. There is a linear relationship between the true response and the independent variable, i.e.

$$\eta = \beta_0 + \beta_1 X_1 \qquad (7.11)$$

   or, the model for the ith observation of the dependent variable Y is

$$Y_i = \beta_0 + \beta_1 X_{1i} + \epsilon_i \qquad i = 1, \ldots, n \qquad (7.12)$$

2. The ϵ_i's are random variables with the properties
   a. $E[\epsilon_i] = 0$ for all $i = 1, \ldots, n$.
   b. $V(\epsilon_i) = \sigma_E^2$, a constant for all values of X_1.
   c. ϵ_i's are statistically uncorrelated; i.e., $E[\epsilon_i \epsilon_{i'}] = 0$ for $i \neq i'$.
3. The observed values of X_1 are measured without error.

Occasionally, our data will consist of several observations of the dependent variable for each value of the independent variable. The sample is

$$(X_{11}, Y_{11}), (X_{11}, Y_{12}), \ldots, (X_{11}, Y_{1n_1}), (X_{12}, Y_{21}), \ldots, (X_{1k}, Y_{kn_k})$$

and the model becomes

$$Y_{ij} = \beta_0 + \beta_1 X_{1i} + \epsilon_{ij} \qquad i = 1, \ldots, k; j = 1, \ldots, n_i \qquad (7.13)$$

where now $n = \sum_{i=1}^{k} n_i$. We assume $E[\epsilon_{ij}] = 0$ and $V(\epsilon_{ij}) = \sigma_E^2$ for all i
and j. Also, the ϵ_{ij}'s are uncorrelated; i.e., $E[\epsilon_{ij} \epsilon_{i'j'}] = 0$, for $i \neq i'$ or
$j \neq j'$.

7.3.2 Least Squares Estimation

Using the method of least squares, the estimates b_0 and b_1 are found by minimizing

$$S(b_0, b_1) = \sum_{i=1}^{n} (Y_i - b_0 - b_1 X_{1i})^2 \tag{7.14}$$

Thus the estimates b_0 and b_1 are solutions of the normal equations,

$$b_0 n + b_1 \sum_{i=1}^{n} X_{1i} = \sum_{i=1}^{n} Y_i$$

$$b_0 \sum_{i=1}^{n} X_{1i} + b_1 \sum_{i=1}^{n} X_{1i}^2 = \sum_{i=1}^{n} X_{1i} Y_i \tag{7.15}$$

The estimate of β_1 is

$$b_1 = \frac{\sum_{i=1}^{n} (X_{1i} - \overline{X}_{1.})(Y_i - \overline{Y})}{\sum_{i=1}^{n} (X_{1i} - \overline{X}_{1.})^2} = \frac{\sum_{i=1}^{n} X_{1i} Y_i - \left(\sum_{i=1}^{n} X_{1i}\right)\left(\sum_{i=1}^{n} Y_i\right)\Big/ n}{\sum_{i=1}^{n} X_{1i}^2 - \left(\sum_{i=1}^{n} X_{1i}\right)^2 \Big/ n} \tag{7.16}$$

where $\overline{X}_{1.}$ denotes the mean $\sum_{i=1}^{n} X_{1i}/n$. Alternatively, we can write b_1 as

$$b_1 = \sum_{i=1}^{n} x_{1i} y_i \Big/ \sum_{i=1}^{n} x_{1i}^2 \tag{7.17}$$

where $x_{1i} = X_{1i} - \overline{X}_{1.}$ and $y_i = Y_i - \overline{Y}$. The estimate of β_0 is

$$b_0 = \overline{Y} - b_1 \overline{X}_{1.} \tag{7.18}$$

Using the estimates b_0 and b_1 in place of β_0 and β_1 respectively in the regression equation leads to the *estimated regression equation*

$$\hat{Y} = b_0 + b_1 X_1 \tag{7.19}$$

Further, under the assumptions stated in Section 7.3.1, the best estimate of the variance σ_E^2 of ϵ is

$$s_E^2 = \sum_{i=1}^{n} (Y_i - \hat{Y}_i)^2/(n - 2) \tag{7.20}$$

where $\hat{Y}_i$ is the value of the estimated regression equation at the ith value of X_1.

7.3.3 An Example

To illustrate the methods of simple linear regression, consider the data in Table 7.1. Following the methods outlined in the preceding

TABLE 7.1–Schopper-Riegler Freeness Test of Paper Pulp during Beating

Hours of Beating (X_1)	Schopper-Riegler (*degrees*) (Y)	Hours of Beating (X_1)	Schopper-Riegler (*degrees*) (Y)
1	17	8	64
2	21	9	80
3	22	10	86
4	27	11	88
5	36	12	92
6	49	13	94
7	56		

Source: Davies (5), p. 161.

sections, we obtain the normal equations

$$13b_0 + 91b_1 = 732 \qquad 91b_0 + 819b_1 = 6485$$

which yield $\hat{Y} = 3.962 + 7.478X_1$ as the estimated regression equation. Also, the estimate of the variance σ_E^2 is $s_E^2 = 28.83$.

The estimated function is pictured in Figure 7.3. Examination of Figure 7.3 suggests that perhaps a cubic equation (i.e., a third-degree polynomial) would have been a better model to fit to the observed data. The fact that our postulated model may be inadequate could lead to an

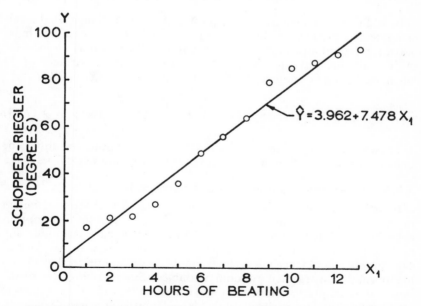

Fig. 7.3–Plot of data in Table 7.1 with the least squares line inserted.

estimate s_E^2 that is badly inflated. Discussion of this and the appropriateness of the model and possible tests of the goodness of the model will be deferred until a later section.

7.3.4 Estimates of Errors in Simple Linear Regression

In Eq. (7.20) we have already estimated the variance σ_E^2 of the random error term ϵ in the regression model. Since the estimates b_0, b_1, and $\hat{Y}$ are functions of the observed Y_i, they are themselves statistics and hence are random variables. The estimates of the variances of these statistics are given without derivation.

1. Estimated variance of b_1

$$s_{b_1}^2 = s_E^2 / \sum_{i=1}^{n} (X_{1i} - \overline{X}_{1.})^2 \tag{7.21}$$

2. Estimated variance of b_0

$$s_{b_0}^2 = s_E^2 \left[(1/n) + (\overline{X}_{1.})^2 / \sum_{i=1}^{n} (X_{1i} - \overline{X}_{1.})^2 \right] \tag{7.22}$$

3. Estimated variance of $\hat{Y}$, when used as an estimate of η (the expected value of Y at $X_1 = X_1'$),

$$s_{\hat{Y}}^2 = s_E^2 \left[(1/n) + (X_1' - \overline{X}_{1.})^2 / \sum_{i=1}^{n} (X_{1i} - \overline{X}_{1.})^2 \right] \tag{7.23}$$

In addition to $\hat{Y}$ being used as an estimate of η, $\hat{Y}$ is also used as an estimate of a predicted value of an individual Y. In this case one must allow for the unknown random error, and this shows up as an increase in the estimated variance of $\hat{Y}$. That is, the estimated variance of $\hat{Y}$, when used as an estimate of the predicted value of Y at $X_1 = X_1'$ is

$$s_E^2 \left[1 + (1/n) + (X_1' - \overline{X}_{1.})^2 / \sum_{i=1}^{n} (X_{1i} - \overline{X}_{1.})^2 \right] \tag{7.24}$$

These estimated variances will be used to test hypotheses about, or to provide interval estimates of, various unknown parameters.

7.3.5 Estimation in Simple Linear Regression

In order to provide more than just point estimates of the parameters of the regression equation, we must make one further assumption beyond those listed in Section 7.3.1. Up to now we have not made any assumption about the type of distribution the random errors follow. To construct confidence intervals, we make the assumption that the ϵ's are normally distributed random variables. Summarizing the distributional assumptions about the ϵ's, we assume that the ϵ's are normal independent random variables with $E[\epsilon] = 0$ and $V(\epsilon) = \sigma_E^2$, a constant. [**Note:** These assumptions are summarized in $\epsilon \sim NI(0, \sigma_E^2)$.] With this assumption we are assuming that, for a given X_{1i}, Y_i is a normal ran-

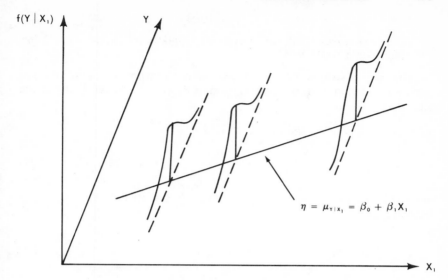

Fig. 7.4–Simple linear regression with $\epsilon \sim NI(0, \sigma_E^2)$.

dom variable with $E[Y_i] = \beta_0 + \beta_1 X_{1i}$ and $V(Y_i) = \sigma_E^2$. That is, the regression equation is a straight line connecting the means of many normal populations. This is illustrated in Figure 7.4.

With this added assumption of the normality of the ϵ's we are led to the distributional facts that b_0, b_1, and $\hat{Y}$ are normal random variables with means, β_0, β_1, and η respectively and variances as given in Eqs. (7.21)–(7.23) with σ_E^2 replacing s_E^2. We now use the techniques of Section 5.4 to construct confidence intervals for the unknown parameters.

In most linear regression problems the parameter of greatest importance is generally the slope β_1. Of course, b_1 is a point estimate of β_1. A 100γ percent confidence interval for β_1 is given by the limits

$$\left.\begin{matrix} L \\ U \end{matrix}\right\} = b_1 \mp t_{[(1+\gamma)/2](n-2)} s_{b_1} \qquad (7.25)$$

where s_{b_1} is defined in Eq. (7.21).

For the intercept β_0, b_0 provides a point estimate and the 100γ percent confidence interval for β_0 is given by the limits

$$\left.\begin{matrix} L \\ U \end{matrix}\right\} = b_0 \mp t_{[(1+\gamma)/2](n-2)} s_{b_0} \qquad (7.26)$$

where s_{b_0} is defined in Eq. (7.22).

Example 7.2

Using the data in Table 7.1 and the results in the previous sections, it may be verified that

$$s_E^2 = 28.83 \qquad s_{b_1}^2 = 0.1584 \qquad s_{b_0}^2 = 9.977$$

Thus 95 percent confidence intervals for β_0 and β_1 are respectively $(-2.989, 10.913)$ and $(6.602, 8.354)$.

It is also possible that we might wish to determine a confidence region for the simultaneous estimation of β_0 and β_1. Making use of the fact that

$$Q = \left[n(b_0 - \beta_0)^2 + 2n\overline{X}_{1.}(b_0 - \beta_0)(b_1 - \beta_1) \right.$$
$$\left. + (b_1 - \beta_1)^2 \sum_{i=1}^{n} X_{1i}^2 \right] \Big/ \sigma_E^2 \qquad (7.27)$$

is distributed as $\chi_{(2)}^2$ and that $(n - 2)s_E^2/\sigma_E^2$ is distributed as $\chi_{(n-2)}^2$, it is seen that

$$F = \frac{n(b_0 - \beta_0)^2 + 2n\overline{X}_{1.}(b_0 - \beta_0)(b_1 - \beta_1) + (b_1 - \beta_1)^2 \sum_{i=1}^{n} X_{1i}^2}{2s_E^2}$$
$$(7.28)$$

is distributed as F with $\nu_1 = 2$ and $\nu_2 = n - 2$ degrees of freedom. The boundary of the 100γ percent confidence region is then determined by solving

$$\left[n(b_0 - \beta_0)^2 + 2n\overline{X}_{1.}(b_0 - \beta_0)(b_1 - \beta_1) + (b_1 - \beta_1)^2 \sum_{i=1}^{n} X_{1i}^2 \right] \Big/ 2s_E^2$$
$$= F_{\gamma(2,n-2)} \qquad (7.29)$$

for β_0 and β_1.

Another estimation problem of importance in simple linear regression is associated with $\hat{Y} = b_0 + b_1 X_1$. As you will remember, $\hat{Y} = b_0 + b_1 X_1$ is an estimate of $\mu_{Y|X_1} = \eta = \beta_0 + \beta_1 X_1$. Further, by the assumptions of Section 7.3.1, η is the mean of a normal population. Thus it should not be surprising that a 100γ percent confidence interval estimate of η is provided by

$$\left.\begin{matrix} L \\ U \end{matrix}\right\} = \hat{Y} \mp t_{[(1+\gamma)/2](n-2)} s_{\hat{Y}} \qquad (7.30)$$

where $s_{\hat{Y}}$ is defined in Eq. (7.23).

It should be noted that $\hat{Y} = b_0 + b_1 X_1$ is also a predictor of $Y = \beta_0 + \beta_1 X_1 + \epsilon$. That is, $\hat{Y}$ can also be used to predict an individual Y value associated with a given X_1 value. (**Note:** This is in contrast to the preceding paragraph where $\hat{Y}$ was used to estimate the mean of a normal population.) When $\hat{Y}$ is used to predict an individual value rather than a mean value, a 100γ percent *prediction interval* is provided by

$$\left.\begin{array}{c} L' \\ U' \end{array}\right\} = \hat{Y} \mp t_{[(1+\gamma)/2](n-2)} s_{\hat{Y}} \tag{7.31}$$

where $s_{\hat{Y}}$ is defined in Eq. (7.24).

The nature of the confidence and prediction intervals specified by Eqs. (7.30) and (7.31) is illustrated in Figure 7.5. The most noticeable feature of Figure 7.5 is the curvature of the confidence and prediction limits. That is, our estimates are most precise at the average value of X_1 and may be almost useless at values of X_1 far removed from $\overline{X}_{1.}$. By "almost useless," we mean that the confidence and prediction intervals may turn out to be so wide as to render them of little value. To state the preceding conclusion in a positive rather than a negative fashion, any estimate of the mean value of Y for a given X_1 or any prediction about an individual Y associated with a given X_1 will be most meaningful for those values of X_1 near $\overline{X}_{1.}$.

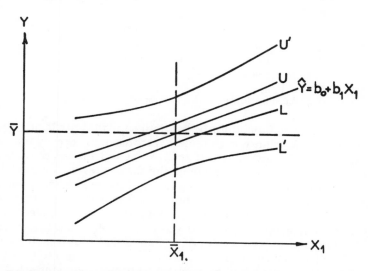

Fig. 7.5–Graphical representation of the confidence and prediction intervals specified by Eqs. (7.30) and (7.31).

As a corollary to the preceding paragraph, it is clear that if the estimation of β_0 is of prime importance, the values of X_1 should be selected (prior to collecting data) so that $\overline{X}_{1.} = 0$. The reason for this statement should be clear. By so choosing the X_1 values, the narrowest confidence and/or prediction interval will occur at $X_1 = 0$, and at this value of X_1, $\hat{Y} = b_0$.

Following up the line of thought in the preceding paragraph, one might wonder if choosing the X_1 values in accordance with the expressed recommendation is best for all purposes. For example, if the estimation of β_1 rather than β_0 is of prime importance, should the values of the controlled variable still be selected so that $\overline{X}_{1.} = 0$? The answer is, "Definitely not." Then if our only interest lies in β_1 (and it frequently does), how should the values of X_1 be chosen? In this case the appropriate recommendation is: select two values of X_1 (as far apart as reasonable) and obtain random observations on the Y variable at only those two X_1 values. By following this rule, the standard error of b_1 will be made as small as possible subject to the (uncontrollable) magnitude of s_E. In other words, if we proceed as indicated, the confidence interval for β_1 should be kept "small." (**Note:** The reader can verify heuristically the wisdom of this approach by noting that widely divergent X_1 values will increase $\sum_{i=1}^{n} (X_{1i} - \overline{X}_{1.})^2 = \sum_{i=1}^{n} x_{1i}^2$, the denominator of Eq. (7.21), and thus decrease the size of $s_{b_1}^2$.)

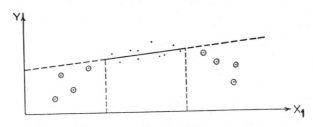

Fig. 7.6–Illustration of the danger of extrapolation.

Another fact that should not be overlooked is that predicting values of Y for a given X_1 value is even more hazardous than already indicated if we attempt such a procedure for an X_1 value outside the range of the chosen values of X_1 used in obtaining the sample regression line. That is, *extrapolation* beyond the observed range of the independent variable is very risky unless we are reasonably certain that the same regression function does exist over a wider range of X_1 values than we have in our sample. A simple illustration will suffice to point out the possible trouble. Suppose we have values of X_1 and Y that plot (see dots) as in Figure 7.6. In the given range of X_1, a straight line appears to be a good fit to the data and we might be tempted to project our regression line farther in both directions. However, it is entirely possible that if we had chosen a wider range of X_1 values and observed the associated Y values (see circled dots), a second degree polynomial might have been indicated as the true form of the regression function rather than the straight line we have drawn. You can readily see that predicting values of Y using an extrapolation of the straight line could lead to serious errors. Therefore the research worker is advised to act with caution whenever he makes predictions that involve going outside the observed range of the independent variable.

Although only two-sided confidence and prediction limits have been discussed in this section, the reader will realize that one-sided limits should be used if the problem calls for such a procedure. If only an upper or lower limit is required, the researcher should make the same changes in procedure as outlined in Chapter 5 but continue to use the statistics specified in this section.

7.3.6 Tests of Hypotheses in Simple Linear Regression

In addition to estimating the parameters in a regression model, we may also be interested in testing hypotheses about the values of the parameters. In particular, suppose we are interested in testing whether the slope β_1 is some hypothesized value (say β_1'). That is, we wish to test $H:\beta_1 = \beta_1'$ against the alternative hypothesis $A:\beta_1 \neq \beta_1'$. Using the distributional facts given in Section 7.3.5, the appropriate test statistic is

$$t = (b_1 - \beta_1')/s_{b_1} \tag{7.32}$$

where s_{b_1} is defined in Eq. (7.21). Since t has a t distribution with $(n - 2)$ degrees of freedom if H is true, the hypothesis H is rejected if $|t| \geq t_{(1-\alpha/2)(n-2)}$. Frequently, the hypothesized value is $\beta_1' = 0$, which reflects the hypothesis that Y is independent of X_1 (in a linear sense). That is, it is hypothesized that X_1 is of no value in predicting Y if a linear approximation is used. In this case when testing $H:\beta_1 = 0$ versus $A:\beta_1 \neq 0$, one can use the t statistic given in Eq. (7.32) with $\beta_1' = 0$. However, an F statistic can also be used, and this will be discussed in the next section.

Other test procedures in simple linear regression are concerned with such hypotheses as: (1) $H:\beta_0 = \beta_0'$, (2) $H:\mu_{Y|X_1=X_{10}} = \mu_0$, and (3) $H:\beta_0 = \beta_0'$ and $\beta_1 = \beta_1'$. These tests are summarized in Table 7.2.

Again we shall do no more than remind the reader of the possibility of one-sided test procedures. By this time the procedure in such cases should be obvious once the two-tailed tests have been specified.

TABLE 7.2–Summary of Test Procedures in Simple Linear Regression

Hypothesis(H)	Test Statistic	Critical Region for a Two-sided Alternative
$\beta_0 = \beta_0'$	$t = (b_0 - \beta_0')/s_{b_0}$	$|t| \geq t_{(1-\alpha/2)(n-2)}$
$\beta_1 = \beta_1'$	$t = (b_1 - \beta_1')/s_{b_1}$	$|t| \geq t_{(1-\alpha/2)(n-2)}$
$\mu_{Y|X_1=X_{10}} = \mu_0$	$t = (b_0 + b_1 X_{10} - \mu_0)/s_{\hat{Y}}$	$|t| \geq t_{(1-\alpha/2)(n-2)}$
$\beta_0 = \beta_0'$ and $\beta_1 = \beta_1'$	$F = \left[n(b_0 - \beta_0')^2 \right.$ $+ 2n\overline{X}_{1.}(b_0 - \beta_0')$ $\cdot (b_1 - \beta_1') + (b_1 - \beta_1')^2$ $\left. \cdot \sum_{i=1}^{n} X_{1i}^2 \right] \bigg/ 2s_E^2$	$F \geq F_{(1-\alpha)[2,(n-2)]}$

Example 7.3

Referring to Example 7.2, test the hypothesis $H:\beta_1 = 0$. Using $\alpha = 0.01$, the tabled value is $t_{0.995(11)} = 3.106$. The calculated value of the t statistic given in Eq. (7.32) is $t = (7.478 - 0)/0.398 = 18.788$. Therefore, we reject the hypothesis and conclude $\beta_1 \neq 0$.

7.3.7 Partitioning the Sum of Squares of the Dependent Variable and the Test of $H:\beta_1 = 0$ vs. $A:\beta_1 \neq 0$

As mentioned in the previous section, a frequent hypothesis tested in regression analysis is $H:\beta_1 = 0$ versus $A:\beta_1 \neq 0$. In terms of the regression model, we can restate the hypothesis as a comparison of two models; i.e., $H:\eta = \beta_0$ and $A:\eta = \beta_0 + \beta_1 X_1$. To develop the test statistic, we consider an analysis of variance under H and A.

Under H the model for the ith observation of Y in the sample is

$$Y_i = \beta_0 + \epsilon_i' \qquad i = 1, \ldots, n \qquad (7.33)$$

Based on the sample, we can express each Y_i as a function of the sample mean and the deviation from the mean. That is,

$$Y_i = \overline{Y} + (Y_i - \overline{Y}) \qquad (7.34)$$

where $\overline{Y}$ is an estimate of β_0 and $Y_i - \overline{Y}$ estimates ϵ_i'. Expressing each observation in this form, we can partition the variation in the data as measured by the total sum of squares of the sample observations into the variation attributable to each of the terms β_0 and ϵ_i' in the model. Squaring both sides of Eq. (7.34) and summing over the n observations, we have the identity

$$\sum_{i=1}^{n} Y_i^2 = n(\overline{Y})^2 + \sum_{i=1}^{n} (Y_i - \overline{Y})^2 \qquad (7.35)$$

since the sum of cross products $\sum_{i=1}^{n} \overline{Y}(Y_i - \overline{Y}) = 0$. In words, the identity is

$$\text{Total } SS = SS \text{ due to } \beta_0 + SS \text{ due to } \epsilon'$$

$$= SS(\beta_0) + RSS(H) \qquad (7.36)$$

where SS = sum of squares, and RSS = residual (error) sum of squares.
Under A the model for the ith observation is

$$Y_i = \beta_0 + \beta_1 X_{1i} + \epsilon_i \qquad i = 1, \ldots, n \qquad (7.37)$$

Similarly, we can express Y_i as

$$Y_i = \hat{Y}_i + (Y_i - \hat{Y}_i) \qquad (7.38)$$

where $\hat{Y}_i = b_0 + b_1 X_{1i}$ is our estimate of $\beta_0 + \beta_1 X_{1i}$, and $Y_i - \hat{Y}_i$ estimates ϵ_i. From Eq. (7.38) we get the analysis of variance identity

$$\sum_{i=1}^{n} Y_i^2 = \sum_{i=1}^{n} \hat{Y}_i^2 + \sum_{i=1}^{n} (Y_i - \hat{Y}_i)^2 \qquad (7.39)$$

TABLE 7.3–General Analysis of Variance for Simple Linear Regression

Source of Variation	Degrees of Freedom	Sum of Squares	Mean Square	Expected Mean Square
Due to β_0	1	$SS(\beta_0)$	...	...
Due to $\beta_1 \mid \beta_0$	1	$SS(\beta_1 \mid \beta_0)$	$SS(\beta_1 \mid \beta_0)$	$\sigma_E^2 + \beta_1^2 \sum_{i=1}^{n} (X_{1i} - \bar{X}_{1.})^2$
Residual	$n - 2$	$RSS(A) = E_{yy}$	s_E^2	σ_E^2
Total	n	$\sum_{i=1}^{n} Y_i^2$	...	...

In words,

$$\text{Total } SS = SS(\beta_0, \beta_1) + RSS(A) \tag{7.40}$$

Comparing Eqs. (7.36) and (7.40), the increase $SS(\beta_0, \beta_1) - SS(\beta_0)$ in the regression sum of squares [equivalently, the reduction $RSS(H) - RSS(A)$ in the residual sum of squares] is the variation accounted for by including the term $\beta_1 X_1$ in the model. This additional sum of squares due to β_1 after fitting β_0 we denote $SS(\beta_1 \mid \beta_0)$. From Eqs. (7.35) and (7.39)

$$SS(\beta_1 \mid \beta_0) = \sum_{i=1}^{n} \hat{Y}_i^2 - n(\bar{Y})^2 = \sum_{i=1}^{n} (\hat{Y}_i - \bar{Y})^2 \tag{7.41}$$

These computations are generally summarized in an analysis of variance table such as Table 7.3. The entries in the mean square column are the corresponding sums of squares divided by the appropriate degrees of freedom. The last column contains the expected values of the corresponding mean squares based on the model in Eq. (7.37). Convenient computing formulas for the sums of squares are

$$SS(\beta_0) = \left(\sum_{i=1}^{n} Y_i\right)^2 \Big/ n = n(\bar{Y})^2 \tag{7.42}$$

$$SS(\beta_1 \mid \beta_0) = b_1 \sum_{i=1}^{n} (X_{1i} - \bar{X}_{1.})(Y_i - \bar{Y}) \tag{7.43}$$

and

$$E_{yy} = RSS(A) = \sum_{i=1}^{n} (Y_i - \hat{Y}_i)^2$$

$$= \sum_{i=1}^{n} Y_i^2 - SS(\beta_0) - SS(\beta_1 \mid \beta_0) \tag{7.44}$$

To test the hypothesis $H: \beta_1 = 0$ versus $A: \beta_1 \neq 0$, we see from the expected mean squares in Table 7.3 that under H both $SS(\beta_1 \mid \beta_0)$ and s_E^2, the residual mean square, estimate σ_E^2. Thus the test statistic to use is

$$F = SS(\beta_1 \mid \beta_0)/s_E^2 \qquad (7.45)$$

It can be demonstrated that under H, F follows an F distribution with $\nu_1 = 1$ and $\nu_2 = n - 2$ degrees of freedom. Thus for significance level α, the hypothesis H is rejected if $F \geq F_{(1-\alpha)(1,n-2)}$.

Example 7.4

Again referring to the example given in Section 7.3.3, the analysis of variance appropriate for testing the hypothesis $H:\beta_1 = 0$ is given in Table 7.4. The value of the F statistic given in Eq. (7.45) is $F = 10{,}177.59/28.83 = 353.02$. Since this exceeds $F_{0.99(1,11)} = 9.65$, the hypothesis is rejected. This is the same conclusion reached in Example 7.3.

TABLE 7.4–Analysis of Variance of the Schopper-Riegler Data of Table 7.1

Source of Variation	Degrees of Freedom	Sum of Squares	Mean Square
Due to β_0	1	41,217.23	41,217.23
Due to $\beta_1 \mid \beta_0$	1	10,177.59	10,177.59
Residual	11	317.18	28.83
Total	13	51,712.00	. . .

7.3.8 Inverse Prediction in Simple Linear Regression

The equation $\hat{Y} = b_0 + b_1 X_1$ may sometimes be used to estimate the unknown value of X_1 associated with an observed Y value. For example, suppose that in addition to the data of Table 7.1, we have a Schopper-Riegler reading of $Y = 60$, but the hours of beating X_1 are unknown. How shall this unknown value be estimated? The procedure is as follows. Compute

$$\hat{X}_1 = (Y_0 - b_0)/b_1 \qquad (7.46)$$

where Y_0 is the observed value of Y for which we desire to estimate the associated X_1 value. A 100γ percent confidence interval for the true but unknown X_1 value is defined by

$$\left.\begin{matrix} L \\ U \end{matrix}\right\} = \bar{X}_{1.} + \frac{b_1(Y_0 - \bar{Y})}{D} \mp \frac{ts_E}{D}\left[B(Y_0 - \bar{Y})^2 + D\left(\frac{n+1}{n}\right)\right]^{1/2} \qquad (7.47)$$

where

$$B = 1\left/\sum_{i=1}^{n}(X_{1i} - \bar{X}_{1.})^2\right. = 1\left/\sum_{i=1}^{n}x_{1i}^2\right. \qquad (7.48)$$

$$D = b_1^2 - t^2 s_E^2 B = b_1^2 - t^2 s_{b_1}^2 \qquad (7.49)$$

and

$$t = t_{[(1+\gamma)/2](n-2)} \qquad (7.50)$$

If, as is frequently the case, one has several (say m) values of Y associated with the unknown X_1, Eqs. (7.46) and (7.47) are modified to read

$$\hat{X}_1 = (\overline{Y}_0 - b_0)/b_1 \qquad (7.51)$$

and

$$\left.\begin{matrix} L \\ U \end{matrix}\right\} = \overline{X}_1. + \frac{b_1(\overline{Y}_0 - \overline{Y})}{D} \mp \frac{ts'_E}{D}\left[B(\overline{Y}_0 - \overline{Y})^2 + D\left(\frac{n+m}{nm}\right)\right]^{1/2} \quad (7.52)$$

where

$$\overline{Y}_0 = \sum_{i=1}^{m} Y_{0i}/m \qquad (7.53)$$

$$(s'_E)^2 = \frac{(n-2)s_E^2 + \sum_{i=1}^{m} (Y_{0i} - \overline{Y}_0)^2}{n+m-3} \qquad (7.54)$$

$$t = t_{[(1+\gamma)/2](n+m-3)} \qquad (7.55)$$

and B and D are the same as before. Since in practice m is usually quite small relative to n, the computational labor may be reduced materially by using s_E^2 rather than $(s'_E)^2$. This leads to an approximate solution, but one that is sufficiently accurate for most situations.

Example 7.5

Consider the problem posed at the beginning of this section. Using Eq. (7.46) and the results of Section 7.3.3, we obtain $\hat{X}_1 = 7.494$. Using Eqs. (7.47)–(7.50), a 95 percent confidence interval estimate is determined to be (5.85, 9.15).

7.4 MULTIPLE LINEAR REGRESSION

In the previous section we assumed a simple linear model involving only one independent variable. We now consider models involving several independent variables. We assume the model involves a linear function of the unknown parameters, thus it can be written in the form

$$\eta = \beta_0 + \beta_1 X_1 + \beta_2 X_2 + \cdots + \beta_p X_p \qquad (7.56)$$

or, if we have n sets of observations $X_{1i}, X_{2i}, \ldots, X_{pi}, Y_i$ ($i = 1, 2, \ldots, n$), the model for the ith observation is

$$Y_i = \beta_0 + \beta_1 X_{1i} + \beta_2 X_{2i} + \cdots + \beta_p X_{pi} + \epsilon_i \qquad i = 1, \ldots, n \qquad (7.57)$$

Again, if we have repeated observations of Y at the same values of the independent variables, we will alter the model notation to allow for this. For example, if we have n_i values of Y at the ith set of values of the X's, the model is written

$$Y_{ij} = \beta_0 + \beta_1 X_{1i} + \cdots + \beta_p X_{pi} + \epsilon_{ij} \qquad i = 1, \ldots, k; j = 1, \ldots, n_i$$
$$(7.58)$$

Finally, we make the usual assumptions about the ϵ's; i.e., we assume that the ϵ_i's are independent random variables with zero mean and constant variance σ_E^2. We also assume that the ϵ_i's are normally distributed when necessary for estimation and testing.

The general model as given in Eqs. (7.56)–(7.58) really encompasses many models. Some examples follow:

$$\eta = \beta_0 + \beta_1 X + \beta_2 X^2 + \cdots + \beta_p X^p \qquad (7.59)$$

is a pth order polynomial in one independent variable X. The simple linear model of Section 7.3 is a special case of such a model with $p = 1$.

$$\eta = \beta_0 + \beta_1 X_1 + \beta_2 X_2 + \beta_3 X_1^2 + \beta_4 X_2^2 + \beta_5 X_1 X_2 \qquad (7.60)$$

is a 2nd-degree polynomial in two independent variables.

$$\eta = \beta_0 + \beta_1 \ln X + \beta_2 X_2 \qquad (7.61)$$

where $X_1 = \ln X$.

$$\eta = \beta_0 + \beta_1 / X \qquad (7.62)$$

where $X_1 = 1/X$. Several special models are dealt with in more detail in later sections.

7.4.1 Estimation

As in the case of simple linear regression, we use the method of least squares to estimate the values of the unknown parameters. Using the model as given in Eq. (7.57), the least squares estimates are those values of $b_0, \ldots, b_p$ such that

$$SS(b_0, \ldots, b_p) = \sum_{i=1}^{n} (Y_i - b_0 - b_1 X_{1i} - \cdots - b_p X_{pi})^2 \qquad (7.63)$$

is minimum. Differentiating as before with respect to the b's, the normal equations are

$$b_0 n + b_1 \sum_{i=1}^{n} X_{1i} + \cdots + b_p \sum_{i=1}^{n} X_{pi} = \sum_{i=1}^{n} Y_i$$

$$b_0 \sum_{i=1}^{n} X_{1i} + b_1 \sum_{i=1}^{n} X_{1i}^2 + \cdots + b_p \sum_{i=1}^{n} X_{1i} X_{pi} = \sum_{i=1}^{n} X_{1i} Y_i$$

$$\vdots$$

$$b_0 \sum_{i=1}^{n} X_{pi} + b_1 \sum_{i=1}^{n} X_{1i} X_{pi} + \cdots + b_p \sum_{i=1}^{n} X_{pi}^2 = \sum_{i=1}^{n} X_{pi} Y_i$$
$$(7.64)$$

We assume the values of the X's are such that the normal equations have a unique solution. Thus we can solve for the estimates b_0, b_1, $\ldots$, b_p.

With regard to the actual mechanics of solving the normal equations, if the model only includes a few terms, we could solve the normal equations as a system of simultaneous linear equations in the b's. The model and normal equations can also be set up using matrices. Solving for the b's in terms of matrices involves inverting a $(p + 1) \times (p + 1)$ matrix. This is discussed in Section 7.4.4. Several methods for solving a set of simultaneous linear equations (or for inverting matrices), appropriate for use on a desk calculator or for programming on a computer, appear in the literature. We present one such method, the abbreviated Doolittle method, in Section 7.4.5. Perhaps the most common way to solve for the values of the b's is through the use of a regression program on a digital computer. In any case, we assume that the estimates b_0, b_1, $\ldots$, b_p are available.

Given the estimates $b_0, b_1, \ldots, b_p$, the estimated regression equation is

$$\hat{Y} = b_0 + b_1 X_1 + \cdots + b_p X_p \qquad (7.65)$$

One additional estimate is the estimate of the variance σ_E^2. As in simple linear regression the estimate is the residual mean square

$$s_E^2 = \sum_{i=1}^{n} (Y_i - \hat{Y}_i)^2 / (n - p - 1) \qquad (7.66)$$

which has $n - p - 1$ degrees of freedom associated with it.

Example 7.6

Consider the data given in Table 7.5. The model proposed is

$$Y_i = \beta_0 + \beta_1 X_{1i} + \beta_2 X_{2i} + \beta_3 X_{3i} + \beta_4 X_{4i} + \epsilon_i \qquad i = 1, \ldots, 32$$

The normal equations for estimating the β's are

$$
\begin{aligned}
32b_0 + 1{,}256b_1 + 133.8b_2 + 7{,}728b_3 + 10{,}627b_4 &= 629.1 \\
1{,}256b_0 + 50{,}282.5b_1 + 5{,}535.68b_2 + 298{,}732.1b_3 + 413{,}189.7b_4 &= 25{,}153.59 \\
133.8b_0 + 5{,}535.68b_1 + 772.22b_2 + 29{,}549.7b_3 + 42{,}746b_4 &= 2{,}964.88 \\
7{,}728b_0 + 298{,}732.1b_1 + 29{,}549.7b_2 + 1{,}910{,}002b_3 + 2{,}599{,}887b_4 &= 147{,}996.6 \\
10{,}627b_0 + 413{,}189.7b_1 + 42{,}746b_2 + 2{,}599{,}887b_3 + 3{,}680{,}003b_4 &= 225{,}418.0
\end{aligned}
$$

Rather than solving these equations by hand, a regression program on the computer was used. The estimates are $b_0 = -6.821$, $b_1 = 0.2272$, $b_2 = 0.5537$, $b_3 = -0.1495$, and $b_4 = 0.1547$. Thus $\hat{Y} = -6.821 + 0.2272 X_1 + 0.5537 X_2 - 0.1495 X_3 + 0.1547 X_4$ is the estimated regression equation. Also, $s_E^2 = 4.9927$.

7.4.2 Estimates of Errors and Confidence Intervals

Under our assumed model we can evaluate the variances and covariances of the b's as a function of the X's and the variance σ_E^2. We

TABLE 7.5–Crude Oil Properties and Actual Gasoline Yields

Crude Oil Gravity, °API (X_1)	Crude Oil Vapor Pressure, PSIA (X_2)	Crude Oil ASTM 10% Point, °F (X_3)	Gasoline End Point, °F (X_4)	Gasoline Yield Percent of Crude Oil (Y)
38.4	6.1	220	235	6.9
40.3	4.8	231	307	14.4
40.0	6.1	217	212	7.4
31.8	0.2	316	365	8.5
40.8	3.5	210	218	8.0
41.3	1.8	267	235	2.8
38.1	1.2	274	285	5.0
50.8	8.6	190	205	12.2
32.2	5.2	236	267	10.0
38.4	6.1	220	300	15.2
40.3	4.8	231	367	26.8
32.2	2.4	284	351	14.0
31.8	0.2	316	379	14.7
41.3	1.8	267	275	6.4
38.1	1.2	274	365	17.6
50.8	8.6	190	275	22.3
32.2	5.2	236	360	24.8
38.4	6.1	220	365	26.0
40.3	4.8	231	395	34.9
40.0	6.1	217	272	18.2
32.2	2.4	284	424	23.2
31.8	0.2	316	428	18.0
40.8	3.5	210	273	13.1
41.3	1.8	267	358	16.1
38.1	1.2	274	444	32.1
50.8	8.6	190	345	34.7
32.2	5.2	236	402	31.7
38.4	6.1	220	410	33.6
40.0	6.1	217	340	30.4
40.8	3.5	210	347	26.6
41.3	1.8	267	416	27.8
50.8	8.6	190	407	45.7

Source: Prater (11), pp. 236–38.

denote the variance of b_h $(h = 0, 1, \ldots, p)$ as

$$V(b_h) = c_{hh}\sigma_E^2 \tag{7.67}$$

and the covariance of b_h and $b_{h'}$ $(h \neq h')$ as

$$\text{Cov}(b_h, b_{h'}) = c_{hh'}\sigma_E^2 \tag{7.68}$$

where the constants c_{hh} and $c_{hh'}$ are functions of the X's. We use the notation c_{hh} and $c_{hh'}$ because, as we point out in Sections 7.4.4 and

7.4.5, these values are elements of the inverse of a certain matrix and hence part of the output of the abbreviated Doolittle method. In general, estimates of the variances and covariances are part of the output of a regression program.

In terms of the estimate s_E^2 of σ_E^2, the estimated variance of b_h is

$$s_{b_h}^2 = c_{hh} s_E^2 \tag{7.69}$$

Also, the estimated covariance of b_h and $b_{h'}$ is

$$s_{b_h b_{h'}} = c_{hh'} s_E^2 \tag{7.70}$$

Likewise, $\hat{Y}$, as an estimator of η at $X_1 = X_1'$, $X_2 = X_2'$, ..., $X_p = X_p'$, has estimated variance

$$s_{\hat{Y}}^2 = s_E^2 \sum_{i=0}^{p} \sum_{j=0}^{p} c_{ij} X_i' X_j' \tag{7.71}$$

where we let $X_0' = 1$.

Using the assumption that the ϵ's are normal random variables, it follows that b_h, as an estimator of β_h, has a normal distribution. That is, $b_h \sim N(\beta_h, c_{hh} \sigma_E^2)$. Also $\hat{Y}$, as an estimator of η at $X_1 = X_1', \ldots, X_p = X_p'$, has a normal distribution; i.e,

$$\hat{Y} \sim N\left(\beta_0 + \sum_{i=1}^{p} \beta_i X_i', \sigma_E^2 \sum_{i=0}^{p} \sum_{j=0}^{p} c_{ij} X_i' X_j'\right)$$

Based on these distributional facts, a 100γ percent confidence interval for β_h is given by

$$\left.\begin{array}{c} L \\ U \end{array}\right\} = b_h \mp t_{[(1+\gamma)/2](n-p-1)} s_{b_h} \tag{7.72}$$

where s_{b_h} is defined in Eq. (7.69). Similarly, a 100γ percent confidence interval for η at $X_1 = X_1', \ldots, X_p = X_p'$ is given by

$$\left.\begin{array}{c} L \\ U \end{array}\right\} = \left(b_0 + \sum_{i=1}^{p} b_i X_i'\right) \mp t_{[(1+\gamma)/2](n-p-1)} s_{\hat{Y}} \tag{7.73}$$

where $s_{\hat{Y}}$ is defined in Eq. (7.71).

Example 7.7

Continuing to use the data in Table 7.5, the value of c_{22} is $c_{22} = 0.02738$. Using this and the estimate of σ_E^2 (i.e., $s_E^2 = 4.9927$), the estimated variance of b_2 is $s_{b_2}^2 = (0.02738)(4.9927) = 0.13670$. Thus, using Eq. (7.72), a 95 percent confidence interval for β_2 is $(-0.2050, 1.3124)$.

7.4.3 Tests of Hypotheses

We could, as we did in simple linear regression, test hypotheses about the individual β's. Thus if we hypothesize $H : \beta_h = \beta_h'$ versus $A : \beta_h \neq$

β'_h, the appropriate test statistic is

$$t = (b_h - \beta'_h)/s_{b_h} = (b_h - \beta'_h)/s_E \sqrt{c_{hh}} \qquad (7.74)$$

and for the significance level α, the critical region is $|t| \geq$ $t_{(1 - \alpha/2)(n - p - 1)}$.

Many other tests can be made in multiple linear regression. We shall discuss only the following types:

1. Test of significance of regression.
2. Test of significance of a subset of the independent variables.
3. Sequential test of significance of each variable.
4. Partial test of significance of each variable.

All these tests are characterized by the property that the independent variables in the model under H constitute a subset of the independent variables in the model under A. In all cases the test statistic is based on the difference in the regression sum of squares for the model in A and the model in H (or, equivalently, the difference in the residual sum of squares). This is best illustrated by looking at the tests.

Test of significance of regression. The purpose of this test is to assess the overall significance of fitting the regression equation. The hypothesis is

$$H:\beta_h = 0 \qquad \text{for all } h = 1, \ldots, p$$
$$A: \text{at least one } \beta_h \neq 0 \qquad h = 1, \ldots, p$$

In terms of the model in Eq. (7.56) the hypothesis is a comparison of the two models

$$H:\eta = \beta_0$$
$$A:\eta = \beta_0 + \beta_1 X_1 + \cdots + \beta_p X_p$$

where in the model in A at least one $\beta_h(h = 1, \ldots, p)$ is not zero.

As outlined for simple linear regression, the analysis is based on partitioning the total sum of squares into sums of squares attributable to the appropriate terms in the model. Under H, the identity is

$$\sum_{i=1}^{n} Y_i^2 = n(\bar{Y})^2 + \sum_{i=1}^{n} (Y_i - \bar{Y})^2 \qquad (7.75)$$

or in words,

$$\text{Total } SS = SS(\beta_0) + RSS(H) \qquad (7.76)$$

Similarly, under A the identity is

$$\sum_{i=1}^{n} Y_i^2 = SS(\beta_0, \beta_1, \ldots, \beta_p) + RSS(A) \qquad (7.77)$$

The sums of squares on the right-hand side of Eq. (7.77) can be computed using the equations

$$SS(\beta_0, \beta_1, \ldots, \beta_p) = b_0 \sum_{i=1}^{n} Y_i + b_1 \sum_{i=1}^{n} X_{1i} Y_i + \cdots + b_p \sum_{i=1}^{n} X_{pi} Y_i$$

$$= n(\overline{Y})^2 + b_1 \sum_{i=1}^{n} (X_{1i} - \overline{X}_{1.})(Y_i - \overline{Y})$$

$$+ \cdots + b_p \sum_{i=1}^{n} (X_{pi} - \overline{X}_{p.})(Y_i - \overline{Y}) \qquad (7.78)$$

and

$$RSS(A) = \sum_{i=1}^{n} (Y_i - \hat{Y}_i)^2 = \sum_{i=1}^{n} Y_i^2 - SS(\beta_0, \beta_1, \ldots, \beta_p) \qquad (7.79)$$

where $\overline{X}_{k.}$ denotes the mean $\sum_{i=1}^{n} X_{ki}/n$ of the kth independent variable. The sums of squares given in Eqs. (7.78) and (7.79) are extensions of the analogous sums of squares for simple linear regression given in Eqs. (7.42)–(7.44).

The difference in the regression sum of squares due to including the independent variables $X_1, \ldots, X_p$ in the model after β_0 or, equivalently, the reduction in the residual sum of squares by including $X_1, \ldots, X_p$ after β_0 is

$$SS(\beta_1, \ldots, \beta_p \mid \beta_0) = SS(\beta_0, \beta_1, \ldots, \beta_p) - SS(\beta_0)$$

$$= b_1 \sum_{i=1}^{n} (Y_i - \overline{Y})(X_{1i} - \overline{X}_{1.})$$

$$+ \cdots + b_p \sum_{i=1}^{n} (Y_i - \overline{Y})(X_{pi} - \overline{X}_{p.}) \qquad (7.80)$$

These computations are summarized in an analysis of variance table like Table 7.6.

The residual sum of squares E_{yy} is $RSS(A)$. To test the hypothesis H the test statistic is

$$F = MS(\beta_1, \ldots, \beta_p \mid \beta_0)/s_E^2 \qquad (7.81)$$

where MS = mean square. Under H, F has an F distribution with $\nu_1 = p$ and $\nu_2 = n - p - 1$ degrees of freedom. The critical region includes large values of F.

TABLE 7.6–Analysis of Variance for Testing the Significance of Regression

Source of Variation	Degrees of Freedom	Sum of Squares	Mean Square
Due to β_0	1	$SS(\beta_0)$	$\cdots$
Due to $\beta_1, \ldots, \beta_p \mid \beta_0$	p	$SS(\beta_1, \ldots, \beta_p \mid \beta_0)$	$MS(\beta_1, \ldots, \beta_p \mid \beta_0)$
Residual	$n - p - 1$	$RSS(A) = E_{yy}$	s_E^2
Total	n	$\sum_{i=1}^{n} Y_i^2$	$\cdots$

TABLE 7.7–Analysis of Variance for Data in Table 7.5

Source of Variation	Degrees of Freedom	Sum of Squares	Mean Square
Due to β_0	1	12,367.713	...
Due to $\beta_1, \ldots, \beta_4 \mid \beta_0$	4	3,429.273	857.318
Residual	27	134.804	4.993
Total	32	15,931.790	...

Example 7.8

Using the same data as in Examples 7.6 and 7.7, the analysis of variance table for testing the significance of regression is given in Table 7.7. The value for the F statistic in Eq. (7.81) is $F = 857.318/4.993 = 171.713$, which exceeds $F_{0.999(4,27)}$, so we reject the hypothesis $H:\beta_1 = \beta_2 = \beta_3 = \beta_4 = 0$.

Test of significance of a subset of the independent variables. Sometimes we are interested in testing to see if the model can be simplified by eliminating some of the independent variables. To indicate the test procedure for doing this, we consider the special case where $p = 3$. Suppose we want to test the significance of β_1 and β_2; i.e., we test $H:\beta_1 = \beta_2 = 0$ versus $A:\beta_1 \neq 0$ and/or $\beta_2 \neq 0$. In terms of the model in Eq. (7.56), we are comparing $H:\eta = \beta_0 + \beta_3 X_3$ and $A:\eta = \beta_0 + \beta_1 X_1 + \beta_2 X_2 + \beta_3 X_3$, where in A, at least one or both β_1 and β_2 are nonzero. The analysis of variance table is given in Table 7.8. In the sum

TABLE 7.8–Analysis of Variance for Testing Significance of β_1 and β_2

Source of Variation	Degrees of Freedom	Sum of Squares	Mean Square
Due to β_0, β_3	2	$SS(\beta_0, \beta_3)$	...
Due to $\beta_1, \beta_2 \mid \beta_0, \beta_3$	2	$SS(\beta_1, \beta_2 \mid \beta_0, \beta_3)$	$MS(\beta_1, \beta_2 \mid \beta_0, \beta_3)$
Residual	$n - 4$	E_{yy}	s_E^2
Total	n	$\sum_{i=1}^{n} Y_i^2$	...

of squares column, $SS(\beta_0, \beta_3)$ is the regression sum of squares evaluated by fitting the model under H and $SS(\beta_1, \beta_2 \mid \beta_0, \beta_3)$ is the difference of the regression sum of squares $SS(\beta_0, \beta_1, \beta_2, \beta_3)$, evaluated by fitting the model under A, and $SS(\beta_0, \beta_3)$.

Again, the residual sum of squares E_{yy} is the residual sum of squares for the model in A. The test statistic to use is

$$F = MS(\beta_1, \beta_2 \mid \beta_0, \beta_3)/s_E^2 \qquad (7.82)$$

which under H has an F distribution with $\nu_1 = 2$ and $\nu_2 = n - 4$ degrees of freedom.

Example 7.9

Again consider the model for the data in Table 7.5,

$$\eta = \beta_0 + \beta_1 X_1 + \beta_2 X_2 + \beta_3 X_3 + \beta_4 X_4$$

and suppose we wish to test $H:\beta_2 = \beta_3 = 0$. The analysis of variance for this test is summarized in Table 7.9. Using the test statistic in Eq. (7.82), $F = 363.571/4.993 = 72.82$. Since $72.82 > F_{0.99(2,27)}$, we reject H and conclude $\beta_2 \neq 0$ and/or $\beta_3 \neq 0$.

TABLE 7.9–Analysis of Variance for Testing $H:\beta_2 = \beta_3 = 0$ Using Data in Table 7.5

Source of Variation	Degrees of Freedom	Sum of Squares	Mean Square
Due to $\beta_0, \beta_1, \beta_4$	3	15,069.843	. . .
Due to $\beta_2, \beta_3 \mid \beta_0, \beta_1, \beta_4$	2	727.143	363.571
Residual	27	134.804	4.993
Total	32	15,931.790	. . .

Sequential test of significance of each variable. Assuming that the independent variables have been included in the model in some meaningful order, it may be of interest to test how significant each term is after all the previous terms have been included. In this context we are testing a sequence of hypotheses:

$$H_1:\beta_1 = 0 \qquad A_1:\beta_1 \neq 0$$
$$H_2:\beta_2 = 0 \qquad A_2:\beta_2 \neq 0$$
$$\vdots \qquad\qquad \vdots$$
$$H_p:\beta_p = 0 \qquad A_p:\beta_p \neq 0$$

To illustrate the use of this test, suppose the variables were ordered according to how important each variable is thought to be, with X_1 being considered the most important. In this case we would begin by testing if X_p can be removed from the model; thus we would test H_p versus A_p. That is, we are comparing the models

$$H_p:\eta = \beta_0 + \beta_1 X_1 + \cdots + \beta_{p-1} X_{p-1}$$
$$A_p:\eta = \beta_0 + \beta_1 X_1 + \cdots + \beta_{p-1} X_{p-1} + \beta_p X_p$$

TABLE 7.10–Analysis of Variance for Sequential Test of Significance

Source of Variation	Degrees of Freedom	Sum of Squares	Mean Square
Due to β_0	1	$SS(\beta_0)$	$SS(\beta_0)$
Due to $\beta_1 \mid \beta_0$	1	$SS(\beta_1 \mid \beta_0)$	$SS(\beta_1 \mid \beta_0)$
Due to $\beta_2 \mid \beta_0, \beta_1$	1	$SS(\beta_2 \mid \beta_0, \beta_1)$	$SS(\beta_2 \mid \beta_0, \beta_1)$
$\vdots$	$\vdots$	$\vdots$	
Due to $\beta_p \mid \beta_0, \beta_1, \ldots, \beta_{p-1}$	1	$SS(\beta_p \mid \beta_0, \beta_1, \ldots, \beta_{p-1})$	$SS(\beta_p \mid \beta_0, \beta_1, \ldots, \beta_{p-1})$
Residual	$n - p - 1$	E_{yy}	s_E^2
Total	n	$\sum_{i=1}^{n} Y_i^2$	$\cdots$

If H_p is accepted we would test if X_{p-1} can be considered nonsignificant. Thus we test H_{p-1} versus A_{p-1} or compare the models

$$H_{p-1}: \eta = \beta_0 + \beta_1 X_1 + \cdots + \beta_{p-2} X_{p-2}$$
$$A_{p-1}: \eta = \beta_0 + \beta_1 X_1 + \cdots + \beta_{p-2} X_{p-2} + \beta_{p-1} X_{p-1}$$

Summarization of the computations for the entire sequence of tests is given in Table 7.10. The test statistic for $H_h: \beta_h = 0$ is

$$F_h = SS(\beta_h \mid \beta_0, \beta_1, \ldots, \beta_{h-1})/s_E^2 \qquad (7.83)$$

which under H_h has an F distribution with $\nu_1 = 1$ and $\nu_2 = n - p - 1$ degrees of freedom. The denominator s_E^2 is the residual mean square based on the model including all the independent variables.

Example 7.10

Using the same data as in Examples 7.7, 7.8, and 7.9, the analysis of variance table useful for sequentially testing the significance of the β_h's is given in Table 7.11. Since $F_{0.99(1,27)} \cong 7.68$, β_1, β_2, and β_4 are significantly different from zero at $\alpha = 0.01$. Also, β_3 is significant at $\alpha = 0.05$.

TABLE 7.11–Analysis of Variance for Sequential Test of Data in Table 7.5.

Source of Variation	Degrees of Freedom	Sum of Squares	Mean Square	F
Due to β_0	1	12,367.713		
Due to $\beta_1 \mid \beta_0$	1	216.256	216.256	43.31
Due to $\beta_2 \mid \beta_0, \beta_1$	1	309.851	309.851	62.06
Due to $\beta_3 \mid \beta_0, \beta_1, \beta_2$	1	29.214	29.214	5.85
Due to $\beta_4 \mid \beta_1, \beta_2, \beta_3$	1	2,873.952	2,873.952	575.63
Residual	27	134.804	4.993	
Total	32	15,931.790	$\cdots$	$\cdots$

Partial test of significance of each variable. Sometimes it is meaningful to test how important any one term is in the model after all the other terms have been included. Consider the special case of $p = 3$. We may be interested in the hypotheses:

$$H : \beta_1 = 0 \qquad A : \beta_1 \neq 0$$

Thus in this example, we are testing if X_1 can be eliminated from the model if X_2 and X_3 are included. In terms of the model, we are comparing

$$H : \eta = \beta_0 + \beta_2 X_2 + \beta_3 X_3$$
$$A : \eta = \beta_0 + \beta_1 X_1 + \beta_2 X_2 + \beta_3 X_3$$

The test statistic is

$$F = SS(\beta_1 \mid \beta_0, \beta_2, \beta_3)/s_E^2 = b_1^2/c_{11} s_E^2 \qquad (7.84)$$

which has an F distribution with $\nu_1 = 1$ and $\nu_2 = n - 4$ degrees of freedom if H is true. In general, $\nu_2 = n - p - 1$. Again, s_E^2 is the residual mean square for the full model. The latter expression for the F statistic in Eq. (7.84) is the square of the t statistic in Eq. (7.74) for $h = 1$ and $\beta_1' = 0$. This follows since the partial test of significance of β_1 is the same as the test discussed in Section 7.4.3 with the hypothesized value $\beta_1' = 0$.

Example 7.11

For the data in Table 7.5 consider the hypotheses

$$H : \beta_3 = 0 \qquad A : \beta_3 \neq 0$$

The test statistic as given in Eq. (7.84) is $F = b_3^2/c_{33} s_E^2$. Since $b_3 = -0.1495$ and $c_{33} = 0.0017$, $F = 2.629 < F_{0.90(1,27)}$. Thus we would not reject H.

Before finishing this section we shall mention by way of a simple example the test procedure for testing a general hypothesis about functions of the β's. Suppose our model is

$$Y_i = \beta_0 + \beta_1 X_{1i} + \beta_2 X_{2i} + \beta_3 X_{3i} + \beta_4 X_{4i} + \epsilon_i \qquad i = 1, \ldots, n$$
$$(7.85)$$

and we are interested in testing the hypotheses

$$H : \beta_1 = 0, \beta_2 + \beta_3 = 0, \beta_4 - \beta_3 = 0$$
$$A : H \text{ is not true}$$

The general procedure is to rewrite the model so that H is satisfied, fit the transformed model, and evaluate the residual sum of squares for the model. The test statistic is based on a comparison of the residual sum of squares for the transformed model and the original model, which we assume true if H is not true.

In our example we can easily solve the equations in H, which yield $\beta_1 = 0, \beta_2 = -\beta_3, \beta_4 = \beta_3$.

Substituting the solutions in Eq. (7.85), the transformed model becomes

$$Y_i = \beta_0 + \beta_3(X_{3i} + X_{4i} - X_{2i}) + \epsilon_i \qquad i = 1, \ldots, n \qquad (7.86)$$

Letting $W_{1i} = (X_{3i} + X_{4i} - X_{2i})$, $\gamma_0 = \beta_0$, and $\gamma_1 = \beta_3$, we can rewrite the transformed model as

$$Y_i = \gamma_0 + \gamma_1 W_{1i} + \epsilon_i \qquad i = 1, \ldots, n \qquad (7.87)$$

which is an ordinary simple linear regression model with W_1 as the independent variable and γ_0 and γ_1 as the unknown parameters. Fitting the model and letting c_0 and c_1 denote the estimates of γ_0 and γ_1 respectively,

$$
\begin{aligned}
RSS(H) &= \sum_{i=1}^{n} (Y_i - c_0 - c_1 W_{1i})^2 \\
&= \sum_{i=1}^{n} Y_i^2 - n(\bar{Y})^2 - c_1 \sum_{i=1}^{n} (W_{1i} - \bar{W}_{1.})(Y_i - \bar{Y}) \qquad (7.88)
\end{aligned}
$$

where the latter expression is analogous to the expression for the residual sum of squares for simple linear regression given in Eq. (7.44). This sum of squares has $n - 2$ degrees of freedom.

To set up the test statistic, recall that the residual sum of squares for the original model when $p = 4$, as given in Eq. (7.79), is

$$
\begin{aligned}
RSS(A) &= \sum_{i=1}^{n} Y_i^2 - SS(\beta_0, \beta_1, \beta_2, \beta_3, \beta_4) \\
&= \sum_{i=1}^{n} Y_i^2 - n(\bar{Y})^2 - b_1 \sum_{i=1}^{n} (X_{1i} - \bar{X}_{1.})(Y_i - \bar{Y}) \\
&\quad - \cdots - b_4 \sum_{i=1}^{n} (X_{4i} - \bar{X}_{4.})(Y_i - \bar{Y}) \qquad (7.89)
\end{aligned}
$$

which has $n - 5$ degrees of freedom. The test statistic for testing H is based on comparing the residual sum of squares for the transformed model and the original model and is

$$
\begin{aligned}
F &= \frac{[RSS(H) - RSS(A)]/[(n - 2) - (n - 5)]}{RSS(A)/(n - 5)} \\
&= \frac{[RSS(H) - RSS(A)]/3}{RSS(A)/(n - 5)} \qquad (7.90)
\end{aligned}
$$

If H is true, F has an F distribution with $\nu_1 = 3$ and $\nu_2 = n - 5$ degrees of freedom. The hypothesis is rejected for large values of F. For a more general discussion of tests of this type, the reader is referred to some of the texts on regression analysis given in the list of references.

7.4.4 Matrix Approach to Multiple Linear Regression

In multiple linear regression we are frequently dealing with large sets of data, and it is sometimes convenient to think of the observations as elements of vectors and matrices. Therefore we shall briefly outline the results of multiple regression analysis in terms of matrices.

Assuming that the model in Eq. (7.57) is applicable; i.e., the model is

$$Y_i = \beta_0 + \beta_1 X_{1i} + \cdots + \beta_p X_{pi} + \epsilon_i \qquad i = 1, \ldots, n \qquad (7.91)$$

we can write the model for all n observations simultaneously by defining the matrices

$$\mathbf{Y} = \begin{bmatrix} Y_1 \\ Y_2 \\ \vdots \\ Y_n \end{bmatrix} \quad \mathbf{X} = \begin{bmatrix} 1 & X_{11} & X_{21} & \ldots & X_{p1} \\ 1 & X_{12} & X_{22} & \ldots & X_{p2} \\ & & \vdots & & \\ 1 & X_{1n} & X_{2n} & \ldots & X_{pn} \end{bmatrix} \quad \beta = \begin{bmatrix} \beta_0 \\ \beta_1 \\ \vdots \\ \beta_p \end{bmatrix} \quad \epsilon = \begin{bmatrix} \epsilon_1 \\ \epsilon_2 \\ \vdots \\ \epsilon_n \end{bmatrix}$$

Then the model, in matrix notation, is

$$\mathbf{Y} = \mathbf{X}\beta + \epsilon \qquad (7.92)$$

with the usual assumptions about the ϵ's stated as

$$\mathbf{E}[\epsilon] = \phi \qquad \mathbf{V}(\epsilon) = \sigma_E^2 \mathbf{I} \qquad (7.93)$$

where ϕ is an $(n \times 1)$ matrix of zeros, $\mathbf{V}(\epsilon)$ denotes the matrix of variances and covariances of the ϵ's, and $\mathbf{I}$ is the $(n \times n)$ identity matrix.

To estimate the parameters, the sum of squares to be minimized is

$$SS(b_0, b_1, \ldots, b_p) = (\mathbf{Y} - \mathbf{XB})'(\mathbf{Y} - \mathbf{XB}) \qquad (7.94)$$

where $\mathbf{B} = (b_0, b_1, \ldots, b_p)'$ and the normal equations can be written as

$$(\mathbf{X}'\mathbf{X})\mathbf{B} = \mathbf{X}'\mathbf{Y} \qquad (7.95)$$

where

$$\mathbf{X}'\mathbf{X} = \begin{bmatrix} n & \sum_{i=1}^{n} X_{1i} & \cdots & \sum_{i=1}^{n} X_{pi} \\ \sum_{i=1}^{n} X_{1i} & \sum_{i=1}^{n} X_{1i}^2 & \cdots & \sum_{i=1}^{n} X_{1i}X_{pi} \\ \sum_{i=1}^{n} X_{2i} & \sum_{i=1}^{n} X_{2i}X_{1i} & \cdots & \sum_{i=1}^{n} X_{2i}X_{pi} \\ & & \vdots & \\ \sum_{i=1}^{n} X_{pi} & \sum_{i=1}^{n} X_{pi}X_{1i} & \cdots & \sum_{i=1}^{n} X_{pi}^2 \end{bmatrix} \quad \mathbf{X}'\mathbf{Y} = \begin{bmatrix} \sum_{i=1}^{n} Y_i \\ \sum_{i=1}^{n} X_{1i}Y_i \\ \sum_{i=1}^{n} X_{2i}Y_i \\ \vdots \\ \sum_{i=1}^{n} X_{pi}Y_i \end{bmatrix}$$

As we mentioned earlier, assuming the normal equations have a unique solution, the estimates **B** can be solved by inverting the $\mathbf{X'X}$ matrix. The least squares estimates are

$$\mathbf{B} = (\mathbf{X'X})^{-1}\mathbf{X'Y} \tag{7.96}$$

Unless the number of independent variables is small, we would not solve for the estimates using Eq. (7.96) by manually inverting $\mathbf{X'X}$. Instead, we can solve for the estimates by using a regression program on the computer. Although many of the regression programs solve for **B** using Eq. (7.96), this is not necessarily the most accurate method of solution. This is particularly true for certain types of large matrices when the computer routine for inverting matrices can introduce inaccuracies into the estimates. We assume a regression program is available that will provide accurate estimates of the regression coefficients. The residual mean square that is an estimate of σ_E^2 is

$$s_E^2 = [\mathbf{Y'Y} - \mathbf{B'X'Y}]/(n - p - 1) \tag{7.97}$$

Example 7.12

Reworking Example 7.5 using matrix notation, the normal equations for estimating β are

$$\begin{bmatrix} 32.00 & 1{,}256.00 & 133.80 & 7{,}728.00 & 10{,}627.00 \\ 1{,}256.00 & 50{,}282.50 & 5{,}535.68 & 298{,}732.10 & 413{,}189.70 \\ 133.80 & 5{,}535.68 & 772.22 & 29{,}549.70 & 42{,}746.00 \\ 7{,}728.00 & 298{,}732.10 & 29{,}549.70 & 1{,}910{,}002.00 & 2{,}599{,}887.00 \\ 10{,}627.00 & 413{,}189.70 & 42{,}746.00 & 2{,}599{,}887.00 & 3{,}680{,}003.00 \end{bmatrix} \begin{bmatrix} b_0 \\ b_1 \\ b_2 \\ b_3 \\ b_4 \end{bmatrix} = \begin{bmatrix} 629.10 \\ 25{,}153.59 \\ 2{,}964.88 \\ 147{,}996.60 \\ 225{,}418.00 \end{bmatrix}$$

The estimate of β is

$$\mathbf{B} = \begin{bmatrix} b_0 \\ b_1 \\ b_2 \\ b_3 \\ b_4 \end{bmatrix} = \begin{bmatrix} -6.82 \\ 0.23 \\ 0.55 \\ -0.15 \\ 0.15 \end{bmatrix}$$

The variances and covariances of the b's are easily expressed in matrix notation. Letting $\mathbf{V(B)}$ denote the matrix of variances and covariances, it follows from the regression model and the assumption about ϵ that

$$\mathbf{V(B)} = (\mathbf{X'X})^{-1}\sigma_E^2 \tag{7.98}$$

If we let

$$\mathbf{C} = (c_{ij})_{\substack{i=0,1,\ldots,p \\ j=0,1,\ldots,p}} = (\mathbf{X'X})^{-1} \tag{7.99}$$

be the inverse of $\mathbf{X'X}$, and use s_E^2 as an estimate of σ_E^2, the estimated variance of b_h is

$$s_{b_h}^2 = c_{hh}s_E^2 \tag{7.100}$$

and the estimated covariance of b_h and $b_{h'}$ is

$$s_{b_h b_{h'}} = c_{hh'} s_E^2 \tag{7.101}$$

These agree with the expressions of the estimated variance and co-variance given in Eqs. (7.69) and (7.70) respectively. Using Eqs. (7.99)–(7.101) the estimated variance of $\hat{Y}$, as an estimate of η, is

$$s_{\hat{Y}}^2 = (X^{*\prime} C X^*) s_E^2 \tag{7.102}$$

where $X^* = (1, X_1', X_2', \ldots, X_p')'$, the particular set of X values for which an estimate is desired.

Example 7.13

Continuing to use the data in Table 7.5, the inverse of the $X'X$ matrix is

$$C = \begin{bmatrix} 20.52545 & -0.13541 & -0.53670 & -0.05387 & 0.00023 \\ -0.13541 & 0.00200 & 0.00027 & 0.00022 & 0.00001 \\ -0.53670 & 0.00027 & 0.02738 & 0.00183 & -0.00009 \\ -0.05387 & 0.00022 & 0.00183 & 0.00017 & -0.00001 \\ 0.00023 & 0.00001 & -0.00009 & -0.00001 & 0.00001 \end{bmatrix}$$

Using this and the estimate of σ_E^2 (i.e., $s_E^2 = 4.9927$), we see, for example, that

$$s_{b_2}^2 = c_{22} s_E^2 = (0.02738)(4.9927) = 0.13670$$

Also, $s_{\hat{Y}}^2 = 1.9954$ for $X^* = (1, 40.0, 6.1, 220, 275)'$.

Finally, when testing hypotheses about the regression coefficients, the test statistic involves the ratio of a mean square that is associated with the β's under test and the residual mean square used to estimate the variance σ_E^2. These mean squares are also easily expressible in terms of matrices. As an example, consider the multiple linear regression model

$$Y = X\beta + \epsilon \tag{7.103}$$

and suppose we are interested in the test of significance of regression. That is, we test the hypothesis $H:\beta_h = 0$ for all $h = 1, 2, \ldots, p$. The sums of squares needed for this test are summarized in Table 7.6. These can be expressed in matrix notation as

$$SS(\beta_0, \beta_1, \ldots, \beta_p) = B'X'Y \tag{7.104}$$

$$\text{Total } SS = Y'Y \tag{7.105}$$

thus

$$RSS = E_{yy} = Y'Y - B'X'Y \tag{7.106}$$

Also

$$SS(\beta_0) = n(\bar{Y})^2 \tag{7.107}$$

thus

$$SS(\beta_1, \ldots, \beta_p \mid \beta_0) = B'X'Y - n(\bar{Y})^2 \tag{7.108}$$

The test statistic for testing H is given in Eq. (7.81) and is

$$F = MS(\beta_1, \ldots, \beta_p \mid \beta_0)/s_E^2 = [\mathbf{B}'\mathbf{X}'\mathbf{Y} - n(\overline{Y})^2]/ps_E^2 \qquad (7.109)$$

The remaining tests can be expressed in terms of matrices as well.

7.4.5 Abbreviated Doolittle Method

The estimation problem in a multiple linear regression model involves solving the normal equations that are a set of simultaneous linear equations in the b's. Several methods for solving such a set of equations (or of inverting matrices) appear in the literature. In this section we shall discuss one method, the *abbreviated Doolittle method.* Among several attractive features, this method is well suited to programming for high-speed computers as well as being useful when only desk calculators are available. Further, the method incorporates self-checking features that permit verification of the accuracy of the arithmetic calculations at each stage.

To illustrate the abbreviated Doolittle method, we consider the problem of estimating

$$\eta = \beta_0 + \beta_1 X_1 + \beta_2 X_2 + \beta_3 X_3 + \beta_4 X_4 \qquad (7.110)$$

The least squares estimate is

$$\hat{Y} = b_0 + b_1 X_1 + b_2 X_2 + b_3 X_3 + b_4 X_4 \qquad (7.111)$$

where the coefficients $b_i (i = 0, 1, 2, 3, 4)$ are found by solving the normal equations given in Eq. (7.64) with $p = 4$. Writing the normal equations in matrix notation, the matrix $\mathbf{B}' = [b_0, b_1, b_2, b_3, b_4]$ is the solution of the equations

$$(\mathbf{X}'\mathbf{X})\mathbf{B} = \mathbf{X}'\mathbf{Y} \qquad (7.112)$$

To simplify the notation, we denote $\mathbf{X}'\mathbf{X}$ by $\mathbf{A}$ and $\mathbf{X}'\mathbf{Y}$ by $\mathbf{G}$. In this notation Eq. (7.112) is

$$\mathbf{AB} = \mathbf{G} \qquad (7.113)$$

and the solution is

$$\mathbf{B} = \mathbf{A}^{-1}\mathbf{G} = \mathbf{CG} \qquad (7.114)$$

where $\mathbf{C} = \mathbf{A}^{-1} = (\mathbf{X}'\mathbf{X})^{-1}$.

The abbreviated Doolittle method can be used to obtain:

1. The b's.
2. The sums of squares associated with the *sequential* fitting of the β's.
3. The estimated variances of the b's.
4. The estimated covariances between pairs of b's.
5. The elements of the inverse of the $\mathbf{X}'\mathbf{X}$ matrix, that is, the elements of $\mathbf{C}$.

The mechanics of the *forward solution* using the abbreviated Doolittle method are summarized in Table 7.12. A discussion of the steps

TABLE 7.12–The Generalized Abbreviated Doolittle Method Illustrated on the Symmetric 5 × 5 Matrix Associated with the Set of Linear Equations Specified by Eq. (7.112)

Row	Front Half					Constant Terms	Back Half					Check
	b_0	b_1	b_2	b_3	b_4							
(0)	a_{00}	a_{01}	a_{02}	a_{03}	a_{04}	g_0	1	0	0	0	0	T_0
(1)		a_{11}	a_{12}	a_{13}	a_{14}	g_1	0	1	0	0	0	T_1
(2)			a_{22}	a_{23}	a_{24}	g_2	0	0	1	0	0	T_2
(3)				a_{33}	a_{34}	g_3	0	0	0	1	0	T_3
(4)					a_{44}	g_4	0	0	0	0	1	T_4
(5) = (0)	A_{00}	A_{01}	A_{02}	A_{03}	A_{04}	A_{0y}	1	0	0	0	0	T_5
(6) = (5)/A_{00}	1	B_{01}	B_{02}	B_{03}	B_{04}	B_{0y}	B'_{00}	0	0	0	0	T_6
(7) = (1) − A_{01}(6)		A_{11}	A_{12}	A_{13}	A_{14}	A_{1y}	A'_{10}	1	0	0	0	T_7
(8) = (7)/A_{11}		1	B_{12}	B_{13}	B_{14}	B_{1y}	B'_{10}	B'_{11}	0	0	0	T_8
(9) = (2) − A_{02}(6) − A_{12}(8)			A_{22}	A_{23}	A_{24}	A_{2y}	A'_{20}	A'_{21}	1	0	0	T_9
(10) = (9)/A_{22}			1	B_{23}	B_{24}	B_{2y}	B'_{20}	B'_{21}	B'_{22}	0	0	T_{10}
(11) = (3) − A_{03}(6) − A_{13}(8) − A_{23}(10)				A_{33}	A_{34}	A_{3y}	A'_{30}	A'_{31}	A'_{32}	1	0	T_{11}
(12) = (11)/A_{33}				1	B_{34}	B_{3y}	B'_{30}	B'_{31}	B'_{32}	B'_{33}	0	T_{12}
(13) = (4) − A_{04}(6) − A_{14}(8) − A_{24}(10) − A_{34}(12)					A_{44}	A_{4y}	A'_{40}	A'_{41}	A'_{42}	A'_{43}	1	T_{13}
(14) = (13)/A_{44}					1	B_{4y}	B'_{40}	B'_{41}	B'_{42}	B'_{43}	B'_{44}	T_{14}

involved is best given in two parts, one associated with the first section of the table and one associated with all the succeeding sections:

First section [rows (0) through (4)]

1. In the front half of the table are entered the elements of the matrix of coefficients defined by $X'X$, omitting those obvious from symmetry. That is, we have entered $a_{ij} = \sum_{k=1}^{n} X_{ik} X_{jk}$ recognizing that $a_{ji} = a_{ij}$.
2. In the column headed "constant terms" are entered the elements of the vector $X'Y$. That is, we have entered $g_i = \sum_{k=1}^{n} X_{ik} Y_k$.
3. In the back half of the table are entered the elements of the identity matrix, again omitting those obvious from symmetry.
4. In the check column are entered the sums of all entries in the corresponding rows, *including those elements omitted because of symmetry.*

Succeeding sections [rows (5) through (14)]

1. Each entry in a given row is generated according to the instruction specified for that row. (See the first column of the table.) This applies to the front half, the constant terms, the back half, and the check column.
2. The sum of all the entries in a row (with the exception of the entry in the check column) should equal (within rounding error) the entry in the check column. The advantage of the checking procedure should be obvious: if an arithmetic error has been made, it will be corrected before calculations are started on the next row.
3. Steps such as those described are continued until a row is reached in which only a single B_{pq} appears. With the calculation of all entries in this row and the satisfaction of the "check," the forward solution is complete.

The next step in the analysis is the completion of the *backward solution*. This will be performed in two parts, one to determine the b's and the other to determine the c_{ij} values.

Determination of the b's

The forward solution of the abbreviated Doolittle method has provided us with the following set of equations:

$$
\begin{aligned}
(1) \quad b_0 + (B_{01})b_1 + (B_{02})b_2 + (B_{03})b_3 + (B_{04})b_4 &= B_{0y} \\
(1) \quad b_1 + (B_{12})b_2 + (B_{13})b_3 + (B_{14})b_4 &= B_{1y} \\
(1) \quad b_2 + (B_{23})b_3 + (B_{24})b_4 &= B_{2y} \\
(1) \quad b_3 + (B_{34})b_4 &= B_{3y} \\
(1) \quad b_4 &= B_{4y}
\end{aligned}
\qquad (7.115)
$$

Solving these in reverse order (hence the name "backward solution"), we obtain:

$$b_4 = B_{4y}$$
$$b_3 = B_{3y} - b_4 B_{34}$$
$$b_2 = B_{2y} - b_4 B_{24} - b_3 B_{23}$$
$$b_1 = B_{1y} - b_4 B_{14} - b_3 B_{13} - b_2 B_{12}$$
$$b_0 = B_{0y} - b_4 B_{04} - b_3 B_{03} - b_2 B_{02} - b_1 B_{01} \tag{7.116}$$

Determination of the c_{ij}

1. Since $\mathbf{C} = \mathbf{A}^{-1} = (\mathbf{X'X})^{-1}$ is the inverse of a symmetric matrix, it will be symmetric. This reduces the number of calculations to be performed since $c_{ji} = c_{ij}$.
2. All c_{ij} values may be calculated using the equation

$$c_{ij} = \sum_{k=0}^{4} A'_{ki} B'_{kj} \tag{7.117}$$

in which some of the A' values may be 0 or 1 and some of the B' values may be 0. It should be noted that some of the c_{ij} values may be read directly from the forward solution, namely,

$$c_{40} = B'_{40} \qquad\qquad c_{42} = B'_{42} \qquad\qquad c_{44} = B'_{44}$$
$$c_{41} = B'_{41} \qquad\qquad c_{43} = B'_{43}$$

3. If we choose to ignore the symmetry mentioned in (1) and calculate all the c_{ij} independently, a check can be made on the arithmetic by comparing c_{ij} and c_{ji}.
4. A final check could be made by seeing if $\mathbf{CA}$ equals $\mathbf{I}$. It should. However, rounding errors may cause minor discrepancies.

Having completed both the forward and backward solutions using the abbreviated Doolittle method, the next step is to indicate how to obtain the analysis of variance table and the standard errors associated with the various statistics. Using the formula

$$SS(\beta_i \mid \beta_0, \beta_1, \ldots, \beta_{i-1}) = A_{iy} B_{iy} \tag{7.118}$$

the various sums of squares may be evaluated easily once the forward solution of the abbreviated Doolittle method has been completed. The analysis of variance may then be recorded as in Table 7.13. If we do not wish to record the reduction in the residual sum of squares associated with the sequential fitting of each additional β, it is proper to note that the

$$SS(\beta_1, \beta_2, \beta_3, \beta_4 \mid \beta_0) = \sum_{i=1}^{4} A_{iy} B_{iy} \tag{7.119}$$

and this pooled sum of squares possesses 4 degrees of freedom.

TABLE 7.13–Analysis of Variance Associated with the Multiple Regression Problem Discussed in Table 7.12

Source of Variation	Degrees of Freedom	Sum of Squares	Mean Square
Due to β_0	1	$A_{0y}B_{0y}$	$A_{0y}B_{0y}$
Due to $\beta_1 \mid \beta_0$	1	$A_{1y}B_{1y}$	$A_{1y}B_{1y}$
Due to $\beta_2 \mid \beta_0, \beta_1$	1	$A_{2y}B_{2y}$	$A_{2y}B_{2y}$
Due to $\beta_3 \mid \beta_0, \beta_1, \beta_2$	1	$A_{3y}B_{3y}$	$A_{3y}B_{3y}$
Due to $\beta_4 \mid \beta_0, \beta_1, \beta_2, \beta_3$	1	$A_{4y}B_{4y}$	$A_{4y}B_{4y}$
Residual	$n-5$	E_{yy}	s_E^2
Total	n	$\sum_{i=1}^{n} Y_i^2$	...

Example 7.14

Consider the data given in Table 7.5. The abbreviated Doolittle method is applied to these data in Table 7.14, where the X's and Y have been coded so that the elements of $\mathbf{X'X}$ and $\mathbf{X'Y}$ are approximately of the same order of magnitude. (This is done to facilitate the computations.) In this case, the coding is as follows: divide X_0 by 10, X_1 by 100, X_2 by 10, X_3 by 1000, X_4 by 1000, and Y by 100. Then Eq. (7.116) is used to obtain $\hat{Y}/100 = -0.681468(X_0/10) + 0.227227(X_1/100) + 0.055349(X_2/10) - 1.495563(X_3/1000) + 1.546520(X_4/1000)$, and Eq. (7.117) yields

$$\mathbf{C} = \begin{bmatrix} 2052.64 & -135.410 & -53.6729 & -538.760 & 2.25605 \\ -135.410 & 20.0032 & 0.27303 & 22.3301 & 0.54994 \\ -53.6729 & 0.27303 & 2.73843 & 18.3161 & -0.92804 \\ -538.760 & 22.3301 & 18.3161 & 171.128 & -11.6657 \\ 2.25605 & 0.54994 & -0.92804 & -11.6657 & 8.32196 \end{bmatrix}$$

where the elements of $\mathbf{C}$ reflect the coding explained above. The analysis of variance of the coded data is presented in Table 7.15.

Example 7.15

The example of Section 7.3.3 is reworked in Table 7.16 using the abbreviated Doolittle method. Again we get $\hat{Y} = 3.962 + 7.478X_1$. Also, $c_{00} = 0.3460$, $c_{01} = -0.0384$, and $c_{11} = 0.0055$. It can be verified that the sums of squares agree (within rounding error) with the values reported in Table 7.4.

7.5 SOME SPECIAL MODELS

In this section we look in more detail at some models that are special cases of the general multiple linear regression model discussed in Section 7.4.

TABLE 7.14–Abbreviated Doolittle Method Applied to the Data of Table 7.5

Row	Front Half					Constant Terms	Back Half					Check
	b_0	b_1	b_2	b_3	b_4							
(0)	0.3200	1.256000	1.338000	0.772800	1.062700	0.629100	1	0	0	0	0	6.378600
(1)		5.028250	5.535680	2.987321	4.131897	2.515359		1	0	0	0	22.454507
(2)			7.722200	2.954970	4.274600	2.964880			1	0	0	25.790330
(3)				1.910002	2.599887	1.479966				1	0	13.704946
(4)					3.680003	2.254180					1	19.003267
(5)	0.3200	1.256000	1.338000	0.772800	1.062700	0.629100	1	0	0	0	0	6.378600
(6)	1	3.925000	4.181250	2.415000	3.320938	1.965938	3.125000	0	0	0	0	19.933125
(7)		0.098450	0.284030	−0.045919	−0.039201	0.046141	− 3.925000	1	0	0	0	− 2.581497
(8)		1	2.885018	−0.466420	−0.398182	0.468674	−39.867953	10.157440	0	0	0	− 26.221402
(9)			1.308256	−0.143823	−0.055719	0.201337	7.142445	−2.885018	1	0	0	6.567474
(10)			1	−0.109935	−0.042590	0.153897	5.459516	−2.205240	0.764376	0	0	5.020022
(11)				0.006461	0.009057	0.004344	− 3.460493	0.149255	0.109935	1	0	− 2.181439
(12)				1	1.401795	0.672342	−535.597121	23.100913	17.015168	154.774802	0	−337.631791
(13)					0.120164	0.185836	0.271096	0.066083	−0.111517	−1.401795	1	0.130072
(14)					1	1.546520	2.256050	0.549940	−0.928040	−11.665682	8.321960	1.082454

TABLE 7.15–Analysis of Variance Associated with
the Regression Analysis of the Data in Table 7.5

Source of Variation	Degrees of Freedom	Sum of Squares	Mean Square
Due to β_0	1	1.236772	1.236772
Due to $\beta_1 \mid \beta_0$	1	0.021625	0.021625
Due to $\beta_2 \mid \beta_0, \beta_1$	1	0.030985	0.030985
Due to $\beta_3 \mid \beta_0, \beta_1, \beta_2$	1	0.002921	0.002921
Due to $\beta_4 \mid \beta_0, \beta_1, \beta_2, \beta_3$	1	0.287399	0.287399
Residual	27	0.013477	0.000499
Total	32	1.593179	. . .

TABLE 7.16–Solution of the Example of Section 7.3.3
by the Abbreviated Doolittle Method

Row	Front Half b_0	Front Half b_1	Constant Terms	Back Half		Check
(0)	13	91	732	1	0	837
(1)		819	6485		1	7396
(2)	13	91	732	1	0	837
(3)	1	7	56.3077	0.0769	0	64.3846
(4)		182	1360.9993	−6.99790	1	1537.0014
(5)		1	7.4780	−0.03845	0.0054945	8.44506

7.5.1 Polynomial Models

We first consider a model in which the regression function is a polynomial in one independent variable X. That is, the model is

$$Y_i = \beta_0 + \beta_1 X + \beta_2 X^2 + \cdots + \beta_p X^p + \epsilon_i \qquad i = 1, \ldots, n \qquad (7.120)$$

where we assume $\epsilon_i \sim \mathrm{NI}(0, \sigma_E^2)$. In terms of the general multiple linear regression model in Eq. (7.57), $X_j = X^j$.

In addition to testing the overall significance of regression, a frequent test of interest for polynomial models is the sequential test of significance. In using the latter test, we are investigating the minimum degree polynomial necessary to adequately describe the behavior of the dependent variable. Generally, the minimum degree polynomial is $k - 1$ if β_k is the first nonsignificant term. Some researchers prefer to conclude that two successive terms β_k and β_{k+1} are nonsignificant before concluding that the $(k - 1)$st degree polynomial is adequate.

If we have more than one variable, the polynomial model usually involves the powers of each of the variables as well as the cross-product terms. For example, the quadratic model involving two independent variables X_1 and X_2 is

$$\eta = \beta_0 + \beta_1 X_1 + \beta_2 X_2 + \beta_3 X_1^2 + \beta_4 X_2^2 + \beta_5 X_1 X_2 \qquad (7.121)$$

We will make use of such models in a later section.

7.5.2 Orthogonal Polynomial Models

When the regression function describing the relationship between two variables X and Y is a polynomial and the values of X (the independent variable) are equally spaced, another method of fitting the polynomial equation has much to recommend it. This is the method of *orthogonal polynomials.* One should observe that, if we are building a polynomial model by adding one term at a time, each time we add a term it is necessary to start the solution from the beginning and solve a new set of normal equations. For example, when fitting a second-order model, we were unable to use the results of fitting the first-order model. However, when the data permit the use of orthogonal polynomial techniques, we can salvage the previous results and simply perform the calculations required to add a new term to the polynomial (of one less degree) determined at the preceding stage.

The method of orthogonal polynomials will be illustrated only for the case where the values of X are equally spaced at unit intervals and where each X has but one Y value associated with it. If the X values are equally spaced at intervals not equal to unity, we may code the X values by dividing through by the length of the common interval and then proceed in the manner to be developed below. If there is more than one Y value associated with each X, the method is not applicable unless we have an *equal* number of Y values associated with each X value. In the latter case, the complete solution may be obtained by introducing the proper divisor into the calculations. If the X values are unequally spaced, a solution may be obtained [see Kendall (9)], but the operation is so cumbersome that it will not be presented in this text. Thus in all but the simpler cases it is usually better for the research worker to use the method described earlier in this chapter. However, if the experimental data are amenable to simple treatment by the method of orthogonal polynomials, the research worker is advised to use that method, for it saves time and also allows him to calculate and evaluate readily, step by step, the contribution made by fitting each additional term in the regression function.

What, then, is the method of orthogonal polynomials? It may be shown that any polynomial, for example,

$$\hat{Y} = b_0 + b_1 X + \cdots + b_p X^p \qquad (7.122)$$

may be rewritten as

$$\hat{Y} = A_0 + A_1 \xi_1' + \cdots + A_p \xi_p' \qquad (7.123)$$

TABLE 7.17–Partial Table of ξ' Values

	Degree of Polynomial									
	$p = 1$	$p = 2$		$p = 3$			$p = 4$			
n	ξ'_1	ξ'_1	ξ'_2	ξ'_1	ξ'_2	ξ'_3	ξ'_1	ξ'_2	ξ'_3	ξ'_4
1	-1	-1	1	-3	1	-1	-2	2	-1	1
2	1	0	-2	-1	-1	3	-1	-1	2	-4
3	...	1	1	1	-1	-3	0	-2	0	6
4	...	...		3	1	1	1	-1	-2	-4
5	...	...		...			2	2	1	1

in which the ξ'_i $(i = 1, \ldots, p)$ are orthogonal polynomials[1] and the $A_i (i = 0, \ldots, p)$ are constants defined by

$$A_0 = \sum Y/n = \overline{Y} \qquad (7.124)$$

and

$$A_i = \sum Y\xi'_i / \sum (\xi'_i)^2 \qquad i = 1, \ldots, p \qquad (7.125)$$

For the case we are considering (i.e., where X takes on the values 1, 2, \ldots, n$), the first three orthogonal polynomials may be expressed as

$$\xi'_1 = \lambda_1 (X - \overline{X}) \qquad (7.126)$$

$$\xi'_2 = \lambda_2 \left[(X - \overline{X})^2 - \frac{n^2 - 1}{12} \right] \qquad (7.127)$$

$$\xi'_3 = \lambda_3 \left[(X - \overline{X})^3 - (X - \overline{X}) \frac{3n^2 - 7}{20} \right] \qquad (7.128)$$

where the λ_i are constants (depending on n) chosen so that the values of the ξ''s are integers reduced to their lowest terms. Abbreviated ξ' values are given in Table 7.17; a more complete table may be found in Anderson and Houseman (1).

In addition to the ease of computing the estimates of the coefficients with orthogonal polynomials, the sum of squares due to each variable being in the model is easily computed. The formulas for these are:

$$SS(\beta_0) = A_0 \sum Y \qquad (7.129)$$

$$SS(\beta_i \mid \beta_0, \ldots, \beta_{i-1}) = SS(\beta_i) = A_i \sum Y\xi'_i \qquad i = 1, \ldots, p \qquad (7.130)$$

The identity included in Eq. (7.130) points out another nice feature of orthogonal polynomials; namely, no matter when the ith term is added in the model (first, in the middle, or last), the additional sum of squares due to that term is the same.

---

1. Two polynomials are said to be orthogonal if, when X takes on a specified set of values, $\sum \xi'_i \xi'_k = 0$ for $i \neq k$, where the summation means that we first compute the product $\xi'_i \xi'_k$ for each value of X and then obtain the sum of these products.

TABLE 7.18–Data for Regression Analysis of the Schopper-Riegler
Freeness Test Using Orthogonal Polynomials

Y	ξ_1'	ξ_2'	ξ_3'	$Y\xi_1'$	$Y\xi_2'$	$Y\xi_3'$
17	−6	22	−11	−102	374	−187
21	−5	11	0	−105	231	0
22	−4	2	6	−88	44	132
27	−3	−5	8	−81	−135	216
36	−2	−10	7	−72	−360	252
49	−1	−13	4	−49	−637	196
56	0	−14	0	0	−784	0
64	1	−13	−4	64	−832	−256
80	2	−10	−7	160	−800	−560
86	3	−5	−8	258	−430	−688
88	4	2	−6	352	176	−528
92	5	11	0	460	1012	0
94	6	22	11	564	2068	1034

Example 7.16

Consider the data given in Table 7.1. The results of the Schopper-Riegler freeness test have been reproduced in Table 7.18 along with the values of the orthogonal polynomials up to ξ_3' for $n = 13$. Using Eqs. (7.124) and (7.125), the estimated orthogonal polynomial model is $\hat{Y} = 56.3 + 7.478\xi_1' - 0.0365\xi_2' - 0.6801\xi_3'$. Applying Eqs. (7.126)–(7.128) with $\lambda_1 = \lambda_2 = 1$ and $\lambda_3 = 1/6$, the estimated regression model in terms of the original X, hours of beating, is $\hat{Y} = 21.7277 - 5.8458X + 2.3449X^2 - 0.1134X^3$. The sum of squares due to including the terms ξ_1', ξ_2', etc., may be calculated using Eqs. (7.129) and (7.130).

7.5.3 The Special Case: $\eta = \beta X$

In some instances it is reasonable to assume that the true regression line passes through the origin. That is, if simple linear regression is being considered, β_0 in

$$\eta = \beta_0 + \beta_1 X \tag{7.131}$$

is *assumed to be* 0 and Eq. (7.131) is rewritten as

$$\eta = \beta X \tag{7.132}$$

where $\beta = \beta_1$. It is clear that such an assumption, *if justified*, will simplify the calculational procedures. It can be verified that for this special case

$$b = \hat{\beta} = \sum XY / \sum X^2 \tag{7.133}$$

Please do not make the mistake of adopting this simpler form just because it is easier to handle. Further, even if the assumption is justified (such as when X = height and Y = weight of men), it may be better to forego the simplifying assumption and consider $\eta = \beta_0 + \beta_1 X$ as being more appropriate for the range of X values being studied in the experiment.

In summary, the mathematical model should only be chosen after proper consideration has been given to all the factors involved.

7.5.4 Pseudolinear Models

Our discussion so far has centered on *linear* regression functions, i.e., functions that are linear with respect to the unknown parameters. Certainly there are many models that do not have this property. Such functions are called *nonlinear* models. A few nonlinear models, frequently encountered in applied work, are

$$\eta = \alpha\beta^X \tag{7.134}$$

$$\eta = \alpha e^{\beta X} \tag{7.135}$$

$$1/\eta = \gamma + \alpha\beta^X \tag{7.136}$$

$$\eta = \gamma(1 - e^{\alpha - \beta X}) \tag{7.137}$$

If the method of least squares is used for these models, the resulting normal equations are not amenable to easy solution. This is discussed in Section 7.13. In this section we indicate an approximate procedure that can be used for certain nonlinear models. Some models, which we call pseudolinear models, can be put into the form of a linear regression model by transforming the variables. For example, consider Eq. (7.134). This model can be made into a linear model by taking logarithms (logarithms to the base 10 are most convenient). The model becomes

$$\log \eta = \log \alpha + (\log \beta)X \tag{7.138}$$

which is a simple linear regression equation

$$\eta' = \alpha' + \beta' X \tag{7.139}$$

with $\eta' = \log \eta$, $\alpha' = \log \alpha$ and $\beta' = \log \beta$. Estimates of α' and β' may be found following the methods described in Section 7.3. This solution, which is equivalent to fitting a straight line by least squares to the data when plotted on semilogarithmic paper, is not identical with a least squares solution of the original problem using Eq. (7.134) and ordinary graph paper. However, the approximation is sufficiently accurate for most problems.

Example 7.17

Consider the data given in Table 7.19. The model proposed is $\eta = \alpha\beta^X$. After transforming the data and using the methods of Section 7.3, the estimated transformed model is $\hat{Z} = 0.9469 + 0.0026X$.

There are many other pseudolinear functions. Consider Eq. (7.135). Taking natural logarithms of both sides, the model becomes

$$\ln \eta = \ln \alpha + \beta X \tag{7.140}$$

TABLE 7.19–Protein Content and Proportion of Vitreous Kernels in Samples of Wheat

Sample	Proportion of Vitreous Kernels (X)	Protein Content (Y)	$Z = \log Y$
1	6	10.3	1.013
2	75	12.2	1.086
3	87	14.5	1.161
4	55	11.1	1.045
5	34	10.9	1.037
6	98	18.1	1.258
7	91	14.0	1.146
8	45	10.8	1.033
9	51	11.4	1.057
10	17	11.0	1.041
11	36	10.2	1.009
12	97	17.0	1.230
13	74	13.8	1.140
14	24	10.1	1.004
15	85	14.4	1.158
16	96	15.8	1.199
17	92	15.6	1.193
18	94	15.0	1.176
19	84	13.3	1.124
20	99	19.0	1.279

Source: Ezekiel and Fox (6), p. 92.

or

$$\eta' = \alpha' + \beta X \tag{7.141}$$

Still another example is the hyperbolic function

$$\eta = X/(\alpha X + \beta) \tag{7.142}$$

Taking the inverse, the model becomes

$$1/\eta = \alpha + \beta/X \tag{7.143}$$

or

$$\eta' = \alpha + \beta X' \tag{7.144}$$

In any case, ordinary least squares can be used to get approximate estimates of the unknown parameters.

7.6 EXAMINATION OF RESIDUALS

When we set up the regression model

$$Y = \beta_0 + \beta_1 X_1 + \cdots + \beta_p X_p + \epsilon \tag{7.145}$$

we always make the assumption that

$$\eta = \beta_0 + \beta_1 X_1 + \cdots + \beta_p X_p \qquad (7.146)$$

describes the true or expected value of Y. Also, we make several assumptions about the characteristics of the error ϵ. Generally, we assume the errors are independent identically distributed normal random variables with zero means and constant variance σ_E^2.

Given a set of observations $Y_1, \ldots, Y_n$ and an estimated regression equation $\hat{Y}$, we are in a position to examine some of these assumptions. We do this by examining the tendencies of the residuals $e_i = Y_i - \hat{Y}_i$. If the assumptions are true, we would expect the e_i's to be consistent with the assumptions. Otherwise, the e_i's will tend to deviate from trends consistent with the model.

Although it is recognized that the residuals e_i are not independent nor do they have constant variance as we assume for the errors ϵ_i, the residuals appear to have approximately the same properties as the true errors. Thus by looking at plots of the residuals, we can get some idea of the consistency of the assumptions for the data being analyzed.

Several plots of the residuals can be used to check on the validity of the model assumptions. We discuss three types of plots.

1. *Normal plots.* One way to examine the assumption of normal errors with constant variance is to plot the ordered residuals on normal probability paper. Such plots should approximately follow a straight line through the origin with slope $1/\sigma_E$ if the normality assumption is valid. Serious deviations from a straight line indicate nonnormal errors and a possible need for a transformation. An individual value far off the line may be an indication of an outlier observation that does not follow the model.

2. *Plots of e versus the estimated regression function $\hat{Y}$.* These should follow a horizontal trend centered at zero. Deviations from such a plot may be an indication of nonconstant variance if the variability of the e_i's changes or an inadequate model if the e_i's follow a trend other than a constant line about zero.

3. *Plots of e versus the independent variables.* These should follow a horizontal trend centered at zero. As in item 2, deviations from the constant line may be an indication of nonconstant variance or a need for perhaps higher power terms of the independent variables.

For a more complete discussion of plots of residuals we particularly refer the reader to the book by Daniel and Wood (2).

7.7 TEST OF LACK OF FIT

In Sections 7.3 and 7.4 the assumption was made that the failure of the model to fit the observations exactly is solely a function of the errors. That is, we assume our model

$$\eta = \beta_0 + \beta_1 X_1 + \cdots + \beta_p X_p \qquad (7.147)$$

is the true model. This assumption is seldom true, although it may be nearly so in many cases; therefore, we should check the adequacy of the model whenever possible. We have already indicated that an examination of plots of the residuals will help us check our model as well as the other assumptions we usually make. We now indicate an additional test for checking the model. Such a test is called a test of *lack of fit*.

To check on the validity of the model, one must have some measure of error other than the residual mean square. It can be shown that when the proposed model is not correct, the residual mean square is a biased estimate of σ_E^2, the error variance. Thus, we need an estimate of σ_E^2 that is independent of the model used. One way to obtain such a measure is to insist that the experiment be repeated some number of times at at least one value of the independent (or controlled) variable. In addition, it is also wise to insist on running the experiment at as many different values of the controlled variable as feasible. In the example considered in Section 7.3.3, the latter recommendation was followed but no repetition of the experiment at any value of the controlled variable was undertaken. This enabled us to make a visual judgment about lack of fit, but no statistical analysis was possible.

To indicate the test of lack of fit, we restrict our discussion to the simple linear regression model. Thus suppose we have n_i observations of the dependent variable for each value of the independent variable. Our assumed model is

$$Y_{ij} = \beta_0 + \beta_1 X_{1i} + \epsilon_{ij} \qquad i = 1, \ldots, k; j = 1, \ldots, n_i \qquad (7.148)$$

where we assume $\epsilon_{ij} \sim \mathrm{NI}(0, \sigma_E^2)$.

The procedure for testing for lack of fit is as follows:

1. Using all the data, fit the assumed regression equation

$$Y_{ij} = \beta_0 + \beta_1 X_{1i} + \epsilon_{ij} \qquad (7.149)$$

to get estimates of β_0 and β_1 and the usual sums of squares.
2. Estimate σ_E^2, independent of the model, by

$$\hat{\sigma}_E^2 = \sum_{i=1}^{k} \sum_{j=1}^{n_i} (Y_{ij} - \overline{Y}_{i.})^2 / (n - k) \qquad (7.150)$$

where $\sum_{i=1}^{k} (n_i - 1) = n - k$ and $\overline{Y}_{i.} = \sum_{j=1}^{n_i} Y_{ij}/n_i$. This is just the pooled estimate of σ_E^2 from k independent populations.
3. Perform the following analysis of variance

$$\sum_{i=1}^{k} \sum_{j=1}^{n_i} Y_{ij}^2 = n(\overline{Y}_{..})^2 + \sum_{i=1}^{k} n_i(\hat{Y}_i - \overline{Y}_{..})^2 + \sum_{i=1}^{k} \sum_{j=1}^{n_i} (Y_{ij} - \hat{Y}_i)^2$$

$$(7.151)$$

or in words

$$\text{Total } SS = SS(\beta_0) + SS(\beta_1 \mid \beta_0) + RSS \qquad (7.152)$$

TABLE 7.20–Analysis of Variance for Test of Lack of Fit

Source of Variation	Degrees of Freedom	Sum of Squares	Mean Square	F test
Due to β_0	1	$SS(\beta_0)$	...	...
Due to $\beta_1 \mid \beta_0$	1	$SS(\beta_1 \mid \beta_0)$	...	...
Residual	$n - 2$	$R_{yy} = RSS$	...	...
Lack of fit	$k - 2$	$L_{yy} = R_{yy} - E_{yy}$	L	L/E
Experimental error	$n - k$	E_{yy}	E	...
Total	n	$\sum\limits_{i=1}^{k} \sum\limits_{j=1}^{n_i} Y_{ij}^2$	...	...

It is possible to partition the residual sum of squares further when we have an independent estimate of σ_E^2 as in item (2). In particular,

$$\sum_{i=1}^{k} \sum_{j=1}^{n_i} (Y_{ij} - \hat{Y}_i)^2 = \sum_{i=1}^{k} \sum_{j=1}^{n_i} (Y_{ij} - \overline{Y}_{i.})^2 + \sum_{i=1}^{k} n_i(\hat{Y}_i - \overline{Y}_{i.})^2 \qquad (7.153)$$

or in words

$$RSS = \text{experimental error } SS + \text{lack of fit } SS \qquad (7.154)$$

We summarize this analysis of variance in Table 7.20 where E_{yy} is the experimental error SS and L_{yy} is the lack of fit SS. The test statistic, as indicated in Table 7.20, is $F = L/E$, which under the hypothesis $H:\eta = \beta_0 + \beta_1 X_1$ has an F distribution with $\nu_1 = k - 2$ and $\nu_2 = n - k$ degrees of freedom.

Example 7.18

Consider the data reported by Hunter (8) and reproduced in Table 7.21. The model used is the simple linear regression model $\eta = \beta_0 + \beta_1 X_1$. The estimated regression equation is $\hat{Y} = 1.76 + 2.86 X_1$, and the analysis of variance is given in Table 7.22.

Since $F = 38.64 > F_{0.99(3,7)} = 8.45$, we reject the hypothesis that the simple linear model is the true model and conclude that the model inadequately describes the data. The researcher is obligated to consider other models, perhaps a higher order polynomial or a nonlinear model.

7.8 COMPARISON OF REGRESSION MODELS FOR SEVERAL SAMPLES

In this section a topic of considerable importance will be discussed: given several samples or groups of observations, may all the data be pooled into one large sample? This sort of problem has arisen earlier and it is not surprising that it also arises when dealing with regression analyses.

Although the problem can exist regardless of the form of the regression function, the discussion here will be limited to the case of simple

TABLE 7.21–Percentage of Impurities at Different Temperatures

Temperature (C)	Coded Temperature (X_1)	Percent of Impurities (Y)
200	1	6.4
200	1	5.6
200	1	6.0
210	2	7.5
210	2	6.5
220	3	8.3
220	3	7.7
230	4	11.7
230	4	10.3
240	5	17.6
240	5	18.0
240	5	18.4

Source: Hunter (8), pp. 16–24. By permission of the author and editor.

TABLE 7.22–Analysis of Variance for the Data of Table 7.21

Source of Variation	Degrees of Freedom		Sum of Squares	Mean Square	F Ratio
Due to β_0	1		1,281.3333	1,281.3333	...
Due to $\beta_1 \mid \beta_0$	1		228.5715	228.5715	...
Residual	10		40.3952	...	...
Lack of fit		3	38.0952	12.6984	38.64
Experimental error		7	2.3000	0.3286	...
Total	12		1,550.3000	...	...

linear regression. If other functional forms are pertinent, a statistician should be consulted.

When several sets of sample data are available, the question most frequently asked is, Can one regression line be used for all the data? Additional questions of interest, particularly if the answer to the first question is no, are:

1. Taking liberties with the system of notation adhered to up to this point and letting β_i denote the true slope of the regression function for the ith sample, does $\beta_1 = \cdots = \beta_k$? That is, do all the regression lines have the same slope?
2. Assuming $\beta_1 = \cdots = \beta_k$, does a simple linear regression function provide an adequate fit of the group means?
3. Assuming $\beta_1 = \cdots = \beta_k$ and the regression of the group means is linear with true regression coefficient β_M, is $\beta_W = \beta_M$ where β_W is the true *pooled* within groups regression coefficient?

To mention one case where it is necessary to know the answers to the questions stated above, we cite the technique known as analysis of co-variance, which we shall study in detail in a later chapter. This technique has as one of its basic assumptions the requirement that the same regression coefficient β apply to all groups. Hence the need for an appropriate test is clear. Let us now outline the general procedure to be followed.

Suppose we have k groups of observations Y_{ij}, X_{ij} $(i = 1, \ldots k;$ $j = 1, \ldots, n_i)$. (**Note:** we deviate from our previous notation for the independent variable; here X_{ij} refers to the jth value of the independent variable in the ith group.) Perhaps the first step in the analysis of this situation is to test if the variances for the k groups of data are all the same. To test this, we assume the model

$$Y_{ij} = \beta_{0i} + \beta_i X_{ij} + \epsilon_{ij} \qquad i = 1, \ldots, k; j = 1, \ldots, n_i \quad (7.155)$$

where $\epsilon_{ij} \sim NI(0, \sigma_i^2)$ and test the hypothesis $H:\sigma_1 = \cdots = \sigma_k = \sigma$.

The appropriate test is Bartlett's test for homogeneity of variance (Section 6.8) where now

$$s_i^2 = \frac{1}{n_i - 2}\left\{\sum_{j=1}^{n_i} (Y_{ij} - \overline{Y}_{i.})^2 - \left[\sum_{j=1}^{n_i} (X_{ij} - \overline{X}_{i.})(Y_{ij} - \overline{Y}_{i.})\right]^2 \middle/ \sum_{j=1}^{n_i} (X_{ij} - \overline{X}_{i.})^2\right\}$$

$$(7.156)$$

is based on $n_i - 2$ degrees of freedom. The pooled estimate of variance is

$$s^2 = \sum_{i=1}^{k} (n_i - 2)s_i^2 \middle/ \sum_{i=1}^{k} (n_i - 2) = \sum_{i=1}^{k} (n_i - 2)s_i^2 \middle/ (n - 2k)$$

$$(7.157)$$

if we let $n = \sum_{i=1}^{k} n_i$. The pooled estimate has $n - 2k$ degrees of freedom.

Assuming H is not rejected, we now indicate the test procedures for answering the questions posed at the beginning of this section.

1. *Can one regression line be used for all the observations?* To answer this question we set up the hypotheses:

$$H:\eta_{ij} = \beta_0 + \beta X_{ij}$$
$$A:\eta_{ij} = \beta_{0i} + \beta_i X_{ij} \qquad i = 1, \ldots, k; j = 1, \ldots, n_i$$

Under H the estimate of β is

$$b = \sum_{i=1}^{k}\sum_{j=1}^{n_i} (X_{ij} - \overline{X}_{..})(Y_{ij} - \overline{Y}_{..}) \middle/ \sum_{i=1}^{k}\sum_{j=1}^{n_i} (X_{ij} - \overline{X}_{..})^2$$

$$(7.158)$$

and the residual sum of squares is:

$$S_T = RSS(H) = \sum_{i=1}^{k} \sum_{j=1}^{n_i} (Y_{ij} - \bar{Y}_{..})^2 - b^2 \sum_{i=1}^{k} \sum_{j=1}^{n_i} (X_{ij} - \bar{X}_{..})^2$$

(7.159)

with $n - 2$ degrees of freedom.

Under A the estimate of β_i is

$$b_i = \sum_{j=1}^{n_i} (X_{ij} - \bar{X}_{i.})(Y_{ij} - \bar{Y}_{i.}) \bigg/ \sum_{j=1}^{n_i} (X_{ij} - \bar{X}_{i.})^2 \qquad i = 1, \ldots, k$$

(7.160)

and the residual sum of squares is

$$S_1 = RSS(A) = s^2(n - 2k) \tag{7.161}$$

where s^2 is given in Eq. (7.157). This has $n - 2k$ degrees of freedom.
The test statistic is

$$F = \frac{(S_T - S_1)/2(k - 1)}{S_1/(n - 2k)} \tag{7.162}$$

which under H has an F distribution with $\nu_1 = 2(k - 1)$ and $\nu_2 = n - 2k$ degrees of freedom.

2. If we reject H in item 1 above, we might be interested in the reason and thus will consider the following question. *Do the individual regression functions have the same slopes?* That is, does $\beta_1 = \beta_2 = \cdots = \beta_k$? The test is based on the hypotheses

$$\begin{aligned} H &: \eta_{ij} = \beta_{0i} + \beta_W X_{ij} \\ A &: \eta_{ij} = \beta_{0i} + \beta_i X_{ij} \end{aligned} \qquad i = 1, \ldots, k; j = 1, \ldots, n_i$$

Under H the estimate of the common slope is

$$b_W = \sum_{i=1}^{k} \sum_{j=1}^{n_i} (X_{ij} - \bar{X}_{i.})(Y_{ij} - \bar{Y}_{i.}) \bigg/ \sum_{i=1}^{k} \sum_{j=1}^{n_i} (X_{ij} - \bar{X}_{i.})^2 \quad (7.163)$$

and the residual sum of squares is

$$S_2 = RSS(H) = \sum_{i=1}^{k} \sum_{j=1}^{n_i} (Y_{ij} - \bar{Y}_{i.})^2 - b_W^2 \sum_{i=1}^{k} \sum_{j=1}^{n_i} (X_{ij} - \bar{X}_{i.})^2$$

(7.164)

which is based on $n - k - 1$ degrees of freedom.

The appropriate test statistic is

$$F = [(S_2 - S_1)/(k - 1)]/[S_1/(n - 2k)] \qquad (7.165)$$

which under H has an F distribution with $\nu_1 = k - 1$ and $\nu_2 = n - 2k$ degrees of freedom.

3. Assuming $\beta_1 = \cdots = \beta_k$, we may ask the question, *Is the regression of the group means linear?* The appropriate test is a goodness of fit test of the hypothesized model

$$H: \overline{Y}_{i.} = \beta_0 + \beta_M \overline{X}_{i.} + \epsilon_{i.} \qquad i = 1, \ldots, k$$

Under H the estimate of β_M is

$$b_M = \sum_{i=1}^{k} n_i(\overline{X}_{i.} - \overline{X}_{..})(\overline{Y}_{i.} - \overline{Y}_{..}) \Big/ \sum_{i=1}^{k} n_i(\overline{X}_{i.} - \overline{X}_{..})^2 \qquad (7.166)$$

and the residual sum of squares is

$$S_3 = RSS(H) = \sum_{i=1}^{k} n_i(\overline{Y}_{i.} - \overline{Y}_{..})^2 - b_M^2 \sum_{i=1}^{k} n_i(\overline{X}_{i.} - \overline{X}_{..})^2$$

$$(7.167)$$

which has $k - 2$ degrees of freedom. If $\beta_1 = \cdots = \beta_k$, $S_2/(n - k - 1)$ is an estimate of σ_E^2. Also, under H, $S_3/(k - 2)$ is an estimate of σ_E^2, so an appropriate test statistic is

$$F = [S_3/(k - 2)]/[S_2/(n - k - 1)] \qquad (7.168)$$

which under H is distributed as F with $\nu_1 = k - 2$ and $\nu_2 = n - k - 1$ degrees of freedom.

4. Finally, if $\beta_1 = \cdots = \beta_k$ and the model for the means is linear and a single model is not appropriate, we would test, *Is $\beta_W = \beta_M$?* To test $H: \beta_W = \beta_M$, the test statistic is

$$F = \frac{(b_W - b_M)^2 \left[\sum_{i=1}^{k} n_i(\overline{X}_{i.} - \overline{X}_{..})^2 \right] \left[\sum_{i=1}^{k} \sum_{j=1}^{n_i} (X_{ij} - \overline{X}_{i.})^2 \right]}{\left[\sum_{i=1}^{k} \sum_{j=1}^{n_i} (X_{ij} - \overline{X}_{..})^2 \right] S_2/(n - k - 1)} \qquad (7.169)$$

which under H is distributed as F with $\nu_1 = 1$ and $\nu_2 = n - k - 1$ degrees of freedom.

It should be clear that the *order* in which these tests are performed is very important since the assumptions necessary for the later tests are tested as hypotheses in the earlier tests. Note also that if a sequence of tests is applied, the critical level (true probability of Type I error) of the sequence is not known, though it is needed for proper interpretation of the results.

7.9 ADJUSTED Y VALUES

Closely related to the reduction in sum of squares, mentioned several times in this chapter, is the technique of adjusting values of the dependent variable to take account of differences among the associated values of the independent variable. For example, if we are concerned with measurements on the gains in weights of certain animals, a valid comparison among the gains does not seem possible unless we adjust for such a value as the initial weight of the animals or the feed consumed. That is, if one animal gains 60 pounds while consuming 300 pounds of feed and another animal gains 40 pounds while consuming 200 pounds of feed, we do not feel justified in making a direct comparison between 60 pounds and 40 pounds. We should first attempt to make some adjustment or correction for the different amounts of feed consumed. One way to make such an adjustment is through regression. If from the present or other data we have an estimated regression function $\hat{Y} = b_0 + b_1 X$, where Y = gain in weight and X = feed consumed, we can adjust the observed gains in weight to some common value for feed consumed. The value of X most commonly selected is the sample mean $\bar{X}$, but any value will do. The reason why the mean is usually adopted as the point of comparison is that in general it is near the center of the range of values of the independent variable.

What then is the procedure for determining adjusted Y values? The formula defining adjusted Y values (i.e., adjusted to $\bar{X}$) is

$$\text{adj } Y = Y - b_1(X - \bar{X}) \qquad (7.170)$$

and the nature of the adjustment is illustrated in Figure 7.7. Here only three sample points have been plotted, since these are sufficient to illustrate the technique. It is seen that all the adjusted Y values (represented by circles) appear on the line erected vertically at $\bar{X}$ because we adjusted to $X = \bar{X}$. Note that it is possible to have adjusted Y_1 > adjusted Y_2 even when $Y_1 < Y_2$. Thus it is readily apparent that the adjustment of a set of measurements based on a concomitant variable may completely change the entire picture of an experiment. As a consequence we might reach much different conclusions based on an analysis of adjusted values than would have been reached if no account were taken of the functional relation existing between the dependent and independent variables.

It should be evident that adjusted Y values may also be determined when dealing with other than simple linear regression. For example, if a multiple linear regression analysis has been performed and the regression equation determined to be

$$\hat{Y} = b_0 + b_1 X_1 + \cdots + b_p X_p \qquad (7.171)$$

then adjusted Y values are defined by

$$\text{adj } Y = Y - b_1(X_1 - \bar{X}_1) - b_2(X_2 - \bar{X}_2) - \cdots - b_p(X_p - \bar{X}_p)$$
$$(7.172)$$

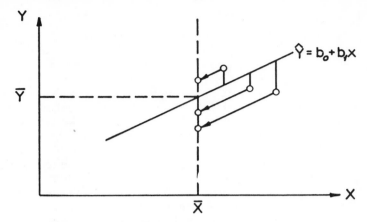

Fig. 7.7–Illustration of adjusted Y values.

Equation (7.172) would be evaluated using the appropriate sample values $Y_i, X_{1i}, \ldots, X_{pi}$ and the calculated mean values.

Rather than dwell on the topic of adjusted values at this time, we shall defer further discussion until later in the book where a more efficient method of analysis, namely, *covariance analysis,* will be introduced.

7.10 SAMPLING FROM A BIVARIATE NORMAL POPULATION

Let us consider as far as practicable the consequences of obtaining a random sample of values of both X and Y from some bivariate population rather than first choosing values of X and then observing random Y values associated with these chosen X values. What effect will such a procedure have on our estimates? As we have stated the problem, it is much too general to permit a satisfactory answer in this book. However, if we make the assumption that our bivariate population is a *bivariate normal population,* then we may examine the effect of obtaining random pairs of X and Y rather than choosing X values and then observing the random values of Y associated with the selected values of X.

In this case two approaches are possible: (1) obtain the best regression equation for estimating a value of Y associated with a specified value of X or (2) obtain the best regression equation for estimating a value of X associated with a specified value of Y. That is, we can obtain

$$\hat{Y} = b_0 + b_1 X \qquad (7.173)$$

as in Section 7.3.2, or we can obtain

$$\hat{X} = c_0 + c_1 Y \qquad (7.174)$$

where

$$c_0 = \overline{X} - c_1 \overline{Y} \qquad (7.175)$$

and

$$c_1 = \sum (X - \overline{X})(Y - \overline{Y})/\sum (Y - \overline{Y})^2 \qquad (7.176)$$

It is to be noted that the above relations assume no "errors of measurement" in X and Y. If, however, our variables are subject to errors in measurement so that we really observe $Z = X + \epsilon$ and $W = Y + \delta$, where ϵ and δ are independently and normally distributed with mean 0 and variances σ_ϵ^2 and σ_δ^2 respectively, what estimation procedure may we use? If in the future we measure Z and wish to estimate Y, the regression of W on Z should be used; if we measure W and wish to estimate X, the regression of Z on W should be calculated and used.

A reasonable question to ask at this point is, What effects do the above-mentioned errors of measurement have on the accuracy and precision of our estimates? Some answers are:

1. If the random errors of measurement are associated only with the dependent variable and are not related to the true values, they will not affect our estimate of the true slope but will cause s_E to overestimate σ_E.
2. If the random errors of measurement are associated only with the independent variable and are not related to the true values, they not only cause s_E to overestimate σ_E but also tend to produce underestimates of the true slope.
3. If both variables are subject to error, the consequences are not so easily determined, and much care should be taken when making predictions based on such data.

Suppose, however, that we want to estimate the *true relationship* between X and Y. To accomplish this, we need further information about σ_ϵ^2 and σ_δ^2. Such information (i.e., estimates s_ϵ^2 and s_δ^2) can sometimes be obtained by making duplicate measurements; however, such a procedure is not always possible. When duplicate measurements are not available, other approaches must be explored. Many scholars have considered the problems associated with regression analyses in which both variables are subject to error, and several solutions have been proposed. However, because no general optimal solution has yet been obtained and the subject may rightly be considered to be beyond the scope of this text, no attempt will be made to illustrate any of the proposed methods of analysis.

7.11 MODEL BUILDING AND THE SELECTION OF INDEPENDENT VARIABLES

In our discussion of regression analysis we have always assumed that the functional relationship is known well enough for an initial regression equation to be stated. Much of the analysis beyond the initial estimation centers on testing this initial model. We now discuss the situation in which the functional relationship is unknown or, even if it is known, is very complex and does not lend itself very well to being expressed in the form of a multiple linear regression model. This situation often arises in trying to model engineering processes involving complex interrelationships between the process parameters and variables. In such situations we may still be interested in constructing a model that can be used for predicting the value of some response. Although the model may not be the true functional relationship, it can be used to describe the behavior of a response and predict future responses. It is in such situations that multiple linear regression analysis can be used to construct predictive models.

7.11.1 Model Building

How does one proceed to construct a predictive model? Initially we must identify the response variable and the independent variables that may affect the response and are measurable. Then the independent variables to be included in the model are selected. This selection could be based on an examination of plots (Section 7.1) of the response variable versus each of the identified independent variables. Also, several procedures are available for a systematic selection of the significant variables. One of these procedures is described in the next section. Once the model is developed, it should be tested both within the range of values of the independent variables used in the analysis and beyond this range for future predictive value. The resulting model would be subject to revision as new information is obtained.

7.11.2 Selection of Independent Variables

An important step in the construction of a predictive model is the selection of the independent variables to be included. Although several procedures can be used, we will describe only one, the *stepwise regression* procedure. The stepwise regression procedure for building a predictive model is as follows: suppose we have a set of independent variables $X_1, \ldots, X_p$, which are thought to significantly affect the response to be predicted, then:

Step 1. Among all the independent variables $X_1, \ldots, X_p$ the first variable to be included in the model is the one that minimizes the residual sum of squares and has a regression coefficient that is significantly different from zero. Thus X_j is the first independent variable included in the model if

$$\sum_{i=1}^{n}(Y_i - b_0 - b_j X_{ji})^2 = \min_{\substack{k \\ k=1,\ldots,p}} \sum_{i=1}^{n}(Y_i - b_0 - b_k X_{ki})^2$$

and the hypothesis $H:\beta_j = 0$ is rejected.

Step 2. Having selected X_j as the first independent variable included in the model, a second variable is selected from the remaining variables such that the residual sum of squares for the selected variable combined with X_j is minimum and the partial test of significance of the coefficient of the selected variable indicates that the coefficient is different from zero. Thus $X_{j'}$ is the second independent variable included in the model if

$$\sum_{i=1}^{n} (Y_i - b_0 - b_j X_{ji} - b_{j'} X_{j'i})^2$$

$$= \min_{\substack{k \\ k=1,\ldots,p \\ k \neq j}} \sum_{i=1}^{n} (Y_i - b_0 - b_j X_{ji} - b_k X_{ki})^2$$

and the hypothesis $H{:}\beta_{j'} = 0$ is rejected.

Step. 3. Once $X_{j'}$ is included in the model, a partial test of significance of the coefficient of X_j is performed to test if X_j should be included, given $X_{j'}$ is in the model. Thus if $H{:}\beta_j = 0$ is rejected, both X_j and $X_{j'}$ are included in the model. If $H{:}\beta_j = 0$ is not rejected, X_j is eliminated and step 2 is followed with $X_{j'}$ as the first independent variable. Of course, X_j is not considered at this stage but can be considered later in the procedure.

The procedure continues, alternating between steps 2 and 3 adding (and occasionally eliminating) variables until none of the remaining variables meet the criteria in step 2 to be included in the model.

Stepwise regression and other procedures are available in packaged computer programs so that the mechanics of the analysis are done. Understanding and interpreting the final model is a critical part of the analysis. For a more comprehensive discussion of this procedure and others we refer the reader to some of the more advanced texts on regression analysis.

7.12 WEIGHTED REGRESSION ANALYSIS

Suppose the data to be considered are of such a nature that the assumption of homoscedasticity (homogeneous variances) is no longer justified. That is, suppose we cannot assume that the variances of the observations are all the same, or in terms of the variances of the ϵ_i's, $V(\epsilon_i) = \sigma_i^2 \neq \sigma_E^2$ for all i. We may also be faced with the situation of having correlated errors; i.e., the ϵ_i's are not independent. In either case the usual estimation procedure is no longer optimal (the estimates are unbiased but not minimum variance as is usually the case for least squares estimates), and we must turn to an alternative analysis.

Suppose we assume that $\sigma_i^2 = \sigma_E^2/w_i$, where w_i is a known constant. That is, restricting ourselves to the simple linear model, we assume the model

$$Y_{ij} = \beta_0 + \beta_1 X_{1i} + \epsilon_{ij} \qquad i = 1, \ldots, k; j = 1, \ldots, n_i \qquad (7.177)$$

where $\epsilon_{ij} \sim NI(0, \sigma_E^2/w_i)$. To use the usual method of least squares to estimate β_0 and β_1, we must transform the observations so that the usual assumption about the ϵ_{ij}'s is satisfied. Doing this for the model in

Eq. (7.177), it may be shown that the resulting normal equations are

$$\left(\sum_{i=1}^{k} n_i w_i\right)b_0 + \left(\sum_{i=1}^{k} n_i w_i X_{1i}\right)b_1 = \sum_{i=1}^{k}\sum_{j=1}^{n_i} w_i Y_{ij}$$

$$\left(\sum_{i=1}^{k} n_i w_i X_{1i}\right)b_0 + \left(\sum_{i=1}^{k} n_i w_i X_{1i}^2\right)b_1 = \sum_{i=1}^{k}\sum_{j=1}^{n_i} w_i X_{1i} Y_{ij} \qquad (7.178)$$

The solutions b_0 and b_1 are the weighted regression estimates of β_0 and β_1 respectively.

In the more general situation, using matrix notation, the model is

$$\mathbf{Y} = \mathbf{X}\beta + \epsilon \qquad (7.179)$$

where $\epsilon \sim N(\phi, \mathbf{V}\sigma_E^2)$. That is, we assume the variance-covariance matrix of ϵ is of the form $\mathbf{V}\sigma_E^2$. Note that the variance-covariance structure is the only change from the model in Eq. (7.92). As indicated earlier, the basic idea of the analysis in this situation is to transform the observations to other variables that satisfy the usual assumptions. Without going through the derivation of this transformation, the least squares method, assuming the model in Eq. (7.179), leads to finding $\mathbf{B}$, which minimizes

$$S(b_0, \dots, b_p) = (\mathbf{Y} - \mathbf{XB})'\mathbf{V}^{-1}(\mathbf{Y} - \mathbf{XB}) \qquad (7.180)$$

The normal equations become

$$\mathbf{X}'\mathbf{V}^{-1}\mathbf{XB} = \mathbf{X}'\mathbf{V}^{-1}\mathbf{Y} \qquad (7.181)$$

and assuming $\mathbf{X}'\mathbf{V}^{-1}\mathbf{X}$ has an inverse, the solution is

$$\mathbf{B} = (\mathbf{X}'\mathbf{V}^{-1}\mathbf{X})^{-1}\mathbf{X}'\mathbf{V}^{-1}\mathbf{Y} \qquad (7.182)$$

The analysis of variance table useful for testing the significance of regression is given in Table 7.23. Also, the variances and covariances of the b's are given in the matrix $\mathbf{V}^*(\mathbf{B})$,

$$\mathbf{V}^*(\mathbf{B}) = (\mathbf{X}'\mathbf{V}^{-1}\mathbf{X})^{-1} \sigma_E^2 \qquad (7.183)$$

where $\mathbf{V}^*(\mathbf{B})$ is used to differentiate from the $\mathbf{V}$ matrix above.

TABLE 7.23–Analysis of Variance for Weighted Regression Analysis

Source of Variation	Degrees of Freedom	Sum of Squares	Mean Square
Due to $\beta_0, \dots, \beta_p$	$p + 1$	$SS(\beta_0, \dots, \beta_p)$ $= \mathbf{B}'\mathbf{X}'\mathbf{V}^{-1}\mathbf{Y}$	$SS(\beta_0, \dots, \beta_p)/(p + 1)$
Residual	$n - p - 1$	$E_{yy} = \mathbf{Y}'\mathbf{V}^{-1}\mathbf{Y}$ $- \mathbf{B}'\mathbf{X}'\mathbf{V}^{-1}\mathbf{Y}$	E
Total	n	$\mathbf{Y}'\mathbf{V}^{-1}\mathbf{Y}$	$\dots$

One of the difficult problems related to this analysis is that the matrix $\mathbf{V}$ is not always known. One approach to overcoming this difficulty is to run an initial regression analysis with $\mathbf{V} = I$ and then look at the plot of the residuals in an attempt to estimate $\mathbf{V}$. An example in which $\mathbf{V}$ does not need to be estimated is given in Example 7.19.

Example 7.19

Consider the simple linear regression model in which the variances are proportional to the value of the independent variable. In particular, consider the model

$$Y_{ij} = \beta_0 + \beta_1 X_{1i} + \epsilon_{ij} \qquad i = 1, \ldots, k; j = 1, \ldots, n_i \qquad (7.184)$$

where $\epsilon \sim N(\phi, \mathbf{V}\sigma_E^2)$, and

$$\mathbf{V} = \begin{bmatrix} X_{11} & 0 & \cdots & 0 \\ 0 & X_{12} & & \vdots \\ \vdots & & \ddots & 0 \\ 0 & \cdots & 0 & X_{1k} \end{bmatrix}$$

The normal equations are

$$\begin{bmatrix} \sum_{i=1}^{k} n_i/X_{1i} & \sum_{i=1}^{k} n_i \\ \sum_{i=1}^{k} n_i & \sum_{i=1}^{k} n_i X_{1i} \end{bmatrix} \begin{bmatrix} b_0 \\ b_1 \end{bmatrix} = \begin{bmatrix} \sum_{i=1}^{k} \sum_{j=1}^{n_i} Y_{ij}/X_{1i} \\ \sum_{i=1}^{k} \sum_{j=1}^{n_i} Y_{ij} \end{bmatrix} \qquad (7.185)$$

and the estimate of σ_E^2 is

$$s_E^2 = \sum_{i=1}^{k} \sum_{j=1}^{n_i} (Y_{ij} - b_0 - b_1 X_{1i})^2 / X_{1i}(n-2) \qquad (7.186)$$

7.13 NONLINEAR REGRESSION ANALYSIS

In Section 7.5 we indicated some examples of nonlinear regression functions and also an appropriate analysis for those models that could be put into the form of a linear regression function by a transformation. We now comment briefly on the analysis for a nonlinear regression model that cannot be linearized.

Suppose the model for the ith observation in the sample $Y_1, \ldots, Y_n$ is

$$Y_i = \phi(X_{1i}, \ldots, X_{pi}; \beta_0, \ldots, \beta_p) + \epsilon_i \qquad i = 1, \ldots, n \qquad (7.187)$$

where $\epsilon_i \sim NI(0, \sigma_E^2)$ and ϕ is the regression function that is nonlinear in some of the β's. An example of a nonlinear regression function is

$$\phi(X_{1i}, X_{2i}; \beta_0, \beta_1, \beta_2) = \beta_0 + \beta_1 X_{1i} + e^{-\beta_2 X_{2i}} \qquad (7.188)$$

This function cannot be linearized by a transformation of variables as the regression functions in Section 7.5.

Using the least squares method of estimation, we find the estimates

$b_0, \ldots, b_p$ that minimize

$$SS(b_0, \ldots, b_p) = \sum_{i=1}^{n} [Y_i - \phi(X_{1i}, \ldots, X_{pi}; b_0, \ldots, b_p)]^2 \quad (7.189)$$

For the regression function in Eq. (7.188) the sum of squares in Eq. (7.189) is

$$SS(b_0, b_1, b_2) = \sum_{i=1}^{n} (Y_i - b_0 - b_1 X_{1i} - e^{-b_2 X_{2i}})^2 \quad (7.190)$$

Differentiating with respect to the b's results in the set of normal equations

$$nb_0 + b_1 \sum_{i=1}^{n} X_{1i} + \sum_{i=1}^{n} e^{-b_2 X_{2i}} = \sum_{i=1}^{n} Y_i$$

$$b_0 \sum_{i=1}^{n} X_{1i} + b_1 \sum_{i=1}^{n} X_{1i}^2 + \sum_{i=1}^{n} X_{1i} e^{-b_2 X_{2i}} = \sum_{i=1}^{n} X_{1i} Y_i$$

$$b_0 \sum_{i=1}^{n} X_{2i} e^{-b_2 X_{2i}} + b_1 \sum_{i=1}^{n} X_{1i} X_{2i} e^{-b_2 X_{2i}} + \sum_{i=1}^{n} X_{2i} e^{-2b_2 X_{2i}} = \sum_{i=1}^{n} Y_i X_{2i} e^{-b_2 X_{2i}}$$

which is a set of nonlinear equations in the unknown b's. Thus to find estimates of the coefficients in a nonlinear regression model involves solving a set of nonlinear equations. This differs from the linear regression case when the normal equations were a set of linear equations in the unknown b's.

Several methods of solution for nonlinear models have been developed. One method based on the linearization of the normal equations is due to Hartley (7). Another solution is based on the method of steepest descent; this is described in Davies (4). A combination of these has been suggested by Marquardt (10). We refer the reader to some of the advanced texts on regression analysis for further discussion and analysis.

In any case most of the techniques available have been programmed for use on the computer, although care is required in use of these programs. A statistician familiar with these techniques should be consulted if a nonlinear regression model must be used.

7.14 SOME USES OF REGRESSION ANALYSIS

The uses to which regression techniques may be put are numerous. A few of the more important are:

1. To reduce the length of a confidence interval when estimating some population mean (or total) by considering the effect of concomitant variables.
2. To eliminate certain "environmental" effects from our estimates of treatment effects; i.e., we may wish to examine adjusted Y values.
3. To predict Y knowing values of $X_1, \ldots, X_k$ (our auxiliary variables) *whether or not* a causal relationship exists.

4. To influence the outcome of the dependent variables assuming that we have a causal relationship.

Many other uses for regression methods might have been listed. We have not attempted to exhaust the possibilities nor to give our examples in any order of importance. The relative importance of the different uses will vary depending on the subject matter being discussed.

PROBLEMS

7.1 Derive the normal equations specified by Eq. (7.15).

7.2 Derive Eq. (7.39).

7.3 Derive Eq. (7.41) from Eqs. (7.35) and (7.38).

7.4 Given the following values, find (a) $\hat{Y} = b_0 + b_1 X$, (b) s_E, (c) s_{b_1}.

$$\sum X^2 = 121 \qquad \sum X = 20 \qquad \sum XY = -82$$
$$\sum Y^2 = 516 \qquad \sum Y = 40 \qquad n = 10$$

7.5 Find the linear regression of Y on X given the values:

$$
\begin{array}{c|ccccc}
X: & 3 & 8 & 4 & 11 & 9 \\
Y: & 5 & 3 & 4 & 1 & 2 \\
\end{array}
$$

7.6 Given that $n = 277$, $\bar{X} = 65$, $\bar{Y} = 72$, $\sum (X - \bar{X})^2 = 1600$, $\sum (Y - \bar{Y})^2 = 3600$, $\sum (X - \bar{X})(Y - \bar{Y}) = 2000$, compute (a) s_E, (b) s_{b_1}, (c) $s_{\hat{y}}$ for $X = 45$.

7.7 Given the abbreviated analysis of variance shown below, perform the following:
(a) Test $H:\beta_1 = 0$, using $\alpha = 0.01$.
(b) Compute the standard error of estimate s_E.

Source of Variation	Degrees of Freedom	Sum of Squares	Mean Square
Due to regression	1	40	40
Deviations about regression	50	200	$\mathcal{L}$
Total	51	240	

7.8 Given $n = 38$, $\bar{X} = 5$, $\bar{Y} = 40$, $\sum (Y - \bar{Y})^2 = 10,000$, $\sum (X - \bar{X})(Y - \bar{Y}) = -800$, $\sum (X - \bar{X})^2 = 100$, answer the following:
(a) Determine $\hat{Y} = b_0 + b_1 X$.
(b) Test $H:\beta_1 = 0$, using $\alpha = 0.05$.
(c) Partition $\sum (Y - \bar{Y})^2$ into two parts, one associated with the slope of the linear regression and the other with the deviations about regression.
(d) For the observation $X = 8$, $Y = 36$, compute the adjusted value of Y.
(e) Interpret both b_1 and β_1.

7.9 Given that $n = 62$, $\bar{X} = 10$, $\bar{Y} = 20$, $\sum (X - \bar{X})^2 = 40$, $\sum (Y - \bar{Y})^2 = 250$, $\sum (X - \bar{X})(Y - \bar{Y}) = -80$, solve the following:
(a) Determine $\hat{Y} = b_0 + b_1 X$.

(b) Compute a 99 percent confidence interval for β_1. State all assumptions.

(c) Estimate the gain in information from using X as a statistical control (the regression of Y on X) in estimating the population mean of Y. (**Note:** *Information* is here used as a synonym for "the reciprocal of the variance.")

7.10   Given that $b_1 = 0.2$ g gain/g feed eaten, find the *net* difference between the gains of two rats where one animal consumed 200 g of feed and gained 60 g while the other animal consumed 300 g and gained 90 g.

7.11   The data in the table given below represent the heights X and the weights Y of several men. We selected the heights in advance and then observed the weights of a random group of men having the selected heights.

$X(in.)$	$Y(lb)$
60	110
61	135
60	120
61	126
62	140
60	130
62	135
65	158
64	145
70	170
72	185
70	180

(a) Plot a scatter diagram.
(b) Obtain the estimated regression line $\hat{Y} = b_0 + b_1 X$.
(c) Calculate and interpret a 90 percent confidence interval for β_1.
(d) Calculate and interpret a 98 percent confidence interval for β_0.
(e) Calculate and interpret a 95 percent confidence interval for $\mu_{Y|X=65}$.
(f) Test the hypothesis $H:\beta_1 = 0$.
(g) Test the hypothesis $H:\beta_1 = 6$.
(h) Test the hypothesis $H:\beta_0 = -30$.
(i) Predict the weight of an individual who is 66 in. in height. Give a "prediction interval."
(j) Estimate the height of a man whose recorded weight is 170 lb. Give both point and interval estimates.
(k) Test for "linearity of regression."
(l) What proportion of the variation in Y is "explained" by the regression of weight on height?

(**Note:** Give all assumptions and use a probability of Type I error equal to 0.05 in each test.)

7.12   Assume the data given in Problem 3.3 to be a random sample from a bivariate normal population.
(a) Calculate the regression for estimating weight from height.
(b) Calculate the regression for estimating height from weight.
(c) Plot a scatter diagram and show both regression lines thereon.

7.13   The *Consumer Market Data Handbook,* 1939 edition, U.S. Department of Commerce, lists consumer market data by states, counties, and cities. Among the types of information listed are "Population and Dwellings,"

"Volume and Type of Business and Industry, 1935," "Employment and Payrolls, 1935," "Retail Distribution by Kinds of Business, 1935," and "Related Indicators of Consumer Purchasing Power." Among the latter are numbers of income tax returns, automobile registrations, radios, telephones, electric meters, and magazine subscribers.

Such information as listed above might be used by national advertising agencies, large sales organizations, and individual retail or manufacturing agencies for various purposes in planning their business activities. The numbers and kinds of analyses that might be considered for such data are large. We have selected only a small portion for study in this problem. The data given here are for a group of Iowa counties.

County	Yearly per Capita Sales of Filling Stations (Y)	Automobile Registrations per 1000 Persons (X)
Adair	$17	206
Adams	25	233
Allamakee	16	237
Appanoose	13	183
Audubon	28	243
Benton	27	230
Black Hawk	20	272
Boone	21	214
Bremer	22	314
Buchanan	16	263
Buena Vista	32	314
Butler	27	295
Calhoun	27	273
Carroll	21	279
Cass	30	283
Cedar	21	276
Cerro Gordo	23	265
Cherokee	43	254
Chickasaw	23	264
Clarke	32	194
Clay	23	285
Clayton	14	255
Clinton	21	232
Crawford	19	238
Dallas	24	271
Davis	18	224
Decatur	12	203
Delaware	22	230

(a) Plot these data on an 8″ × 11″ sheet of graph paper. On the *abscissa* or X axis place automobile registrations and on the *ordinate* or Y axis plot sales per capita. Plot the point ($\bar{X}, \bar{Y}$) from the results to be obtained below.

(b) Calculate the means $\bar{X}$ and $\bar{Y}$ and the standard deviations for X and Y.

(c) Fit a straight line to the plotted points by obtaining the regression

226 CHAPTER 7

of Y on X as a least squares fit. What is the model in this case, i.e., for a single observation, county per capita sales by filling stations? What parameters do the statistics b_0 and b_1 estimate? Explain in words the meaning of b_0 and b_1; interpret the results of your analysis.

(d) Plot $\hat{Y} = b_0 + b_1 X$ on the scattergram.
(e) Calculate $\hat{Y}$ and $Y - \hat{Y}$ for each X.
(f) Calculate $(Y - \hat{Y})^2$ for each X and thus obtain $\sum (Y - \hat{Y})^2$. Compare this value with $\sum (Y - \overline{Y})^2 - b_1 \sum (X - \overline{X})(Y - \overline{Y})$. Then obtain s_E and explain its relationship to the values of $Y - \hat{Y}$.
(g) Determine 95 percent confidence limits for β_1.

7.14 On the basis of the following tabulations comparing years of service with ratings, the management seeks to discover whether or not there is a distinct tendency to rate old employees higher than more recent additions to the working force.

Employee*	Service (years)	Rating	Employee*	Service (years)	Rating
A	1	5	K	6	9
B	9	6	L	7	4
C	8	8	M	1	2
D	3	8	N	1	3
E	3	6	O	3	8
F	2	7	P	1	6
G	4	5	Q	2	5
H	5	6	R	2	3
I	5	4	S	4	4
J	6	5	T	2	7

*Source: Davies and Yoder (3), p. 313.

(a) Plot a scatter diagram (X = service, Y = rating).
(b) Obtain the regression line $\hat{Y} = b_0 + b_1 X$.
(c) Compute s_E^2.
(d) Compute $s_{b_1}^2$.
(e) Set out your results in an analysis of variance table.
(f) Test $H:\beta_1 = 0$; using (1) a t test and (2) an F test.
(g) Estimate the *average* rating that might be given an employee with (1) 4 years of service, (2) 15 years of service. Give both point and interval estimates. Discuss the validity of these estimates.
(h) Estimate by interval what rating an individual employee with 4 years of service might be expected to receive.

(Note: Whenever necessary, state all the assumptions made in order to use the techniques involved.)

7.15 Assume you are an investment counselor for a large insurance company. As one of your duties you would need to have some idea of the amount of policy loans per year, i.e., loans to policyholders, using their life insurance policies as collateral. Suppose you wish to estimate the total amount of policy loans your company will make during the coming year. Assume the date to be Jan. 1, 1948. You are given the data sheet shown below.

(a) What methods of estimation might you use and what would your estimates be?

(b) What further information might you request in order to do a better job?

(c) Give reasons for the answers you make to (b).

Year	National Income* ($ million)	Estimated Population of U.S.† (thousands)	Policy Loans Made by U.S. Life Insurance Companies‡ ($ million)
1929	87,355	121,770	2,379
1930	75,003	123,077	2,807
1931	58,873	124,040	3,369
1932	41,690	124,840	3,806
1933	39,584	125,579	3,769
1934	48,613	126,374	3,658
1935	56,789	127,250	3,540
1936	64,719	128,053	3,411
1937	73,627	128,825	3,399
1938	67,375	129,825	3,389
1939	72,532	130,880	3,248
1940	81,347	131,970	3,091
1941	103,834	133,203	2,919
1942	136,486	134,665	2,683
1943	168,262	136,497	2,373
1944	182,407	138,083	2,134
1945	181,731	139,586	1,962
1946	179,289	141,235	1,891
1947	202,500	144,034	1,937
1948	224,400§	146,571	...

*Statistical Abstract of the United States, 1949, p. 281.
†Ibid., p. 7.
‡Life Insurance Fact Book, 1949, p. 67.
§Estimated.

7.16 Let us assume that one of your duties is that of preparing reports for the managing director of a firm that is engaged in manufacturing. He is interested in the average cost per unit of production. Units of production are easily measured, but average cost requires lengthy and difficult computations. If some relationship between these two quantities can be determined empirically, an estimation procedure may be employed. From past records the following data are available:

Average cost (Y) (cents)	Units Produced (X) (thousands)
1.1	9
1.9	13
3.5	5
5.9	17
7.4	18
1.4	8

(Continued)

Average cost (Y) (cents)	Units Produced (X) (thousands)
2.6	14
1.4	12
1.9	7
3.5	15
1.0	10
1.1	11
4.6	4
17.9	23

(a)   What methods would you employ to have available a means of estimating values of Y if X were known?

(b)   Describe briefly what devices you would use to determine the type of curve that would best fit a given set of data.

7.17   An advertising concern is interested in prorating sales by counties for Maryland. In hopes of using magazine circulation per 1000 population to aid them, they obtained the following data:

Magazine Circulation/ 1000 Population (X)	Per Capita Sales (Y)
159	279
114	184
67	137
79	126
112	213
124	184
129	181
58	133
85	161
127	228
64	129
131	182
75	142
116	199
141	268
133	189
76	161
48	105
68	102
127	235
150	259
136	232
114	216

$$\sum X = 2433 \qquad\qquad \sum Y = 4245$$

$$\sum X^2 = 281{,}019 \qquad \sum Y^2 = 841{,}133 \qquad \sum XY = 482{,}786$$

(a) Determine the regression equation.
(b) If the circulation in County A was 90, what would you estimate the per capita sales to be? What is the standard error of $\hat{Y}$?
(c) Is the regression coefficient significant?
(d) What are your assumptions? Are they justified?

7.18 Given the following data satisfying the normality and homogeneous variance assumptions, do you believe that the true regression is actually linear?

			X		
	1	2	3	4	5
	4	18	26	38	44
Y	3	19	25	35	43
	6	18	24	28	39
	7	13	21	31	38

7.19 For the following data test the hypothesis that $\beta_1 = \beta_2 = \beta_3 = \beta_4$, where we assume normality, etc., as required for simple linear regression.

Sample	Degrees of Freedom	$\sum(X - \bar{X})^2$	$\sum(X - \bar{X})(Y - \bar{Y})$	$\sum(Y - \bar{Y})^2$
A	67	300	312	550
B	75	500	515	758
C	115	200	216	375
D	34	200	300	500

7.20 Using the data given below, fit a second-degree polynomial (parabola) for gross profit per farm against months of labor to obtain a curve for Iowa farms.

Farm No.*	Gross Profit (Y)	Months of Labor (X)	Farm No.*	Gross Profit (Y)	Months of Labor (X)
1	16.7	20	15	11.2	14
2	17.9	19	16	9.4	15
3	17.4	24	17	8.7	12
4	14.9	15	18	12.2	17
5	16.2	24	19	7.7	14
6	14.0	15	20	11.5	14
7	15.1	24	21	7.3	13
8	18.3	24	22	11.8	16
9	11.3	16	23	15.1	23
10	18.3	26	24	10.5	33
11	17.1	24	25	17.0	29
12	12.0	16	26	15.6	30
13	15.2	25	27	13.2	31
14	16.2	28	28	17.2	22

(Continued)

Farm No.*	Gross Profit (Y)	Months of Labor (X)	Farm No.*	Gross Profit (Y)	Months of Labor (X)
29	14.9	24	32	14.6	32
30	10.5	15	33	12.2	34
31	16.5	27	34	9.8	36

*Source: Selected values from Iowa farm records plus some supplementary hypothetical observations.

7.21 Given the linear regression: $\hat{Y} = 4 + 10X$ with $\bar{Y} = 244$, $s_{b_1} = 2$, $\sum X^2 = 58{,}000$, and $n = 100$:
   (a) What is the standard error of $\hat{Y} = 254$?
   (b) For what $\hat{Y}$ value is its variance a minimum?
   (c) Given that information is the reciprocal of the variance, how may we maximize our information about the unknown parameter β_1 in estimating a linear regression similar to the above?

7.22 In a regression study the following preliminary calculations were made: $\bar{X} = 20$, $\bar{Y} = 22$, $\sum(X - \bar{X})^2 = 225$, $\sum(Y - \bar{Y})^2 = 414$, $\sum(X - \bar{X})(Y - \bar{Y}) = 180$, $n = 32$.
   (a) What is the estimate for the population regression coefficient?
   (b) How do you interpret the population regression coefficient β_1?
   (c) Obtain the regression equation in the form $\hat{Y} = b_0 + b_1 X$.
   (d) Test the hypothesis $H : \beta_1 = 1$.

7.23 An economist from the University of Hawaii and an economist from the University of Chicago were comparing their studies of income and the consumption of various goods. Among the items studied was gasoline for use in private automobiles. Each had used a sample of about 100 university employees chosen to cover the range of salaries and wages. The Chicago economist reported that he had observed an increase in gas consumption of 10 gallons per $100 increase in income, while the Hawaii economist noted an increase of only 4 gallons per $100 increase in income. They then looked at the variances of their regression coefficients and gave these figures, $\hat{V}(b_C) = 2.41$ and $\hat{V}(b_H) = 2.0$.
   (a) Could the observed difference between the regression coefficients be expected to occur more than once in 20 times by chance if we consider the necessary assumptions for such a test to be fulfilled?
   (b) Would your conclusion be changed if the change in gas consumption had been reported as 0.1 gallon per $1 increase for Chicago and 0.04 gallon per $1 increase for Hawaii? Or what would the variance of b_H be per dollar increase in income if the regression coefficients had been reported in the latter unit?
   (c) When the Hawaii economist tested the hypothesis $\beta_H = 0$, he obtained a t value of $2.828 = (4 - 0)/\sqrt{2}$. Approximately what F value would he have obtained if he had examined the reduction in sum of squares due to linear regression by preparing an analysis of variance?
   [**Note:** $\hat{V}(\hat{\theta})$ is another way of expressing $s_{\hat{\theta}}^2$, e.g., $\hat{V}(b_C) = s_{b_C}^2$.]

7.24 The performance of a tensile test on a specific metal yielded the following results:

Brinell Hardness Number (X)	Tensile Strength (Y) (1000 psi)
104	38.9
106	40.4
106	39.9
106	40.8
102	33.7
104	39.5
102	33.0
104	37.0
102	33.2
102	33.9
101	29.9
105	39.5
106	40.6
103	35.1

(a) Determine the best linear regression equation by least squares and obtain confidence limits for estimating the mean tensile strength associated with a specified Brinell number.

(b) Is any functional form other than a linear equation indicated by these data? Make the appropriate test and discuss your results.

7.25 Using the data of Table 7.5, obtain the following regression equations:

(a) $\hat{Y} = b_0 + b_1 X_1$
(b) $\hat{Y} = b_0 + b_2 X_2$
(c) $\hat{Y} = b_0 + b_3 X_3$
(d) $\hat{Y} = b_0 + b_4 X_4$

For each regression equation, perform a complete analysis. Comment on the four different values of b_0. Also, compare the results of this problem with the multiple regression analysis obtained in Example 7.5.

7.26 The solubility of nitrous oxide in nitrogen dioxide was investigated with the results below. Perform a complete regression analysis and interpret your results.

Reciprocal Temperature (= 1000/degrees absolute)	3.801	3.731	3.662	3.593	3.533
Solubility (percent by weight)	1.28	1.21	1.11	0.81	0.65
	1.33	1.27	1.04	0.82	0.59
	1.52	. . .	. . .	. . .	0.63

7.27 A Rockwell hardness test is fairly simple to perform. However, the determination of abrasion loss is difficult. In an attempt to find a way of predicting abrasion loss from a measurement of hardness, an experiment was run and data collected on 30 samples. The following results were obtained: $\bar{X} = 70.27$, $\bar{Y} = 175.4$, $\sum(X - \bar{X})^2 = 4300$, $\sum(Y - \bar{Y})^2 = 225{,}011$, $\sum(X - \bar{X})(Y - \bar{Y}) = -22{,}946$, $s_E^2 = 3663$, and $\hat{Y} = 550.4 - 5.336X$. Estimate the abrasion loss when hardness is 70. Discuss the usefulness of the prediction equation.

7.28 A gauge is to be calibrated using dead weights. If X represents the standard and Y the gauge reading, perform a linear regression analysis based on the following results from 10 observations: $\overline{X} = 230$, $\overline{Y} = 226$, $\sum (X - \overline{X})(Y - \overline{Y}) = 1532$, $\sum (X - \overline{X})^2 = 1561$, $\sum (Y - \overline{Y})^2 = 1539$. Test $H:\beta_1 = 1$, using $\alpha = 0.01$.

7.29 Elongation of steel plate Y is related to the applied force in psi, X. Given the following data, perform a complete regression analysis and interpret your results.

X	Y
1.33	26
2.68	51
3.50	66
4.40	84
5.35	101
6.27	117
7.11	133
8.93	150
9.76	182
10.81	202

7.30 It is desired to determine the relationship of a twisting movement to the amount of strain imposed on a piece of test metal. Eight samples were obtained and the following data observed. Determine the "best" relationship between X and Y. Interpret your results.

Twisting Movement (X)	Strain (Y)
100	112
300	330
500	546
700	770
900	1010
1000	1100
1200	1323
1300	1515

7.31 The data given below and identified as Y, X_1, and X_2 represent annual figures for 1919 to 1943, a 25-year period, for three adjacent counties in the semiarid central area of South Dakota. Y_i is the average yield of oats in the ith year, X_{1i} is preseason precipitation in inches, e.g., 9.82 for X_{11} is the rainfall from August 1918, to March 31, 1919, etc. X_{2i} is the growing season precipitation in inches. This rainfall covers the period April 1 to July 31 for each crop year listed. Due to the nature of weather and yield data we may assume that these data fulfill our necessary assumptions for multiple linear regression. Do a complete analysis and interpretation of the data. The reader should note, however, that these are time series data, and thus an ordinary multiple linear regression analysis may be of doubtful validity.

Year	Y	X_1	X_2
1919	30.8	9.82	14.85
1920	34.2	9.12	17.30
1921	14.3	6.24	9.92
1922	34.5	14.06	9.33
1923	32.7	5.29	12.01
1924	36.0	7.74	10.87
1925	33.8	9.40	11.78
1926	3.7	4.22	7.14
1927	26.1	8.11	14.44
1928	18.6	6.30	8.95
1929	15.0	10.58	6.15
1930	23.8	8.62	8.63
1931	4.4	10.53	6.19
1932	23.5	7.05	8.86
1933	0.1	7.75	7.97
1934	0.0	4.41	4.93
1935	19.7	7.05	11.27
1936	0.0	6.90	5.37
1937	4.5	7.97	8.78
1938	14.4	5.41	10.37
1939	13.4	7.30	8.78
1940	11.8	5.94	7.06
1941	22.2	6.77	10.44
1942	42.9	11.23	14.58
1943	24.6	8.55	9.57

7.32   A study of 18 regions gives the following data on suicide rate, age, percentage male, and business failures. Fit an equation for the linear regression of Y on X_1, X_2, and X_3, where Y = suicide rate, X_1 = age, X_2 = percentage male, X_3 = business failures, and analyze completely. The summary of the data follows:

$$\sum Y = 285.3 \qquad \sum X_3^2 = 199843.52$$
$$\sum X_1 = 531.09 \qquad \sum YX_1 = 8536.6165$$
$$\sum X_2 = 911.95 \qquad \sum YX_2 = 14500.1161$$
$$\sum X_3 = 1800 \qquad \sum YX_3 = 29644.847$$
$$\sum Y^2 = 4905.6904 \qquad \sum X_1 X_2 = 26913.822$$
$$\sum X_1^2 = 15731.2223 \qquad \sum X_1 X_3 = 53614.575$$
$$\sum X_2^2 = 46218.4473 \qquad \sum X_2 X_3 = 91131.630$$

7.33   Do a complete multiple linear regression analysis of the following data. Interpret your results.

Rabbit No.	Choles- terol Dosage (g/day) (X_1)	Average Blood Total Choles- terol (mg) (X_2)	Initial Weight (kg) (X_3)	Ratio of Final Weight to Initial Weight (X_4)	Average Food Intake /kg Initial Weight (g/day) (X_5)	Degree of Athero- sclerosis (Y)
1	30	424	2.46	0.90	18	2
2	30	313	2.39	0.91	10	0
3	35	243	2.75	0.95	30	2
4	35	365	2.19	0.95	21	2
5	43	396	2.67	1.00	39	3
6	43	356	2.74	0.79	19	2
7	44	346	2.55	1.26	56	3
8	44	156	2.58	0.95	28	0
9	44	278	2.49	1.10	42	4
10	44	349	2.52	0.88	21	1
11	44	141	2.36	1.29	56	1
12	44	245	2.36	0.97	24	1
13	45	297	2.56	1.11	45	3
14	45	310	2.62	0.94	20	2
15	45	151	3.39	0.96	35	3
16	45	370	.3.57	0.88	15	4
17	45	379	1.98	1.47	64	4
18	45	463	2.06	1.05	31	3
19	45	316	2.45	1.32	60	4
20	45	280	2.25	1.08	36	4
21	44	395	2.15	1.01	27	1
22	49	139	2.20	1.36	59	0
23	49	245	2.05	1.13	37	4
24	49	373	2.15	0.88	25	1
25	51	224	2.15	1.18	54	3
26	51	677	2.10	1.16	33	4
27	51	424	2.10	1.40	59	4
28	51	150	2.10	1.05	30	0

7.34   You are presented with farm records for one year for a sample of 89 dairy farms located in a fairly homogeneous area in the same milk shed. The records contain the following information: Y = milk sold per cow (lb), X_1 = amount of concentrates fed per cow, X_2 = silage fed per cow, X_3 = pasture cost per cow, X_4 = amount of other roughage fed. You first decide to fit a multiple linear regression of Y, milk sold, on the four independent variates, the X's given above. Thus the regression equation is of the form $\hat{Y} = b_0 + b_1X_1 + b_2X_2 + b_3X_3 + b_4X_4$.

    (a)   List the numerical quantities and statistics you would compute to obtain this regression equation for $\hat{Y}$. You need not give detailed formulas. In particular, you will wish to compare β_2 and β_4, or silage with other roughage fed in effect on milk production. Also, pasture is quite homogeneous in the area, so you suspect β_3 may not be different from zero. Include in your list such items as needed for examination of the indicated regression coefficients.

(b)   Supposing you obtain $b_1 = +0.30$, what interpretation would you make of this statistic?

(c)   Can you suggest any other form for this regression function, using only the given X's? If so, write it out.

7.35   Using the data given below for 25 Iowa counties, obtain a multiple linear regression equation. (Do a complete analysis.) Then, consider other possible analyses and comment on the "best" functional relationship.

Observation Number	County	Corn Yield/ Acre 1910–19 (X_1)	Percentage Farmland in Small Grain (X_2)	No. Improved Acres/ Farm (X_3)	No. Brood Sows/ 1000 Acres (X_4)	Percentage Farmland in Corn (X_5)	Value/ Acre of Land Jan. 1, 1920 (Y)	Sum (W)
1	Allamakee	40	11	103	42	14	$ 87	297
2	Bremer	36	13	102	58	30	133	372
3	Butler	34	19	137	53	30	174	447
4	Calhoun	41	33	160	49	39	285	607
5	Carroll	39	25	157	74	33	263	591
6	Cherokee	42	23	166	85	34	274	624
7	Dallas	40	22	130	52	37	235	516
8	Davis	31	9	119	20	20	104	303
9	Fayette	36	13	106	53	27	141	376
10	Fremont	34	17	137	59	40	208	495
11	Howard	30	18	136	40	19	115	358
12	Ida	40	23	185	95	31	271	645
13	Jefferson	37	14	98	41	25	163	378
14	Johnson	41	13	122	80	28	193	477
15	Kossuth	38	24	173	52	31	203	521
16	Lyon	38	31	182	71	35	279	636
17	Madison	34	16	124	43	26	179	422
18	Marshall	45	19	138	60	34	244	540
19	Monona	34	20	148	52	30	165	449
20	Pocahontas	40	30	164	49	38	257	578
21	Polk	41	22	96	39	35	252	485
22	Story	42	21	132	54	41	280	570
23	Wapello	35	16	96	41	23	167	378
24	Warren	33	18	118	38	24	168	399
25	Winneshiek	36	18	113	61	21	115	364
	Sum	937	488	3342	1361	745	4955	11828
	Mean	37.48	19.52	133.68	54.44	29.80	198.20	473.12

Source: H.A. Wallace and G.W. Snedecor (12).

REFERENCES

1. Anderson, R. L.; and Houseman, E. E. 1942. Tables of orthogonal polynomial values extended to $N = 104$. Agr. Exp. Sta. Res. Bull. 297, Iowa State University, Ames.
2. Daniel, C.; and Wood, F. S. 1971. *Fitting Equations to Data.* Wiley, New York.
3. Davies, G. R.; and Yoder, D. 1937. *Business Statistics,* Wiley, New York.
4. Davies, O. L. (ed.). 1956. *Design and Analysis of Industrial Experiments,* 2nd ed. Oliver and Boyd, Edinburgh.
5. ———. 1957. *Statistical Methods in Research and Production,* 3rd ed. Oliver and Boyd, Edinburgh.
6. Ezekiel, M.; and Fox, K. A. 1959. *Methods of Correlation and Regression Analysis,* 3rd ed. Wiley, New York.

7. Hartley, H. O. 1961. The modified Gauss-Newton method for fitting of non-linear regression functions by least squares. *Technometrics*, vol. 3.

8. Hunter, J. S. 1958. Determination of optimum operating conditions by experimental methods. Part II-1. Models and methods. *Ind. Qual. Control* 15(Dec.):16–24.

9. Kendall, M. G. 1946. *The Advanced Theory of Statistics,* Vols. I and II. Charles Griffin, London.

10. Marquardt, D. W. 1963. An algorithm for least squares estimation of nonlinear parameters. *J. SIAM,* vol. 2.

11. Prater, N. H. 1956. Estimate gasoline yields from crudes. *Petroleum Refiner* 35(May): 236–38.

12. Wallace, H. A.; and Snedecor, G. W. 1931. *Correlation and Machine Calculations,* rev. ed. Iowa State College Press, Ames.

FURTHER READING

Bowker, A. H.; and Lieberman, G. J. 1972. *Engineering Statistics,* 2nd ed. Prentice-Hall, Englewood Cliffs, N.J.

Brownlee, K. A. 1965. *Statistical Theory and Methodology in Science and Engineering,* 2nd ed. Wiley, New York.

Daniel C.; and Wood, F. S. 1971. *Fitting Equations to Data.* Wiley, New York.

Dixon, W. J.; and Massey, F. J., Jr. 1969. *Introduction to Statistical Analysis,* 3rd ed. McGraw-Hill, New York.

Draper, N. R.; and Smith, H. 1966. *Applied Regression Analysis.* Wiley, New York.

Dunn, O. J.; and Clark, V. A. 1974. *Applied Statistics: Analysis of Variance and Regression.* Wiley, New York.

Neter, J.; and Wasserman, W. 1974. *Applied Linear Statistical Models: Regression, Analysis of Variance and Experimental Design.* Irwin, Homewood, Ill.

Snedecor, G. W.; and Cochran, W. G. 1967. *Statistical Methods,* 6th ed. Iowa State University Press, Ames.

Sprent, P. 1969. *Models in Regression and Related Topics.* Methuen, London.

Williams, E.J. 1959. *Regression Analysis.* Wiley, New York.

CHAPTER 8

CORRELATION ANALYSIS

In Chapter 7 methods of estimating functional relationships among variables were presented. Such methods have many uses in experimental work; however, a related matter also deserves attention when the joint variation of two or more variables is discussed. This is: How closely are the variables associated? Or in other words, what is the degree (or intensity) of association among the variables?

8.1 MEASURES OF ASSOCIATION

The techniques developed to provide measures of the degree of association between variables are known as *correlation methods*. This name reflects the universal practice of speaking about "measures of correlation" rather than about "measures of the degree (or intensity) of association." Consequently, when an analysis is performed to determine the amount of correlation, it is referred to as a *correlation analysis*. The resulting measure of correlation is usually called a *correlation coefficient*.

In this chapter some of the more frequently used measures of correlation will be presented. However, because of the close ties between this and some of the preceding chapters (particularly Chapter 7), a minimum of discussion will be sufficient.

8.2 INTUITIVE APPROACH TO CORRELATION

Because of the nature of the concept of correlation, it is clear that (in most cases) it is closely related to the concept of regression. In fact, for a given regression equation, it seems reasonable to expect that a correlation coefficient will measure how well the regression equation fits the data or, stating this in reverse fashion, how closely the sample points hug the regression curve. Thus a correlation coefficient will undoubtedly be related to the standard error of estimate (s_E), which measures the dispersion of the points about the regression curve.

Pursuing this idea and denoting the correlation coefficient by the symbol R, we express R as a function of s_E,

$$R = f(s_E) \tag{8.1}$$

If R is to perform satisfactorily as a measure of correlation, it is desirable that it exhibit two characteristics: (1) it should be large when the degree of association is high and small when the degree of association is low and (2) it should be independent of the units in which the variables are measured.

One way to achieve the desired properties is to (approximately) define R by

$$R^2 \cong 1 - s_E^2/s_Y^2 \qquad (8.2)$$

where

$$s_E^2 = \sum_{i=1}^{n} (Y_i - \hat{Y}_i)^2/(n - p) \qquad (8.3)$$

$$s_Y^2 = \sum_{i=1}^{n} (Y_i - \overline{Y})^2/(n - 1) \qquad (8.4)$$

and p is the number of parameters in the true regression function, estimated by the regression equation symbolized by $\hat{Y}$. If n is large relative to p, another approximation is

$$R^2 \cong 1 - \sum_{i=1}^{n} (Y_i - \hat{Y}_i)^2 / \sum_{i=1}^{n} (Y_i - \overline{Y})^2 \qquad (8.5)$$

Since $\sum_{i=1}^{n} (Y_i - \hat{Y}_i)^2 \leq \sum_{i=1}^{n} (Y_i - \overline{Y})^2$, it is clear that $0 \leq R^2 \leq 1$. Further, if the sample points hug the regression curve closely (i.e., the correlation is high), R^2 will be close to 1. Similarly, if the regression curve is a poor fit, the sample points will be widely dispersed about the estimated regression and R^2 will be close to 0, reflecting a low correlation.

Having given the foregoing intuitive approach to correlation, it is necessary that a more precise approach be formulated. This will now be done. It is hoped that the remarks given earlier in this section will aid the reader in appreciating the discussions to follow.

8.3 CORRELATION INDEX

Rewriting Eq. (8.5) as

$$R^2 = \left[\sum_{i=1}^{n} (Y_i - \overline{Y})^2 - \sum_{i=1}^{n} (Y_i - \hat{Y}_i)^2 \right] \bigg/ \sum_{i=1}^{n} (Y_i - \overline{Y})^2 \qquad (8.6)$$

and referring to Sections 7.3 and 7.4, it is seen that

$$R^2 = SS \text{ due to regression/total (corrected) } SS \qquad (8.7)$$

where SS = sum of squares. Since the ratio defined by Eq. (8.7) may be calculated for any estimated regression equation, it is a most general and useful measure of correlation and is referred to as the *correlation index*. In succeeding sections special cases will be examined in detail.

8.4 CORRELATION IN SIMPLE LINEAR REGRESSION

In Section 7.3.7, the partitioning of the sum of squares of the dependent variable was discussed and the results presented in Table 7.3. Referring to Table 7.3 and using Eqs. (7.43) and (7.44), R^2 in Eq. (8.6) can be written in the form

$$r^2 = \frac{b_1 \sum_{i=1}^{n} (X_{1i} - \bar{X}_{1.})(Y_i - \bar{Y})}{\sum_{i=1}^{n} (Y_i - \bar{Y})^2} = \frac{\left[\sum_{i=1}^{n} (X_{1i} - \bar{X}_{1.})(Y_i - \bar{Y})\right]^2}{\sum_{i=1}^{n} (X_{1i} - \bar{X}_{1.})^2 \sum_{i=1}^{n} (Y_i - \bar{Y})^2}$$

(8.8)

since $b_1 = \sum_{i=1}^{n} (X_{1i} - \bar{X}_{1.})(Y_i - \bar{Y})/\sum_{i=1}^{n} (X_{1i} - \bar{X}_{1.})^2$. The symbol r^2 is used instead of R^2 to conform with standard practice. If we let $x_{1i} = X_{1i} - \bar{X}_{1.}$ and $y_i = Y_i - \bar{Y}$, Eq. (8.8) can be written as

$$r^2 = b_1 \sum_{i=1}^{n} x_{1i} y_i \Big/ \sum_{i=1}^{n} y_i^2 = \left(\sum_{i=1}^{n} x_{1i} y_i\right)^2 \Big/ \left(\sum_{i=1}^{n} x_{1i}^2\right)\left(\sum_{i=1}^{n} y_i^2\right)$$

(8.9)

It is customary to talk about r rather than r^2. Thus we have

$$r = \sum_{i=1}^{n} x_{1i} y_i \Big/ \left[\left(\sum_{i=1}^{n} x_{1i}^2\right)\left(\sum_{i=1}^{n} y_i^2\right)\right]^{1/2}$$

(8.10)

which assumes the same sign as $\sum_{i=1}^{n} x_{1i} y_i$ and hence the same sign as b_1

It is readily seen that the correlation coefficient associated with simple linear regression is easily obtained once a regression analysis has been performed. Further, it is clear that

$$-1 \le r \le 1$$

(8.11)

where -1 represents *perfect negative linear association in the sample* and $+1$ represents *perfect positive linear association in the sample.* A value of 0 is interpreted to mean that no linear association between X and Y exists in the sample. Since r is only a sample value, any inference to the sampled population must be carefully stated. More will be said concerning this a little later.

Example 8.1

Referring to Tables 7.1 and 7.4, the coefficient of linear correlation between X and Y for the Schopper-Riegler data is determined as follows:

$$r^2 = 10,177.59/(51,712.00 - 41,217.23) = 0.9698$$
$$r = \sqrt{0.9698} = 0.98$$

Example 8.2

For the following data,

X	Y
-2	4
-1	1
0	0
1	1
2	4

it may be verified that $r = 0$, indicating no *linear* association. Please note carefully the word "linear," for a moment's reflection will reveal that X and Y are perfectly associated, the relationship being $Y = X^2$. What we calculated was a measure of linear correlation when the indicated relationship is actually quadratic. This simple example should call to your attention one of the greatest potential trouble spots in correlation analysis, namely, the use of an inappropriate measure of correlation.

The preceding discussion and interpretation of r, or perhaps we should say of r^2, is most valuable in regression analyses. Examination of Eqs. (8.7) and (8.8) reminds us that $100r^2$ is the percentage of the corrected sum of squares that is "explained by" the fitting of the simple linear regression $\hat{Y} = b_0 + b_1 X_1$. If this percentage is not large enough to satisfy us, a better fitting regression equation should be found.

Some terms associated with the coefficient of correlation that are sometimes encountered are:

$$r^2 = \text{coefficient of determination} \qquad (8.12)$$

$$1 - r^2 = \text{coefficient of nondetermination} \qquad (8.13)$$

and

$$\sqrt{1 - r^2} = \text{coefficient of alienation} \qquad (8.14)$$

Example 8.3

For the Schopper-Riegler data, $r^2 = 0.9698$ and $1 - r^2 = 0.0302$. Thus 96.98 percent of the variation in Y (Schopper-Riegler rating) is "explained by" the linear regression of Y on X (hours of beating).

8.5 SAMPLING FROM A BIVARIATE NORMAL POPULATION

The interpretation of r given in the preceding section is valid for any simple linear regression regardless of what assumptions are made concerning the variables X and Y. However, if a random sample is drawn from a bivariate normal population, r [defined by Eq. (8.10)] is a sample estimate of the population parameter

$$\rho_{XY} = \rho = \sigma_{XY}/\sigma_X \sigma_Y \qquad (8.15)$$

The reader should note that this is the same correlation coefficient defined in Eq. (2.69), and thus it is not surprising that r is sometimes referred to as the *sample product–moment correlation coefficient*.

When sampling from a bivariate normal population, it is natural to want to test hypotheses about the true value of ρ. Since such tests are simply further examples of the general techniques introduced in Chapter 6, only a brief explanation will be given.

To test $H:\rho = 0$ versus the alternative $A:\rho \neq 0$, we calculate

$$t = (r - 0)/s_r = r\sqrt{n - 2}/\sqrt{1 - r^2} \qquad (8.16)$$

and reject H if $t \geq t_{(1 - \alpha/2)(n - 2)}$ or if $t \leq -t_{(1 - \alpha/2)(n - 2)}$. However, a minimum amount of simple algebra will show that

$$t = r/s_r = b_1/s_{b_1} \tag{8.17}$$

and thus the test just detailed is identically equivalent to the test of $H:\beta_1 = 0$ versus $A:\beta_1 \neq 0$ as given in Section 7.3.6. A review of that section will remind you that the hypothesis might also be tested using an F ratio [see Eq. (7.45)]. Consequently, three equivalent methods of testing are available, the choice being determined by the form of the analysis.

Example 8.4

Given the sample observations

X	3	8	4	11	9
Y	5	3	4	1	2

it is easily verified that $r = -0.98$. Using Eq. (8.16), we obtain $t = (-0.98)\sqrt{3}/\sqrt{0.0413} = -8.24$. Since $t = -8.24 < -t_{0.995(3)} = -5.841$, the hypothesis $H:\rho = 0$ is rejected in favor of the alternative $A:\rho \neq 0$. Clearly, a 1 percent significance level was used. It is suggested that the reader consider $H:\beta_1 = 0$ versus $A:\beta_1 \neq 0$ and compare the resulting test statistic with that computed above.

If the hypothesis to be tested is $H:\rho = \rho_0$ versus $A:\rho \neq \rho_0$, where $\rho_0 \neq 0$, the test procedure is more complicated. The complication arises because $(r - \rho_0)/s_r$ is not distributed as "Student's" t unless $\rho_0 = 0$. When $\rho_0 \neq 0$, an approximate test is provided by

$$\begin{aligned} z_r &= (1/2)[\ln (1 + r) - \ln (1 - r)] \\ &= (1.1513)[\log_{10}(1 + r) - \log_{10}(1 - r)] \end{aligned} \tag{8.18}$$

Fisher (1) has shown that z_r is approximately normally distributed with mean z_{ρ_0} and variance $\sigma_z^2 = 1/(n - 3)$. The approximate test procedure is to calculate

$$z = (z_r - z_{\rho_0})/\sigma_z \tag{8.19}$$

and compare this quantity with fractiles of the standard normal distribution. The hypothesis $H:\rho = \rho_0$ would be rejected if

$$z \geq z_{(1-\alpha/2)} \tag{8.20}$$

or if

$$z \leq -z_{(1-\alpha/2)} \tag{8.21}$$

The research worker may also be interested in obtaining a confidence interval estimate of ρ. This may be obtained by calculating

$$\left.\begin{matrix} L \\ U \end{matrix}\right\} = z_r \mp z_{[(1+\gamma)/2]}\sigma_z \tag{8.22}$$

and then using Eq. (8.18) to solve for r_L and r_U.

Quite frequently the research worker has several independent samples, each randomly selected from a bivariate normal population, from which estimates $r_1, \ldots, r_k$ are obtained. If the research worker can accept the hypothesis $H:\rho_1 = \cdots = \rho_k$, it is permissible to obtain a pooled estimate of the common population correlation coefficient, and this pooled estimate should be more reliable than any of the individual estimates. If calculations are carried out as in Table 8.1 and the observed chi-square is not judged significant at the 100α percent significance level, a pooled estimate of ρ (corresponding to the "average z") may be found.

TABLE 8.1–Calculations for Testing the Hypothesis $\rho_1 = \cdots = \rho_k$ $(k = 3)$

Sample	Size of Sample (n)	$n - 3$	r	z	$(n - 3)z$	$(n - 3)z^2$
1	102	99	0.63245	0.74551	73.80549	55.02273
2	102	99	0.77459	1.03168	102.13632	105.37200
3	102	99	0.67082	0.81223	80.41077	65.31204
Total	306	297	...	...	256.35258	225.70677

$$\text{ave } z = \sum_{i=1}^{k} (n_i - 3)z_i \bigg/ \sum_{i=1}^{k} (n_i - 3) = 256.35258/297 = 0.86314$$

$$\chi^2 = \sum_{i=1}^{k} (n_i - 3)z_i^2 - \left[\sum_{i=1}^{k} (n_i - 3)z_i\right]^2 \bigg/ \sum_{i=1}^{k} (n_i - 3)$$

$$= 225.70677 - 221.26817 = 4.4386$$

Reject $H:\rho_1 = \cdots = \rho_k$ if $\chi^2 \geq \chi^2_{(1-\alpha/2)(k-1)}$

8.6 CORRELATION IN MULTIPLE LINEAR REGRESSION

When a multiple linear regression equation has been fitted to a set of data as in Section 7.4, it is natural to seek a measure of correlation that reflects the "goodness of the fit." The correlation index defined in Section 8.3 may be used to give us what we desire. Referring to Eq. (7.80), we see that

$$R^2 = \frac{SS \text{ due to regression}}{\text{total (corrected) } SS} = \frac{SS(\beta_1, \ldots, \beta_p \mid \beta_0)}{\sum_{i=1}^{n} (Y_i - \overline{Y})^2} \tag{8.23}$$

This may also be expressed as

$$R^2 = \frac{b_1 \sum_{i=1}^{n} (X_{1i} - \overline{X}_{1.})(Y_i - \overline{Y}) + \cdots + b_p \sum_{i=1}^{n} (X_{pi} - \overline{X}_{p.})(Y_i - \overline{Y})}{\sum_{i=1}^{n} (Y_i - \overline{Y})^2}$$

$$= \sum_{j=1}^{p} b_j \left(\sum_{i=1}^{n} x_{ji} y_i \right) \Big/ \sum_{i=1}^{n} y_i^2 \tag{8.24}$$

which is analogous to the expression for r^2 given in Eq. (8.9). If we calculate $R = \sqrt{R^2}$, where R^2 is defined by Eq. (8.23) or Eq. (8.24), then R is known as the *multiple correlation coefficient*. The significance of R may be assessed by the F test specified in Eq. (7.81). No example will be given at this time since nothing new and different is involved. However, some of the problems at the end of the chapter will require the calculation and interpretation of the coefficient of multiple correlation.

It is also worth noting that R, as defined by Eq. (8.23), may be thought of as a simple linear correlation between Y and $\hat{Y}$ where $\hat{Y} = b_0 + b_1 X_1 + \cdots + b_p X_p$.

Closely allied to the topic of multiple correlation is that of *partial correlation*. By partial correlation is meant the correlation between two variables in a multivariable problem under the restriction that any common association with the remaining variables (or some of them) has been "eliminated." Clearly, many partial correlation coefficients may be calculated. For example, a *first-order partial correlation coefficient* is one that measures the degree of linear association between two variables after taking into account their common association with a third variable. Symbolically,

$$r_{12.3} = (r_{12} - r_{13}r_{23})/(\sqrt{1 - r_{13}^2}\,\sqrt{1 - r_{23}^2}) \tag{8.25}$$

where the subscripts refer to the three variables X_1, X_2, and X_3. Here, of course, $r_{12.3}$ is attempting to measure the correlation between X_1 and X_2 independent of X_3. It should also be clear that $r_{ij}(i,j = 1, 2, 3)$ are simple linear correlation coefficients measuring the correlation between X_i and X_j. A *second-order partial correlation coefficient* may be illustrated by

$$r_{12.34} = (r_{12.3} - r_{14.3}r_{24.3})/(\sqrt{1 - r_{14.3}^2}\,\sqrt{1 - r_{24.3}^2}) \tag{8.26}$$

which measures the correlation between X_1 and X_2 independent of X_3 and X_4.

Before proceeding to another topic, it will be worth digressing for a moment to discuss a related matter (related to partial correlation, that is) in regression. In Section 7.4, the equation

$$\hat{Y} = b_0 + b_1 X_1 + \cdots + b_p X_p \tag{8.27}$$

was discussed for the case $p = 4$. At that time, had we so desired, it would have been appropriate to call attention to a different system of notation sometimes encountered. For $p = 4$, Eq. (8.27) would appear as

$$\hat{Y} = b_0 + b_1 X_1 + b_2 X_2 + b_3 X_3 + b_4 X_4 \tag{8.28}$$

An alternative notation is

$$\hat{Y} = b_0 + b_{Y1.234}X_1 + b_{Y2.134}X_2 + b_{Y3.124}X_3 + b_{Y4.123}X_4 \quad (8.29)$$

and in this form the analogy with partial correlation is evident. Strictly speaking, the coefficients should be called *partial regression coefficients* where, for example, $b_{Y1.234}$ represents how Y would vary per unit change in X_1 if $X_2, X_3,$ *and* X_4 *were all held fixed.* Thus $b_{Y1.234}$ (or, as we usually denote it, b_1) gives only a partial picture of what happens to Y as X_1 changes. Hence the adjective "partial." It should be clear that the less cumbersome notation was used (at the risk of not clearly defining the meaning) solely to simplify the writing of the equations.

8.7 CORRELATION RATIO

Closely related to the correlation index is a quantity known as the *correlation ratio.* Denoted by E^2, it is defined by

$$E^2 = \sum_{i=1}^{k} n_i(\overline{Y}_i - \overline{Y})^2 \bigg/ \sum_{i=1}^{k} \sum_{j=1}^{n_i} (Y_{ij} - \overline{Y})^2 \quad (8.30)$$

where $\overline{Y}_i$ is the mean of the ith group consisting of n_i observations and $\overline{Y}$ is the mean of all observations. Expressing Eq. (8.30) in words,

$$E^2 = \text{among groups } SS/\text{total (corrected) } SS \quad (8.31)$$

where the quantity labeled "among groups sum of squares" is most easily found using the identity

$$\sum_{i=1}^{k} n_i(\overline{Y}_i - \overline{Y})^2 = \sum_{i=1}^{k} G_i^2/n_i - T^2/n \quad (8.32)$$

where

$$G_i = \sum_{j=1}^{n_i} Y_{ij} = n_i\overline{Y}_i$$

$$= \text{total of the observations in the } i\text{th group} \quad (8.33)$$

$$T = \sum_{i=1}^{k} G_i = n\overline{Y} = \text{total of all observations} \quad (8.34)$$

and $n = \sum_{i=1}^{k} n_i$ = total number of observations. Also, the total (corrected) SS can be found using the identity

$$\sum_{i=1}^{k} \sum_{j=1}^{n_i} (Y_{ij} - \overline{Y})^2 = \sum_{i=1}^{k} \sum_{j=1}^{n_i} Y_{ij}^2 - T^2/n \quad (8.35)$$

A moment's reflection will indicate that the value of E^2 is highly dependent on the choice of groups. For example, if there is only one observation in each group, the value of E^2 is unity; if all the observations are in one group, the value of E^2 is 0. Great care must be exercised when grouping the observations.

Another point of interest is the following: once the observations are assembled in groups, the value of E^2 is determined solely from the values of the "dependent" variable. Consequently, the "independent" variable need not be quantitative but can be a *qualitative* variable. Thus, subject to the dangers implicit in the grouping, the correlation ratio may be used to measure the correlation between a quantitative variable and a qualitative variable.

Since grouping is so important, some guidance is necessary. One rule of thumb is to have three to five groups, each containing a large number of observations (say 100). Strict rules of procedure are hard to define, but the preceding rule may prove helpful. Denoting the population correlation ratio by η^2, Woo (9) gives tables for use in testing the hypothesis $H:\eta = 0$ when we are willing to assume that the Y_{ij} are normally and independently distributed (with common variance) in each group.

TABLE 8.2–Analysis of Variance Associated with the
Calculation of a Correlation Ratio

Source of Variation	Degrees of Freedom	Sum of Squares	Mean Square
Mean	1	$M_{yy} = T^2/n$	$M_{yy}/1$
Among groups	$k - 1$	$G_{yy} = \displaystyle\sum_{i=1}^{k} G_i^2/n_i - T^2/n$	$G_{yy}/(k - 1)$
Within groups	$\displaystyle\sum_{i=1}^{k} (n_i - 1)$	$W_{yy} = \displaystyle\sum_{i=1}^{k} \sum_{j=1}^{n_i} Y_{ij}^2 - M_{yy} - G_{yy}$	$W_{yy}\Big/\displaystyle\sum_{i=1}^{k} (n_i - 1)$
Total	n	$\displaystyle\sum_{i=1}^{k} \sum_{j=1}^{n_i} Y_{ij}^2$	$\ldots$

Because the analysis of variance form of presenting results is so often encountered, it should not be surprising to find it helpful in the situation under discussion. Referring to Table 8.2, it is seen that the sums of squares needed in Equation (8.31) are easily accessible. (**Note:** Now that the opportunity has presented itself, we shall take a moment to review the symbolism introduced in Section 6.5 and used in Table 8.2. It seems almost unnecessary to remark that the letters M, G, and W in the symbols M_{yy}, G_{yy}, and W_{yy} (M_{xx}, G_{xx} and W_{xx} in Section 6.5) were chosen to stand for the words "Mean, Groups, and Within," respectively. However, since this abbreviated method of representing various sums of squares will be used extensively in later chapters, it is a good idea to become well acquainted with the notation as early as possible.)

8.8 BISERIAL CORRELATION

A measure of correlation encountered frequently in such areas of specialization as education, psychology, and public health is the *biserial correlation coefficient.* Only a brief discussion will be given in this text. Those persons interested in more detail are referred to McNemar (4), Pearson (7), and Treloar (8).

The biserial correlation coefficient, usually denoted by r_b, is used where one variable Y is quantitatively measured while the second variable X is dichotomized, i.e., defined by two groups. The assumptions necessary for a meaningful interpretation of r_b are:

1. Y is normally distributed and suffers little due to broad grouping (if grouping is necessary).
2. The true distribution underlying the dichotomized variable X should be of normal form.
3. The regression of Y on X is linear.
4. The mean value of Y in the minor or smaller category as specified by X, denoted by $\overline{Y}_1$, is to be on the regression line. This assumption implies a large number of observations in the minor segment.

If we define:

p = proportion of observations in the major category
q = proportion of observations in the minor category
z = ordinate of the standard normal curve at the point cutting off a tail of that distribution with area equal to q
$\overline{Y}_2$ = mean of the Y values in the major category
s_Y = standard deviation of all the Y values

then

$$r_b = (\overline{Y}_2 - \overline{Y}_1)pq/zs_Y \tag{8.36}$$

and this gives us a measure of the degree of linear association between X and Y.

It should be mentioned that in a manner analogous to the way in which we developed the correlation ratio, Pearson (7) introduced the concept of a *biserial correlation ratio,* denoted by E_b, which extends the biserial correlation concept to cover any postulated regression function. We shall not go into detail here, but the reader is referred to Pearson (7) and Treloar (8) if he is interested in such problems.

8.9 TETRACHORIC CORRELATION

Another measure frequently encountered in some areas of research is the *tetrachoric correlation coefficient.* This is generally denoted by r_t and is used to measure the degree of linear association between two variables X and Y, where *both* are dichotomized and the true underlying distributions are assumed to be normal. That is, if we have samples from a bivariate normal population but the measurements are *not* available (we know only to which cell of a 2 × 2 contingency table each observation belongs), we can obtain a measure of the correlation between X and Y. It is not feasible to present a formula for r_t, but reference to McNemar (4), Treloar (8), and other works will indicate calculational methods for those interested in this particular statistic.

8.10 COEFFICIENT OF CONTINGENCY

Of some interest also is a measure of the degree of association between two characteristics, where our observational data are classified in an $r \times c$ contingency table. In Chapter 6 we gave a method for testing the hypothesis that these two characteristics, or classifications, were independent of one another. Suppose, however, that we are more interested in estimating the degree of association between them than testing the hypothesis of independence. How may we do this? Pearson (6) proposed for this purpose a measure known as the *coefficient of contingency* defined by

$$C = [\chi^2/(n + \chi^2)]^{1/2} \tag{8.37}$$

where n is the total number of observations and

$$\chi^2 = \sum_{i=1}^{r} \sum_{j=1}^{c} (O_{ij} - E_{ij})^2/E_{ij}$$

as given in Chapter 6. In the case of a 2×2 table, this may seem to be analogous to a tetrachoric correlation coefficient, but the coefficient of contingency is of wider generality because we no longer require the assumption of normality of the underlying distributions. Any distribution, discrete or continuous, is acceptable. However, there is a disadvantage to this measure of association; its maximum possible value varies with the number of rows and columns, and thus two different values of C are not directly comparable unless computed from tables of the same size. For further remarks on this measure the reader is referred to McNemar (4) and Treloar (8).

8.11 RANK CORRELATION

Let us now consider a slightly different problem but one that arises quite frequently in certain areas of research. The problem is as follows: n individuals are ranked from 1 to n according to some specified characteristic by m observers, and we wish to know if the m rankings are substantially in agreement with one another. How may we answer such a query? Kendall and Smith (3) have proposed a measure known as the *coefficient of concordance* W for answering this question, which is defined by

$$W = 12S/m^2(n^3 - n) \tag{8.38}$$

where S equals the sum of the squares of the deviations of the total of the ranks assigned to each individual from $m(n + 1)/2$. The quantity $m(n + 1)/2$ is, of course, the average value of the totals of the ranks, and hence S is the usual sum of squares of deviations from the mean. W varies from 0 to 1, 0 representing no community of preference, while unity represents perfect agreement. The hypothesis that the observers have no community of preference may be tested, using tables given in Kendall (2) or, more simply (for $n > 7$), by calculating

$$\chi^2 = m(n-1)W = 12S/mn(n+1) \tag{8.39}$$

which is approximately distributed as chi-square with $\nu = n - 1$ degrees of freedom. If there are "ties" in some of the rankings, it may be necessary to modify our formulas somewhat; if such a case is encountered, the researcher is referred to Kendall (2).

If we find W to be significant, the next step is to estimate the true ranking of the n individuals. This is done by ranking them according to the sum of the ranks assigned to each, the one with the smallest sum being ranked first, the one with the next smallest sum being ranked second, and so on. If two sums are equal, we rank these two individuals by the sum of the *squares* of the ranks assigned to them, the one with the smaller sum of squares obviously being ranked ahead of the other. If W is not significant, we are not justified in attempting to find an "average," or "pooled," estimate of a true ranking, for we are not at all certain that such a true ranking even exists.

When $m = 2$ (i.e., when only two rankings are available) a slightly different approach is often used. In this case a measure known as Spearman's rank correlation coefficient is computed. Spearman's rank correlation coefficient, denoted by r_S, is defined by

$$r_S = 1 - 6 \sum_{i=1}^{n} d_i^2/(n^3 - n) \tag{8.40}$$

where d_i equals the difference between the two ranks assigned to the ith individual. It can easily be seen that r_S varies from -1 to $+1$, whereas W varied only from 0 to 1, -1 signifying perfect disagreement and $+1$ signifying perfect agreement between the two rankings. A test of the null hypothesis $H: \rho_S = 0$ may be made using tables provided by Olds (5). We must remember, however, that the same conclusion (namely, to accept or reject H) could be reached by computing W and comparing with the tabulated values for $m = 2$. Incidentally, Kendall (2) does not tabulate W itself but only the associated value of S. This cuts down the amount of arithmetic required since it is not necessary actually to compute the value of W to perform our statistical test. Similarly, Olds (5) only tabulates $\sum_{i=1}^{n} d_i^2$.

TABLE 8.3–Preferences for Six Lemonades as Expressed by Two Judges

Lemonade	Ranking Given by Judge No. 1	Ranking Given by Judge No. 2	Difference in Ranks = d
A	4	4	0
B	1	2	-1
C	6	5	1
D	5	6	-1
E	3	1	2
F	2	3	-1

Example 8.5

Consider the data of Table 8.3. Calculations yield $r_S = 0.771$ with $\sum_{i=1}^{6} d_i^2 = 8$. Using $\alpha = 0.05$, the hypothesis $H: \rho_S = 0$ is rejected. [Note: This conclusion was reached after consulting the tables provided by Olds (5).] Thus it is concluded that the two judges are in quite good agreement.

Example 8.6

Consider the data of Table 8.4. It may be verified that $m(n + 1)/2 = 10.5$, $S = 25.5$, and $W = 0.162$. Examination of the tables in Kendall (2) leads us to accept the hypothesis of no community of preference among our three judges, and thus we shall not attempt to estimate any "true order of preference."

TABLE 8.4–Preferences for Six Lemonades as Expressed by Three Judges

Lemonade	Ranking Given by Judge No. 1	Ranking Given by Judge No. 2	Ranking Given by Judge No. 3	Sum of Ranks
A	5	2	4	11
B	4	3	1	8
C	1	1	6	8
D	6	5	3	14
E	3	6	2	11
F	2	4	5	11

8.12 INTRACLASS CORRELATION

The measure of correlation to be discussed in this section was devised to assess the degree of association (or similarity) among individuals within classes or groups. For this reason, the measure is known as the *intraclass correlation coefficient*. (**Note:** Some authors have referred to the intraclass correlation coefficient as the *coefficient of homotypic correlation* but the former term is more common.)

As an example of a situation in which the intraclass correlation coefficient is the proper measure, consider the problem of measuring the correlation between heights of brothers. Because all that is desired is a measure of similarity between heights of brothers, any attempt to label one as X and the other Y (e.g., by age) would introduce a spurious element into the correlation. This would be that an ordinary (simple linear) correlation would measure the correlation between the heights of older brothers and the heights of younger brothers rather than simply assess the "sameness" of heights of brothers.

The intraclass correlation coefficient, denoted by r_I, is most easily calculated using analysis of variance techniques. Given the data of Table 8.5, the variation among the kn observations may be summarized

TABLE 8.5–Symbolic Representation of Data to Be Used in Calculating
the Intraclass Correlation Coefficient

	Group			
	1	2	$\cdots$	k
Observations*	Y_{11} Y_{12} . . . Y_{1n}	Y_{21} Y_{22} . . . Y_{2n}		Y_{k1} Y_{k2} . . . Y_{kn}
Total	G_1	G_2		G_k

*Each observation is assumed to be of the form $Y_{ij} = \mu + g_i + \epsilon_{ij}$ where μ is a constant, g_i is a random variable with mean 0 and variance σ_G^2, and ϵ_{ij} is a random variable with mean 0 and variance σ^2. That is, a linear model has been postulated which states that any observation is a linear combination of three contributing factors: an overall mean effect, an effect due to the particular group to which the observation belongs, and an "error" effect representing all extraneous sources of variation.

TABLE 8.6–General Analysis of Variance for Calculating the Intraclass
Correlation Coefficient Using the Data of Table 8.5

Source of Variation	Degrees of Freedom	Sum of Squares	Mean Square	Expected Mean Square
Mean	1	M_{yy}	$\cdots$	$\cdots$
Among groups	$k - 1$	G_{yy}	$s^2 + ns_G^2$	$\sigma^2 + n\sigma_G^2$
Within groups	$k(n - 1)$	W_{yy}	s^2	σ^2
Total	kn	$\sum_{i=1}^{k} \sum_{j=1}^{n} Y_{ij}^2$	$\cdots$	$\cdots$

as in Table 8.6, where

$$T = \sum_{i=1}^{k} G_i \tag{8.41}$$

$$M_{yy} = T^2/kn \tag{8.42}$$

$$G_{yy} = \sum_{i=1}^{k} G_i^2/n - M_{yy} \tag{8.43}$$

and

$$W_{yy} = \sum_{i=1}^{k} \sum_{j=1}^{n} Y_{ij}^2 - M_{yy} - G_{yy} \tag{8.44}$$

Since the population intraclass correlation coefficient is defined by

$$\rho_I = \sigma_G^2/(\sigma^2 + \sigma_G^2) \qquad (8.45)$$

a sample estimate is provided by

$$r_I = \frac{s_G^2}{s^2 + s_G^2} = \frac{MS_a - MS_w}{MS_a + (n-1)MS_w} \qquad (8.46)$$

where

$$MS_a = \text{mean square among groups}$$
$$= s^2 + ns_G^2$$
$$= G_{yy}/(k-1) \qquad (8.47)$$

and

$$MS_w = \text{mean square within groups}$$
$$= s^2$$
$$= W_{yy}/k(n-1) \qquad (8.48)$$

It will be seen that if $n = 2$, the analysis would fit the situation described earlier, namely, the correlation between the heights of brothers. (**Note:** Once again we have availed ourselves of the opportunity to introduce some new notation. This time the concept of *components of variance*, denoted by s^2 and s_G^2, has been used as an alternative way of expressing mean squares. The relationship between "expected mean squares" and "mean squares" is simply the familiar relationship between "population parameters" and "sample statistics." The determination of the form of the various expected mean squares will be examined in detail in succeeding chapters, where linear models will be the main topic of discussion. Those who desire more information on this topic may jump ahead to the appropriate sections.)

If we are willing to assume that the individuals within groups are random samples from normal populations (one population per group) and that each population has the same variance, then the hypothesis $H:\rho_I = 0$ is equivalent to the hypothesis $H:\sigma_G^2 = 0$ and this may be tested using

$$F = MS_a/MS_w \qquad (8.49)$$

with degrees of freedom $\nu_1 = k - 1$ and $\nu_2 = k(n - 1)$.

Example 8.7

Given the data in Table 8.7, calculations will lead to the analysis of variance shown in Table 8.8. From this we obtain $r_I = 0.6974$. To test $H:\rho_I = 0$, we calculate $F = 30.857/5.500 = 5.61$ with $\nu_1 = 7$ and $\nu_2 = 8$ degrees of freedom. Since $F = 5.61 > F_{0.95(7,8)} = 3.5$, the hypothesis $H:\rho_I = 0$ is rejected.

TABLE 8.7–Heights of Eight Pairs of Brothers

Pair	Heights (in.)
A	71; 71
B	69; 72
C	59; 65
D	65; 64
E	66; 60
F	73; 72
G	68; 67
H	70; 68

TABLE 8.8–Analysis of Variance for Data of Table 8.7

Source of Variation	Degrees of Freedom	Sum of Squares	Mean Square	Expected Mean Square
Mean	1	72,900.0	72,900.0	...
Among groups (among pairs of brothers)	7	216.0	30.857	$\sigma^2 + 2\sigma_G^2$
Within groups (between brothers within pairs)	8	44.0	5.500	σ^2
Total	16	73,160.0	...	...

8.13 CORRELATIONS OF SUMS AND DIFFERENCES

Reference to Section 2.16 reminds us that for any constants a_i and any variables X_i the linear combination specified by

$$U = \sum_{i=1}^{n} a_i X_i \tag{8.50}$$

has mean

$$\mu_U = E[U] = \sum_{i=1}^{n} a_i \mu_i \tag{8.51}$$

and variance

$$\sigma_U^2 = E[(U - \mu_U)^2] = \sum_{i=1}^{n} a_i^2 \sigma_i^2 + 2\sum_{i<j} a_i a_j \sigma_{ij} \tag{8.52}$$

where μ_i is the mean of X_i, σ_i^2 is the variance of X_i, and σ_{ij} is the covariance of X_i and X_j. Thus if $U = X_1 \pm X_2$,

$$\mu_U = \mu_1 \pm \mu_2 \tag{8.53}$$

and

$$\sigma_U^2 = \sigma_1^2 + \sigma_2^2 \pm 2\sigma_{12} \qquad (8.54)$$

Utilizing Eq. (2.69), it is easily verified that Eq. (8.54) may be rewritten as

$$\sigma_U^2 = \sigma_1^2 + \sigma_2^2 \pm 2\rho_{12}\sigma_1\sigma_2 \qquad (8.55)$$

Rearranging terms, we obtain

$$\rho_{12} = (\sigma_U^2 - \sigma_1^2 - \sigma_2^2)/2\sigma_1\sigma_2 \qquad \text{if } U = X_1 + X_2 \qquad (8.56)$$

or

$$\rho_{12} = (\sigma_1^2 + \sigma_2^2 - \sigma_U^2)/2\sigma_1\sigma_2 \qquad \text{if } U = X_1 - X_2 \qquad (8.57)$$

This leads to an alternative method of obtaining r_{12} (the sample estimate of ρ_{12}), namely:

$$r_{12} = (s_U^2 - s_1^2 - s_2^2)/2s_1s_2 \qquad \text{if } U = X_1 + X_2 \qquad (8.58)$$

or

$$r_{12} = (s_1^2 + s_2^2 - s_U^2)/2s_1s_2 \qquad \text{if } U = X_1 - X_2 \qquad (8.59)$$

Before terminating our discussion of the correlation of sums and differences, attention must be directed to the relationship between the contents of this section and the "method of paired observations" examined in Sections 5.5 and 6.4. Noting that $D = X - Y$ is analogous to $U = X_1 - X_2$, we recognize that a legitimate pairing of related observations will yield a smaller standard error of the mean difference if a positive correlation exists. Such a reduction in the standard error represents a gain in efficiency (relative to nonpairing), which will be reflected in a shorter confidence interval, an easier establishment of statistical significance, or a smaller sample size. Clearly, the success of pairing in any situation depends upon the extent to which the researcher can introduce positive correlation into an experiment.

PROBLEMS

8.1   Using the data of Example 8.4, test $H:\beta_1 = 0$ versus $A:\beta_1 \neq 0$ using (a) a t test, (b) an F test. In both tests let $\alpha = 0.01$.

8.2   If $U = a + bX$ and $V = c + dY$, show that $r_{UV} = r_{XY}$.

8.3   Verify Eq. (8.17).

8.4   Interpret a simple linear correlation coefficient of -0.8.

8.5   If the simple linear (product-moment) correlation coefficient between X and Y is $r_{XY} = 0.8$, what are the values of (a) r_{xy}, (b) $r_{X\hat{Y}}$, and (c) $r_{Y\hat{Y}}$?

8.6   Using the data of Problem 3.3 and the results of Problem 7.12, compute and interpret the appropriate measure of correlation.

8.7   Using the data of the problem indicated, compute and interpret the appropriate measure of correlation:

(a)	7.4	(c)	7.6	(e)	7.8	(g)	7.11
(b)	7.5	(d)	7.7	(f)	7.9	(h)	7.13

(i)	7.14	(n)	7.21	(r)	7.27	(v)	7.31
(j)	7.15	(o)	7.22	(s)	7.28	(w)	7.32
(k)	7.16	(p)	7.24	(t)	7.29	(x)	7.33
(l)	7.17	(q)	7.26	(u)	7.30	(y)	7.35
(m)	7.20						

8.8   The following table gives hypothetical data for the covariates X and Y, selected at random from a bivariate normal distribution.

X	Y	X	Y
12	74	18	149
20	170	16	142
17	147	13	144
11	75	18	173
8	46	11	101
8	59	16	140
4	20	15	132
12	90	5	35
9	74	14	96
12	77	6	50
16	144	3	24
11	110	5	26
10	99	8	95
13	109	6	73
15	109	17	159

(a) Compute the means, the standard deviations, and the standard errors of the means of X and Y.

(b) Make a scatter diagram to show the relation between these two series. Also, draw one line through the plotted data showing the mean of X and another showing the mean of Y.

(c) Fit a straight line to the points on the scatter diagram in order to express mathematically the average relationship between these two variables. The required equation is $\hat{Y} = b_0 + b_1 X$. This calls for the computation of: (1) the regression coefficient $b_1 = \sum xy / \sum x^2$ and (2) the Y-intercept $b_0 = \bar{Y} - b_1 \bar{X}$. The regression equation may be written $\hat{Y} = \bar{Y} + b_1(X - \bar{X})$. Find b_0 and b_1 geometrically from the graph.

(d) Calculate the estimated value of Y for each of the 30 values of X from the equation $\hat{Y} = b_0 + b_1 X$. Also, compute the errors of estimate $Y - \hat{Y}$ for each X.

(e) Interpret the constants b_0 and b_1 obtained for $\hat{Y} = b_0 + b_1 X$.

(f) Compute and interpret the standard error of estimate from the formula

$$s_E = \left[\sum (Y - \hat{Y})^2 / (n - 2) \right]^{1/2}$$

(g) Compute the sum of squares of the errors of estimate (deviations from regression) with the formula

$$\sum (Y - \hat{Y})^2 = \sum y^2 - \frac{\left(\sum xy\right)^2}{\sum x^2}$$

(h) Test the regression coefficient b_1 for significance.

(i) Compute the correlation coefficient using the formula

$$r = \sum xy \Big/ \left[\left(\sum x^2 \right) \left(\sum y^2 \right) \right]^{1/2}$$

(j) Compute and interpret the coefficient of determination r^2.

(k) Partition $\sum y^2$ into two parts, that associated with regression and that attributed to errors of estimate.

(l) Compute the correlation coefficient between X and $\hat{Y}$.

(m) Compute the correlation coefficient between Y and $\hat{Y}$.

(n) Compute the correlation coefficient between x and y.

(o) Compute and interpret the 95 percent confidence limits of β_1.

(p) Compute the standard errors for the estimated values of Y for each of the following: (1) the mean of all Y's whose X value is equal to 10 and (2) particular Y's whose X value is equal to 10.

(q) Compute the sum of squares attributed to regression using the formula $\sum (\hat{Y} - \bar{Y})^2$. The shortcut formula is $\left(\sum xy \right)^2 \big/ \sum x^2$ or $r^2 \sum y^2$. Show computationally that the three formulas give the same sum of squares.

(r) Show computationally that $(1 - r^2) \sum y^2 = \sum (Y - \hat{Y})^2$.

(s) Compute the regression of X on Y; i.e., compute the constants in the equation $\hat{X} = b_0' + b_1' Y$, where $b_1' = \sum xy \big/ \sum y^2$ and $b_0' = \bar{X} - b_1' \bar{Y}$. Plot the regression on the same sheet on which the regression $\hat{Y} = b_0 + b_1 X$ was plotted.

(t) Show that $r^2 = b_1 b_1'$, where b_1 and b_1' are the two regression coefficients.

(u) Compute $r^2 = 1 - \sum (Y - \hat{Y})^2 \big/ \sum y^2$.

(v) Show logically, algebraically, or geometrically that $|r|$ cannot be less than 0 nor greater than 1.

8.9 We have this sample of X and Y values:

Y	X
9	4
11	2
7	5
10	1
8	3

(a) Compute the product-moment correlation between Y and X for this sample.

(b) What assumptions are required for testing the significance of a sample value of r? What parameter is estimated by the sample correlation?

(c) Indicate or describe three methods for testing the hypothesis that the true value of the correlation is 0 in the bivariate population from which the above sample was taken. (Exact formulas are not required.)

8.10 Management seeks to discover a measure of correlation between length of service on the part of a certain type of machine and the annual repair bills on such machines. From the following data:

Machine	Years of Service	Annual Repair Cost
A	1	$2.00
B	3	1.50
C	4	2.50
D	2	2.00
E	5	3.00
F	8	4.00
G	9	4.00
H	10	5.00
I	13	8.00
J	15	8.00

(a) Make a scatter diagram, designating years of service as the X series and annual repair costs as the Y series.

(b) Find the correlation coefficient r.

(c) Is the measure of correlation significant?

(d) What are your assumptions?

8.11   Given that $\sum y^2 = 1000$ and that the sum of squares due to regression is 640, compute the value of r showing all your steps. What assumptions are necessary if r is to be interpreted as a sample estimate of a population correlation coefficient?

8.12   The correlation coefficient between the CAVD and the graduate record verbal tests was 0.60 for a sample of 67 men students and 0.50 for a sample of 39 women students. With a risk of Type 1 error of 5 percent, is this evidence that the two groups are random samples from bivariate normal populations of the same correlation?

8.13   Given the following data and statistics for a random sample from a bivariate normal distribution:

$$\bar{X} = 6 \qquad \sum x^2 = 100 \qquad b_1 = -4 \qquad s_E = 6.708$$
$$\bar{Y} = 20 \qquad \sum y^2 = 2500 \qquad r = -0.8 \qquad s_{b_1} = 0.6708$$
$$n = 22 \qquad \sum xy = -400 \qquad b_0 = 44$$

(a) Give a detailed interpretation of the linear regression of Y on X. Include all inferences that can be made about the population regression. Also, interpret all inferences made.

(b) Interpret the above correlation coefficient.

(c) What assumptions are implicit in the use of the regression in (a)?

8.14   Using the following regression and correlation data for seven types of sheeting, test the hypothesis $H:\rho_i = \rho(i = 1, \ldots, 7)$. Also test the hypothesis $H:\beta_i = \beta(i = 1, \ldots, 7)$. State the assumptions made in each case.

Fabric	Degrees of Freedom	$\sum x^2$	$\sum xy$	$\sum y^2$	Correlation Coefficient	Regression Coefficient	Degrees of Freedom	Sum of Squares	Mean Square
1	139	60357.14	− 989.64	1965.89	−0.0909	−0.0164	138	1949.66	...
2	139	60357.14	− 1970.43	2351.43	−0.1654	−0.0326	138	2287.10	...
3	139	60357.14	− 1647.50	3190.85	−0.1186	−0.0273	138	3145.88	...
4	139	60357.14	− 192.86	3258.61	−0.0138	−0.0032	138	3257.99	...
5	139	60357.14	− 5482.14	2804.04	−0.4214	−0.0908	138	2306.11	...
6	139	60357.14	− 7605.00	2276.79	−0.6487	−0.1260	138	1318.56	...
7	139	60357.14	− 12458.50	4375.60	−0.7666	−0.2064	138	1804.00	...
							966	16069.30	16.63
Total	973	422499.98	− 30346.07	20223.21		−0.5028	972	20201.41	...
			Difference for testing among regression coefficients				6	4132.11	688.68
			$F = 688.68/16.63 = 41.41$						

REFERENCES

1. Fisher, R. A. 1921. On the probable error of a coefficient of correlation deduced from a small sample. *Metron* 1(4):3.
2. Kendall, M. G. 1970. *Rank Correlation Methods,* 4th ed. Griffin, London.
3. Kendall, M. G.; and Smith, B. B. 1939. The problem of *m* rankings. *Ann. Math. Stat.* 10:275.
4. McNemar, Q. 1969. *Psychological Statistics,* 4th ed. Wiley, New York.
5. Olds, E. G. 1938. Distributions of sums of squares of rank differences for small numbers of individuals. *Ann. Math. Stat.* 9:133.
6. Pearson, K. 1904. *Mathematical Contributions to the Theory of Evolution. XIII. On the Theory of Contingency and Its Relation to Association and Normal Correlation.* Drapers' Co., Res. Mem., Biometric Series I. Cambridge University Press, London.
7. Pearson, K. 1910. On a new method of determining correlation when one variable is given by alternative and the other by multiple categories. *Biometrika* 7:248.
8. Treloar, A. E. 1942. *Correlation Analysis.* Burgess, Minneapolis.
9. Woo, T. L. 1929. Tables for ascertaining the significance or nonsignificance of association measured by the correlation ratio. *Biometrika* 21:1.

FURTHER READING

DuBois, P. H. 1957. *Multivariate Correlational Analysis.* Harper, New York.
Ferguson, G. A. 1971. *Statistical Analysis in Psychology and Education,* 3rd ed. McGraw-Hill, New York.
Garrett, H. E. 1967. *Statistics in Psychology and Education,* 6th ed. McKay, New York.
Guilford, J. P.; and Fruchter, B. 1973. *Fundamental Statistics in Psychology and Education,* 5th ed. McGraw-Hill, New York.

CHAPTER 9

DESIGN OF EXPERIMENTAL INVESTIGATIONS

Before proceeding to the introduction and discussion of further techniques of statistical analysis, time will be taken to examine certain aspects of data acquisition. Such a digression is justified because *the analysis of any set of data is dictated to a large extent by the manner in which the data were obtained.* The truth of the foregoing statement will be illustrated many times throughout the remainder of this book.

9.1 SOME GENERAL REMARKS

It has been well demonstrated in the preceding chapters that statistics (as a science) deals with the development and application of methods and techniques for the collection, tabulation, analysis, and interpretation of data so that the uncertainty of conclusions based upon the data may be evaluated by means of the mathematics of probability. However, it should also be evident that there is something more to statistics than the routine analysis of data using standard techniques. For example, the reader should realize that the analyses are exact only if all the underlying assumptions are satisfied. Since this is rarely true, much depends on the skill of the researcher in selecting the method of analysis that best fits the circumstances of the experimental situation being studied. Thus it seems safe to say that statistics is an art as well as a science.

9.2 MEANING OF "THE DESIGN OF AN EXPERIMENT"

Designing an experiment simply means *planning* an experiment so that information will be collected that is relevant to the problem under investigation. All too often data collected are of little or no value in any attempted solution of the problem. The *design of an experiment* is, then, the complete sequence of steps taken ahead of time to insure that the appropriate data will be obtained in a way that permits an objective analysis leading to valid inferences with respect to the stated problem. Such a definition of designing an experiment implies that the person formulating the design clearly understands the objectives of the proposed investigation.

9.3 NEED FOR AN EXPERIMENTAL DESIGN

That some sort of design is necessary before any experiment is performed may be demonstrated by considering an example.

[258]

Example 9.1

It is desired to determine the effect of gasoline and oil additives on carbon and gum formation of engines (see ref. 12, p. 79). Twenty additives are to be tested in combination with a "control" gasoline and oil mixture. Eighty similar engines are available for use in the experimental program.

As the problem is now stated it is far too general to permit the selection of a particular design. Many questions must be asked (and answers obtained) before the statistician can propose a suitable design. Typical questions are:

1. How is the effect to be measured? That is, what are the characteristics to be analyzed?
2. What factors influence the characteristics to be analyzed?
3. Which of these factors will be studied in this investigation?
4. How many times should the basic experiment be performed?
5. What should be the form of the analysis?
6. How large an effect will be considered important?

When we recognize that the foregoing questions are only a small sample of those that might be asked, it is evident that much thought should be given to the planning stage in any experimental investigation. In fact, the importance of this recommendation cannot be overemphasized.

9.4 PURPOSE OF AN EXPERIMENTAL DESIGN

The purpose of any experimental design is to provide a maximum amount of information relevant to the problem under investigation. However, it is also important that the design, plan, or test program, be kept as simple as possible. Further, the investigation should be conducted as efficiently as possible; i.e., every effort should be made to conserve time, money, personnel, and experimental material. Fortunately, most of the simple statistical designs are not only easy to analyze but also efficient in both the economic and statistical senses. For this reason, a statistician should be consulted in the early stages of any proposed research project. He can often recommend a simple design that is both economical and efficient.

Having said that the purpose of any experimental design is to provide a maximum amount of information at minimum cost, it is evident that the design of experiments is a subject that involves both statistical methodology and economic analysis. A person planning an experiment should incorporate both of these features into his design; i.e., he should strive for *statistical efficiency* and *resource economy*. However, an examination of books on statistical methods and the design of experiments will seldom reveal many explicit references to the cost aspects of the problem. This is unfortunate. On the other hand, the subject of cost is implicit in most discussions of experimental design. We have only to note the continual attempts to plan experiments using the smallest size sample possible to realize that the cost aspect has not been overlooked.

Fortunately, most simple designs are both economical and efficient, and thus the statistician's efforts to achieve statistical efficiency usually lead also to economy of experimentation.

9.5 BASIC PRINCIPLES OF EXPERIMENTAL DESIGN

It has been stated many times that there are three basic principles of experimental design: *replication, randomization,* and *local control.* Because of the fundamental nature of these concepts, each will be discussed separately. Further, it is recommended that the reader strive for as complete an understanding and appreciation of these ideas as possible, for they will play a very important role in much of the remainder of this book.

9.6 REPLICATION

By *replication* we mean the *repetition* of the basic experiment. The reasons why replication is desirable are:

1. It provides an estimate of experimental error that acts as a "basic unit of measurement" for assessing the significance of observed differences or for determining the length of a confidence interval.
2. Since under certain assumptions experimental error may be estimated in the absence of replication, it is also fair to state that replication sometimes provides a more accurate estimate of experimental error.
3. It enables us to obtain a more precise estimate of the mean effect of any factor since $\sigma_{\bar{Y}}^2 = \sigma^2/n$, where σ^2 represents the true experimental error and n the number of replications.

It must be emphasized that multiple readings do not necessarily represent true replication. This statement may best be substantiated by an example.

Example 9.2

Two manufacturing processes are used to produce thermal batteries. Sample batteries are obtained from each of two production lots, one lot being produced by process A and the other by process B. The batteries are then tested and the activated life of each battery is recorded.

If an analysis of the above experiment were attempted, it would be discovered that no valid estimate of error is available for testing the difference between processes. The variation among batteries within lots yields a valid estimate of error for assessing only the lot-to-lot variability. True replication would require that batteries be tested from each of several lots manufactured by each process. (**Note:** In the example just given, the effects of lots and processes are said to be *confounded.* This term will be discussed more fully a little later.)

Sometimes the absence of true replication is more easily recognized than in Example 9.2. For instance, if multiple measurements of activated life had been obtained by connecting several clocks to a single battery, the researcher would easily have recognized that the observed data were not true replications but only repeated measurements on the same experimental unit. Another example of the same type of spurious replication (i.e., multiple measurements rather than true replication) would be multiple determinations of the silicon content of a particular batch of pig iron where the variability among processes was to be assessed.

9.7 EXPERIMENTAL ERROR AND EXPERIMENTAL UNITS

In the preceding discussion of replication, the terms experimental error and experimental unit were used. Because of their wide usage, it is necessary to have a clear understanding of their meanings. An *experimental unit* is that unit to which a single treatment (which may be a combination of many factors) is applied in one replication of the basic experiment. The term *experimental error* describes the failure of two identically treated experimental units to yield identical results.

In one respect the term "experimental error" is unfortunate, especially the word "error." This word is probably a legacy from the physical sciences, particularly astronomy, where the investigators (*observers*) were concerned with errors in both measurement and observation. However, the influence of *experimenters* in both the biological and physical sciences should not be discounted entirely. The adoption of the word "error" could just as easily be attributed to them, for they clearly recognized the existence of errors of technique in the performance of their experiments. But whatever the history of the word "error," a thoughtful examination of the definition of the term "experimental error" will reveal that its meaning to the statistician is much more general. In each particular situation, it reflects: (1) errors of experimentation, (2) errors of observation, (3) errors of measurement, (4) the variation of the experimental material (i.e., among experimental units), and (5) the combined effects of all extraneous factors that could influence the characteristics under study but have not been singled out for attention in the current investigation.

Another item related to the term experimental error is sometimes confusing to the statistical novice. This is the practice of the professional statistician of referring to "the experimental error for testing a particular effect." Such a phrase suggests that in a given experiment more than one experimental error may exist even though examination of the assumed statistical model will reveal only one such term. As confusing as this practice may be to the uninitiated, it serves a useful purpose. As the reader progresses through the book, he will become more familiar with the way the expression is used and thus, we hope, become more tolerant of what seems at the moment to be an unwise use of words that have been carefully defined. In an attempt to give a somewhat more specific defense, let us say that all the statistician is really doing is reminding you of the fact that every statistic has its own standard error. Perhaps his choice of words is not the best, but it is a firmly entrenched part of the language of experimental design. Thus, we strongly recommend that you forgive the statistician his choice of words and concentrate on the more important task of learning *how* and *when* to use statistical methods.

Before terminating this discussion of experimental error, ways of reducing its magnitude should be indicated. The following are general statements, for specific recommendations can be made only when a particular design problem is being considered. Experimental error may usually be reduced by adoption of one or more of the following techniques: (1) using more homogeneous experimental material or by careful stratification of available material, (2) utilizing information provided by related variates, (3) using more care in conducting the experiment, and (4) using a more efficient experimental design.

9.8 CONFOUNDING

In Section 9.6 the word "confounded" was introduced to describe a certain phenomenon that is fairly common in experimentation. Since this phenomenon is so important in the design of experiments, it is appropriate that time be taken to investigate and describe it more thoroughly. This will best be done through the use of examples.

Example 9.3

A chemist has developed a new synthetic fertilizer and wishes to compare it with an established product. He contacts a nearby university and they agree to run an experiment on two available experimental plots. The established product will be applied to one plot of ground and the experimental product to the other. The characteristic to be measured and used as the index of performance will be the yield (converted to bushels per acre) of a specified cereal crop. However, when the two yields are compared, we are unable to say how much of the difference is due to fertilizers and how much is due to inherent differences (in fertility, soil type, etc.) between the two plots. That is, any comparison of fertilizers is said to be *confounded* with a comparison of plots or, in slightly different words, the effects of fertilizers and plots are confounded.

Example 9.4

An analyst is engaged in determining the percentage of iron in chemical compounds. Two different procedures are to be compared. The analyst takes a sample of the first chemical compound and makes a determination of the iron content using procedure A. Then he makes a determination using procedure B. This sequence (i.e., first A and then B) of steps is repeated several times, each time on a new sample from a different compound. But here again, as in Example 9.3, we are troubled by the existence of confounding. Any comparison of the two procedures (A and B) will be *confounded* with a comparison of the first and second determinations made (on each compound) by the analyst. That is, if there is any improvement in technique (due to a learning process) from the first to the second determination, this effect will be confounded with the difference between procedures.

Examination of the preceding examples will show that the word "confounded" is simply a synonym for "mixed together." That is, two or more effects are said to be *confounded* in an experiment if it is impossible to separate the effects when the subsequent statistical analysis is performed.

Since one of the purposes of experimental design is to provide unambiguous results, it would seem almost obvious that a good design should avoid confounding. It is, therefore, disconcerting to the uninitiated to learn that the statistician frequently deliberately introduces confounding into a design. However, as you will see later, such a procedure is not followed indiscriminately. When confounding is introduced into a design it is done for a good reason, as often as not to achieve economy through reduction of the size of the experiment.

9.9 RANDOMIZATION

It was noted in Section 9.6 that replication provides an estimate of experimental error that can be used for assessing the significance of observed differences. That is, replication makes a test of significance possible. But what makes such a test valid? We have seen that every test procedure has certain underlying assumptions that must be satisfied if the test is to be valid. Perhaps the most frequently invoked assumption is the one stating that the observations (or the errors therein) are independently distributed. How can we be certain this assumption is true? We cannot; but by insisting on a random sample from a population or on a random assignment of treatments to the experimental units, we can proceed as though the assumption is true. That is, *randomization makes the test valid* by making it appropriate to analyze the data as though the assumption of independent errors is true. Note that we have not said that randomization guarantees independence, but only that randomization permits us to proceed as though independence is a fact. The reason for this distinction should be clear: errors associated with experimental units adjacent in space or time will tend to be correlated, and all randomization does is to assure us that the effect of this correlation on any comparison among treatments will be made as small as possible. Some degree of correlation will still remain, for no amount of randomization can ever eliminate it entirely. That is, in any experiment true and complete independence of errors is an ideal that can never be achieved. However, such independence should be sought, and randomization is the best technique devised so far to attain the desired end.

Sometimes the concept of randomization is introduced as a device for "eliminating" bias. To illustrate the thinking behind this approach, consider again Example 9.4. There, any comparison of procedures A and B would be *biased* in favor of B if a learning effect existed. However, if each time a new compound was to be investigated, the analyst had decided *at random* which procedure to use first, the bias would have been reduced, perhaps even eliminated; but even more would have been accomplished. If other biases were operating, these would also have had their effects eliminated (or at least reduced) by the randomization. That is, by randomly assigning treatments to the experimental units, we try to make certain that treatments will not be continually favored or handicapped by extraneous sources of variation over which the experimenter has no control or chooses not to control. In other words, randomization is like insurance: it is always a good idea, and sometimes even better than we expect.

Regardless of the foregoing arguments in favor of randomization, some persons have spoken out in favor of systematic (nonrandom) designs. "Can we not," they ask, "obtain a more accurate measurement of differences among treatments if such treatments are applied to the experimental units in a systematic manner?" The only honest answer to this query is, "Possibly." Why, then, does the statistician insist on randomization? The reason is the same as expressed earlier—because the statistician wishes to make certain inferences from the observed data and desires to attach a measure of reliability to these inferences. If randomization is not employed, the quoted measure of reliability may be biased. Further, any inference would be unsupported by a meaningful probability statement. (**Note:** The reader is reminded of the discussion of judgment versus random samples presented in Section 3.3.)

There are situations in which complete randomization is either impossible or uneconomical. The statistician should not, therefore, adopt the unrelenting position of insisting on complete randomization in every case. On the other hand, neither should he agree to the use of a completely systematic design, for the experimenter must reconcile himself to the fact that some degree of randomization is required for the valid application of most statistical analyses. Some intermediate position between the two extremes of complete randomization or a strictly systematic design is often most realistic. Once the experimenter and the statistician recognize one another's problems, a compromise plan can usually be found that is mutually satisfactory. [**Note:** The question of which is better, a systematic or a randomized design, has never been completely settled; most likely it never will be. Most designs in common use today involve both systematic and random elements, and this seems a reasonable state of affairs. For the person who wishes to pursue this point further, the literature offers many papers discussing the argument, both pro and con. See (1), (9), (13), (14), (17).]

9.10 LOCAL CONTROL

In Section 9.5, it was stated that the three basic principles of experimental design are replication, randomization, and local control; the first two have already been discussed. In one sense, local control is synonymous with experimental design. However, this interpretation of experimental design is very narrow, and not consistent with our earlier definition. If we agree, then, that experimental design is as defined in Section 9.2, then local control is only a part of the total complex. In this sense, *local control* refers to the amount of balancing, blocking, and grouping of the experimental units that is employed in the adopted statistical design. It was observed earlier (Section 9.9) that replication and randomization make a valid test of significance possible. What then is the function of local control? The function or purpose of local control is to make the experimental design more efficient. That is, local control makes any test of significance more sensitive or (in the language of Section 6.1) the test procedure more powerful. This increase in efficiency (or sensitivity or power) results because a proper use of local control will reduce the magnitude of the estimate of experimental error. (**Note:** The reader should recognize that local control can be exerted in several ways. The more common methods have been suggested above and in the last paragraph of Section 9.7.)

9.11 BALANCING, BLOCKING, AND GROUPING

In the preceding section, the terms *balancing, blocking,* and *grouping* were introduced in connection with the principle of local control. A few sentences of explanation will be given so that the researcher will understand what is implied. Actually, it is possible to say that the three terms are synonymous. However, in this text we shall use them to describe different aspects of design philosophy. It is hoped this will not lead to confusion when other references are consulted.

By *grouping* will be meant the placing of a set of homogeneous experimental units into groups in order that the different groups may be subjected to different treatments. These groups may consist of different numbers of experimental units.

Example 9.5

A pharmaceutical company is investigating the comparative effects of three proposed compounds. The experiment will consist of injecting rats with the compounds and recording the pertinent reaction. A litter consisting of 11 rats (experimental units) is available. Each of the 11 rats is assigned at random to one of three groups subject only to the restriction that the groups contain 4, 4, and 3 rats respectively. The animals in the first group are then injected with compound A, those in the second group with compound B, and those in the third group with compound C.

By *blocking* will be meant the allocation of the experimental units to blocks in such a manner that the units within a block are relatively homogeneous, while the greater part of the predictable variation among units has been confounded with the effect of blocks. That is, using the researcher's prior knowledge concerning the nature of the experimental units, the statistician can design the experiment in such a way that much of the anticipated variation will not be a part of experimental error. In this way a more efficient design is provided.

Example 9.6

Consider again the problem outlined in Example 9.5. This time, however, let us assume that 12 rats are available and the pedigrees show 6 are from litter X, 3 from litter Y, and 3 from litter Z. Since it may well be expected that rats in the same litter will perform more nearly alike than rats from different litters (due to inherited characteristics), it would seem natural to form three blocks. The first block would contain the 6 rats from litter X, the second would contain the 3 rats from litter Y, and the third would contain the 3 rats from litter Z. The three treatments ($A, B,$ and C) would then be assigned at random to the rats within blocks. Since each rat is subjected to only one treatment, the block containing 6 rats would undoubtedly end up with 2 rats seeing treatment A, 2 seeing treatment B, and 2 seeing treatment C. The other two blocks would have single rats seeing each treatment.

By *balancing* will be meant the obtaining of the experimental units, the grouping, the blocking, and the assignment of the treatments to the experimental units in such a way that a balanced configuration results. (Circular though the preceding definition is, it projects the thought we wish to impart. Consequently, we hope you will forgive the poor logic.) It should be clear that we can have little or no balance, partial balance, approximate balance, or complete balance in any particular design. For instance, Example 9.5 illustrates a case of approximate balance, while Example 9.6 might be construed as an illustration of partial balancing. Rather than manufacturing further examples at this time, let us defer the matter until later. As you progress through the following chapters on various designs, it will become clear that the statistician continually strives for balanced designs.

9.12 TREATMENTS AND TREATMENT COMBINATIONS

Several times in the preceding sections, the word "treatments" has been used with little or no explanation. Just what is meant by this word? Like so many other terms in statistics, the word "treatments" entered the literature because of its use in agronomic experimentation. However, the word "treatments" (like "blocks" and "plots") has long since lost its strict agronomic connotation. In fact, the three phrases mentioned in the preceding sentence are now an accepted part of the language of statistics, regardless of the area of application.

To the statistician, *treatment* or *treatment combination* implies the particular set of experimental conditions that will be imposed on an experimental unit within the confines of the chosen design. By way of explanation several illustrations are given:

1. In agronomic experimentation, a treatment might refer to (a) a brand of fertilizer, (b) an amount of fertilizer, (c) a depth of seeding, or (d) a combination of (b) and (c). The last of these would more properly be termed a treatment combination.
2. In animal nutrition experimentation, a treatment might refer to (a) the breed of sheep, (b) the sex of the animals, (c) the sire of the experimental animal, or (d) the particular ration fed to an animal.
3. In psychological and sociological studies, a treatment might refer to (a) age, (b) sex, or (c) amount of education.
4. In an investigation of the effects of various factors on the efficiency of washing clothes in the home, the treatments were various combinations of (a) the type of water (hard or soft), (b) temperature of

water, (c) length of wash time, (d) type of washing machine, and (e) kind of cleansing agent.

5. In an experiment to study the yield of a certain chemical process, the treatments might be all combinations of (a) the temperature at which the process was operated and (b) the amount of catalyst used.

6. In a research and development study concerned with batteries, the treatments could be various combinations of (a) the amount of electrolyte and (b) the temperature at which the battery was activated.

Many more examples could be cited from every field in which experimentation is performed. However, later chapters will abound with such examples.

9.13 FACTORS, FACTOR LEVELS, AND FACTORIALS

In any discussion of experimental design, the word "factorial" is almost certain to be heard. Frequently, the reference is to a "factorial design." However, this is actually a misnomer; there is no such thing as a factorial design. The adjective "factorial" refers to a special way in which treatment combinations are formed and not to any basic type of design. Thus if a randomized complete block design (see Chapter 11 for definition) has been selected and the treatment combinations are of a factorial nature, a more correct expression would be "a randomized complete block design involving a factorial treatment arrangement." Some writers, such as Yates (18), have recognized this situation and speak of factorial experiments rather than factorial designs. This shift in terminology, while in the proper direction, does not completely resolve the difficulty, since the word "experiment" seems to imply that survey data are to be excluded. To avoid any such implication, we shall not speak of factorial designs nor factorial experiments but simply of *factorials*. It is to be understood that this is only an abbreviation for a more lengthy expression describing the nature of the treatments.

Having introduced the subject of factorials, it is desirable that specific terms be defined in an explicit manner. In most investigations, the researcher is concerned with more than one independent variable and in the changes that occur in the dependent variable as one or more of the independent variables are permitted to vary. In the language of experimental design, an independent variable is referred to as a *factor*. Referring to the illustrations in the preceding section, it is noted that five factors were listed for the home washing study, while the battery study involved only two factors. The reader can easily find many more examples of investigations involving several factors by consulting various technical journals.

Before proceeding to the definition of the next term arising in connection with factorials, it will be wise to indicate the generally accepted notation used to represent factors. Most writers use *lowercase Latin letters* to represent factors. As an illustration, the five factors in the home washing experiment might be represented by

$$m = \text{type of washing machine}$$

$$a = \text{kind of cleansing agent}$$
$$b = \text{type of water}$$
$$c = \text{temperature of water}$$
$$d = \text{length of wash time}$$

A second illustration is provided by an investigation conducted by Ratner (16). His experiment involved a study of how long it took to perform a certain move, and the factors investigated were

$$d = \text{distance}$$
$$w = \text{weight}$$
$$o = \text{operator-pair}$$

The researcher is generally interested in experimental results (observations on the dependent variables) as one or more factors are allowed to vary. It can be seen in Ratner's study that he considered 3 distances (d_1, d_2, d_3), 10 weights $(w_1, \ldots, w_{10})$, and 4 operator-pairs (o_1, o_2, o_3, o_4). In the home washing experiment, the investigator used 2 types of machine, 2 kinds of cleansing agent, 2 types of water, 2 temperatures of water, and 2 lengths of wash time. These various values, or classifications of the factors, are known as the *levels* of the factors. Thus there were 10 levels of weight, 3 levels of distance, and 4 levels of operator-pairs in Ratner's experiment. In the home washing study, each factor appeared at 2 levels. These two examples should indicate that the word "level" is a very general term, which may be applied in many varied situations. Ratner's investigation of move times provides an excellent example of this diversity, for the 3 levels of distance (6, 12, and 18 inches) are values of a continuous variable, while the 4 levels of operator-pairs (i.e., 4 distinct pairs of operators formed from 8 individuals) are classifications of a qualitative variable.

Since so many experiments involve factorial treatment arrangements, some notation must be adopted to represent the various treatment combinations. Unfortunately, several systems of notation appear in the literature. These are summarized in Table 9.1 for a case involving 12 treatment combinations where the 12 combinations were formed from 2 levels of factor a, 2 levels of factor b, and 3 levels of factor c. In this representation, using Method I as an example, the symbol $a_i b_j c_k$ ($i = 1, 2; j = 1, 2; k = 1, 2, 3$) represents the treatment combination formed by using the ith level of factor a, the jth level of factor b, and the kth level of factor c.

Another item of terminology should be mentioned in the present context. This item is best explained by example. The factorial arrangement of the treatments used in Table 9.1 would be referred to by the statistician as a $2 \times 2 \times 3$ factorial. Similarly, Ratner's investigation would be termed a $3 \times 10 \times 4$ factorial, while the home washing study was a $2 \times 2 \times 2 \times 2 \times 2 = 2^5$ factorial.

Before leaving (for the time being) the subject of factorials, it is only fair that the reader be warned of a double use of certain symbols that

TABLE 9.1–Illustrations of Notation Used to
Represent Factorial Treatment Combinations

Treatment Combination	Method				
	I	II	III	IV	V*
1	$a_1b_1c_1$	111	$a_0b_0c_0$	000	(1)
2	$a_1b_1c_2$	112	$a_0b_0c_1$	001	c
3	$a_1b_1c_3$	113	$a_0b_0c_2$	002	c^2
4	$a_1b_2c_1$	121	$a_0b_1c_0$	010	b
5	$a_1b_2c_2$	122	$a_0b_1c_1$	011	bc
6	$a_1b_2c_3$	123	$a_0b_1c_2$	012	bc^2
7	$a_2b_1c_1$	211	$a_1b_0c_0$	100	a
8	$a_2b_1c_2$	212	$a_1b_0c_1$	101	ac
9	$a_2b_1c_3$	213	$a_1b_0c_2$	102	ac^2
10	$a_2b_2c_1$	221	$a_1b_1c_0$	110	ab
11	$a_2b_2c_2$	222	$a_1b_1c_1$	111	abc
12	$a_2b_2c_3$	223	$a_1b_1c_2$	112	abc^2

*In this representation the absence of a letter implies that the factor it represents is at the lowest level. In general, the exponents on the letters agree with the subscripts used in Method III. Thus $a_0b_1c_2$ becomes $a^0b^1c^2 = bc^2$. The symbol (1) is used to signify that each factor is at its lowest level; i.e., $a_0b_0c_0$ is equivalent to $a^0b^0c^0 = (1)$.

could (but should not) lead to confusion. The situation is as follows: it is common practice to use the letters $a, b, c, \ldots$ to denote not only the various factors but also the number of levels of the factors. For example, a statistical model might be written as

$$Y_{ij} = \mu + \alpha_i + \beta_j + \epsilon_{ij} \qquad i = 1, \ldots, a; j = 1, \ldots, b \qquad (9.1)$$

where μ = mean effect

α_i = effect of the ith level of factor a

β_j = effect of the jth level of factor b

ϵ_{ij} = experimental error

and

$$\sum_{i=1}^{a} \alpha_i = \sum_{j=1}^{b} \beta_j = 0$$

while the ϵ_{ij} are $NI(0, \sigma^2)$. In this and similar situations, the decision to use a and b to denote not only the factors but also the number of levels of each factor should not lead to any confusion. The sense in which a letter is being used in any particular instance should always be perfectly clear from the context.

9.14 EFFECTS AND INTERACTIONS

Whenever a statistician undertakes the design of an experiment, he must first ascertain the objectives of the researcher. Frequently, the objectives may be very simple. For example, the researcher may wish to determine the effect on the yield of a chemical reaction of changing the operating temperature while all other factors (variables) are held constant at predetermined levels. On the other hand, he may have no interest whatsoever in temperature; his concern might be only with pH. In this case an experiment would be planned to determine the effect of pH under the restriction that all other factors (including temperature) are held constant.

Experiments such as these are fine if the effects of pH and temperature on the response variable are independent. However, if we know that the factors are interdependent or if we are doubtful of the validity of an assumption of independence, an experiment that estimates both main effects and interactions should be recommended. Such an experiment would utilize a factorial arrangement of the treatments.

Example 9.7

It is suggested that the effects of pH and temperature on the yield of a certain chemical reaction are not independent. It is therefore recommended that a design be adopted that utilizes treatment combinations formed by combining different levels of the two factors involved. It is decided that two levels of each factor will be investigated. Denote pH by a and temperature by b, then the four treatment combinations might be:

$$a_0 b_0 = \text{pH of 4.0 and a temperature of 30 C}$$
$$a_0 b_1 = \text{pH of 4.0 and a temperature of 40 C}$$
$$a_1 b_0 = \text{pH of 4.4 and a temperature of 30 C}$$
$$a_1 b_1 = \text{pH of 4.4 and a temperature of 40 C}$$

Before we can say how the performance of an experiment involving a factorial set of treatment combinations will help answer our questions concerning independence of the factors, it will be necessary to define certain terms. These terms (effect, main effect, and interaction) have already been used without explanation. Now specific definitions must be given.

We shall consider first a 2^2 factorial such as the one used in Example 9.7. If we agree that the symbols $a_i b_j$ ($i = 0, 1; j = 0, 1$) can represent not only the treatment combinations but also the average yields from all experimental units subjected to the similarly designated treatment combinations, it is possible to define *effect, main effect,* and *interaction* as noted below. (**Note:** To avoid complicating the discussion, it has been assumed that each average yield was obtained from the same number of experimental units.)

$$\text{Effect of } a \text{ at level } b_0 \text{ of } b = a_1 b_0 - a_0 b_0 \tag{9.2}$$

$$\text{Effect of } a \text{ at level } b_1 \text{ of } b = a_1 b_1 - a_0 b_1 \tag{9.3}$$

$$\text{Main effect of } a = [(a_1 b_0 - a_0 b_0) + (a_1 b_1 - a_0 b_1)]/2$$

$$= (a_1 - a_0)(b_1 + b_0)/2 = A \tag{9.4}$$

Similarly,

$$\text{Effect of } b \text{ at level } a_0 \text{ of } a = a_0 b_1 - a_0 b_0 \qquad (9.5)$$

$$\text{Effect of } b \text{ at level } a_1 \text{ of } a = a_1 b_1 - a_1 b_0 \qquad (9.6)$$

$$\text{Main effect of } b = [(a_0 b_1 - a_0 b_0) + (a_1 b_1 - a_1 b_0)]/2$$

$$= (a_1 + a_0)(b_1 - b_0)/2 = B \qquad (9.7)$$

If a and b were acting independently, the effect of a at b_0 and the effect of a at b_1 should be the same. (A similar statement holds for the effects of b at a_0 and a_1.) Thus any difference in these two effects is a measure of the degree of interdependence between the factors, i.e., of the extent to which a and b interact. Accordingly, we define the *interaction* between a and b by

$$AB = [(a_1 b_1 - a_0 b_1) - (a_1 b_0 - a_0 b_0)]/2$$

$$= [(a_1 b_1 - a_1 b_0) - (a_0 b_1 - a_0 b_0)]/2$$

$$= (a_1 - a_0)(b_1 - b_0)/2 \qquad (9.8)$$

If the symbols used in the preceding definitions are simplified by replacing a_0 and b_0 by unity, and a_1 and b_1 by a and b, the effects and interactions may be defined by

$$4M = (a + 1)(b + 1) \qquad (9.9)$$

$$2A = (a - 1)(b + 1) \qquad (9.10)$$

$$2B = (a + 1)(b - 1) \qquad (9.11)$$

$$2AB = (a - 1)(b - 1) \qquad (9.12)$$

where M represents the mean effect (i.e., the mean yield of all experimental units).

Example 9.8

Let us assume that an experiment has been performed involving treatments such as described in Example 9.7. To illustrate the computation of main effects and interactions, three hypothetical cases will be examined.

	I				II				III	
	a_0	a_1			a_0	a_1			a_0	a_1
b_0	63	67		b_0	63	67		b_0	63	67
b_1	69	73		b_1	69	78		b_1	69	70

Case   I: $M = 68, A = 4, B = 6,$ and $AB = 0$
Case II: $M = 69.25, A = 6.5, B = 8.5,$ and $AB = 2.5$
Case III: $M = 67.25, A = 2.5, B = 4.5,$ and $AB = -1.5$

Having defined and illustrated (for a 2^2 factorial) the concepts of effects, main effects, and interactions, it is appropriate that an attempt be made to put these ideas into words rather than symbols. However, the reader is reminded again that the understanding of a concept is much more important than the memorization of any definition, whether it is in words or in mathematical symbolism. With that reminder, let us now attempt definitions of "interaction" and "main effect." Utilizing the earlier definitions and the illustrations in Example 9.8, we may say that:

1. *Interaction* is the differential response to one factor in combination with varying levels of a second factor applied simultaneously. That is, interaction is an additional effect due to the combined influence of two (or more) factors.
2. The *main effect* of a factor is a measure of the change in the response variable to changes in the level of the factor averaged over all levels of all the other factors.

It should be clear that the concepts described as effects and interactions will also be present in situations involving more than two factors. For example, in a case involving four factors, there would be four main effects, six two-factor interactions involving the combined effect of two factors averaged over the other two factors, four three-factor interactions involving the combined effect of three factors averaged over the one remaining factor, and one four-factor interaction involving the combined effect of all four factors. Extensive discussion of these ideas will be deferred until a later chapter.

Before terminating the discussion of effects and interactions, however, two additional topics will be mentioned. One is a convenient method of determining the effects in 2^n factorials; the other is the definition of effects and interactions for 3^n factorials.

To illustrate the method of calculating effects in 2^n factorials, let us consider a 2^3 factorial. Using the abbreviated notation for treatment combinations given in Table 9.1 and letting these symbols also represent the average yields of experimental units subjected to the similarly designated treatment combinations, the main effects and interactions may be found by adding and subtracting yields according to the signs given in Table 9.2. It can easily be verified that this procedure is simply

TABLE 9.2–Schematic Representation of Effects and Interactions in a 2^3 Factorial

Treatment Combination								Effect or Interaction
(1)	a	b	ab	c	ac	bc	abc	
+	+	+	+	+	+	+	+	$8M$
−	+	−	+	−	+	−	+	$4A$
−	−	+	+	−	−	+	+	$4B$
+	−	−	+	+	−	−	+	$4AB$
−	−	−	−	+	+	+	+	$4C$
+	−	+	−	−	+	−	+	$4AC$
+	+	−	−	−	−	+	+	$4BC$
−	+	+	−	+	−	−	+	$4ABC$

a tabular device for calculating the effects and interactions defined by

$$X = (a \pm 1)(b \pm 1)(c \pm 1)/2^2 \qquad (9.13)$$

where the sign in each set of parentheses is plus if the corresponding capital letter is not contained in X and minus if it is contained in X, and the right-hand side is to be expanded and the yields substituted for the appropriate treatment combination symbols. Equation (9.13) may be extended to the 2^n factorial case by simply adding more multiplicative factors as shown in Eq. (9.14),

$$X = (a \pm 1)(b \pm 1)(c \pm 1)(d \pm 1) \ldots /2^{n-1} \qquad (9.14)$$

When factors are investigated at only two levels, the best the researcher can do (apart from a simple test of significance) is to determine (1) whether the effect of a factor is positive or negative and (2) whether the factors are independent. However, when factors are investigated at more than two levels, the researcher can probe more deeply. He now has the opportunity to see if the effect of a factor is linear or nonlinear. In most experimental work this is a very important item of information, and thus the researcher should give serious consideration to factorials involving more than two levels of the factors when planning an investigation.

If an experiment is designed involving two factors, each at three levels, the main effects and interactions may be used to study the non-linearity of the response variable. Rather than going into excessive detail at this time, only the pertinent formulas will be presented. In these formulas we have again used the symbols $a_i b_j$ ($i = 0, 1, 2$; $j = 0, 1, 2$) to represent both the treatment combinations and the yields from the treatment combinations.

Linear effect of $a = A_L = (a_2 - a_0)(b_0 + b_1 + b_2)/3 \qquad (9.15)$

Quadratic effect of $a = A_Q = (a_2 - 2a_1 + a_0)(b_0 + b_1 + b_2)/6 \quad (9.16)$

Linear effect of $b = B_L = (a_0 + a_1 + a_2)(b_2 - b_0)/3 \qquad (9.17)$

Quadratic effect of $b = B_Q = (a_0 + a_1 + a_2)(b_2 - 2b_1 + b_0)/6 \quad (9.18)$

Linear $\times$ linear interaction $= A_L B_L = (a_2 - a_0)(b_2 - b_0)/2 \qquad (9.19)$

Linear $\times$ quadratic interaction

$$= A_L B_Q = (a_2 - a_0)(b_2 - 2b_1 + b_0)/4 \qquad (9.20)$$

Quadratic $\times$ linear interaction

$$= A_Q B_L = (a_2 - 2a_1 + a_0)(b_2 - b_0)/4 \qquad (9.21)$$

Quadratic $\times$ quadratic interaction

$$= A_Q B_Q = (a_2 - 2a_1 + a_0)(b_2 - 2b_1 + b_0)/8 \qquad (9.22)$$

Example 9.9

Consider an experiment similar to that described in Example 9.7 but involving three levels of pH and three levels of temperature. As in Example 9.8 three cases will be considered.

I

	a_0	a_1	a_2
b_0	10	13	16
b_1	13	16	19
b_2	16	19	22

II

	a_0	a_1	a_2
b_0	22	10	14
b_1	25	13	17
b_2	30	18	22

III

	a_0	a_1	a_2
b_0	10	12	11
b_1	14	17	21
b_2	19	25	35

Case   I: $A_L = 6$, $A_Q = 0$, $B_L = 6$, $B_Q = 0$, $A_L B_L = 0$, $A_L B_Q = 0$,
$A_Q B_L = 0$, and $A_Q B_Q = 0$

Case II: $A_L = -8$, $A_Q = 8$, $B_L = 8$, $B_Q = 1$, $A_L B_L = 0$, $A_L B_Q = 0$,
$A_Q B_L = 0$, and $A_Q B_Q = 0$

Case III: $A_L = 8$, $A_Q = 1/3$, $B_L = 46/3$, $B_Q = 4/3$, $A_L B_L = 15/2$,
$A_L B_Q = 3/4$, $A_Q B_L = 7/4$, and $A_Q B_Q = -1/8$

It should be noted that the "no interaction" result in Cases I and II could have been predicted by observing that the pattern of differences between yields at varying levels of b is the same for each level of a. (**Note:** We could just as easily have examined the differences between yields at varying levels of a for each level of b).

From the preceding discussion it should be evident that much can be said about effects and interactions. As a matter of fact, what began as a short section exposing the reader to general concepts has grown into a rather detailed discussion of the topic. On the other hand, the surface has only been scratched. Some additional material will be discussed in later chapters, while the remainder will be left to books devoted to experimental design. For those who wish to read further on these topics, the following references are recommended: Cochran and Cox (3), Cox (4), Davies (5), Federer (6), Finney (7 and 8), Kempthorne (10), Quenouille (15), and Yates (18).

9.15 TREATMENT COMPARISONS

In most experiments involving several treatments, the researcher will be interested in certain specific comparisons among the treatment means. To aid in making such comparisons, the statistician finds it convenient to talk in terms of "contrasts." Algebraically, a *contrast* among the quantities $T_1, \ldots, T_k$ (where T_i is the sum of n_i observations) is defined by

$$C_j = c_{1j}T_1 + c_{2j}T_2 + \cdots + c_{kj}T_k \qquad (9.23)$$

where

$$\sum_{i=1}^{k} n_i c_{ij} = 0 \tag{9.24}$$

If each $n_i = n$; i.e., if each T_i is the sum of the same number of observations, then the necessary condition for a contrast reduces to

$$\sum_{i=1}^{k} c_{ij} = 0 \tag{9.25}$$

Example 9.10

Consider an experiment involving batteries in which four treatments are to be investigated. The four treatments happen to be four different electrolytes. However, it is noted that electrolytes 1 and 2 are quite similar in composition, that 3 and 4 are also similar, but that 1 and 2 differ considerably from 3 and 4. It would, then, be reasonable to plan comparisons of: (1) treatments 1 and 2 versus 3 and 4, (2) treatment 1 versus 2, and (3) treatment 3 versus 4. Assuming that 20 batteries (experimental units) are used and they are allocated to the treatments in the ratio 4:2:5:9, what would be the form of the contrasts for the selected treatment comparisons? Denoting the treatment totals by $T_i (i = 1, 2, 3, 4)$, the desired contrasts are:

$$C_1 = (7)(T_1 + T_2) + (-3)(T_3 + T_4) = 7T_1 + 7T_2 - 3T_3 - 3T_4$$
$$C_2 = (1)T_1 + (-2)T_2 + (0)T_3 + (0)T_4 = T_1 - 2T_2$$
$$C_3 = (0)T_1 + (0)T_2 + (9)T_3 + (-5)T_4 = 9T_3 - 5T_4$$

One might ask how we obtained the coefficients $c_{ij} (i = 1, 2, 3, 4; j = 1, 2, 3)$ used in the above comparisons. A short explanation at this moment should serve to clear up any difficulties. Consider the case of comparison C_1. We are attempting to compare the mean of 6 observations $(4 + 2)$ with the mean of 14 observations $(5 + 9)$. It is necessary to adjust for the spurious weighting given by our comparison of treatment totals based on unequal numbers of observations. Since the smallest integer that may be divided evenly by both 6 and 14 is 42, we see that 7 and 3 are the indicated weights to be used if our comparison is to be unaffected by the differing numbers of observations associated with the various treatments. The remaining coefficients are found in a like manner.

Example 9.11

Consider a research situation similar to that described in Example 9.10 but involving five treatments. Suppose that four batteries are allocated to each treatment. If treatment 2 represents a commonly used electrolyte, while 1, 3, 4, and 5 are newly developed electrolytes in which 1 and 3 are of type A and 4 and 5 are of type B, the contrasts specified in Table 9.3 are appropriate for the obvious treatment comparisons.

TABLE 9.3–Symbolic Representation of the Contrasts for the
Treatment Comparisons Specified in Example 9.11

Contrast	Electrolyte				
	1	2	3	4	5
C_1	-1	4	-1	-1	-1
C_2	1	0	1	-1	-1
C_3	1	0	-1	0	0
C_4	0	0	0	1	-1

Let us now note another item of importance. If two contrasts,

$$C_p = c_{1p}T_1 + c_{2p}T_2 + \cdots + c_{kp}T_k \tag{9.26}$$

and

$$C_q = c_{1q}T_1 + c_{2q}T_2 + \cdots + c_{kq}T_k \tag{9.27}$$

are such that

$$\sum_{i=1}^{k} n_i c_{ip} c_{iq} = 0 \qquad p \neq q \tag{9.28}$$

then contrast C_p is *orthogonal* to contrast C_q. (**Note:** It is common practice to speak of orthogonal contrasts or orthogonal treatment comparisons.) If $n_i = n$ (for all i), the orthogonality condition reduces to

$$\sum_{i=1}^{k} c_{ip} c_{iq} = 0 \qquad p \neq q \tag{9.29}$$

The reader can easily verify that the contrasts specified in Examples 9.10 and 9.11 are orthogonal in each case. In addition, the perceptive reader will have noted that the effects and interactions discussed in Section 9.14 were also orthogonal contrasts.

The question might well be asked, "Are orthogonal contrasts better than nonorthogonal contrasts?" Intuitively, orthogonal contrasts seem to be preferable. (**Note:** Actually, they are preferable if one wishes the estimates derived from the different contrasts to be uncorrelated.) However, occasionally it is desirable to design an experiment with the expressed intent of analyzing a set of nonorthogonal contrasts. In such cases, the probability statements accompanying the associated tests of significance are of an ambiguous nature (due to the correlation between the contrasts), and much care should be exercised in interpreting the experimental results.

One final remark needs to be made: regardless of the desirability of orthogonal contrasts, the statistician should not let his preference for such override the needs of the researcher. By this is meant that, as nice as it is to have a set of orthogonal contrasts, only those contrasts that are meaningful to the researcher should be analyzed.

9.16 STEPS IN DESIGNING AN EXPERIMENT

Each statistician has his own list of steps that he follows when designing an experiment. However, a comparison of various lists reveals that they all cover essentially the same points.

According to Kempthorne (10) a statistically designed experiment consists of the following steps:

1. Statement of the problem.
2. Formulation of hypotheses.
3. Devising of the experimental technique and design.
4. Examination of possible outcomes and reference back to the reasons for the inquiry to be sure the experiment provides the required information to an adequate extent.
5. Consideration of the possible results from the point of view of the statistical procedures that will be applied to them, to ensure that the conditions necessary for these procedures to be valid are satisfied.
6. Performance of the experiment.
7. Application of statistical techniques to the experimental results.
8. Drawing conclusions with measures of the reliability of estimates of any quantities that are evaluated, careful consideration being given to the validity of the conclusions for the population of objects or events to which they are to apply.
9. Evaluation of the whole investigation, particularly with other investigations on the same or similar problems.

In a later section these steps will be illustrated through the consideration of some design problems.

Since the designing of an experiment or the planning of a test program is such an important part of any investigation, the statistician must make every effort to obtain all the relevant information. This will usually require one or more conferences with the researcher and the asking of many questions. It has been our experience that the amount of time consumed in this phase can be materially reduced if, at the preliminary meeting between the researcher (e.g., a development engineer) and the statistician, time is taken to explore the relationship between research and/or development experimentation and the statistical design of experiments. (**Note:** Frequently, there is a formidable communications barrier that must be overcome.) One of the best ways to convince the researcher of the need for the multitude of questions posed by the statistician is to give him (in the first meeting) a check list that specifies various stages in the planning of a test program. (An even more efficient arrangement if you are the statistician in an industrial organization is to distribute copies of such a list to all persons who may at some time have need of your services.) One such list, prepared by Bicking (2), is reproduced below.

Check List for Planning Test Programs

A. *Obtain a clear statement of the problem.*

 1. Identify the new and important problem area.
 2. Outline the specific problem within current limitations.

3. Define the exact scope of the test program.
4. Determine the relationship of the particular problem to the whole research or development program.

B. *Collect available background information.*

1. Investigate all available sources of information.
2. Tabulate data pertinent to planning the new program.

C. *Design the test program.*

1. Hold a conference of all parties concerned.
   a. State the propositions to be proved.
   b. Agree on the magnitude of differences considered worthwhile.
   c. Outline the possible alternative outcomes.
   d. Choose the factors to be studied.
   e. Determine the practical range of these factors and the specific levels at which tests will be made.
   f. Choose the end measurements that are to be made.
   g. Consider the effect of sampling variability and precision of test methods.
   h. Consider possible interrelationships (or "interactions") of the factors.
   i. Determine limitations of time, cost, materials, manpower, instrumentation and other facilities and of extraneous conditions such as weather.
   j. Consider human relation angles of the program.
2. Design the program in preliminary form.
   a. Prepare a systematic and inclusive schedule.
   b. Provide for stepwise performance or adaptation of the schedule if necessary.
   c. Eliminate the effect of variables not under study by controlling, balancing, or randomizing them.
   d. Minimize the number of experimental runs.
   e. Choose the method of statistical analysis.
   f. Arrange for orderly accumulation of data.
3. Review the design with all concerned.
   a. Adjust the program in line with comments.
   b. Spell out in unmistakable terms the steps to be followed.

D. *Plan and carry out the experimental work.*

1. Develop methods, materials, and equipment.
2. Apply the methods or techniques.
3. Attend to and check details; modify methods if necessary.
4. Record any modifications of program design.
5. Take precautions in collection of data.
6. Record progress of the program.

E. *Analyze the data.*

1. Reduce the recorded data to numerical form if necessary.
2. Apply proper mathematical statistical techniques.

F. *Interpret the results.*

1. Consider all the observed data.
2. Confine conclusions to strict deductions from the evidence at hand.
3. Test questions suggested by the data by independent experiments.
4. Arrive at conclusions as to the technical meaning of results as well as their statistical significance.
5. Point out implications of the findings for application and further work.
6. Account for any limitations imposed by the methods used.
7. State results in terms of verifiable probabilities.

G. *Prepare the report.*

1. Describe work clearly giving background, pertinence of the problems, and meaning of results.
2. Use tabular and graphic methods of presenting data in good form for future use.
3. Supply sufficient information to permit the reader to verify results and draw his own conclusions.
4. Limit conclusions to an objective summary of evidence so that the work recommends itself for prompt consideration and decisive action.

The reader should realize that the two lists presented in this section are only guides. Very seldom will the various steps be tackled and settled in the particular order given. The statistician does not operate in such a mechanical and routine fashion. Questions will be asked and answers received which will trigger new lines of thought, and thus the planning conference will find itself jumping from one step to another in a seemingly haphazard manner. Furthermore, it is not surprising to find, as the conference progresses and new information is brought forth, the same step being considered several times. Regardless of the inherent repetition, such a procedure is good.

In summary, then, the designing of an experiment can be a time-consuming and occasionally painful process. Thus the use of check lists such as those presented earlier can be most helpful (as a supplement to common sense) in making relatively certain that nothing has been overlooked.

9.17 ILLUSTRATIONS OF THE STATISTICIAN'S APPROACH TO DESIGN PROBLEMS

To illustrate the manner in which a statistician approaches a design problem, a series of examples will be considered. The first of these will demonstrate the application of Kempthorne's nine steps, while the remainder will illustrate various topics discussed in Sections 9.1 through 9.16.

Example 9.12

Suppose a machine is constructed for the purpose of generating a random series of 0's and 1's. If the machine is truly a generator of random binary elements, it should, among other things, yield 0's 50 percent of the time and 1's 50 percent of the time. It is proposed that an experiment be devised to check on this particular aspect of the randomness of the machine.

The preceding paragraph illustrates Kempthorne's step 1, the statement of the problem. If we formulate $H:p_0 = 1/2$ (where p_0 stands for the probability of a 0) and $A:p_0 \neq 1/2$, we have taken care of step 2. The devising of an experimental technique and design (step 3) is fairly simple. In this case we shall operate the device a certain number of times (say n), record the proportion of 0's ($\hat{p}_0$), and see if this is in close enough agreement with the hypothesis H. If the agreement is good, we accept H; if the agreement is poor, we reject H and accept A, the alternative hypothesis. The only remaining part of step 4 to be taken care of is the determination of the number of operations of the device that are required before we feel safe in making a decision. Suppose it is desired that the probability of rejecting $H:p_0 = 1/2$ (when it is really true) should be no greater than $\alpha = 0.05$. This implies $n \geq 6$, as can easily be shown. Note carefully the concept of rejecting a true hypothesis. The value of n would also be influenced by fixing the probability of accepting a false hypothesis, but we choose to ignore this in the present example. Step 5 consists, in this case, of recognizing that the results will be analyzed using the binomial distribution, and thus we should make certain that the repeated events (operations of the device) are statistically independent. Step 6 is evident, though sometimes troublesome. When discussing steps 3 and 4, the content of step 7 was alluded to, and all that remains is the formalizing of the analysis. Step 8 implies that we should produce a confidence interval estimate of the true probability of producing a 0 with our device; i.e., a point estimate, $\hat{p}_0$, is not sufficient. We must also be very careful to state that our conclusions only hold for the particular device operated, *unless* this device was randomly selected from a larger group (or population) of devices. Had other devices of a similar nature been investigated, the results of our experiment should be evaluated along with all pertinent information from the allied studies (step 9).

The reader will probably have recognized the similarity of this illustration to Example 6.5. It is, of course, the same. All we have done here is "dress up" the problem and use it to illustrate the various steps in the design of an experiment.

Example 9.13

Consider the problem of an engineer who wishes to assess the relative effects of eight treatments (for the moment undefined) on the activated life of a particular type of thermal battery. Assume that 64 relatively homogeneous batteries are available for experimentation. With only this much information, the most efficient design would be to randomly assign the batteries to the eight treatments (groups) subject to the restriction that 8 batteries be allocated to each treatment. Such an assignment is illustrated in Table 9.4. The reader should note that the major design decisions reached in this example were concerned with *balancing* and *grouping*. (**Note:** The type of design described above is known as a *completely randomized design*, and is discussed further in Chapter 10.)

TABLE 9.4–Random Assignment of Batteries to
Treatments as Described in Example 9.13

			Treatments				
A	B	C	D	E	F	G	H
9	58	37	18	14	21	48	43
22	53	36	38	1	15	63	56
64	26	30	33	50	3	60	41
34	11	5	29	27	45	57	23
17	52	6	61	16	47	25	10
4	51	13	40	49	32	59	12
31	8	2	35	46	19	7	20
28	14	54	39	44	62	55	42

Note: Numbers in the table represent serial numbers of units; a random order of testing would also be determined.

Example 9.14

As a second illustration of a completely randomized design, consider the agronomist who has 28 homogeneous experimental plots available for testing the relative effects of four different fertilizers on the yield of a particular variety of oats. A reasonable design would be to allocate at random a different fertilizer to each plot. If the restriction is imposed that 7 of the experimental plots be allocated to each fertilizer (treatment), complete *balance* will have been achieved.

Example 9.15

Referring to Example 9.13, suppose you are now advised that the 64 batteries consist of 8 batteries from each of 8 different production lots. How will this additional information affect the design? If it is suspected that real differences exist among the lots, the precision of the experiment can be improved by removing the lot-to-lot variation from the estimate of experimental error. Such an improvement in design may be accomplished by assigning the treatments to the batteries *at random within each lot.* Such a restricted randomization is illustrated in Table 9.5. (**Note:** The type of design described above is known as a *randomized complete block design,* and is discussed further in Chapter 11.)

The major benefit resulting from this type of *blocking* is a gain in efficiency in analysis. That is, more sensitive tests of significance for treatment differences can be made and shorter confidence interval estimates of treatment effects can be obtained.

Example 9.16

Another illustration of a randomized complete block design is provided by the following problem in nutrition research. A nutritionist wishes to assess the relative effects of four newly developed rations on the weight-gaining ability of rats. He has 20 rats available for experimentation. Examination of the pedigrees of the experimental animals indicates that the 20 rats consist of 4 rats from each of 5 litters. The statistician would, under these circumstances, recommend that the rations be assigned to the rats at random within each litter (block).

TABLE 9.5–Random Assignment of Treatments to Batteries
within Lots as Described in Example 9.15

			Lots				
1	2	3	4	5	6	7	8
1-*H*	9-*H*	17-*E*	25-*C*	33-*E*	41-*G*	49-*B*	57-*D*
2-*C*	10-*E*	18-*A*	26-*D*	34-*F*	42-*H*	50-*E*	58-*A*
3-*F*	11-*D*	19-*B*	27-*E*	35-*D*	43-*B*	51-*G*	59-*F*
4-*B*	12-*F*	20-*G*	28-*B*	36-*G*	44-*C*	52-*F*	60-*G*
5-*E*	13-*G*	21-*C*	29-*H*	37-*C*	45-*E*	53-*H*	61-*C*
6-*G*	14-*C*	22-*F*	30-*G*	38-*A*	46-*A*	54-*A*	62-*H*
7-*D*	15-*B*	23-*D*	31-*F*	39-*B*	47-*F*	55-*D*	63-*B*
8-*A*	16-*A*	24-*H*	32-*A*	40-*H*	48-*D*	56-*C*	64-*E*

Note: Numbers in the table represent serial numbers of units; the letters represent treatments. It will be observed that we have assumed lot 1 contains batteries 1 to 8, lot 2 contains batteries 9 to 16, etc.

Example 9.17

Consider next a somewhat more complex problem. Assume that we are again concerned with testing batteries, but this time the problem arises during the development phase. The development engineer has to reach a decision about three things: (1) how much electrolyte should be incorporated in this particular model, (2) what weight of heat paper should be used in the construction of the batteries, and (3) what effect will the temperature at which the batteries are activated have on the activated life of the batteries?

Denoting electrolyte by *a*, heat paper by *b*, and temperature by *c* and assuming that two levels of each factor are to be investigated, the eight treatment combinations might be as shown in Table 9.6.

It is decided that 16 batteries will be built to each of the four "electrolyte–heat paper" specifications, providing a total of 64 batteries for testing. As a precaution against bias being introduced because the last bat-

TABLE 9.6–Factors, Factor Levels, and Treatment Combinations for
the Experiment Described in Example 9.17

Treatment Combination	Factors and Factor Levels		
	Amount of electrolyte (gm/cell)	Weight of heat paper (gm/cell)	Test temperature (F)
(1)	1	4	−50
a	2	4	−50
b	1	6	−50
ab	2	6	−50
c	1	4	100
ac	2	4	100
bc	1	6	100
abc	2	6	100

teries built might be better than the first, the 64 batteries will be built in a random order. Next, in each set of 16 batteries, 8 will be randomly selected for testing at low temperature, and the remaining 8 will be reserved for testing at high temperature.

When this stage is reached (i.e., once each battery has been built and assigned a test temperature), the 64 batteries will be arranged in random order for individual testing. As you can probably anticipate, this last restriction frequently proves to be unpopular, especially if only one temperature chamber is available. (**Note:** The design that has been formulated is a completely randomized design involving a 2^3 factorial with 8 experimental units per treatment.)

Example 9.18

Suppose that we now consider a slightly different problem. Like many of us engaged in research, the development engineer is often hard pressed for funds. If this were the case in the situation described in Example 9.17, the development engineer might place a preliminary order for 8 batteries. Of the 8, 2 would be assembled to each of the 4 "electrolyte–heat paper" combinations. His plan would be to test 1 battery in each pair at low temperature and 1 at high temperature. This testing would naturally take place in a random order.

Next, assume that after the first 8 batteries are built and tested, funds are made available for the building and testing of 8 additional batteries. These would be ordered without delay to provide some *replication* of the experiment. However, due to the way the batteries were produced and tested (i.e., first 8 and then 8 more), it is clear that the combined analysis of all 16 batteries must take into account the *blocking* implicit in the data. (**Note:** In this example we have a randomized complete block design consisting of two blocks and involving a 2^3 factorial set of treatment combinations.)

Example 9.19

Referring again to the problem described in Example 9.17, suppose that two additional complications arise: (1) only 8 batteries can be tested in a normal work day and (2) in the interest of economy the test engineer wishes to place 4 batteries in the temperature chamber at the same time. Under these restrictions, we have a natural set of *blocks,* namely, days. Further, within each block it would be desirable to test 1 battery corresponding to each of the 8 treatment combinations. Because of the temperature chamber restriction we would decide, randomly for each day, whether to first test batteries at high temperature and then test batteries at low temperature or vice versa. Once this decision is made, the random order of testing batteries within temperatures must be specified. As might be expected, the eventual analysis of the data will take due cognizance of all restrictions placed on the test program. (**Note:** The type of design illustrated in this example is known as a *split plot design,* and is discussed further in Chapter 12.)

9.18 ADVANTAGES AND DISADVANTAGES OF STATISTICALLY DESIGNED EXPERIMENTS

It is now appropriate that the advantages and disadvantages of statistically designed experiments be considered. These will be expressed in different ways by different people. However, as was true for the steps involved in designing experiments, an examination of various lists will show that they cover essentially the same points.

Advantages of Statistically Designed Experiments

Bicking (2, p. 22) has listed the advantages of statistical designs as follows:

1. Close teamwork is required between the statisticians and the research or development scientists with consequent advantages in the analysis and interpretation stages of the program.
2. Emphasis is placed on anticipating alternatives and systematic pre-planning, yet permitting stepwise performance and producing only data useful for analysis in later combinations.
3. Attention is focused on interrelationships and identifying and measuring sources of variability in results.
4. The required number of tests is determined reliably and often may be reduced.
5. Comparison of effects of changes is more precise because of grouping of results.
6. The correctness of conclusions is known with definite mathematical preciseness.

If these advantages truly exist, and we believe they do, the value of statistical aid in planning experiments is evident and should always be sought.

Disadvantages of Statistically Designed Experiments

Happily, there are more advantages than disadvantages associated with statistically designed experiments. However, a careful reading of Mandelson (11), together with a realistic appraisal of the implementation of certain statistically designed experiments, yielded the following possible disadvantages:

1. Such designs and their analyses are usually accompanied by statements couched in the technical language of statistics. It would be much better if the statistician would translate such statements into terms that are meaningful to the nonstatistician. In addition, the statistician should not overlook the value of presenting the results in graphical form. As a matter of fact, he should always consider plotting the data as a preliminary step to a more analytical approach.
2. Many statistical designs, especially when first formulated, are criticized as being too expensive, complicated, or time consuming. Such criticisms, when valid, must be accepted in good grace and an honest attempt made to improve the situation, provided that the solution of the problem is not compromised.

Before terminating this discussion, some mention should be made of particular advantages and disadvantages associated with factorials.

This is deemed necessary because of the important role that factorials play in the design and analysis of experiments. (**Note:** There will undoubtedly be some overlap between the advantages and disadvantages given for statistically designed experiments in general and those about to be given for factorials.)

Advantages of factorials:

1. Greater efficiency in the use of available experimental resources is achieved.
2. Information is obtained about the various interactions.
3. The experimental results are applicable over a wider range of conditions; i.e., due to the combining of the various factors in one experiment, the results are of a more comprehensive nature.
4. There is a gain due to the hidden replication arising from the factorial arrangement.

Disadvantages of factorials:

1. The experimental setup and the resulting statistical analysis are more complex.
2. With a large number of treatment combinations the selection of homogeneous experimental units becomes more difficult.
3. Certain of the treatment combinations may be of little or no interest; consequently, some of the experimental resources may be wasted.

9.19 SUMMARY

In this chapter several extremely important topics have been discussed. It is recommended that they be reexamined from time to time as the reader progresses through the book. Such a periodic reappraisal will prove beneficial for a number of reasons; e.g., (1) a thorough understanding of the concepts, principles, and techniques involved is essential for a fruitful study of the experimental designs that are the subjects of the next three chapters and (2) an appreciation of these important principles should manifest itself in improved experimentation.

PROBLEMS

9.1   Choosing practical situations from your special field of interest, describe three problems whose solutions must be determined experimentally.

9.2   With reference to Problem 9.1, discuss the need for an experimental design in each of the three illustrations.

9.3   It is sometimes said that experimental design is a subject consisting of two almost distinct parts: (a) the choice of treatments, experimental units, and characteristics to be observed and (b) the choice of the number of experimental units and the method of assigning the treatments to the experimental units. Discuss this classification from the points of view of the researcher and the statistician.

9.4   Define "systematic error" and discuss the relationship between this factor and the statistical design of experiments.

9.5   Some terms that occur rather frequently in the literature are: (a) accuracy, (b) precision, (c) validity, (d) reliability, and (e) bias. Restricting your remarks to the theory of statistics or to applications of statistical methods, define and discuss each of these terms.

9.6   Cox (4) uses "Designs for the Reduction of Error" as the title of one of his chapters. What does this title suggest to you?

9.7   With reference to factorials, what is meant by the phrase "hidden replication"?

9.8   Discuss the use of concomitant information in experimental design.

9.9   Choosing practical situations from your own special field of interest, illustrate the concept of confounding. Give examples of (a) unavoidable confounding, (b) unintentional confounding, and (c) intentional confounding.

9.10 Choosing practical situations from your own special field of interest, illustrate the concept of randomization.

9.11 With reference to Problem 9.10, discuss the difficulties, if any, associated with the randomization process.

9.12 Give your interpretation of the phrase "restricted randomization."

9.13 What would you do if, in the planning of a randomized complete block design, the same order of treatments occurred randomly in each block?

9.14 Discuss the following ways treatments can be assigned to experimental units: (a) randomly, (b) subjectively, and (c) systematically. Give illustrations that show the benefits, dangers, and difficulties involved in each of the three approaches.

9.15 Cox (4) uses the phrase "Randomization as a Device for Concealment" as the heading of one of the sections in his book. Without referring to his discussion, what do you believe he has in mind?

9.16 Cox (4) makes a distinction between factors that represent a treatment applied to the experimental units (treatment factors) and factors that correspond to a classification of the experimental units into two or more types (classification factors). Give illustrations of each of these from situations in your special field of interest.

9.17 The statement has been made that an uncontrolled and unmeasured variable may be of sufficient importance to lead to the conclusion that two controlled factors interact to a significant degree. Discuss this idea, including all possible implications. What safeguards do we have against such a result occurring?

9.18 Cox (4) also states that it is sometimes convenient to classify factors as follows: (a) specific qualitative factors, (b) quantitative factors, (c) ranked qualitative factors, and (d) sampled qualitative factors. How would you define each of these? Compare your ideas with those expressed by Cox.

9.19 Show graphically what is meant by an interaction. Illustrate your ideas using the data of Examples 9.8 and 9.9.

9.20 Explain the relationship, if any, between regression functions (i.e., response functions) and the concepts of effects and interactions.

9.21 How would you go about selecting the factors to be investigated in an experiment? Illustrate with examples from your special field of interest.

9.22 Assuming that the factors have been decided upon, how would you go about selecting the factor levels? Illustrate with examples from your special field of interest.

9.23 Choosing practical situations from your special field of interest, illustrate completely randomized, randomized complete block, and split plot designs.

9.24 Indicate how the examples provided in answer to Problem 9.23 attempted to "control error."

9.25 What is meant by the precision of an experiment? of a contrast?

9.26 What is meant by a sequential experiment? Is there any other kind? Please discuss.

9.27 In Problem 9.3, reference was made to "... the choice of treatments, experimental units, and characteristics to be observed." Illustrate each of these with examples from your special field of interest.

9.28 Discuss the following items relative to the selection of experimental units: (a) number of units, (b) size of units, (c) shape of units, (d) independence of units.

9.29 What is meant by a "control" treatment?

9.30 Cox (4) classifies observations into six groups: (a) primary observations, (b) substitute primary observations, (c) explanatory observations, (d) supplementary observations for increasing precision, (e) supplementary observations for detecting interactions, and (f) observations for checking the application of the treatments. Please try to define and illustrate each of these. Then compare your ideas with those expressed by Cox.

9.31 Building on the samples given in Section 9.16, construct your own list of "steps in designing an experiment."

9.32 Contrast the one-factor-at-a-time method of experimentation with the factorial approach. Construct a table that shows and compares the advantages and disadvantages of each.

9.33 Define (a) absolute experiments and (b) comparative experiments. Give examples of each. With which type is this book mainly concerned?

9.34 Consider the following "elements" of experimental method:
   (a) control, or the elimination of the effects of extraneous variables
   (b) accuracy of instruments and data acquisition
   (c) reduction of the number of variables to be investigated
   (d) planning of the test sequence in advance of the start of experimentation
   (e) detection of malfunctions
   (f) testing for reasonableness of results
   (g) analysis and interpretation of results
   Evaluate the foregoing list by comparing it with the ideas expressed in this chapter.

9.35 Choosing practical situations from your special field of interest, give three examples of statistically designed experiments. For each of these point out where and how the concepts of this chapter were employed.

9.36 Choose a practical problem in your area of specialization. Following where practicable the philosophy expressed in this chapter, design an experiment to provide data relevant to the problem. Justify all your decisions and relate them to the discussion in the text. If feasible, perform the experiment and then analyze and interpret the results.

REFERENCES

1. Barbacki, S.; and Fisher, R. A. 1936. A test of the supposed precision of systematic arrangements. *Ann. Eugen.* 7:189.

2. Bicking, C. A. 1954. Some uses of statistics in the planning of experiments. *Ind. Qual. Control* 10(Jan.):20–24.
3. Cochran, W. G.; and Cox, G. M. 1957. *Experimental Designs,* 2nd ed. Wiley, New York.
4. Cox, D. R. 1958. *Planning of Experiments.* Wiley, New York.
5. Davies, O. L. (ed.). 1956. *The Design and Analysis of Industrial Experiments,* 2nd ed. Oliver and Boyd, Edinburgh.
6. Federer, W. T. 1955. *Experimental Design.* Macmillan, New York.
7. Finney, D. J. 1955. *Experimental Design and Its Statistical Basis.* University of Chicago Press.
8. ———. 1960. *An Introduction to the Theory of Experimental Design.* University of Chicago Press.
9. Jeffreys, H. 1939. Random and systematic arrangements. *Biometrika* 31:1.
10. Kempthorne, O. 1952. *The Design and Analysis of Experiments.* Wiley, New York.
11. Mandelson, J. 1957. The relation between the engineer and the statistician. *Ind. Qual. Control* 13(May):31–34.
12. National Bureau of Standards. 1949. Projects and publications of the national applied mathematics laboratories. April–June.
13. Pearson, E. S. 1938. Some aspects of the problem of randomization. *Biometrika* 29:53.
14. ———. 1938. An illustration of "Student's" inquiry into the effect of balancing in agricultural experiments. *Biometrika* 30:159.
15. Quenouille, M. H. 1953. *The Design and Analysis of Experiment.* Charles Griffin, London.
16. Ratner, R. A. 1951. Effect of variations in weight upon move times. M.S. thesis, Iowa State University, Ames.
17. "Student" (W. S. Gosset). 1938. Comparison between balanced and random arrangements of field plots. *Biometrika* 29:363.
18. Yates, F. 1937. The design and analysis of factorial experiments. Tech. Comm. 35, Imperial Bureau of Soil Science.

FURTHER READING

Chapin, F. S. 1955. *Experimental Designs in Sociological Research,* rev. ed. Harper, New York.
Chew, V. (ed.). 1958. *Experimental Designs in Industry.* Wiley, New York.
Cochran, W. G.; and Cox, G. M. 1957. *Experimental Designs,* 2nd ed. Wiley, New York.
Cox, D. R. 1958. *Planning of Experiments.* Wiley, New York.
Davies, O. L. (ed.). 1956. *The Design and Analysis of Industrial Experiments,* 2nd ed. Oliver and Boyd, Edinburgh.
Federer, W. T. 1955. *Experimental Design.* Macmillan, New York.
Finney, D. J. 1960. *An Introduction to the Theory of Experimental Design.* University of Chicago Press.
Fisher, R. A. 1966. *The Design of Experiments,* 8th ed. Oliver and Boyd, Edinburgh.
Kempthorne, O. 1952. *The Design and Analysis of Experiments.* Wiley, New York.
LaClerg, E. L.; Leonard, W. H.; and Clark, A. G. 1962. *Field Plot Technique,* 2nd ed. Burgess, Minneapolis.
Yates, F. 1970. *Experimental Design: Selected Papers of Frank Yates.* Griffin, London.

CHAPTER 10

COMPLETELY RANDOMIZED DESIGN

In Chapter 9 several experimental designs were illustrated. In this chapter we propose to discuss the simplest of these designs, the completely randomized design, in considerable detail. Much attention will be given to methods of analyzing data arising from such a design, and it will be observed that analysis of variance (frequently abbreviated as AOV or ANOVA) is the method most widely used.

10.1 DEFINITION OF A COMPLETELY RANDOMIZED DESIGN

A *completely randomized* (CR) *design* is a design in which the treatments are assigned *completely at random* to the experimental units, or vice versa. That is, it imposes no restrictions, such as blocking, on the allocation of the treatments to the experimental units. As in Examples 9.13 and 9.14, some degree of balance may be sought.

Because of its simplicity the completely randomized design is widely used. However, the researcher is cautioned that its use should be restricted to those cases in which homogeneous experimental units are available. If such units cannot be obtained, some blocking should be utilized to increase the efficiency of the design.

Example 10.1

Given 4 fertilizers, we wish to test the null hypothesis that there are no differences among the effects of these fertilizers on the yield of corn. We shall assume there are 20 experimental plots available to the research worker. A sound procedure would be to place each fertilizer on an equal number of experimental plots so that our estimates of the mean effect of each fertilizer will have equal weight. Then we insist that the fertilizers be assigned to the plots at *random*. This may be accomplished by numbering our plots from 1 through 20 and then drawing tickets at random from a hat, 5 tickets being identified by coloring or code mark with each of the 4 fertilizers. The first one drawn specifies the treatment for plot 1, the second for plot 2, etc.

Example 10.2

If in the preceding example only 17 plots were available, some lack of balance would be inevitable. Assuming that more precise information is desired on fertilizer 1, the randomization procedure could be modified so that, for example, 8 plots would be treated with fertilizer 1, and 3 plots each with fertilizer 2, 3, and 4.

TABLE 10.1–Symbolic Representation of Data in a Completely
Randomized Design (unequal numbers of observations per treatment)

	Treatment				Total
	1	2	$\ldots$	t	
Observations	Y_{11} Y_{12} $\vdots$ Y_{1n_1}	Y_{21} Y_{22} $\vdots$ Y_{2n_2}	$\ldots$ $\ldots$ $\ldots$	Y_{t1} Y_{t2} $\vdots$ Y_{tn_t}	
Total	$T_1 = Y_1.$	$T_2 = Y_2.$	$\ldots$	$T_t = Y_t.$	$T = Y_{..} = \sum_{i=1}^{t} T_i$ $= \sum_{i=1}^{t} \sum_{j=1}^{n_i} Y_{ij}$
Number of observations	n_1	n_2	$\ldots$	n_t	$N = \sum_{i=1}^{t} n_i$
Mean	$\overline{Y}_1.$	$\overline{Y}_2.$	$\ldots$	$\overline{Y}_t.$	$\overline{Y}_{..} = T/N$

10.2 COMPLETELY RANDOMIZED DESIGN WITH ONE OBSERVATION PER EXPERIMENTAL UNIT

Suppose in an experiment where the completely randomized design was used, n_i experimental units were subjected to the ith treatment $(i = 1, \ldots, t)$ and only one observation per experimental unit was obtained. If we let Y_{ij} represent the observation obtained from the jth experimental unit subjected to the ith treatment, the data would appear as in Table 10.1. We use the notation T_i to denote the total of the observations for treatment i, i.e., $T_i = Y_i. = \sum_{j=1}^{n_i} Y_{ij}$.

Before discussing the statistical analysis associated with the data obtained from a completely randomized experiment, certain assumptions must be made about the observations. In general, the assumptions underlying the analysis are the same as those usually associated with regression analysis. These are additivity, linearity, normality, independence, and homogeneous variance. That is, the statistical model most frequently assumed in a completely randomized experiment is a linear model to which has been appended certain restrictions about independent observations from normal distributions.

10.2.1 Completely Randomized Design Model

The basic assumption for a completely randomized experiment with one observation per experimental unit is that the observations may be represented by the linear statistical model

$$Y_{ij} = \mu + \tau_i + \epsilon_{ij} \qquad i = 1, \ldots, t; j = 1, \ldots, n_i \quad \text{(unequal numbers)}$$

$$j = 1, \ldots, n \quad \text{(equal numbers)}$$

$$(10.1)$$

where μ is the true mean effect, τ_i is the true effect of the ith treatment, and ϵ_{ij} is the experimental error associated with the jth experimental unit subjected to the ith treatment. In addition, we assume $\epsilon_{ij} \sim NI(0, \sigma^2)$. An important part of the assumed model is the latter statement that the experimental errors ϵ_{ij} are normal, are independent, and have homogeneous variance for each treatment and experimental unit. Since all the statistical analysis is based on this assumption, it is necessary for the researcher to be sure the data to be analyzed at least approximately satisfy these assumptions. If not, it may be necessary to transform the data prior to the analysis. (**Note:** This problem is more fully discussed in Section 10.13.) Although we have ascribed only the effect of experimental error to ϵ_{ij}, we also include the nonmeasurable effects of any other extraneous factors in ϵ_{ij}. However, we rely on the process of randomization to prevent these factors from contaminating our results.

To use the results of the analysis associated with this model, we need to make one additional assumption to completely specify the model. We distinguish between two alternative views of the parameters τ_i in the model. These alternate views are pertinent to our choice of test statistic and our interpretation of the test results. We distinguish between two models.

Model I: Analysis of variance (fixed effects) model

This model reflects the experimenter's decision that he is only concerned with the t treatments present in the experiment. This assumption is expressed by the identity $\sum_{i=1}^{t} \tau_i = 0$.

Model II: Component of variance (random effects) model

This model reflects the decision that the experimenter is concerned with a population of treatments of which only a random sample (the t treatments) are present in the experiment. This assumption is expressed by the statement $\tau \sim NI(0, \sigma_\tau^2)$.

The model to be used is determined by the experimenter's view of his experiment. Either the results are pertinent to only the treatments present (Model I) or inferences are to be made to a larger population of treatments (Model II). This completes the specification of the model.

10.2.2 Analysis of Variance: Equal Number of Experimental Units per Treatment

Now that we have set up the model, we are in a position to analyze the data. We first do this for the case where an equal number of experimental units are subjected to each treatment; i.e., $n_i = n$ $(i = 1, \ldots, t)$.

Just what analysis will be done depends on the purpose for which the experiment was conducted.

Several problems are associated with experiments of this type; one is that of comparing the relative effects of the treatments. In terms of the model parameters the problem is one of testing the hypotheses:

Model I	Model II
$H: \tau_i = 0 \ (i = 1, \ldots, t)$	$H: \sigma_\tau^2 = 0$
$A: \tau_i$'s are not all equal	$A: \sigma_\tau^2 \neq 0$

That is, the hypothesis tested is that there is no difference among the effects of the treatments (**Note:** In Model I the hypothesis is referring to the t treatments used in the experiment, and in Model II the hypothesis is about the population of treatment effects from which the t treatments in the experiment are a sample).

In either case the test statistic is an F statistic based on the ratio of two mean squares. The computations associated with this test are based on the following ANOVA identity:

$$\sum_{i=1}^{t} \sum_{j=1}^{n} Y_{ij}^2 = tn(\overline{Y}_{..})^2 + n \sum_{i=1}^{t} (\overline{Y}_{i.} - \overline{Y}_{..})^2 + \sum_{i=1}^{t} \sum_{j=1}^{n} (Y_{ij} - \overline{Y}_{i.})^2$$

(10.2)

where $\overline{Y}_{..} = \sum_{i=1}^{t} \sum_{j=1}^{n_i} Y_{ij}/tn$ is the overall mean and $\overline{Y}_{i.} = \sum_{j=1}^{n_i} Y_{ij}/n_i$ is the ith treatment mean. This identity is a partition of the total variation in the data into meaningful sums of squares that can be associated with various sources of variation. The first term on the right-hand side of Eq. (10.2) is the variation attributable to the overall mean, the second term is a measure of the variation caused by differences

TABLE 10.2–ANOVA for Data of Table 10.1 Showing Certain Expected Mean Squares (equal numbers of observations per treatment)

Source of Variation	Degrees of Freedom	Sum of Squares	Mean Square*	Expected Mean Square
Mean	1	M_{yy}	**M**	$\cdots$
Among treatments	$t - 1$	T_{yy}	**T**	$\begin{cases} \sigma^2 + n \sum_{i=1}^{t} \tau_i^2/(t-1) \\ \sigma^2 + n\sigma_\tau^2 \end{cases}$
Experimental error	$t(n - 1)$	E_{yy}	**E**	σ^2
Total	tn	$\sum Y^2$	$\cdots$	$\cdots$

*To avoid confusion with symbols for effects, interactions, totals, etc., the symbols for mean squares will always be set in boldface type. This procedure will be adhered to throughout the remainder of the book.

among the treatments, while the last term is an estimate of the variation due to experimental error. A summary of the sums of squares associated with the various sources of variation is generally given in a table called an ANOVA table. Table 10.2 is the ANOVA table for a completely randomized design with an equal number of experimental units per treatment and one observation per unit. Equations useful for computing the sums of squares included in Table 10.2 are:

$$\sum Y^2 = \sum_{i=1}^{t} \sum_{j=1}^{n} Y_{ij}^2 \tag{10.3}$$

$$M_{yy} = (Y_{..})^2/tn = T^2/tn \tag{10.4}$$

$$T_{yy} = n \sum_{i=1}^{t} (\bar{Y}_{i.} - \bar{Y}_{..})^2 = \sum_{i=1}^{t} T_i^2/n - M_{yy} \tag{10.5}$$

$$E_{yy} = \sum_{i=1}^{t} \sum_{j=1}^{n} (Y_{ij} - \bar{Y}_{i.})^2 = \sum Y^2 - T_{yy} - M_{yy} \tag{10.6}$$

where $T_i = \sum_{j=1}^{n} Y_{ij}$ is the ith treatment total and $T = Y_{..} = \sum_{i=1}^{t} \sum_{j=1}^{n} Y_{ij}$ is the sum of all the observations.

So far we have not said anything about the column labeled "expected mean square." This column is usually included in an ANOVA table for designed experiments and indicates the expected value of the mean squares given in the column of mean squares. These expected mean squares can be of valuable assistance to the researcher, for they indicate the proper procedure to be used in estimating parameters and/or testing hypotheses about parameters within the framework of the assumed model. As set up in Table 10.2, the expected value of the treatment mean square T is given for both Model I and Model II. The upper term in the parenthesis refers to Model I, whereas the lower term refers to Model II. Thus for Model I

$$E[T] = \sigma^2 + n \sum_{i=1}^{t} \tau_i^2/(t-1) \tag{10.7}$$

and for Model II,

$$E[T] = \sigma^2 + n\sigma_\tau^2 \tag{10.8}$$

We do not stop now to say why the expected mean squares are as given nor how one derives the expected values. Further discussion about the expected mean squares is given in Section 10.5.

We are now ready to indicate the test statistic useful for testing the hypotheses indicated at the beginning of the section. Examination of the expected mean squares in Table 10.2 indicates that if the hypothesis

$$H: \tau_i = 0 \ (i = 1, \ldots, t) \qquad \text{(Model I)}$$

or

$$H: \sigma_\tau^2 = 0 \qquad\qquad \text{(Model II)}$$

is true, both the experimental error mean square and the among treatments mean square are estimates of σ^2. Thus if H is true, the ratio

$$F = T/E = \frac{\text{treatment mean square}}{\text{experimental error mean square}} \qquad (10.9)$$

is distributed as F with $\nu_1 = t - 1$ and $\nu_2 = t(n - 1)$ degrees of freedom because of the assumption that the $\epsilon_{ij} \sim NI(0, \sigma^2)$. For a 100α percent significance level we reject H if $F \geq F_{(1-\alpha)[(t-1),t(n-1)]}$, and we would conclude there are significant differences among the treatments.

Example 10.3

Consider that an experiment similar to that described in Example 9.13 has been performed. However, only 4 treatments were investigated and only 20 batteries were available for testing. The data in Table 10.3 resulted. Following Eqs. (10.3)–(10.6), the appropriate calculations are:

$$\sum Y^2 = 104,352$$
$$M_{yy} = (1444)^2/20 = 104,256.8$$
$$T_{yy} = [(369)^2 + (371)^2 + (345)^2 + (359)^2]/5 - 104,256.8 = 84.8$$
$$E_{yy} = 104,352 - 84.8 - 104,256.8 = 10.4$$

These lead to the ANOVA shown in Table 10.4. The expected mean square for treatments has been given for both Model I and Model II. Since $F = 43.49 > F_{0.99(3,16)} = 5.29$, the hypothesis $H:\tau_i = 0$ ($i = 1, 2, 3, 4$) or $H:\sigma_\tau^2 = 0$, whichever applies, is rejected. Since the number of observations per treatment is the same for each treatment, the standard error of a treatment mean is $\sqrt{0.65/5} = \sqrt{0.13} = 0.36$ second [see Eq. (10.21)].

TABLE 10.3–Activated Lives of Twenty Thermal Batteries Resulting from Experiment Described in Example 10.3

	Treatment				
	1	2	3	4	Total
Observations (sec)	73	74	68	71	
	73	74	69	71	
	73	74	69	72	
	75	74	69	72	
	75	75	70	73	
Total	369	371	345	359	1,444
Number of observations	5	5	5	5	20
Mean	73.8	74.2	69.0	71.8	72.2

TABLE 10.4–ANOVA for Data of Table 10.3

Source of Variation	Degrees of Freedom	Sum of Squares	Mean Square	Expected Mean Square	F Ratio
Mean	1	104,256.8	104,256.8	...	...
Treatments	3	84.8	28.27	$\begin{cases} \sigma^2 + (5/3)\sum_{i=1}^{4} \tau_i^2 \\ \sigma^2 + 5\sigma_\tau^2 \end{cases}$	43.49
Experimental error	16	10.4	0.65	σ^2	...
Total	20	104,352.0	...	...	...

10.2.3 Analysis of Variance: Unequal Number of Experimental Units per Treatment

If the number of experimental units per treatment is not the same for all treatments, the ANOVA is similar to that for the equal sample size except for a few modifications in the computations of the sums of squares. The model as given in Eq. (10.1) is

$$Y_{ij} = \mu + \tau_i + \epsilon_{ij} \qquad i = 1, \ldots, t; j = 1, \ldots, n_i \qquad (10.10)$$

where, as in Eq. (10.1), we assume $\epsilon_{ij} \sim NI(0, \sigma^2)$.

The ANOVA table is given in Table 10.5. Equations for computing the sums of squares in Table 10.5 are

$$\sum Y^2 = \sum_{i=1}^{t} \sum_{j=1}^{n_i} Y_{ij}^2 \qquad (10.11)$$

$$M_{yy} = (Y_{..})^2/N = T^2/N \qquad (10.12)$$

where $N = \sum_{i=1}^{t} n_i$ is the total number of observations.

$$T_{yy} = \sum_{i=1}^{t} n_i(\overline{Y}_{i.} - \overline{Y}_{..})^2 = \sum_{i=1}^{t} T_i^2/n_i - M_{yy} \qquad (10.13)$$

$$E_{yy} = \sum_{i=1}^{t} \sum_{j=1}^{n_i} (Y_{ij} - \overline{Y}_{i.})^2 = \sum Y^2 - T_{yy} - M_{yy} \qquad (10.14)$$

To test the hypothesis

$$H:\tau_i = 0 \, (i = 1, \ldots, t) \qquad \text{(Model I)}$$

or

$$H:\sigma_\tau^2 = 0 \qquad \qquad \text{(Model II)}$$

TABLE 10.5–ANOVA for a Completely Randomized Design with n_i Experimental Units for Treatment i and One Observation per Unit

Source of Variation	Degrees of Freedom	Sum of Squares	Mean Square	Expected Mean Square
Mean	1	M_{yy}	M	
Among treatments	t	T_{yy}	T	$\begin{cases} \sigma^2 + \sum\limits_{i=1}^{t} n_i \tau_i^2/(t-1) \\ \sigma^2 + n_0 \sigma_\tau^2 \,(*) \end{cases}$
Experimental error	$\sum\limits_{i=1}^{t} (n_i - 1)$	E_{yy}	E	σ^2
Total	$N = \sum\limits_{i=1}^{t} n_i$	$\sum Y^2$	...	...

*The constant n_0 is a sort of an average n_i and is defined by

$$n_0 = \left[\sum_{i=1}^{t} n_i - \sum_{i=1}^{t} n_i^2 \bigg/ \sum_{i=1}^{t} n_i \right] \bigg/ (t-1)$$

the test statistic is

$$F = T/E \tag{10.15}$$

which, when H is true, is distributed as F with $\nu_1 = t - 1$ and $\nu_2 = \sum_{i=1}^{t}(n_i - 1) = N - t$ degrees of freedom. Again we reject H and conclude there is a significant difference among treatments if F is greater than $F_{(1-\alpha)(\nu_1, \nu_2)}$ for a 100α percent significance level.

Example 10.4

Consider an experiment to study the effect of storage conditions on the moisture content of white pine lumber. Five storage methods were investigated, with varying numbers of experimental units (sample boards) being stored under each condition. The data in Table 10.6 were obtained. Following Eqs. (10.11)–(10.14), the appropriate calculations are:

$$\sum Y^2 = 863.36$$

$$M_{yy} = (108.8)^2/14 = 845.53$$

$$T_{yy} = \left[\frac{(39.9)^2}{5} + \frac{(19.9)^2}{3} + \frac{(14.5)^2}{2} + \frac{(27.4)^2}{3} + \frac{(7.1)^2}{1} \right] - 845.53$$

$$= 10.66$$

$$E_{yy} = 863.36 - 10.66 - 845.53 = 7.17$$

This leads to the ANOVA shown in Table 10.7. Again, the expected mean square for treatments is given for both Models I and II. Since $F = 3.34 < F_{0.95(4,9)} = 3.63$, we are unable to reject the hypothesis $H: \tau_i = 0$

TABLE 10.6–Moisture Contents of Fourteen White Pine Boards Stored under Different Conditions

| | Storage Condition | | | | | |
	1	2	3	4	5	Total
Observations (percent)	7.3	5.4	8.1	7.9	7.1	
	8.3	7.4	6.4	9.5	...	
	7.6	7.1	...	10.0	...	
	8.4	...	...	...	...	
	8.3	...	...	...	...	
Total	39.9	19.9	14.5	27.4	7.1	108.8
Number of observations	5	3	2	3	1	14
Mean	8.0	6.6	7.3	9.1	7.1	7.8
Standard error	0.4	0.5	0.6	0.5	0.9	...

TABLE 10.7–ANOVA for Data of Table 10.6

Source of Variation	Degrees of Freedom	Sum of Squares	Mean Square	Expected Mean Square	F Ratio
Mean	1	845.53	845.53		...
Storage conditions	4	10.66	2.67	$\begin{cases} \sigma^2 + \sum_{i=1}^{5} n_i \tau_i^2/4 \\ \sigma^2 + 2.64\sigma_\tau^2 \end{cases}$	3.34
Experimental error	9	7.17	0.80	σ^2	...
Total	14	863.36	...	...	...

$(i = 1, \ldots, 5)$ or $H:\sigma_\tau^2 = 0$. The standard errors of the treatment means, presented in Table 10.6 for convenience, were calculated using Eq. (10.21) where $s^2 = 0.80$.

10.2.4 Estimation of the Model Parameters

We now consider some estimates that are of interest in a completely randomized design model. Of course, one parameter for which we have already indicated an estimater is the experimental error variance σ^2. An unbiased estimator for σ^2 is

$$s^2 = \text{E} = \sum_{i=1}^{t} \sum_{j=1}^{n} (Y_{ij} - \bar{Y}_{i.})^2/t(n-1) \qquad \text{(equal size)}$$

$$= \sum_{i=1}^{t} \sum_{j=1}^{n_i} (Y_{ij} - \bar{Y}_{i.})^2 / \sum_{i=1}^{t} (n_i - 1) \qquad \text{(unequal size)}$$

$$(10.16)$$

If Model II is the assumed model, the parameter one is most likely to be interested in estimating is the treatment variance component σ_τ^2. Examination of the expected mean squares in Tables 10.2 and 10.5 indicates that the estimate

$$
\begin{aligned}
s_\tau^2 &= (T - E)/n && \text{(equal size)} \\
&= (T - E)/n_0 && \text{(unequal size)}
\end{aligned} \tag{10.17}
$$

is unbiased.

On the other hand, if Model I is assumed, the parameters to be estimated are the treatment effects τ_i or the individual treatment means μ_i, where, in terms of the model parameters,

$$
\mu_i = \mu + \tau_i \qquad i = 1, \ldots, t \tag{10.18}
$$

It should be obvious that a point estimate of μ_i is $\overline{Y}_{i.}$, the sample mean of the observations obtained for the ith treatment. Thus a point estimate for τ_i is $\overline{Y}_{i.} - \overline{Y}_{..}$. Since $\overline{Y}_{i.}$ is a sample mean, we know that

$$
V(\overline{Y}_{i.}) = \sigma^2/n_i \tag{10.19}
$$

where n_i is the sample size for the ith treatment. Consequently, the estimated variance of the mean of the ith treatment in a completely randomized design with one observation per experimental unit is

$$
\hat{V}(\overline{Y}_{i.}) = s_{\overline{Y}_{i.}}^2 = E/n_i = s^2/n_i \tag{10.20}
$$

where E is the error mean square. The standard error of the mean of the ith treatment is

$$
s_{\overline{Y}_{i.}} = \sqrt{E/n_i} = \sqrt{s^2/n_i} = s/\sqrt{n_i} \tag{10.21}
$$

If each $n_i = n$, the same standard error would be attached to each sample mean. The limits of a 100γ percent confidence interval for μ_i are

$$
\left. \begin{aligned} L \\ U \end{aligned} \right\} = \overline{Y}_{i.} \mp t_{[(1+\gamma)/2]\,(\nu)} s_{\overline{Y}_{i.}} \tag{10.22}
$$

where ν is the degrees of freedom associated with E.

In addition to the individual treatment means, one is generally interested in specific comparisons among them. Suppose we are interested in a linear combination of the treatment means,

$$
\theta = \sum_{i=1}^{t} c_i \mu_i \tag{10.23}
$$

A point estimate of θ is

$$
\hat{\theta} = \sum_{i=1}^{t} c_i \overline{Y}_{i.} = \sum_{i=1}^{t} c_i T_i/n_i \tag{10.24}
$$

Also, the estimated variance of $\hat{\theta}$ is

$$\hat{V}(\theta) = s_{\hat{\theta}}^2 = s^2 \sum_{i=1}^{t} c_i^2/n_i \qquad (10.25)$$

so a 100γ percent confidence interval for θ is given by the limits

$$\left.\begin{array}{c} L \\ U \end{array}\right\} = \hat{\theta} \mp t_{[(1+\gamma)/2]\,(\nu)} s_{\hat{\theta}} \qquad (10.26)$$

If several linear combinations θ_k ($k = 1, \ldots, m$) are to be estimated simultaneously, a 100γ percent confidence region for the θ_k's is given by the limits

$$\left.\begin{array}{c} L_k \\ U_k \end{array}\right\} = \theta_k \mp \left[(t - 1)F_{\gamma(\nu_1,\nu_2)} s^2 \sum_{i=1}^{t} c_{ik}^2/n_i \right]^{1/2} \qquad (10.27)$$

where $\nu_1 = t - 1$ and $\nu_2 = \sum_{i=1}^{t}(n_i - 1)$. Treatment comparisons and contrasts are discussed further in Sections 10.8 and 10.9.

Example 10.5

Consider the data presented in Example 10.3. Suppose the researcher considered the four treatments used in the experiment to be a sample from a larger population; i.e., Model II was assumed. The parameter to be estimated is the treatment variance component σ_τ^2. Using Eq. (10.17), the estimate is $s_\tau^2 = (28.27 - 0.65)/5 = 5.524$.

On the other hand, if Model I was assumed, the expected life for treatment 3, (i.e., μ_3), is estimated to be $\overline{Y}_{3.} = 69.0$ seconds. Since there are $n = 5$ observations per treatment, the standard error of a treatment mean is $\sqrt{0.65/5} = \sqrt{0.13} = 0.36$ seconds. A 95 percent confidence interval for μ_3 is (68.24, 69.76). Further, if we want to estimate the difference in the battery life under treatments 2 and 3, the estimate is $\overline{Y}_{2.} - \overline{Y}_{3.} = 5.2$ seconds. A 95 percent confidence interval for the difference $\mu_2 - \mu_3$ is (3.67, 6.73).

10.2.5 Completely Randomized Design Model as a Regression Model

Having indicated the analysis associated with the data from a completely randomized experiment, we point out that the model in Eq. (10.1) is a special case of a multiple linear regression model. This is more easily seen if we restate the model in the form

$$Y_{ij} = \mu + \tau_1 X_{1i} + \tau_2 X_{2i} + \cdots + \tau_t X_{ti} + \epsilon_{ij}$$
$$i = 1, \ldots, t; j = 1, \ldots, n_i \qquad (10.28)$$

where

$$\begin{aligned} X_{ki} &= 1 \quad \text{if } i = k \\ &= 0 \quad \text{otherwise} \end{aligned}$$

Example 10.6

Suppose we have data from an experiment involving the comparison of three kinds of fertilizers on the yield of corn. If for each fertilizer three independent plots of corn were observed and the yield (bushels/acre) was observed, the data are

$$
\begin{array}{ccc}
Y_{11} & Y_{21} & Y_{31} \\
Y_{12} & Y_{22} & Y_{32} \\
Y_{13} & Y_{23} & Y_{33}
\end{array}
$$

and the equation for Y_{ij}, written as a regression equation, is

$$
Y_{ij} = \mu + \tau_1 X_{1i} + \tau_2 X_{2i} + \tau_3 X_{3i} + \epsilon_{ij} \qquad i = 1, 2, 3; j = 1, 2, 3
$$

Following the discussion of multiple regression models in Chapter 7, we can restate the model in terms of matrices as

$$
\mathbf{Y} = \mathbf{X}\boldsymbol{\beta} + \boldsymbol{\epsilon} \tag{10.29}
$$

where now, if $t = 3$ and $n = 3$, the matrices are

$$
\mathbf{Y} =
\begin{bmatrix}
Y_{11} \\
Y_{12} \\
Y_{13} \\
Y_{21} \\
Y_{22} \\
Y_{23} \\
Y_{31} \\
Y_{32} \\
Y_{33}
\end{bmatrix}
\quad
\mathbf{X} =
\begin{bmatrix}
1 & 1 & 0 & 0 \\
1 & 1 & 0 & 0 \\
1 & 1 & 0 & 0 \\
1 & 0 & 1 & 0 \\
1 & 0 & 1 & 0 \\
1 & 0 & 1 & 0 \\
1 & 0 & 0 & 1 \\
1 & 0 & 0 & 1 \\
1 & 0 & 0 & 1
\end{bmatrix}
\quad
\boldsymbol{\beta} =
\begin{bmatrix}
\mu \\
\tau_1 \\
\tau_2 \\
\tau_3
\end{bmatrix}
\quad
\boldsymbol{\epsilon} =
\begin{bmatrix}
\epsilon_{11} \\
\epsilon_{12} \\
\epsilon_{13} \\
\epsilon_{21} \\
\epsilon_{22} \\
\epsilon_{23} \\
\epsilon_{31} \\
\epsilon_{32} \\
\epsilon_{33}
\end{bmatrix}
$$

If one sets up the normal equations and tries to solve them, perhaps by evaluating $(\mathbf{X}'\mathbf{X})^{-1}$, one finds the inverse does not exist; i.e., a unique solution does not exist. This is because the parameters in the model are not independent. A common way to overcome this difficulty, and one that leads to a convenient analysis, is to introduce the restriction

$$
\sum_{i=1}^{t} \tau_i = 0 \tag{10.30}
$$

Solving for τ_t as a function of the other τ_i's, we can further restate Eq. (10.28) as

$$
\begin{aligned}
Y_{ij} &= \mu + \tau_1(X_{1i} - X_{ti}) + \cdots + \tau_{t-1}(X_{t-1i} - X_{ti}) + \epsilon_{ij} \\
&= \mu + \tau_1 x_{1i} + \cdots + \tau_{t-1} x_{t-1i} + \epsilon_{ij} \qquad i = 1, \ldots, t; j = 1, \ldots, n_i
\end{aligned} \tag{10.31}
$$

where

$$x_{ki} = 1 \quad \text{if } i = k; k = 1, \ldots, t-1$$
$$= -1 \quad \text{if } i = t$$
$$= 0 \quad \text{otherwise}$$

In matrix notation the matrices become

$$
Y = \begin{bmatrix} Y_{11} \\ Y_{12} \\ Y_{13} \\ Y_{21} \\ Y_{22} \\ Y_{23} \\ Y_{31} \\ Y_{32} \\ Y_{33} \end{bmatrix}
\quad
X = \begin{bmatrix} 1 & 1 & 0 \\ 1 & 1 & 0 \\ 1 & 1 & 0 \\ 1 & 0 & 1 \\ 1 & 0 & 1 \\ 1 & 0 & 1 \\ 1 & -1 & -1 \\ 1 & -1 & -1 \\ 1 & -1 & -1 \end{bmatrix}
\quad
\beta = \begin{bmatrix} \mu \\ \tau_1 \\ \tau_2 \end{bmatrix}
\quad
\epsilon = \begin{bmatrix} \epsilon_{11} \\ \epsilon_{12} \\ \epsilon_{13} \\ \epsilon_{21} \\ \epsilon_{22} \\ \epsilon_{23} \\ \epsilon_{31} \\ \epsilon_{32} \\ \epsilon_{33} \end{bmatrix}
$$

We can set up the normal equations

$$(X'X)\beta = X'Y \tag{10.32}$$

and since now $(X'X)^{-1}$ exists, there is a unique solution.

Having set up the model for the data from a completely randomized experiment as a regression model, we could use the usual multiple linear regression techniques for estimating the parameters and for testing hypotheses about the parameters. In particular, the regression sum of squares due to the τ_i's is the among treatment sum of squares and, similarly, the residual sum of squares in regression is equal to the experimental error sum of squares. That is, using the notation of Tables 7.6 and 10.2,

$$SS(\tau_1, \ldots, \tau_t \mid \mu) = T_{yy} \tag{10.33}$$

$$RSS = E_{yy} \tag{10.34}$$

Although using regression analysis is a feasible way of analyzing completely randomized experiments, it may not be the most efficient, particularly if we have a large number of treatments and an equal number of experimental units per treatment. Many package computer programs [e.g., Statistical Analysis System (3), Biomedical Computer Programs (10)] will do all the necessary computations for analyzing completely randomized and other designed experiments.

10.3 THE RELATION BETWEEN A COMPLETELY RANDOMIZED DESIGN AND "STUDENT'S" t-TEST OF $H:\mu_1 = \mu_2$ VERSUS $A:\mu_1 \neq \mu_2$

In Section 6.5 it was mentioned that the analysis of variance technique could be used as an alternative to "Student's" t test when examining the hypothesis $H:\mu_1 = \mu_2$. Clearly, this same relationship exists when we have a completely randomized design involving only two treatments. In this instance the hypothesis (under Model I) of $H:\tau_1 = \tau_2 = 0$ is equivalent to $H:\mu_1 = \mu_2$, where $\mu_1 = \mu + \tau_1$ and $\mu_2 = \mu + \tau_2$.

10.4 SUBSAMPLING; COMPLETELY RANDOMIZED DESIGN WITH SEVERAL OBSERVATIONS PER EXPERIMENTAL UNIT

In many experimental situations, several observations may be obtained on each experimental unit. If these observations are all on the same characteristic (i.e., on the same variable), the process of obtaining the observations is often referred to as subsampling. Some examples of subsampling are:

1. In the battery experiment of Example 10.3 several observations per battery might have been obtained by connecting several clocks to each battery. These several observations per battery would be referred to as "samples within experimental units."
2. In a field experiment the researcher may not have time to harvest each experimental plot totally. Thus he might randomly select several quadrats per plot and harvest the grain in each selected quadrat. Again, we would describe these observations as "samples within experimental units."
3. In a food technology experiment involving the storage of frozen strawberries, 10 pints (experimental units) were stored at each of 5 lengths of storage time (treatments). When ascorbic acid determinations were made after storage, two determinations were made on each pint (samples within experimental units).

As you can well imagine, the addition of subsampling to the experimental program will have an effect on the eventual analysis. First, let us see what changes are required in the assumed statistical model. With subsampling, the model for a completely randomized design is

$$Y_{ijk} = \mu + \tau_i + \epsilon_{ij} + \eta_{ijk} \qquad i = 1, \ldots, t; j = 1, \ldots, n_i; k = 1, \ldots, n_{ij}$$

$$(10.35)$$

where μ is the true overall mean, τ_i is the effect of the ith treatment, ϵ_{ij} is the error associated with the jth experimental unit subjected to the ith treatment, and η_{ijk} is the error associated with the kth sample taken from the jth experimental unit subjected to the ith treatment. We assume $\epsilon_{ij} \sim \mathrm{NI}(0, \sigma^2)$ and $\eta_{ijk} \sim \mathrm{NI}(0, \sigma_\eta^2)$.

As before, we still need to specify whether Model I or Model II is appropriate. This will depend on the manner in which the researcher views the treatments included in the experiment. If inferences are to be made only about the treatments present in the experiment, Model I is appropriate and we assume $\sum_{i=1}^{t} \tau_i = 0$. If the treatments used are viewed as a sample from a population of treatments, Model II is appropriate and we assume $\tau \sim \mathrm{NI}(0, \sigma_\tau^2)$.

TABLE 10.8–Generalized ANOVA for a Completely Randomized Design with Subsampling (unequal numbers: Model I)

Source of Variation	Degrees of Freedom	Sum of Squares	Mean Square	Expected Mean Square
Mean	1	M_{yy}	M	$\ldots$
Treatments	$t - 1$	T_{yy}	T	$\sigma_\eta^2 + c_2\sigma^2 + \sum_{i=1}^{t}\left(\sum_{j=1}^{n_i} n_{ij}\right)^2 \tau_i^2\big/(t - 1)$
Experimental error	$\sum_{i=1}^{t}(n_i - 1)$	E_{yy}	E	$\sigma_\eta^2 + c_1\sigma^2$
Sampling error	$\sum_{i=1}^{t}\sum_{j=1}^{n_i}(n_{ij} - 1)$	S_{yy}	S	σ_η^2
Total	$N = \sum_{i=1}^{t}\sum_{j=1}^{n_i} n_{ij}$	ΣY^2	$\ldots$	$\ldots$

The summary of the computations for this model is given in ANOVA form in Table 10.8, in which constants c_1 and c_2 are defined by

$$c_1 = \frac{N - \sum_{i=1}^{t}\left(\sum_{j=1}^{n_i} n_{ij}^2 \bigg/ \sum_{j=1}^{n_i} n_{ij}\right)}{\sum_{i=1}^{t}(n_i - 1)} \tag{10.36}$$

and

$$c_2 = \frac{\sum_{i=1}^{t}\left(\sum_{j=1}^{n_i} n_{ij}^2 \bigg/ \sum_{j=1}^{n_i} n_{ij}\right) - \sum_{i=1}^{t}\sum_{j=1}^{n_i} n_{ij}^2/N}{t - 1} \tag{10.37}$$

Formulas for computing the sums of squares in Table (10.8) are

$$\Sigma Y^2 = \sum_{i=1}^{t}\sum_{j=1}^{n_i}\sum_{k=1}^{n_{ij}} Y_{ijk}^2 \tag{10.38}$$

$$M_{yy} = N(\overline{Y}_{...})^2 = T^2/N \tag{10.39}$$

$$T_{yy} = \sum_{i=1}^{t} N_i(\overline{Y}_{i..} - \overline{Y}_{...})^2 = \sum_{i=1}^{t} T_i^2/N_i - M_{yy} \tag{10.40}$$

$$E_{yy} = \sum_{i=1}^{t}\sum_{j=1}^{n_i}(\overline{Y}_{ij.} - \overline{Y}_{i..})^2 = \sum_{i=1}^{t}\sum_{j=1}^{n_i} E_{ij}^2/n_{ij} - T_{yy} - M_{yy} \tag{10.41}$$

$$S_{yy} = \sum_{i=1}^{t}\sum_{j=1}^{n_i}\sum_{k=1}^{n_{ij}}(Y_{ijk} - \overline{Y}_{ij.})^2 = \Sigma Y^2 - E_{yy} - T_{yy} - M_{yy} \tag{10.42}$$

where

$$T_i = Y_{i..} = \sum_{j=1}^{n_i} \sum_{k=1}^{n_{ij}} Y_{ijk}$$

is the ith treatment total, $N_i = \sum_{j=1}^{n_i} n_{ij}$ is the number of observations on the ith treatment, $E_{ij} = Y_{ij.} = \sum_{k=1}^{n_{ij}} Y_{ijk}$ is the total of the observations in the jth experimental unit subjected to the ith treatment, and $T = Y_{...}$ is the total of all the observations.

Had Model II been assumed, the ANOVA presented in Table 10.8 would be exactly the same except for the "expected mean square for treatments," which would appear as $\sigma_\eta^2 + c_2\sigma^2 + c_3\sigma_\tau^2$, where

$$c_3 = \frac{N - \sum_{i=1}^{t} \left(\sum_{j=1}^{n_i} n_{ij} \right)^2 \Big/ N}{t - 1} \tag{10.43}$$

Example 10.7

Consider an experiment to investigate the fermentative conversion of sugar to lactic acid. We wish to compare the abilities of two microorganisms to carry out this conversion. A quantity of substrate is prepared and divided into two unequal portions. Each portion is then divided into a number of 100 ml subportions (experimental units) as follows: No. 1, 4 units; No. 2, 3 units. Each of the 100 ml units is inoculated with one or the other of the two microorganisms, the 4 units being inoculated with microorganism 1 and the 3 units with microorganism 2. The fermentation is allowed to proceed for 24 hours, and then each experimental unit (100 ml subportion) is examined for the amount of residual sugar, expressed as mg/5 cc, to determine the amount of change produced by each microorganism, the converted sugar having been shown previously to occur as lactic acid. Varying numbers of determinations are made on each sample. The data are recorded in Table 10.9

TABLE 10.9–Amount of Unconverted Sugar in the Substrate Following a 24-Hour Fermentation Due to Two Different Microorganisms (coded data for easy calculation)

Determination	Microorganism 1				Microorganism 2		
	Sample number				Sample number		
	1	2	3	4	1	2	3
1	5.6	5.0	5.4	5.3	7.6	7.4	7.5
2	5.7	5.0	5.4	5.5	7.6	7.0	7.6
3	...	5.1	5.4	...	7.8	7.2	7.5
4	...	...	5.5	...	...	...	7.4
5	...	...	5.4	...	...	...	...
Total	11.3	15.1	27.1	10.8	23.0	21.6	30.0
n_{ij}	2	3	5	2	3	3	4

Following the calculational procedure outlined, we obtain

$n_{11} = 2$	$E_{11} = 11.3$	$n_1 = 4$
$n_{12} = 3$	$E_{12} = 15.1$	$n_2 = 3$
$n_{13} = 5$	$E_{13} = 27.1$	$N = 22$
$n_{14} = 2$	$E_{14} = 10.8$	$t = 2$
$n_{21} = 3$	$E_{21} = 23.0$	$T_1 = 64.3$
$n_{22} = 3$	$E_{22} = 21.6$	$T_2 = 74.6$
$n_{23} = 4$	$E_{23} = 30.0$	$T = 138.9$

and hence

$$\sum Y^2 = 902.07$$

$$M_{yy} = (138.9)^2/22 = 876.9641$$

$$T_{yy} = \left[\frac{(64.3)^2}{12} + \frac{(74.6)^2}{10}\right] - 876.9641 = 901.0568 - 876.9641$$

$$= 24.0927$$

$$E_{yy} = \left[\frac{(11.3)^2}{2} + \frac{(15.1)^2}{3} + \frac{(27.1)^2}{5} + \frac{(10.8)^2}{2} + \frac{(23.0)^2}{3} + \frac{(21.6)^2}{3}\right.$$

$$\left. + \frac{(30.0)^2}{4}\right] - 24.0927 - 876.9641$$

$$= 901.9036 - 901.0568 = 0.8468$$

$$S_{yy} = 902.07 - 0.8468 - 24.0927 - 876.9641 = 0.1664$$

These results are presented in ANOVA form in Table 10.10.

On examination of the expected mean squares in Tables 10.8 and 10.10, it is seen that an exact test of the hypothesis $H:\tau_i = 0$ ($i = 1, \ldots, t$) is impossible. This unfortunate circumstance results from the fact that $c_1 \neq c_2$, and this is so because of the unequal numbers of

TABLE 10.10–ANOVA for Fermentation Data of Table 10.9

Source of Variation	Degrees of Freedom	Sum of Squares	Mean Square	Expected Mean Square
Mean	1	876.9641	876.9641	$\cdots$
Micro-organisms	1	24.0927	24.0927	$\sigma_\eta^2 + c_2\sigma^2 + \sum\limits_{i=1}^{2}\left(\sum\limits_{j=1}^{n_i} n_{ij}\right)^2 \tau_i^2$
Experimental error	5	0.8468	0.1694	$\sigma_\eta^2 + c_1\sigma^2$
Sampling error	15	0.1664	0.0111	σ_η^2
Total	22	902.0700	$\cdots$	$\cdots$

samples per experimental unit and the unequal numbers of experimental units per treatment. This result clearly attests to the desirability of equal numbers of observations in the various subclasses, and for this reason the statistician always recommends "equal frequencies" when he is consulted at the design stage of any research project.

What then can be done in a situation such as described above? That is, since unequal frequencies are sometimes inevitable, is there any approximate test procedure that can be used? There is. However, discussion of this approximation will be deferred until Section 10.7.

An exact test exists if there are an equal number of observations for all treatments. We consider that case briefly. If in a completely randomized design there are t treatments, n experimental units per treatment, and m samples per experimental unit, the appropriate linear statistical model is

$$Y_{ijk} = \mu + \tau_i + \epsilon_{ij} + \eta_{ijk} \qquad i = 1,\ldots,t; j = 1,\ldots,n; k = 1,\ldots,m \tag{10.44}$$

where all terms are defined as before. Again we assume $\epsilon_{ij} \sim NI(0,\sigma^2)$ and $\eta_{ijk} \sim NI(0,\sigma_\eta^2)$. The ANOVA is summarized in Table 10.11.

TABLE 10.11–Generalized ANOVA for a Completely Randomized Design with Subsampling (equal numbers)

Source of Variation	Degrees of Freedom	Sum of Squares	Mean Square	Expected Mean Square
Mean	1	M_{yy}	M	$\cdots$
Treatments	$t-1$	T_{yy}	T	$\begin{cases} \sigma_\eta^2 + m\sigma^2 + nm\sum_{i=1}^{t} \tau_i^2/(t-1) \\ \sigma_\eta^2 + m\sigma^2 + nm\sigma_\tau^2 \end{cases}$
Experimental error	$t(n-1)$	E_{yy}	E	$\sigma_\eta^2 + m\sigma^2$
Sampling error	$tn(m-1)$	S_{yy}	S	σ_η^2
Total	tnm	$\sum Y^2$	$\cdots$	$\cdots$

Formulas for computing the sums of squares included in Table 10.11 are:

$$\sum Y^2 = \sum_{i=1}^{t} \sum_{j=1}^{n} \sum_{k=1}^{m} Y_{ijk}^2 \tag{10.45}$$

$$M_{yy} = tnm(\bar{Y}_{...})^2 = T^2/tnm \tag{10.46}$$

$$T_{yy} = nm \sum_{i=1}^{t} (\bar{Y}_{i..} - \bar{Y}_{...})^2 = \sum_{i=1}^{t} T_i^2/nm - M_{yy} \tag{10.47}$$

$$E_{yy} = m \sum_{i=1}^{t} \sum_{j=1}^{n} (\overline{Y}_{ij.} - \overline{Y}_{i..})^2 = \sum_{i=1}^{t} \sum_{j=1}^{n} E_{ij}^2/m - T_{yy} - M_{yy} \quad (10.48)$$

$$S_{yy} = \sum_{i=1}^{t} \sum_{j=1}^{n} \sum_{k=1}^{m} (Y_{ijk} - \overline{Y}_{ij.})^2 = \sum Y^2 - E_{yy} - T_{yy} - M_{yy} \quad (10.49)$$

Examination of the expected mean squares in Table 10.11 indicates that, because of the equal frequencies, there will be no difficulty in testing $H: \tau_i = 0$ $(i = 1, \ldots, t)$ or $H: \sigma_\tau^2 = 0$. In addition, the components of variance are easily estimated by

$$s_\eta^2 = S \quad (10.50)$$

and

$$s^2 = (E - S)/m \quad (10.51)$$

And, finally, the standard error of a treatment mean is given by

$$\sqrt{\hat{V}(\overline{Y}_{i..})} = s_{\overline{Y}_{i..}} = \sqrt{E/nm} \quad (10.52)$$

(Note: Although not explicitly stated, it should be clear that σ^2 and σ_η^2 could also have been estimated when unequal frequencies occur.)

Example 10.8

An agronomist conducted a field trial to compare the relative effects of 5 particular fertilizers on the yield of Trebi barley. Thirty homogeneous experimental plots were available and 6 were assigned at random to each fertilizer treatment. At harvest time 3 sample quadrats were taken at random from each experimental plot, and the yield was obtained for each of the 90 quadrats. The data in coded form are given in Table 10.12. Using Eqs. (10.45)–(10.49), we obtain:

$$\sum Y^2 = 646,285$$
$$M_{yy} = (7187)^2/90 = 573,921.88$$
$$T_{yy} = [(650)^2 + (1140)^2 + (1608)^2 + (1797)^2 + (1992)^2]/18$$
$$- 573,921.88$$
$$= 639,168.72 - 573,921.88 = 65,246.84$$
$$E_{yy} = [(131)^2 + \cdots + (344)^2]/3 - 639,168.72$$
$$= 641,001.67 - 639,168.72 = 1,832.95$$
$$S_{yy} = 5,283.33 \text{ (by subtraction)}$$

These are summarized in Table 10.13. It is easily verified that $F = 222.47$, with $\nu_1 = 4$ and $\nu_2 = 25$ degrees of freedom, is highly significant, and thus the hypothesis $H: \tau_i = 0$ $(i = 1, \ldots, 5)$ is rejected. (Note: An experienced analyst could probably have predicted this result on examination of the data, but the analysis and the statistical test make the conclusion an objective one rather than a subjective one.) In case a confidence interval estimate of a treatment mean is desired, the standard error of a treatment mean is calculated. Its value is $\sqrt{E/nm} = \sqrt{73.32/18} = \sqrt{4.07} = 2.02$. It is also clear that components of variance may be estimated in a simple manner. For example, $s_\eta^2 = 88.06$. However, when an estimate of

TABLE 10.12–Coded Values of Yields from Ninety Sample Quadrats

		Fertilizer Treatment		
1	2	3	4	5
57	67	95	102	123
46	72	90	88	101
28	66	89	109	113
26	44	92	96	93
38	68	89	89	110
20	64	106	106	115
39	57	91	102	112
39	61	82	93	104
43	61	98	98	112
23	74	105	103	120
36	47	85	90	101
18	69	85	105	111
48	61	78	99	113
35	60	89	87	109
48	75	95	113	111
50	68	85	117	124
37	65	74	93	102
19	61	80	107	118

Note: Under each treatment the 18 observations are arranged in six groups of three. Each group consists of the observed yields on the three quadrats taken from a single experimental plot.

TABLE 10.13–ANOVA for Data of Table 10.12

Source of Variation	Degrees of Freedom	Sum of Squares	Mean Square	Expected Mean Square	F Ratio
Mean	1	573,921.88	573,921.88	. . .	. . .
Fertilizers	4	65,246.84	16,311.71	$\sigma_\eta^2 + 3\sigma^2 + (18/4)\sum_{i=1}^{5} \tau_i^2$	222.47
Experimental error	25	1,832.95	73.32	$\sigma_\eta^2 + 3\sigma^2$	. . .
Sampling error	60	5,283.33	88.06	σ_η^2	. . .
Total	90	646,285.00	. . .	. . .	. . .

σ^2 is sought, the calculations yield $s^2 = (73.32 - 88.06)/3 < 0$. Since σ^2 is positive by definition, it is unreasonable to quote a negative estimate. Thus in the present situation the "best" estimate of σ^2 will be taken to be *zero*, even though this is a biased estimate. More will be said about the implications of this in a later section. For the moment we shall be content with observing that apparently the variation among the true effects of different experimental units is small, and thus the researcher might consider less replication (fewer experimental units per treatment) in a future experiment of this type.

The reader will no doubt have realized that the concept of subsampling may be extended to many stages. That is, we can have "samples within samples within samples...," and the resulting ANOVA would reflect such multistage subsampling by partitioning the total sum of squares into many more parts. Rather than continue the discussion in general terms, we shall rely on problems at the end of the chapter to illustrate not only the principles involved but also the mechanics of the appropriate calculations.

10.5 EXPECTED MEAN SQUARES, COMPONENTS OF VARIANCE, VARIANCES OF TREATMENT MEANS, AND RELATIVE EFFICIENCY

In Sections 10.2 and 10.4 the reader was introduced to the concepts of components of variance, expected mean squares, and variances of treatment means. In those sections no reasons were given as to why the expected mean squares contained the indicated components of variance or why the coefficients of the components of variance were as given. We now propose to remove this deficiency. In addition, a scheme will be proposed that permits the estimation of the relative efficiency of different proposed designs involving various degrees of subsampling.

Reference to Tables 10.11 and 10.13 shows that the expected mean square for sampling error contains only one component of variance. This is so because the only factor that affects (or causes or produces) the variation "among samples within experimental units" is the η_{ijk} factor. However, the expected mean square for experimental error contains two components of variance since this source of variation reflects the variation among the means of the samples taken from each experimental unit, and these means will vary not only because of the variation from experimental unit to experimental unit but also because of the variation among the samples taken from each experimental unit. To discuss the expected mean square for treatments, it is appropriate to consider first the sum of squares. The treatment sum of squares reflects the variation among the means of all the observations or samples recorded for each treatment. Now, these means will vary because of three contributing factors: (1) variation among treatments (fertilizers in Table 10.13), (2) variation among experimental units (plots) within treatments, and (3) variation among samples (quadrats) within experimental units. Thus the expected mean square involves three components of variance if Model II is assumed, or two components of variance and one sum of squares if Model I is assumed. (**Note:** The reader may verify the reasonableness of the foregoing remarks by substituting the assumed linear statistical model for Y_{ijk} in the expressions for the various sums of squares.)

How were the various coefficients in the expected mean squares determined? The coefficient of σ_η^2 is 1 (and thus not shown) because this reflects the variation among individual samples. The coefficient of σ^2 is m ($m = 3$ in Table 10.13) because there were m observations (samples) per experimental unit. The coefficient of σ_τ^2 when Model II is assumed, or of $\sum_{i=1}^{t} \tau_i^2/(t - 1)$ when Model I is assumed, is nm because there were nm observations (m samples on each of n experimental units) per treatment. In Table 10.13, $n = 6$ and $m = 3$. We might note that another way of expressing the justification of the coefficients described above is to say that each treatment mean is the average of nm observations, while each experimental unit mean is the average of m observations.

The estimation of the various components of variance has been well illustrated in the preceding sections. However, a recapitulation will be made to summarize the procedure. Since $\mathbf{S}$ is an unbiased estimater of σ_η^2, it is reasonable to write

$$s_\eta^2 = \mathbf{S} \tag{10.53}$$

Similarly, $\mathbf{E}$ is an unbiased estimater of $\sigma_\eta^2 + m\sigma^2$, and thus we write

$$s_\eta^2 + ms^2 = \mathbf{E} \tag{10.54}$$

If, then, we combine Eqs. (10.53) and (10.54) as shown in Eq. (10.55), an unbiased estimater of σ^2 is

$$s^2 = (\mathbf{E} - \mathbf{S})/m = [(s_\eta^2 + ms^2) - s_\eta^2]/m \tag{10.55}$$

Now that the preceding estimates are available, it is possible to determine (subject to sampling variation) which factor is contributing the most to the observed variation. Then perhaps an improvement can be made in experimental technique, or the design layout (configuration) can be changed to better control the variation in future experiments of the same type. To pursue this aspect of analysis, the concept of "relative efficiency" of one design compared to another of the same type but involving different numbers of experimental units and/or samples will be investigated.

Before such a comparison can be made, a criterion for measuring efficiency must be established. The criterion adopted in this book will involve the estimated variance of a treatment mean. We will say that a design that provides a smaller estimated variance of a treatment mean than does some other design is the more efficient of the two.

With reference to Table 10.11, and in agreement with the definition given earlier, the estimated variance of a treatment mean is

$$\hat{V}(\overline{Y}_{i..}) = \frac{\text{estimated variance of individual items contributing to mean}}{\text{number of items (observations) averaged to get the mean}}$$

$$= \mathbf{E}/nm = (s_\eta^2 + ms^2)/nm = (s_\eta^2/nm) + (s^2/n) \tag{10.56}$$

Examination of Eq. (10.56) leads to the following conclusions:

1. If the estimates of the components of variance s^2 and s_η^2 remain relatively constant, an increase in n or m (or both) will result in a smaller estimated variance of a treatment mean.
2. An increase in n (the number of experimental units per treatment) will have more of an effect than an increase in m (the number of samples per experimental unit) in reducing $\hat{V}(\bar{Y}_{i.})$. This supports the statement made in Section 9.16 to the effect that "It (replication) enables us to obtain a more precise estimate of the mean effect of any factor. . . ."
3. If either s^2 or s_η^2 (or both) can be made smaller, $\hat{V}(\bar{Y}_{i.})$ can be made smaller. This could be accomplished by choosing more homogeneous experimental units or by improving the experimental technique.

Let us now return to the problem of estimating the efficiency of a proposed design relative to the design used. To do this, we must first estimate what the variance of a treatment mean would be if the proposed design v ere used. Assuming that (1) the proposed design would involve n' experimental units per treatment and m' samples per experimental unit and (2) the estimates of σ^2 and σ_η^2 would remain unchanged, the *new* estimated variance of a treatment mean would be

$$\hat{V}'(\bar{Y}_{i.}) = (s_\eta^2 + m's^2)/n'm' \qquad (10.57)$$

If $\hat{V}'(\bar{Y}_{i.}) < \hat{V}(\bar{Y}_{i.})$, the proposed design is said to be more efficient than the present design; if $\hat{V}'(\bar{Y}_{i.}) > \hat{V}(\bar{Y}_{i.})$, the proposed design is said to be less efficient than the present design. Thus, as a measure of *relative efficiency* (RE), we use the ratio of $\hat{V}(\bar{Y}_{i.})$ and $\hat{V}'(\bar{Y}_{i.})$. If the *efficiency of the proposed (new) design relative to the present (old) design* is desired, one calculates (in percent)

$$\text{RE of new to old} = 100[\hat{V}(\bar{Y}_{i.})/\hat{V}'(\bar{Y}_{i.})] \qquad (10.58)$$

while if the *efficiency of the present (old) design relative to the proposed (new) design* is desired, one calculates (in percent)

$$\text{RE of old to new} = 100[\hat{V}'(\bar{Y}_{i.})/\hat{V}(\bar{Y}_{i.})] \qquad (10.59)$$

Some texts use the concept of "relative information," and it would be wise for us to see what relationship this bears to relative efficiency. If *information* is defined as the *reciprocal of the variance*, then it is only a matter of simple algebra to show that *relative information* (RI) is the same as relative efficiency. For example,

$$\text{RI of old to new} = \frac{[1/\hat{V}(\bar{Y}_{i.})]}{[1/\hat{V}'(\bar{Y}_{i.})]} \times 100 = \text{RE of old to new} \qquad (10.60)$$

Similarly,

$$\text{RI of new to old} = \text{RE of new to old} \qquad (10.61)$$

It should be noted that other definitions of relative information to be found in the literature [e.g., Yates (34)] differ from relative efficiency. However, if we define our terms as above, the two concepts may be used interchangeably.

Example 10.9

The experiment on frozen strawberries discussed in item 3 in the first paragraph of Section 10.4 was performed. However, all that is available is the abbreviated ANOVA of Table 10.14. The estimates of the components of variance are $s_\delta^2 = 5$ and $s^2 = (20 - 5)/2 = 7.5$ where the symbol δ is used to denote determinations (rather than η to denote samples). The estimated variance of a treatment mean is

$$\hat{V}(\overline{Y}_{i..}) = (s_\delta^2 + 2s^2)/10(2) = [5 + 2(7.5)]/20 = 1$$

The question is then asked, "Is the present design more or less efficient than a similar design employing 6 pints per storage time and 3 determinations per pint?" Calculating

$$\hat{V}'(\overline{Y}_{i..}) = [(5 + 3(7.5)]/6(3) = 1.53$$

the answer is, "The present design is more efficient than the proposed design." In fact, the efficiency of the present design relative to the proposed design is: RE of old to new $= 100(1.53/1) = 153$ percent.

TABLE 10.14–Abbreviated ANOVA of Ascorbic Acid Content of Frozen Strawberries

Source of Variation	Degrees of Freedom	Sum of Squares	Mean Square	Expected Mean Square
Among storage times	4	400	100	$\sigma_\delta^2 + 2\sigma^2 + \dfrac{20}{4}\displaystyle\sum_{i=1}^{5} \tau_i^2$
Among pints treated alike	45	900	20	$\sigma_\delta^2 + 2\sigma^2$
Between determinations on pints treated alike	50	250	5	σ_δ^2

10.6 SOME REMARKS CONCERNING F RATIOS THAT ARE LESS THAN UNITY

In all the examples considered so far, the calculated F values have been greater than unity. Thus in each of these cases the only decision to be made by the analyst was whether the calculated value should be termed statistically significant or nonsignificant. If significant, the hypothesis $H : \tau_i = 0 (i = 1, \ldots, t)$ or $H : \sigma_\tau^2 = 0$ was rejected; if not significant, the appropriate hypothesis was not rejected (perhaps even accepted). However, it is possible (and quite probable) that a calculated F value will turn out to be less than unity. What should our conclusion be in such a situation?

We can simply say that F was not significant, and thus the hypothesis cannot be rejected. However, such an easy dismissal of the question is not wise, for it could cause us to ignore a valuable warning sign. Suppose (as might happen) that F, with ν_1 and ν_2 degrees of freedom, is so small that $F' = 1/F$, with ν_2 and ν_1 degrees of freedom, is significant. What should our conclusion be in this case? It appears as though *something* should be rejected; but what is it? In this situation *it seems reasonable to reject the postulated statistical model.*

If the statistical model is rejected because of a significant F' value, what are the steps that should then be taken? Some of these are:

1. The experimental procedure should be reviewed to see if the various assumptions are satisfied. For example, if the proper randomization was not employed, the validity of the independence assumption is doubtful.
2. If sufficient observations are available, the assumption of normality could be checked by plotting the data either on regular graph paper or on normal probability paper.
3. The assumption of homogeneous variances might be checked, but this would require a large number of observations within subclasses.
4. The underlying phenomenon should be restudied to see if the assumed linear model is a good approximation to the true state of affairs. If as a result the assumed model is rejected, a search should be made for a new model that better describes the observed data and the phenomenon under investigation.

10.7 SATTERTHWAITE'S APPROXIMATE TEST PROCEDURE

When discussing the analysis of a completely randomized design involving subsampling, it was noted that no exact test of $H:\tau_i = 0$ $(i = 1,\ldots,t)$ was possible when the experiment involved unequal frequencies at the various stages of subsampling. At that time it was noted that an approximate test procedure would be explained later; we are now ready to do so.

The proposed *approximation*, due to Satterthwaite (24), proceeds as follows: *Using estimates of the components of variance, mean squares will be synthesized that will have the same expected value if the hypothesis to be tested is true. These synthetic mean squares will then be used to form a ratio approximately distributed as F.*

How are the synthetic mean squares formed? If we denote the actual mean squares existing in an ANOVA by $MS_1, MS_2,\ldots,MS_k$, then a synthetic mean square may be obtained by forming a linear combination such as

$$L = a_1 MS_1 + a_2 MS_2 + \cdots + a_k MS_k \qquad (10.62)$$

where the a_i are constants. The degrees of freedom association with L are then *estimated* by

$$\hat{\nu} = \frac{(a_1 MS_1 + a_2 MS_2 + \cdots + a_k MS_k)^2}{(a_1 MS_1)^2/\nu_1 + (a_2 MS_2)^2/\nu_2 + \cdots + (a_k MS_k)^2/\nu_k} \qquad (10.63)$$

where ν_i represents the degrees of freedom associated with MS_i

$(i = 1, \ldots, k)$. Sometimes both the numerator and denominator mean squares (in the approximate F ratio) will be synthesized. However, it is more likely that only one synthetic mean square will be used in any given situation.

Because of the lack of uniqueness of the approximate F ratio (different F ratios could result from the use of different synthetic mean squares) and because of the necessity of approximating the degrees of freedom, the procedure is of limited usefulness. However, if used with care, it can be of value to the researcher and/or statistician. The reader is referred to Cochran (8) for a further discussion of this problem.

Example 10.10

Referring to Example 10.7, we recall that an exact test of $H : \tau_1 = \tau_2 = 0$ was impossible. This was so because $c_1 \neq c_2$ in Table 10.10. It is decided to form a "synthetic experimental error mean square" that will have an expected value of $\sigma_\eta^2 + c_2 \sigma^2$. This could be done by calculating

$$
\begin{aligned}
L &= a_1 E + a_2 S \\
&= (c_2/c_1)E + [1 - (c_2/c_1)]S \\
&= (c_2/c_1)(s_\eta^2 + c_1 s^2) + [1 - (c_2/c_1)]s_\eta^2 \\
&= (c_2/c_1)(0.1694) + [1 - (c_2/c_1)](0.0111)
\end{aligned}
$$

The approximate F ratio would then be $F = 24.0927/L$ with degrees of freedom $\nu_1 = 1$ and $\nu_2 = \hat{\nu}$, where

$$
\hat{\nu} = \frac{L^2}{[a_1(0.1694)]^2/5 + [a_2(0.0111)]^2/15}
$$

The details of the numerical calculations are left as an exercise for the reader.

10.8 SELECTED TREATMENT COMPARISONS: GENERAL DISCUSSION

In Section 9.15 the idea of making specific comparisons among treatment means was introduced. At that time also, the concept of an orthogonal contrast was presented, and it was suggested that orthogonal contrasts were to be preferred over nonorthogonal contrasts. However, the researcher was warned not to let the statistician's desire for orthogonality override his needs.

In this section some general comparisons among treatments will be examined, not to illustrate the concept of a contrast, but to demonstrate the manner in which the ANOVA is modified to provide the proper analysis. Because of the many possibilities, this will best be done by discussing a few illustrative cases.

For example, consider an experiment involving t treatments and n experimental units per treatment in which no subsampling occurred. If treatment 1 were a "control" treatment, it would be of interest to make the following specific comparisons: (1) treatment 1 versus the rest and (2) among the rest. The sums of squares for these two comparisons would be determined as follows:

$$(C_1)_{yy} = SS(1 \text{ vs. rest})$$
$$= [T_1^2/n + (T_2 + \cdots + T_t)^2/(t-1)n] - T^2/tn$$
$$(C_2)_{yy} = SS(\text{among the rest})$$
$$= (T_2^2 + \cdots + T_t^2)/n - (T_2 + \cdots + T_t)^2/(t-1)n$$

where the T_i's are the treatment totals and T is the overall total. These results, when coupled with the basic ANOVA, would be presented as in Table 10.15, where the sums of squares (degrees of freedom) for the selected comparisons are offset to indicate that they are portions of the treatment sum of squares (degrees of freedom). (**Note:** In this example the sums of squares for the two comparisons add up to the treatment sum of squares. The reader is warned that this will not always be the case.)

TABLE 10.15–Generalized ANOVA Showing Two Selected Treatment Comparisons

Source of Variation	Degrees of Freedom	Sum of Squares	Mean Square	F Ratio
Mean	1	M_{yy}	M	...
Treatments	$t-1$	T_{yy}	T	T/E
1 vs. rest	1	$(C_1)_{yy}$	C_1	C_1/E
Among the rest	$t-2$	$(C_2)_{yy}$	C_2	C_2/E
Experimental error	$t(n-1)$	E_{yy}	E	...
Total	tn	$\sum Y^2$	...	...

A second illustration based on the same type of design would be the case in which the t treatments segregate into k groups containing $t_1, t_2, \ldots, t_k$ treatments, respectively, where $\sum_{i=1}^{k} t_i = t$. In such a case, the natural comparisons would be (1) among groups and (2) among treatments within the ith group ($i = 1, \ldots, k$). The sum of squares for the first of these $k + 1$ comparisons would be calculated as

$$G_{yy} = SS(\text{among groups}) = \sum_{i=1}^{k} G_i^2/t_i n - T^2/tn$$

where G_i is the total of all observations in the ith group. The sum of squares among treatments in the first group is given by

$$(W_1)_{yy} = \sum_{i=1}^{t_1} T_i^2/n - \left(\sum_{i=1}^{t_1} T_i\right)^2 \bigg/ t_1 n = \sum_{i=1}^{t_1} T_i^2/n - G_1^2/t_1 n$$

The sums of squares among treatments in each of the remaining $k - 1$ groups would be found in a similar manner. The results would then be presented in ANOVA form as in Table 10.16. (**Note:** Once again the

TABLE 10.16–Generalized ANOVA showing $k + 1$
Selected Treatment Comparisons

Source of Variation	Degrees of Freedom	Sum of Squares	Mean Square	F Ratio
Mean	1	M_{yy}	**M**	...
Treatments	$t - 1$	T_{yy}	**T**	**T/E**
Among groups	$k - 1$	G_{yy}	**G**	**G/E**
Within group 1	$t_1 - 1$	$(W_1)_{yy}$	$\mathbf{W_1}$	$\mathbf{W_1/E}$
Within group 2	$t_2 - 1$	$(W_2)_{yy}$	$\mathbf{W_2}$	$\mathbf{W_2/E}$
$\vdots$	$\vdots$	$\vdots$	$\vdots$	$\vdots$
Within group k	$t_k - 1$	$(W_k)_{yy}$	$\mathbf{W_k}$	$\mathbf{W_k/E}$
Experimental error	$t(n - 1)$	E_{yy}	**E**	...
Total	tn	$\sum Y^2$	...	...

sums of squares for the various comparisons add up to the treatment sum of squares.)

One more general illustration will be given. In this instance, assume again that one treatment is a "control." However, the researcher wishes to do more than compare (1) control versus rest and (2) among the rest. He also wishes to compare separately each noncontrol treatment versus the control. Thus, in addition to the sums of squares indicated in the first illustration, he would also compute

$$(C_3)_{yy} = SS(1 \text{ vs. } 2) = (T_1^2 + T_2^2)/n - (T_1 + T_2)^2/2n$$
$$(C_4)_{yy} = SS(1 \text{ vs. } 3) = (T_1^2 + T_3^2)/n - (T_1 + T_3)^2/2n$$
$$\vdots$$
$$(C_{t+1})_{yy} = SS(1 \text{ vs. } t) = (T_1^2 + T_t^2)/n - (T_1 + T_t)^2/2n$$

These results would then be presented as in Table 10.17. (**Note:** This time neither the degrees of freedom nor the sums of squares for the comparisons will add up to the treatment sum of squares.)

Example 10.11

The experiment described in Example 9.10 was performed and we wish to investigate the specified comparisons. Assuming that the data given in Table 6.10 were the results of this experiment, it is seen that

$$(C_1)_{yy} = [(184 + 68)^2/6 + (170 + 378)^2/14] - (800)^2/20 = 34.3$$
$$(C_2)_{yy} = [(184)^2/4 + (68)^2/2] - (252)^2/6 = 192.0$$
$$(C_3)_{yy} = [(170)^2/5 + (378)^2/9] - (548)^2/14 = 205.7$$

Combining these figures with those of Table 6.11, we get Table 10.18. Examination of the F values in Table 10.18 indicates that all the treatments (electrolytes) differ significantly in their effects on the characteristic (of the batteries) being studied. (**Note:** See Problem 10.30 for the expected mean squares.)

TABLE 10.17–Generalized ANOVA Showing $t + 1$
Selected Treatment Comparisons

Source of Variation	Degrees of Freedom	Sum of Squares	Mean Square	F Ratio
Mean	1	M_{yy}	M	...
Treatments	$t - 1$	T_{yy}	T	T/E
Control vs. rest	1	$(C_1)_{yy}$	C_1	C_1/E
Among rest	$t - 2$	$(C_2)_{yy}$	C_2	C_2/E
Control vs. 2	1	$(C_3)_{yy}$	C_3	C_3/E
Control vs. 3	1	$(C_4)_{yy}$	C_4	C_4/E
⋮	⋮	⋮	⋮	⋮
Control vs. t	1	$(C_{t+1})_{yy}$	C_{t+1}	C_{t+1}/E
Experimental error	$t(n - 1)$	E_{yy}	E	...
Total	tn	$\sum Y^2$	...	...

TABLE 10.18–ANOVA for Experiment of Example 10.11
(data in Table 6.10) Showing the Analysis of a
Specified Set of Treatment Comparisons

Source of Variation	Degrees of Freedom	Sum of Squares	Mean Square	F Ratio
Mean	1	32,000	32,000	...
Treatments (electro- lytes)	3	432	144	72
C_1 (1 and 2 vs. 3 and 4)	1	34.3	34.3	17.15
C_2 (1 vs. 2)	1	192.0	192.0	96
C_3 (3 vs. 4)	1	205.7	205.7	102.85
Experimental error	16	32	2	...
Total	20	32,464	...	...

10.9 SELECTED TREATMENT COMPARISONS: ORTHOGONAL AND NONORTHOGONAL CONTRASTS

Having spent considerable time discussing treatment comparisons in general, let us now concentrate on the subject of contrasts, and particularly on orthogonal contrasts.

It may be verified that the sum of squares associated with a particular contrast is given by

$$(C_j)_{yy} = C_j^2 \bigg/ \sum_{i=1}^{t} n_i c_{ij}^2 = \left(\sum_{i=1}^{t} c_{ij} T_i \right)^2 \bigg/ \sum_{i=1}^{t} n_i c_{ij}^2 \qquad (10.64)$$

where all symbols except t are defined as in Section 9.15. The symbol t is used here, rather than k as in Section 9.15, to comform to the notation being used in the present chapter. If each treatment total T_i is the sum of the same number of observations (i.e., if $n_i = n$ for $i = 1, \ldots, t$), Eq. (10.64) simplifies to

$$(C_j)_{yy} = C_j^2/n \sum_{i=1}^{t} c_{ij}^2 = \left(\sum_{i=1}^{t} c_{ij} T_i \right)^2 \bigg/ n \sum_{i=1}^{t} c_{ij}^2 \qquad (10.65)$$

The results would then be presented in ANOVA form in agreement with the format adopted in the preceding section. [**Note:** If a set of $t - 1$ orthogonal contrasts among t treatments is investigated, the individual sums of squares (one for each contrast) will add up to the treatment sum of squares.]

Example 10.12

Consider again the experiment described in Example 9.10 and analyzed in Example 10.11. The sums of squares associated with the three contrasts could also have been calculated as follows:

$$(C_1)_{yy} = \frac{[(7)(184) + (7)(68) + (-3)(170) + (-3)(378)]^2}{4(7)^2 + 2(7)^2 + 5(-3)^2 + 9(-3)^2}$$

$$(C_2)_{yy} = \frac{[(1)(184) + (-2)(68) + (0)(170) + (0)(378)]^2}{4(1)^2 + 2(-2)^2 + 5(0)^2 + 9(0)^2}$$

$$(C_3)_{yy} = \frac{[(0)(184) + (0)(68) + (9)(170) + (-5)(378)]^2}{4(0)^2 + 2(0)^2 + 5(9)^2 + 9(-5)^2}$$

The ANOVA will be the same as in Table 10.18.

Example 10.13

The experiment described in Example 9.11 was performed, and the data shown in Table 10.19 were recorded. The appropriate calculations are

$$\sum Y^2 = 32,378$$
$$M_{yy} = (800)^2/20 = 32,000$$
$$T_{yy} = [(180)^2 + (160)^2 + (160)^2 + (164)^2 + (136)^2]/4 - 32,000$$
$$= 248$$
$$E_{yy} = 32,378 - 32,000 - 248 = 130$$
$$(C_1)_{yy} = \frac{[(-1)(180) + (4)(160) + (-1)(160) + (-1)(164) + (-1)(136)]^2}{4[(-1)^2 + (4)^2 + (-1)^2 + (-1)^2 + (-1)^2]}$$
$$= 0$$
$$(C_2)_{yy} = \frac{[(1)(180) + (0)(160) + (1)(160) + (-1)(164) + (-1)(136)]^2}{4[(1)^2 + (0)^2 + (1)^2 + (-1)^2 + (-1)^2]}$$
$$= 100$$
$$(C_3)_{yy} = \frac{[(1)(180) + (0)(160) + (-1)(160) + (0)(164) + (0)(136)]^2}{4[(1)^2 + (0)^2 + (-1)^2 + (0)^2 + (0)^2]}$$
$$= 50$$

$$(C_4)_{yy} = \frac{[(0)(180) + (0)(160) + (0)(160) + (1)(164) + (-1)(136)]^2}{4[(0)^2 + (0)^2 + (0)^2 + (1)^2 + (-1)^2]}$$
$$= 98$$

These results are then summarized as in Table 10.20. Using $\alpha = 0.05$, all contrasts except C_1 are judged to be statistically significant. (**Note:** See Problem 10.31 for the expected mean squares.)

TABLE 10.19–Data from Experiment Described in Example 9.11 and Discussed in Example 10.13

Electrolyte				
1	2	3	4	5
40	38	44	41	34
45	40	42	43	35
46	38	40	40	34
49	44	34	40	33

TABLE 10.20–ANOVA for Experiment Described in Example 9.11 (data in Table 10.19; discussion in Example 10.13)

Source of Variation	Degrees of Freedom	Sum of Squares	Mean Square	F Ratio
Mean	1	32,000	32,000	...
Electrolytes	4	248	62	7.15
C_1	1	0	0	0
C_2	1	100	100	11.53
C_3	1	50	50	5.77
C_4	1	98	98	11.30
Experimental error	15	130	8.67	...
Total	20	32,378	...	...

Up to this point the discussion of contrasts has centered on (1) ANOVA techniques for isolating the sums of squares associated with each contrast and (2) the use of the corresponding mean squares to test the hypothesis that the true effects estimated by the contrasts are 0. However, the problem of estimation should not be overlooked.

If the true effect estimated by a contrast C_j is denoted by the symbol ϕ_j, it is desirable to construct a confidence interval estimate of ϕ_j. That is, two numbers L and U are sought such that we can be 100γ percent confident that ϕ will be between L and U. To determine L and U, the standard error of a contrast is needed. Defining the estimated variance of a contrast by

$$\hat{V}(C_j) = \hat{V}\left(\sum_{i=1}^{t} c_{ij} T_i\right) = \sum_{i=1}^{t} c_{ij}^2 \hat{V}(T_i) \qquad (10.66)$$

the *standard error of a contrast* is given by $\sqrt{\hat{V}(C_j)}$.

The nature of $V(T_i)$ will depend on whatever assumptions are made concerning the observations. If we are dealing with a completely randomized design involving t treatments and n experimental units per treatment in which no subsampling has been performed and if the usual assumptions (see Section 10.1) have been made, then

$$\hat{V}(C_j) = ns^2 \sum_{i-1}^{t} c_{ij}^2 \qquad (10.67)$$

and

$$\left.\begin{array}{c} L \\ U \end{array}\right\} = C_j \mp t_{[(1+\gamma)/2](\nu)} \sqrt{\hat{V}(C_j)} \qquad (10.68)$$

where ν is the number of degrees of freedom associated with s^2 in Eq. (10.67).

Example 10.14

Consider the experiment discussed in Examples 9.11 and 10.13. The data were presented in Table 10.19 and the ANOVA in Table 10.20. For this case we have

$$\hat{V}(C_1) = 4(8.67)[(-1)^2 + (4)^2 + (-1)^2 + (-1)^2 + (-1)^2]$$
$$\hat{V}(C_2) = 4(8.67)[(1)^2 + (0)^2 + (1)^2 + (-1)^2 + (-1)^2]$$
$$\hat{V}(C_3) = 4(8.67)[(1)^2 + (0)^2 + (-1)^2 + (0)^2 + (0)^2]$$
$$\hat{V}(C_4) = 4(8.67)[(0)^2 + (0)^2 + (0)^2 + (1)^2 + (-1)^2]$$

Since $s^2 = 8.67$ had 15 degrees of freedom, confidence intervals for $\phi_i\,(i = 1, 2, 3, 4)$ may easily be constructed using Eqs. (9.23) and (10.68).

10.10 ALL POSSIBLE COMPARISONS AMONG TREATMENT MEANS

In Sections 10.8 and 10.9, the usual method of analyzing comparisons among treatment means was discussed in considerable detail. However, one very important restriction on the use of the described method was not mentioned: *The comparisons to be studied should be selected in advance of any analysis of the data.* That is, the method of analyzing contrasts described in the preceding sections would not in general be valid if the comparisons were decided upon after a perusal of the data and perhaps a preliminary ANOVA. In other words, the comparisons should have been decided upon during the planning stage.

The restriction stated in the preceding paragraph can, however, work a hardship on the researcher. Much experimentation is of a purely exploratory nature and little if any idea of which comparisons might be of interest is available prior to the collection and analysis of the data. In such cases the researcher would like to gain more from the analysis than a simple statement that the treatment means are or are not statistically significant. He would also like to know, for example, if some of the treatments might be considered equivalent and which treatment is "best."

How can the researcher attain the goals stated in the preceding paragraph? This problem has received much attention from statisticians; e.g., see Bechhofer (5); Duncan (12, 13); Dunnett (14); Hartley (18); Keuls (20); Kramer (21, 22); Newman (23); Scheffé (25); and Tukey (28, 29, 30, 31). Incidentally, the methods of Duncan, Scheffé, and Tukey (the major protagonists) are discussed in detail in Federer (16), and numerical illustrations are given for each method.

Before proceeding to discuss the method we favor, we will mention an associated technique that has been widely used by researchers for many years. This involves what is known as a *least significant difference* or LSD, which is defined by

$$LSD = t_{(1-\alpha)(\nu)} \sqrt{\hat{V}(\overline{Y}_{i.} - \overline{Y}_{j.})} = t_{(1-\alpha)(\nu)} s_{\overline{Y}_{i.} - \overline{Y}_{j.}} \qquad (10.69)$$

where ν represents the degrees of freedom associated with the variance estimate used in $\hat{V}(\overline{Y}_{i.} - \overline{Y}_{j.})$. The LSD technique operates as follows: If the absolute value of the difference between any two treatment means exceeds the LSD, the effects of the two treatments are judged to be significantly different; if the absolute value of the difference does not exceed the LSD, no such conclusion is reached. The reader is warned that indiscriminate use of the LSD technique is dangerous, for if we have enough treatments, the probability is high that at least one of the $t(t - 1)/2$ differences will, due to chance alone, be judged significantly different. Thus the use of the LSD is to be discouraged. (**Note:** When $t = 2$, the LSD is a legitimate but redundant device.)

The method to be used in this book for making (when desirable) all possible comparisons among treatment means is that proposed by Scheffé (25). While this method has not been the one most widely adopted, it does have certain advantages. These advantages are: (1) it is closely related to the concept of a contrast, (2) it uses tables that are widely available (namely, F tables), and (3) it is easy to use. Let us now see how the technique works.

Recalling that a contrast is defined by

$$C_j = \sum_{i=1}^{t} c_{ij} T_i \qquad (10.70)$$

the procedure is to calculate

$$A \sqrt{\hat{V}(C_j)} = A [\hat{V}(C_j)]^{1/2} \qquad (10.71)$$

where

$$A^2 = (t - 1)F_{(1-\alpha)(\nu_1, \nu_2)} \qquad (10.72)$$

$$\hat{V}(C_j) = \sum_{i=1}^{t} c_{ij}^2 \hat{V}(T_i) \qquad (10.73)$$

$$\nu_1 = t - 1 \qquad (10.74)$$

and ν_2 stands for the degrees of freedom associated with the denominator mean square used in the F test of $H: \tau_1 = \tau_2 = \cdots = \tau_t$. Then if $|C_j| > A[\hat{V}(C_j)]^{1/2}$, hypothesis $H: \phi_j = 0$ will be rejected. (See Section 10.9 for the definition of ϕ_j.) That is, if the absolute value of C_j exceeds $A[\hat{V}(C_j)]^{1/2}$, the contrast C_j will be said to differ significantly from 0. [**Note:** The original F test rejects $H:\tau_i = 0$ $(i = 1,\ldots,t)$ if and only if *at least one* C_j is significantly different, by Scheffé's technique, from 0. The application of Scheffé's procedure permits us, then, to determine which of the C_j are significant.]

Example 10.15

Consider the experiment described in Examples 9.10 and 10.11. The data were presented in Table 6.10 and the analyses in Tables 6.11 and 10.18. The value of A to be used in making any desired comparison is found to be 3.98 since $A^2 = (t - 1)F_{(1-\alpha)(\nu_1,\nu_2)} = (3)F_{0.99(3,16)} = (3)(5.29) = 15.87$.

Let us examine contrast C_2 described in Example 9.10, namely,

$$C_2 = (1)T_1 + (-2)T_2 + (0)T_3 + (0)T_4 = (1)(184) + (-2)(68) = 48$$

The estimated variance of C_2 is given by

$$\hat{V}(C_2) = (1)^2(4s^2) + (-2)^2(2s^2) = 12s^2 = 12(2) = 24$$

Therefore, $A[\hat{V}(C_j)]^{1/2} = (3.98)\sqrt{24} = (3.98)(4.899) = 19.498$.

Since $|C_j| = 48 > 19.498$, we conclude that the difference between the effects of treatment 1 and treatment 2 is statistically significant. Incidentally, this agrees with the conclusion reached in Example 10.11. Other comparisons among the treatment effects could be made in a like manner.

Example 10.16

Consider the experiment described in Examples 9.11 and 10.13. The data were presented in Table 10.19 and the analysis in Table 10.20. In this illustration, $A^2 = 4(4.89) = 19.56$ and thus $A = 4.42$. If we are interested in C_2 as defined in Table 9.3, it may be verified that $C_2 = 40$, $\hat{V}(C_2) = 16s^2 = 16(8.67) = 138.72$, $[\hat{V}(C_2)]^{1/2} = 11.78$, and $A[\hat{V}(C_2)]^{1/2} = 52.07$. Since $|C_2| = 40 < A[\hat{V}(C_2)]^{1/2} = 52.07$, we conclude that C_2 is not significantly different from 0.

It is noted, however, that this conclusion is the opposite of that reached in Example 10.13. Why is this? The reason may be explained as follows: Scheffé's method will not lead to significant results (if the appropriate null hypothesis is true) as frequently as will the classical approach of orthogonal comparisons because we have been permitted to examine the data *before* deciding on the analysis. This obviously should lead to fewer cases of claiming significance when no real differences exist. This is as it should be, for if we can look at the data before deciding on the comparisons to be investigated, we should be able to lessen our chances of making errors. From the point of view of estimation, this decrease in the "frequency of errors" takes the form of longer confidence intervals (i.e., our estimates are less precise) than those provided by the classical approach.

10.11 RESPONSE CURVES: A REGRESSION ANALYSIS OF TREATMENT MEANS WHEN THE VARIOUS TREATMENTS ARE DIFFERENT LEVELS OF ONE QUANTITATIVE FACTOR

The reader may be wondering why the subject of this section is under discussion at this time. Did we not discuss regression analyses completely enough in Chapter 7? Of course we did, but now we wish to utilize the techniques of regression to make more complete and informative analyses of data arising from completely randomized designs in which the treatments are different levels of a single quantitative factor.

How is this possible? Let us suppose that the treatments being examined are (1) different levels or rates of application of the same fertilizer, (2) different weights of an object being moved in a time and motion study project, or (3) different intensities of a given stimulus in a psychological experiment. If situations such as these arise, it seems reasonable to investigate *how* the measured characteristic varies with changes in the level of the treatment. That is, we would like to know if the change in the measured characteristic takes place in a linear, quadratic, ... fashion as the level of the treatment is increased or decreased. In other words, we wish to gain some idea of the shape of the *response curve* so that an estimate may be made of the optimum level of the treatment.

Just how will the type of analysis indicated above be carried out? The first step is to plot the treatment means, thus gaining some idea as to the general shape of the response curve. Once this has been done the researcher will be ready to undertake a more rigorous analysis of his sample data.

Equations for various possible response curves could be determined using the techniques of Chapter 7. However, the determination of the equation of the response curve is not the immediate aim of our analysis; it is to reach an objective decision (based on more than a simple plotting of the means) as to the nature of the regression function that will best describe the effect of the treatment on the response variable.

Perhaps the most convenient way of reaching the goal stated in the preceding paragraph is to determine how much of the treatment sum of squares would be associated with each of the terms (linear, quadratic, ...) in a polynomial regression. If the various levels of the treatment being studied are *equally spaced*, this analysis can best be carried out using the method of orthogonal polynomials introduced in Section 7.5.2. (**Note:** The assumption of equal spacing will in general present no problem, for both the researcher and the statistician will ordinarily plan the experiment in such a way as to insure that the assumption will be satisfied. That is, in most applications equal spacing is the usual state of affairs.)

If each treatment total T_i is the sum of n observations, the desired sums of squares are found using

SS due to the kth degree term

$$= \left(\sum_{i=1}^{t} \xi'_{ik} T_i \right)^2 \bigg/ n \sum_{i=1}^{t} (\xi'_{ik})^2 \qquad k = 1, \ldots, t-1 \qquad (10.75)$$

where the ξ'_{ik} are orthogonal polynomial coefficients. Extensive tables of orthogonal polynomial coefficients are given in Anderson and Houseman (1); for your convenience an abbreviated tabulation is provided in

TABLE 10.21–Partial Table of Orthogonal Polynomial Coefficients

	$t = 2$	$t = 3$		$t = 4$			$t = 5$			
i	$k = 1$	$k = 1$	$k = 2$	$k = 1$	$k = 2$	$k = 3$	$k = 1$	$k = 2$	$k = 3$	$k = 4$
1	-1	-1	1	-3	1	-1	-2	2	-1	1
2	1	0	-2	-1	-1	3	-1	-1	2	-4
3	...	1	1	1	-1	-3	0	-2	0	6
4	...	...		3	1	1	1	-1	-2	-4
5	...	...			...		2	2	1	1

Table 10.21. In agreement with the notation previously adopted, the sums of squares associated with the linear, quadratic, cubic,... terms will be denoted by $(T_L)_{yy}$, $(T_Q)_{yy}$, $(T_C)_{yy}$,..... In addition, since it is unlikely that the researcher will wish to isolate more than a few terms when studying the treatment sum of squares, the balance (if any) will be represented by $(T_{dev})_{yy}$. For example, if the linear, quadratic, and cubic effects were isolated, the sum of the squares of the deviations from regression would be given by

$$(T_{dev})_{yy} = T_{yy} - (T_L)_{yy} - (T_Q)_{yy} - (T_C)_{yy} \qquad (10.76)$$

The results of the foregoing calculations may then be summarized as in Table 10.22.

TABLE 10.22–Generalized ANOVA for a Completely Randomized Design Showing the Isolation of the Linear, Quadratic, and Cubic Components of the Treatment Sum of Squares

Source of Variation	Degrees of Freedom	Sum of Squares	Mean Square	F Ratio
Mean	1	M_{yy}	M	...
Treatments	$t - 1$	T_{yy}	T	T/E
T_L	1	$(T_L)_{yy}$	T_L	T_L/E
T_Q	1	$(T_Q)_{yy}$	T_Q	T_Q/E
T_C	1	$(T_C)_{yy}$	T_C	T_C/E
T_{dev}	$t - 4$	$(T_{dev})_{yy}$	T_{dev}	T_{dev}/E
Experimental error	$t(n - 1)$	E_{yy}	E	...
Total	tn	$\sum Y^2$	...	...

Example 10.17

Consider the data in Table 10.23. Although an examination of the treatment totals suggests that a linear response function may be appropriate, the quadratic effect will also be isolated for illustrative purposes. The following sums of squares were obtained:

$$\sum Y^2 = 29,560$$

TABLE 10.23–Yields (converted to bushels/acre) of a
Certain Grain Crop in a Fertilizer Trial

		Level of Fertilizer			
	No treatment	10 lb/plot	20 lb/plot	30 lb/plot	40 lb/plot
	20	25	36	35	43
	25	29	37	39	40
	23	31	29	31	36
	27	30	40	42	48
	19	27	33	44	47
Total	114	142	175	191	214
Mean	22.8	28.4	35	38.2	42.8

$$M_{yy} = (836)^2/25 = 27{,}955.84$$

$$T_{yy} = [(114)^2 + (142)^2 + (175)^2 + (191)^2 + (214)^2]/5 - 27{,}955.84$$
$$= 1256.56$$

$$E_{yy} = 29{,}560 - 27{,}955.84 - 1256.56 = 347.60$$

$$(T_L)_{yy} = \frac{[(-2)(114) + (-1)(142) + (0)(175) + (1)(191) + (2)(214)]^2}{5[(-2)^2 + (-1)^2 + (0)^2 + (1)^2 + (2)^2]}$$
$$= (249)^2/50 = 1240.02$$

$$(T_Q)_{yy} = \frac{[(2)(114) + (-1)(142) + (-2)(175) + (-1)(191) + (2)(214)]^2}{5[(2)^2 + (-1)^2 + (-2)^2 + (-1)^2 + (2)^2]}$$
$$= (-27)^2/70 = 10.41$$

$$(T_{dev})_{yy} = 1256.56 - 1240.02 - 10.41 = 6.13$$

These are summarized in Table 10.24. Examination of the F ratios confirms our subjective judgment that the response of yield to rate of application of the fertilizer is linear within the range of the levels of fertilizer applied. This suggests that the rate of application of the fertilizer might be increased even more, with an accompanying increase in the yield. How-

TABLE 10.24–ANOVA for Data of Table 10.23

Source of Variation	Degrees of Freedom	Sum of Squares		Mean Square	F Ratio
Mean	1	27,955.84		27,955.84	...
Fertilizer levels	4	1,256.56		314.14	18.07
T_L	1		1,240.02	1,240.02	71.35
T_Q	1		10.41	10.41	0.60
T_{dev}	2		6.13	3.07	0.18
Experimental error	20	347.60		17.38	...
Total	25	29,560.00		...	...

ever, the reader is warned that extrapolation of the linear relationship much beyond 40 lb per plot could possibly lead to erroneous conclusions. Another way of putting this is to say that the optimum level of fertilizer application has probably not yet been reached, and further experimentation should be carried out along these lines.

10.12 ANALYSIS OF A COMPLETELY RANDOMIZED DESIGN INVOLVING FACTORIAL TREATMENT COMBINATIONS

The discussion of the analysis of a completely randomized experiment so far has involved the general case of t treatments. We now consider experiments in which the treatments are combinations of levels of several factors.

Example 10.18

A problem of interest to chemical engineers is the study of mixing characteristics of powder in a fluidized bed (a bed of material subject to gas aeration). Several factors that affect the mixing time of powders in such situations are (1) gas velocity, (2) speed of the stirrer mixing the bed, and (3) depth of the material bed. Suppose an experiment is undertaken in which the following levels of these factors are used in the experimental setup—(1) gas velocity (ft/sec): 0.1, 0.12, 0.15, 0.2; (2) stirrer speed (rpm): 25, 50, 100, 150; and (3) bed depth (in.): 12, 30. The experiment consists of putting the material (e.g., flour) in a column with tracer particles, subjecting it to the experimental conditions specified, and periodically drawing some off to observe the degree of mixing. In this experiment the first two factors have 4 levels, while the latter factor has 2 levels. We would call this a 4 × 4 × 2 or 4^2 × 2 factorial experiment. Combining all the levels of the three factors results in $t = (4)(4)(2) = 32$ treatments. Treatment 1 would be (0.1, 25, 12), treatment 2 is (0.1, 25, 30), etc.

With factorial treatment combinations the treatment sum of squares can be partitioned into meaningful sums of squares that are useful for testing the effects of each factor and also the interactions between the factors. To outline the methods of analysis relevant to experiments of this type, we distinguish between three types of factorial models: (1) crossed classification models, (2) nested classification models, and (3) mixed classification models.

10.12.1 Crossed Classification Model

The crossed classification model arises in factorial experiments when all levels of all factors are combined to form the treatments. Thus the treatments are formed by taking all combinations of the a levels of factor a, with the b levels of factor $b, \ldots$, with the l levels of factor l. In all there are $t = a \times b \times \cdots \times l$ treatments included in the experiment. The experiment in Example 10.18 is a cross-classified factorial experiment involving three factors—two factors having four levels and the third factor having two levels. Thus there are 32 treatments in the experiment.

We begin the analysis of factorial experiments by looking first at experiments involving two factors. We are assuming that the experimental design is a completely randomized design. Let Y_{ijk} denote the observation from the kth experimental unit subjected to the treatment formed by combining the ith level of factor a with the jth level of factor b. The linear statistical model for the observation is

$$Y_{ijk} = \mu + \alpha_i + \beta_j + (\alpha\beta)_{ij} + \epsilon_{ijk}$$

$$i = 1,\ldots,a; j = 1,\ldots,b; k = 1,\ldots,n \qquad (10.77)$$

where μ is the true mean effect, α_i is the true effect of the ith level of factor a, β_j is the true effect of the jth level of factor b, $(\alpha\beta)_{ij}$ is the true interaction of the ith level of factor a with the jth level of factor b, ϵ_{ijk} is the error associated with the kth experimental unit subjected to the ijth treatment combination. As usual, it is assumed that μ is a constant and $\epsilon_{ijk} \sim NI(0, \sigma^2)$. Notice also that we are assuming that the same number n of experimental units exists for all treatments. The analysis for unequal numbers of units per treatment is considered in Chapter 12.

TABLE 10.25–ANOVA for a Two-Factor Factorial in a Completely Randomized Design

Source of Variation	Degrees of Freedom	Sum of Squares	Mean Square
Mean	1	M_{yy}	M
Treatments			
$\quad A$	$a - 1$	A_{yy}	A
$\quad B$	$b - 1$	B_{yy}	B
$\quad AB$	$(a - 1)(b - 1)$	$(AB)_{yy}$	AB
Experimental error	$ab(n - 1)$	E_{yy}	E
Total	abn	$\sum Y^2$	...

The analysis of variance is summarized in Table 10.25 and is based on the sum of squares identity

$$\sum_{i=1}^{a}\sum_{j=1}^{b}\sum_{k=1}^{n} Y_{ijk}^2 = abn(\bar{Y}_{...})^2 + bn\sum_{i=1}^{a}(\bar{Y}_{i..} - \bar{Y}_{...})^2$$

$$+ an\sum_{j=1}^{b}(\bar{Y}_{.j.} - \bar{Y}_{...})^2 + n\sum_{i=1}^{a}\sum_{j=1}^{b}(\bar{Y}_{ij.} - \bar{Y}_{i..} - \bar{Y}_{.j.} + \bar{Y}_{...})^2$$

$$+ \sum_{i=1}^{a}\sum_{j=1}^{b}\sum_{k=1}^{n}(Y_{ijk} - \bar{Y}_{ij.})^2 \qquad (10.78)$$

Formulas for computing the sums of squares included in Table 10.25 are:

$$\sum Y^2 = \sum_{i=1}^{a}\sum_{j=1}^{b}\sum_{k=1}^{n} Y_{ijk}^2 \qquad (10.79)$$

$$M_{yy} = abn(\overline{Y}_{...})^2 = T^2/abn \qquad (10.80)$$

$$A_{yy} = bn \sum_{i=1}^{a} (\overline{Y}_{i..} - \overline{Y}_{...})^2 = \sum_{i=1}^{a} A_i^2/bn - M_{yy} \quad (10.81)$$

$$B_{yy} = an \sum_{j=1}^{b} (\overline{Y}_{.j.} - \overline{Y}_{...})^2 = \sum_{j=1}^{b} B_j^2/an - M_{yy} \quad (10.82)$$

$$(AB)_{yy} = n \sum_{i=1}^{a} \sum_{j=1}^{b} (\overline{Y}_{ij.} - \overline{Y}_{i..} - \overline{Y}_{.j.} + \overline{Y}_{...})^2$$

$$= \sum_{i=1}^{a} \sum_{j=1}^{b} T_{ij}^2/n - A_{yy} - B_{yy} - M_{yy} \qquad (10.83)$$

$$E_{yy} = \sum_{i=1}^{a} \sum_{j=1}^{b} \sum_{k=1}^{n} (Y_{ijk} - \overline{Y}_{ij.})^2$$

$$= \sum Y^2 - A_{yy} - B_{yy} - (AB)_{yy} - M_{yy} \qquad (10.84)$$

where we have adopted the following notation:

$$A_i = Y_{i..} = \sum_{j=1}^{b} \sum_{k=1}^{n} Y_{ijk} \qquad (10.85)$$

is the total of all observations for the ith level of factor a.

$$B_j = Y_{.j.} = \sum_{i=1}^{a} \sum_{k=1}^{n} Y_{ijk} \qquad (10.86)$$

is the total of all observations for the jth level of factor b.

$$T_{ij} = Y_{ij.} = \sum_{k=1}^{n} Y_{ijk} \qquad (10.87)$$

is the total of all observations for the ijth treatment (i.e., ith level of factor a and jth level of factor b).

$$T = Y_{...} = \sum_{i=1}^{a} \sum_{j=1}^{b} \sum_{k=1}^{n} Y_{ijk} \qquad (10.88)$$

is the total of all observations.

Having explained the calculation of the various sums of squares, your attention is now directed to the assumptions associated with the α_i, β_j, and $(\alpha\beta)_{ij}$. Four possible sets of assumptions can be made with respect to the true treatment effects. These are:

Model I: Analysis of variance (fixed effects) model

This model is assumed when the researcher is concerned only with the a levels of factor a and the b levels of factor b present in the experiment. Mathematically, these assumptions are summarized by

$$\sum_{i=1}^{a} \alpha_i = \sum_{j=1}^{b} \beta_j = \sum_{i=1}^{a} (\alpha\beta)_{ij} = \sum_{j=1}^{b} (\alpha\beta)_{ij} = 0$$

Model II: Component of variance (random effects) model

This model is assumed when the researcher is concerned with (1) a population of levels of factor a of which only a random sample (the a levels) are present in the experiment and (2) a population of levels of factor b of which only a random sample (the b levels) are present in the experiment. Mathematically, these assumptions are summarized as follows:

$$\alpha_i \sim \text{NI}(0, \alpha_\alpha^2) \qquad \beta_j \sim \text{NI}(0, \sigma_\beta^2) \qquad (\alpha\beta)_{ij} \sim \text{NI}(0, \sigma_{\alpha\beta}^2)$$

Model III: Mixed model (a fixed, b random)

This model is assumed when the researcher is concerned with: (1) only the a levels of factor a present in the experiment and (2) a population of levels of factor b of which only a random sample (the b levels) are present in the experiment. Mathematically, these assumptions are summarized as follows:

$$\sum_{i=1}^{a} \alpha_i = \sum_{i=1}^{a} (\alpha\beta)_{ij} = 0 \qquad \beta_j \sim \text{NI}(0, \sigma_\beta^2)$$

Please note that $\sum_{j=1}^{b} (\alpha\beta)_{ij}$ was *not* assumed to be 0.

Model III: Mixed model (a random, b fixed)

This model is assumed when the researcher is concerned with (1) a population of levels of factor a of which only a random sample (the a levels) are present in the experiment and (2) only the b levels of factor b present in the experiment. Mathematically, these assumptions are summarized as follows:

$$\alpha_i \sim \text{NI}(0, \sigma_\alpha^2) \qquad \sum_{j=1}^{b} \beta_j = \sum_{j=1}^{b} (\alpha\beta)_{ij} = 0$$

Please note that $\sum_{i=1}^{a} (\alpha\beta)_{ij}$ was *not* assumed to be 0.

While the logic underlying the preceding mathematical formulations is beyond the scope of this text, it is hoped that the validity of the expressions will be substantiated by the arguments that will accompany the specification of the several F tests. Thus it is requested that the reader accept the expressions in good faith and concentrate on learning the methods of analysis. In the long run this will prove most beneficial.

Based on the foregoing assumptions, the expected mean squares may now be derived. As in the preceding examples the derivations will be omitted and only the results tabulated. The expected mean squares for each of the four cases are shown in Table 10.26.

Examination of the expected mean squares in Table 10.26 will indicate the proper F tests for such hypotheses as $H_1 : \alpha_i = 0 (i = 1, \ldots, a)$;

TABLE 10.26–Expected Mean Squares for a Two-Factor Factorial in a Completely Randomized Design (see Table 10.25 for the ANOVA)

Source of Variation	Expected Mean Square			
	Model I	Model II	Model III (a fixed, b random)	Model III (a random, b fixed)
Mean	...	...	...	...
Treatments				
A	$\sigma^2 + nb\sum_{i=1}^{a}\alpha_i^2/(a-1)$	$\sigma^2 + n\sigma_{\alpha\beta}^2 + nb\sigma_\alpha^2$	$\sigma^2 + n\sigma_{\alpha\beta}^2 + nb\sum_{i=1}^{a}\alpha_i^2/(a-1)$	$\sigma^2 + nb\sigma_\alpha^2$
B	$\sigma^2 + na\sum_{j=1}^{b}\beta_j^2/(b-1)$	$\sigma^2 + n\sigma_{\alpha\beta}^2 + na\sigma_\beta^2$	$\sigma^2 + na\sigma_\beta^2$	$\sigma^2 + n\sigma_{\alpha\beta}^2 + na\sum_{j=1}^{b}\beta_j^2/(b-1)$
AB	$\sigma^2 + n\sum_{i=1}^{a}\sum_{j=1}^{b}(\alpha\beta)_{ij}^2/(a-1)(b-1)$	$\sigma^2 + n\sigma_{\alpha\beta}^2$	$\sigma^2 + n\sigma_{\alpha\beta}^2$	$\sigma^2 + n\sigma_{\alpha\beta}^2$
Experimental error	σ^2	σ^2	σ^2	σ^2
Total	...	...	...	...

TABLE 10.27–F Ratios for Testing the Appropriate Hypotheses When
Dealing with a Two-Factor Factorial in a Completely Randomized
Design (see Table 10.25 for the ANOVA and Table 10.26 for
the expected mean squares)

Source of Variation	F Ratio			
	Model I	Model II	Model III (a fixed, b random)	Model III (a random, b fixed)
Mean	...	...	...	...
Treatments				
A	A/E	A/AB	A/AB	A/E
B	B/E	B/AB	B/E	B/AB
AB	AB/E	AB/E	AB/E	AB/E
Experimental error	...	...	...	...
Total	...	...	...	...

$H_2:\beta_j = 0 (j = 1,\ldots,b)$; $H_3:(\alpha\beta)_{ij} = 0(i = 1,\ldots,a; j = 1,\ldots,b)$;
$H_4:\sigma_\alpha^2 = 0$; $H_5:\sigma_\beta^2 = 0$; and $H_6:\sigma_{\alpha\beta}^2 = 0$. For your convenience these
are specified in Table 10.27.

Example 10.19

Consider a 4×3 factorial in a completely randomized design with
three experimental units per treatment combination. The data are given
in Table 10.28. Proceeding as indicated, the following sums of squares
were calculated:

$$\sum Y^2 = 564,389$$
$$M_{yy} = (4023)^2/36 = 449,570.2$$
$$A_{yy} = [(726)^2 + (991)^2 + (1022)^2 + (1284)^2]/9 - 449,570.2$$
$$= 17,351.7$$
$$B_{yy} = [(1624)^2 + (1500)^2 + (899)^2]/12 - 449,570.2 = 25,061.2$$

TABLE 10.28–Hypothetical Data for Illustrating the ANOVA for a
4 × 3 Factorial in a Completely Randomized Design

a_1			a_2			a_3			a_4		
b_1	b_2	b_3	b_1	b_2	b_3	b_1	b_2	b_3	b_1	b_2	b_3
128	34	16	152	40	118	76	102	132	180	220	60
42	134	18	128	88	80	158	96	60	90	220	48
136	172	46	216	76	93	168	162	68	150	156	160

$$(AB)_{yy} = [(306)^2 + \cdots + (268)^2]/3 - 25,061.2 - 17,351.7$$
$$- 449,570.2$$
$$= 516,731.0 - 491,983.1 = 24,747.9$$
$$E_{yy} = 564,389 - 24,747.9 - 25,061.2 - 17,351.7 - 449,570.2$$
$$= 47,658.0$$

These results are summarized in ANOVA form in Table 10.29, where the expected mean squares are shown for each of the four cases illustrated in Table 10.26. Since the data were hypothetical, no F tests will be performed. Such tests and the resulting inferences will be illustrated in succeeding examples in which actual experimental data will be examined.

For a two-factor cross-classification factorial experiment using a completely randomized design, the regression model for the observation Y_{ijk} can be written in the form

$$Y_{ijk} = \mu + \alpha_1 x_{1i} + \cdots + \alpha_{a-1} x_{a-1i} + \beta_1 w_{1j} + \cdots + \beta_{b-1} w_{b-1j}$$
$$+ (\alpha\beta)_{11} x_{1i} w_{1j} + \cdots + (\alpha\beta)_{a-1b-1} x_{a-1i} w_{b-1j} + \epsilon_{ijk}$$
$$i = 1, \ldots, a; \, j = 1, \ldots, b$$
$$k = 1, \ldots, n \qquad (10.89)$$

where two sets of independent variables x and w defined as

$$x_{mi} = 1 \quad \text{if } i = m; m = 1, \ldots, a - 1$$
$$= -1 \quad \text{if } i = a$$
$$= 0 \quad \text{otherwise}$$

and

$$w_{mj} = 1 \quad \text{if } j = m; m = 1, \ldots, b - 1$$
$$= -1 \quad \text{if } j = b$$
$$= 0 \quad \text{otherwise}$$

are associated with factor a and factor b, respectively. To write the model in the form given in Eq. (10.89), certain restrictions are imposed on the parameters to ensure that the resulting normal equations have a unique solution (i.e., $X'X$ has an inverse). The restrictions imposed were

$$\sum_{i=1}^{a} \alpha_i = \sum_{j=1}^{b} \beta_j = \sum_{i=1}^{a} (\alpha\beta)_{ij} = \sum_{j=1}^{b} (\alpha\beta)_{ij} = 0$$

To illustrate the regression model for a two-factor cross-classification model, consider the case where $a = 3$, $b = 2$, and $n = 2$. The regression equations, written in matrix notation as

$$Y = X\beta + \epsilon \qquad (10.90)$$

are

TABLE 10.29–ANOVA for Data of Table 10.28

Source of Variation	Degrees of Freedom	Sum of Squares	Mean Square	Expected Mean Square			
				Model I	Model II	Model III (a fixed, b random)	Model III (a random, b fixed)
Mean Treatments	1	449,570.2	449,570.20	$\cdots$	$\cdots$	$\cdots$	$\cdots$
A	3	17,351.7	5,783.90	$\sigma^2 + (9/3)\sum_{i=1}^{4}\alpha_i^2$	$\sigma^2 + 3\sigma_{\alpha\beta}^2 + 9\sigma_\alpha^2$	$\sigma^2 + 3\sigma_{\alpha\beta}^2 + (9/3)\sum_{i=1}^{4}\alpha_i^2$	$\sigma^2 + 9\sigma_\alpha^2$
B	2	25,061.2	12,530.60	$\sigma^2 + (12/2)\sum_{j=1}^{3}\beta_j^2$	$\sigma^2 + 3\sigma_{\alpha\beta}^2 + 12\sigma_{\beta\cdot}^2$	$\sigma^2 + 12\sigma_\beta^2$	$\sigma^2 + 3\sigma_{\alpha\beta}^2 + (12/2)\sum_{j=1}^{3}\beta_j^2$
AB	6	24,747.9	4,124.65	$\sigma^2 + (3/6)\sum_{i=1}^{4}\sum_{j=1}^{3}(\alpha\beta)_{ij}^2$	$\sigma^2 + 3\sigma_{\alpha\beta}^2$	$\sigma^2 + 3\sigma_{\alpha\beta}^2$	$\sigma^2 + 3\sigma_{\alpha\beta}^2$
Experimental error	24	47,658.0	1,985.75	σ^2	σ^2	σ^2	σ^2
Total	36	564,389.0	$\cdots$		$\cdots$	$\cdots$	$\cdots$

$$
\begin{bmatrix} Y_{111} \\ Y_{112} \\ Y_{121} \\ Y_{122} \\ Y_{211} \\ Y_{212} \\ Y_{221} \\ Y_{222} \\ Y_{311} \\ Y_{312} \\ Y_{321} \\ Y_{322} \end{bmatrix} =
\begin{bmatrix}
1 & 1 & 0 & 1 & 1 & 0 \\
1 & 1 & 0 & 1 & 1 & 0 \\
1 & 1 & 0 & -1 & -1 & 0 \\
1 & 1 & 0 & -1 & -1 & 0 \\
1 & 0 & 1 & 1 & 0 & 1 \\
1 & 0 & 1 & 1 & 0 & 1 \\
1 & 0 & 1 & -1 & 0 & -1 \\
1 & 0 & 1 & -1 & 0 & -1 \\
1 & -1 & -1 & 1 & -1 & -1 \\
1 & -1 & -1 & 1 & -1 & -1 \\
1 & -1 & -1 & -1 & 1 & 1 \\
1 & -1 & -1 & -1 & 1 & 1
\end{bmatrix}
\begin{bmatrix} \mu \\ \alpha_1 \\ \alpha_2 \\ \beta_1 \\ (\alpha\beta)_{11} \\ (\alpha\beta)_{21} \end{bmatrix} +
\begin{bmatrix} \epsilon_{111} \\ \epsilon_{112} \\ \epsilon_{121} \\ \epsilon_{122} \\ \epsilon_{211} \\ \epsilon_{212} \\ \epsilon_{221} \\ \epsilon_{222} \\ \epsilon_{311} \\ \epsilon_{312} \\ \epsilon_{321} \\ \epsilon_{322} \end{bmatrix}
\qquad (10.91)
$$

Notice that the columns in X corresponding to μ, the α's, the β's, and the $(\alpha\beta)$'s are mutually orthogonal. Thus the regression sums of squares are easily partitioned into the sum of squares due to μ, the α's (factor a), the β's (factor b), and the $(\alpha\beta)$'s (interaction) exactly as in Table 10.25. Thus

$$
\begin{aligned}
M_{yy} &= SS(\mu) \\
A_{yy} &= SS(\alpha_1, \ldots, \alpha_{a-1} \mid \mu) \\
B_{yy} &= SS(\beta_1, \ldots, \beta_{b-1} \mid \mu) \\
(AB)_{yy} &= SS[(\alpha\beta)_{11}, \ldots, (\alpha\beta)_{a-1\,b-1} \mid \mu]
\end{aligned}
$$

We now look at the analysis of a three-factor factorial experiment. When a three-factor factorial is associated with a completely randomized design involving n experimental units per treatment combination, the appropriate statistical model is

$$
\begin{aligned}
Y_{ijkl} = \mu &+ \alpha_i + \beta_j + (\alpha\beta)_{ij} + \gamma_k + (\alpha\gamma)_{ik} + (\beta\gamma)_{jk} \\
&+ (\alpha\beta\gamma)_{ijk} + \epsilon_{ijkl} \quad i = 1, \ldots, a; j = 1, \ldots, b \\
&\qquad\qquad\qquad\qquad\quad k = 1, \ldots, c; l = 1, \ldots, n \qquad (10.92)
\end{aligned}
$$

in which all terms are defined in a manner analogous to the definitions accompanying Eq. (10.77). The appropriate ANOVA for the three-factor factorial is given in Table 10.30.

Formulas that can be used to compute the sums of squares in Table 10.30 are:

$$
\sum Y^2 = \sum_{i=1}^{a} \sum_{j=1}^{b} \sum_{k=1}^{c} \sum_{l=1}^{n} Y_{ijkl}^2 \qquad (10.93)
$$

$$
M_{yy} = abcn(\overline{Y}_{....})^2 = T^2/abcn \qquad (10.94)
$$

$$
A_{yy} = bcn \sum_{i=1}^{a} (\overline{Y}_{i...} - \overline{Y}_{....})^2 = \sum_{i=1}^{a} A_i^2/bcn - M_{yy} \qquad (10.95)
$$

TABLE 10.30–ANOVA for a Three-Factor Factorial in a Completely Randomized Design

Source of Variation	Degrees of Freedom	Sum of Squares	Mean Square
Mean	1	M_{yy}	M
Treatments			
A	$a - 1$	A_{yy}	A
B	$b - 1$	B_{yy}	B
C	$c - 1$	C_{yy}	C
AB	$(a - 1)(b - 1)$	$(AB)_{yy}$	AB
AC	$(a - 1)(c - 1)$	$(AC)_{yy}$	AC
BC	$(b - 1)(c - 1)$	$(BC)_{yy}$	BC
ABC	$(a - 1)(b - 1)(c - 1)$	$(ABC)_{yy}$	ABC
Experimental error	$abc(n - 1)$	E_{yy}	E
Total	$abcn$	$\sum Y^2$	...

$$B_{yy} = acn \sum_{j=1}^{b} (\overline{Y}_{.j.} + \overline{Y}_{....})^2 = \sum_{j=1}^{b} B_j^2/acn - M_{yy} \qquad (10.96)$$

$$C_{yy} = abn \sum_{k=1}^{c} (\overline{Y}_{..k} - \overline{Y}_{....})^2 = \sum_{k=1}^{c} C_k^2/abn - M_{yy} \qquad (10.97)$$

$$(AB)_{yy} = cn \sum_{i=1}^{a} \sum_{j=1}^{b} (\overline{Y}_{ij.} - \overline{Y}_{i...} - \overline{Y}_{.j.} + \overline{Y}_{....})^2$$

$$= \sum_{i=1}^{a} \sum_{j=1}^{b} T_{ij}^2/cn - A_{yy} - B_{yy} - M_{yy} \qquad (10.98)$$

$$(AC)_{yy} = bn \sum_{i=1}^{a} \sum_{k=1}^{c} (\overline{Y}_{i.k} - \overline{Y}_{i...} - \overline{Y}_{..k} + \overline{Y}_{....})^2$$

$$= \sum_{i=1}^{a} \sum_{k=1}^{c} T_{ik}^2/bn - A_{yy} - C_{yy} - M_{yy} \qquad (10.99)$$

$$(BC)_{yy} = an \sum_{j=1}^{b} \sum_{k=1}^{c} (\overline{Y}_{.jk} - \overline{Y}_{.j.} - \overline{Y}_{..k} + \overline{Y}_{....})^2$$

$$= \sum_{j=1}^{b} \sum_{k=1}^{c} T_{jk}^2/an - B_{yy} - C_{yy} - M_{yy} \qquad (10.100)$$

$$(ABC)_{yy} = n \sum_{i=1}^{a} \sum_{j=1}^{b} \sum_{k=1}^{c} (\overline{Y}_{ijk} - \overline{Y}_{ij.} - \overline{Y}_{i.k} - \overline{Y}_{.jk}$$

$$+ \overline{Y}_{i...} + \overline{Y}_{.j.} + \overline{Y}_{..k} - \overline{Y}_{....})^2$$

$$= \sum_{i=1}^{a} \sum_{j=1}^{b} \sum_{k=1}^{c} T_{ijk}^2/n - A_{yy} - B_{yy} - C_{yy} - (AB)_{yy}$$

$$- (AC)_{yy} - (BC)_{yy} - M_{yy} \qquad (10.101)$$

$$E_{yy} = \sum_{i=1}^{a} \sum_{j=1}^{b} \sum_{k=1}^{c} \sum_{l=1}^{n} (Y_{ijkl} - \bar{Y}_{ijk.})^2$$

$$= \sum Y^2 - \sum_{i=1}^{a} \sum_{j=1}^{b} \sum_{k=1}^{c} T_{ijk}^2/n \qquad (10.102)$$

where

$$A_i = Y_{i...} = \sum_{j=1}^{b} \sum_{k=1}^{c} \sum_{l=1}^{n} Y_{ijkl} \qquad (10.103)$$

$$B_j = Y_{.j..} = \sum_{i=1}^{a} \sum_{k=1}^{c} \sum_{l=1}^{n} Y_{ijkl} \qquad (10.104)$$

$$C_k = Y_{..k.} = \sum_{i=1}^{a} \sum_{j=1}^{b} \sum_{l=1}^{n} Y_{ijkl} \qquad (10.105)$$

$$T_{ij} = Y_{ij..} = \sum_{k=1}^{c} \sum_{l=1}^{n} Y_{ijkl} \qquad (10.106)$$

$$T_{ik} = Y_{i.k.} = \sum_{j=1}^{b} \sum_{l=1}^{n} Y_{ijkl} \qquad (10.107)$$

$$T_{jk} = Y_{.jk.} = \sum_{i=1}^{a} \sum_{l=1}^{n} Y_{ijkl} \qquad (10.108)$$

$$T = Y_{....} \qquad \text{(the total of all observations)} \qquad (10.109)$$

As in the case of a two-factor factorial, the assumptions concerning the true treatment effects can take several forms. In fact, for a three-factor factorial there are eight different situations. Rather than discussing all of these, only four representative cases will be presented.

Model I: Analysis of variance (fixed effects) model

This model is assumed when the researcher is concerned only with the a levels of factor a, the b levels of factor b, and the c levels of factor c present in the experiment. Mathematically, these assumptions are summarized by

$$\sum_{i=1}^{a} \alpha_i = \sum_{j=1}^{b} \beta_j = \sum_{k=1}^{c} \gamma_k = \sum_{i=1}^{a} (\alpha\beta)_{ij} = \sum_{j=1}^{b} (\alpha\beta)_{ij} = \sum_{i=1}^{a} (\alpha\gamma)_{ik}$$

$$= \sum_{k=1}^{c} (\alpha\gamma)_{ik} = \sum_{j=1}^{b} (\beta\gamma)_{jk} = \sum_{k=1}^{c} (\beta\gamma)_{jk} = \sum_{i=1}^{a} (\alpha\beta\gamma)_{ijk}$$

$$= \sum_{j=1}^{b} (\alpha\beta\gamma)_{ijk} = \sum_{k=1}^{c} (\alpha\beta\gamma)_{ijk} = 0$$

Model II: Component of variance (random effects) model

This model is assumed when the researcher is concerned with (1) a population of levels of factor a of which only a random sample (the a levels) are present in the experiment, (2) a population of levels of factor b of which only a random sample (the b levels) are present in the experiment, and (3) a population of levels of factor c of which only a random sample (the c levels) are present in the experiment. Mathematically, these assumptions are summarized as follows:

$$\alpha_i \sim NI(0, \sigma_\alpha^2)$$
$$\beta_j \sim NI(0, \sigma_\beta^2)$$
$$\gamma_k \sim NI(0, \sigma_\gamma^2)$$
$$(\alpha\beta)_{ij} \sim NI(0, \sigma_{\alpha\beta}^2)$$
$$(\alpha\gamma)_{ik} \sim NI(0, \sigma_{\alpha\gamma}^2)$$
$$(\beta\gamma)_{jk} \sim NI(0, \sigma_{\beta\gamma}^2)$$
$$(\alpha\beta\gamma)_{ijk} \sim NI(0, \sigma_{\alpha\beta\gamma}^2)$$

Model III: Mixed model (a and b fixed, c random)

This model is assumed when the researcher is concerned with (1) only the a levels of factor a present in the experiment; (2) only the b levels of factor b present in the experiment; and (3) a population of levels of factor c of which only a random sample (the c levels) are present in the experiment. Mathematically, these assumptions are summarized as follows:

$$\sum_{i=1}^{a} \alpha_i = \sum_{j=1}^{b} \beta_j = \sum_{i=1}^{a} (\alpha\beta)_{ij} = \sum_{j=1}^{b} (\alpha\beta)_{ij} = \sum_{i=1}^{a} (\alpha\gamma)_{ik}$$
$$= \sum_{j=1}^{b} (\beta\gamma)_{jk} = \sum_{i=1}^{a} (\alpha\beta\gamma)_{ijk} = \sum_{j=1}^{b} (\alpha\beta\gamma)_{ijk} = 0$$
$$\gamma_k \sim NI(0, \sigma_\gamma^2)$$

Please note that

$$\sum_{k=1}^{c} (\alpha\gamma)_{ik}, \sum_{k=1}^{c} (\beta\gamma)_{jk}, \text{ and } \sum_{k=1}^{c} (\alpha\beta\gamma)_{ijk}$$

were *not* assumed to be 0.

Model III: Mixed model (a fixed, b and c random)

This model is assumed when the researcher is concerned with (1) only the a levels of factor a present in the experiment; (2) a population of levels of factor b of which only a random sample (the b levels) are present in the experiment; and (3) a population of levels of factor c of

TABLE 10.31–Expected Mean Squares for Four Selected Cases of a Three-Factor Factorial in a Completely Randomized Design (see Table 10.30 for the ANOVA)

Source of Variation	Expected Mean Square			
	Model I	Model II	Model III (a and b fixed, c random)	Model III (a fixed, b and c random)
Mean Treatments	…	…	…	…
A	$\sigma^2 + nbc \sum_{i=1}^{a} \alpha_i^2/(a-1)$	$\sigma^2 + n\sigma_{\alpha\beta\gamma}^2 + nc\sigma_{\alpha\beta}^2 + nb\sigma_{\alpha\gamma}^2 + nbc\sigma_\alpha^2$	$\sigma^2 + nb\sigma_{\alpha\gamma}^2 + nbc \sum_{i=1}^{a} \alpha_i^2/(a-1)$	$\sigma^2 + n\sigma_{\alpha\beta\gamma}^2 + nb\sigma_{\alpha\gamma}^2 + nc\sigma_{\alpha\beta}^2 + nbc \sum_{i=1}^{a} \alpha_i^2/(a-1)$
B	$\sigma^2 + nac \sum_{j=1}^{b} \beta_j^2/(b-1)$	$\sigma^2 + n\sigma_{\alpha\beta\gamma}^2 + nc\sigma_{\alpha\beta}^2 + na\sigma_{\beta\gamma}^2 + nac\sigma_\beta^2$	$\sigma^2 + na\sigma_{\beta\gamma}^2 + nac \sum_{j=1}^{b} \beta_j^2/(b-1)$	$\sigma^2 + na\sigma_{\beta\gamma}^2 + nac\sigma_\beta^2$
C	$\sigma^2 + nab \sum_{k=1}^{c} \gamma_k^2/(c-1)$	$\sigma^2 + n\sigma_{\alpha\beta\gamma}^2 + nb\sigma_{\alpha\gamma}^2 + na\sigma_{\beta\gamma}^2 + nab\sigma_\gamma^2$	$\sigma^2 + nab\sigma_\gamma^2$	$\sigma^2 + na\sigma_{\beta\gamma}^2 + nab\sigma_\gamma^2$
AB	$\sigma^2 + nc \sum_{i=1}^{a}\sum_{j=1}^{b} (\alpha\beta)_{ij}^2/(a-1)(b-1)$	$\sigma^2 + n\sigma_{\alpha\beta\gamma}^2 + nc\sigma_{\alpha\beta}^2$	$\sigma^2 + n\sigma_{\alpha\beta\gamma}^2 + nc \sum_{i=1}^{a}\sum_{j=1}^{b} (\alpha\beta)_{ij}^2/(a-1)(b-1)$	$\sigma^2 + n\sigma_{\alpha\beta\gamma}^2 + nc\sigma_{\alpha\beta}^2$
AC	$\sigma^2 + nb \sum_{i=1}^{a}\sum_{k=1}^{c} (\alpha\gamma)_{ik}^2/(a-1)(c-1)$	$\sigma^2 + n\sigma_{\alpha\beta\gamma}^2 + nb\sigma_{\alpha\gamma}^2$	$\sigma^2 + nb\sigma_{\alpha\gamma}^2$	$\sigma^2 + n\sigma_{\alpha\beta\gamma}^2 + nb\sigma_{\alpha\gamma}^2$
BC	$\sigma^2 + na \sum_{j=1}^{b}\sum_{k=1}^{c} (\beta\gamma)_{jk}^2/(b-1)(c-1)$	$\sigma^2 + n\sigma_{\alpha\beta\gamma}^2 + na\sigma_{\beta\gamma}^2$	$\sigma^2 + na\sigma_{\beta\gamma}^2$	$\sigma^2 + na\sigma_{\beta\gamma}^2$
ABC	$\sigma^2 + n \sum_{i=1}^{a}\sum_{j=1}^{b}\sum_{k=1}^{c} (\alpha\beta\gamma)_{ijk}^2 /(a-1)(b-1)(c-1)$	$\sigma^2 + n\sigma_{\alpha\beta\gamma}^2$	$\sigma^2 + n\sigma_{\alpha\beta\gamma}^2$	$\sigma^2 + n\sigma_{\alpha\beta\gamma}^2$
Experimental error	σ^2	σ^2	σ^2	σ^2

which only a random sample (the c levels) are present in the experiment. Mathematically, these assumptions are summarized as follows:

$$\sum_{i=1}^{a} \alpha_i = \sum_{i=1}^{a} (\alpha\beta)_{ij} = \sum_{i=1}^{a} (\alpha\gamma)_{ik} = \sum_{i=1}^{a} (\alpha\beta\gamma)_{ijk} = 0$$

$$\beta_j \sim NI(0, \sigma_\beta^2)$$

$$\gamma_k \sim NI(0, \sigma_\gamma^2)$$

$$(\beta\gamma)_{jk} \sim NI(0, \sigma_{\beta\gamma}^2)$$

Please note that

$$\sum_{j=1}^{b} (\alpha\beta)_{ij}, \sum_{k=1}^{c} (\alpha\gamma)_{ik}, \sum_{j=1}^{b} (\alpha\beta\gamma)_{ijk}, \text{ and } \sum_{k=1}^{c} (\alpha\beta\gamma)_{ijk}$$

were *not* assumed to be 0.

Based on the foregoing assumptions, the expected mean squares are derived and the results presented in Table 10.31. The proper F tests for various hypotheses are shown in Table 10.32.

TABLE 10.32–F Ratios for Testing the Appropriate Hypotheses When Dealing with a Three-Factor Factorial in a Completely Randomized Design (See Table 10.30 for the ANOVA and Table 10.31 for the expected mean squares)

Source of Variation	Model I	Model II	Model III (a and b fixed, c random)	Model III (a fixed, b and c random)
			F Ratio	
Mean	...	...	...	...
Treatments				
A	A/E	no exact test	A/AC	no exact test
B	B/E	no exact test	B/BC	B/BC
C	C/E	no exact test	C/E	C/BC
AB	AB/E	AB/ABC	AB/ABC	AB/ABC
AC	AC/E	AC/ABC	AC/E	AC/ABC
BC	BC/E	BC/ABC	BC/E	BC/E
ABC	ABC/E	ABC/E	ABC/E	ABC/E
Experimental error	...	...	...	...
Total	...	...	...	...

Example 10.20

Consider a $3 \times 4 \times 3$ factorial in a completely randomized design with six experimental units per treatment combination. The data are given in Table 10.33. Using Eqs. (10.93)–(10.102) the following sums of squares

TABLE 10.33–Hypothetical Data for Illustrating the ANOVA for a
$3 \times 4 \times 3$ Factorial in a Completely Randomized Design

	a_1				a_2				a_3			
	b_1	b_2	b_3	b_4	b_1	b_2	b_3	b_4	b_1	b_2	b_3	b_4
c_1	3	10	9	8	24	8	9	3	2	8	9	8
	2	10	9	8	29	16	11	3	2	7	5	3
	8	10	2	8	27	16	15	8	2	15	7	14
	1	6	8	14	14	13	8	5	9	30	9	2
	7	8	9	6	18	10	2	16	14	7	6	11
	8	1	10	12	3	8	8	4	11	2	2	9
c_2	4	12	3	8	22	7	16	2	2	2	7	2
	7	10	5	8	28	18	10	6	6	6	5	9
	7	9	2	7	27	15	12	7	7	16	1	13
	14	5	7	15	34	11	9	5	13	11	8	3
	7	9	8	2	19	9	12	12	13	6	6	12
	7	6	12	3	3	15	8	4	12	3	2	10
c_3	5	10	5	8	23	9	17	3	2	8	6	3
	9	10	27	8	28	16	11	7	8	9	8	15
	15	7	6	15	30	14	12	5	11	18	3	8
	8	6	4	18	16	12	13	15	17	8	7	16
	7	17	3	10	17	10	20	9	9	8	6	17
	3	2	10	5	3	7	8	6	11	7	3	14

were calculated:

$$\sum Y^2 = 27,981 \qquad C_{yy} = 84.93$$
$$M_{yy} = 19,703.56 \qquad (AB)_{yy} = 1507.69$$
$$T_{yy} = 3283.27 \qquad (AC)_{yy} = 38.60$$
$$E_{yy} = 4994.17 \qquad (BC)_{yy} = 122.11$$
$$A_{yy} = 941.79 \qquad (ABC)_{yy} = 124.36$$
$$B_{yy} = 463.79$$

For convenience, tables of the totals T_{ijk}, T_{ij}, T_{ik}, and T_{jk} are given in
Tables 10.34–10.37 respectively.

These results are summarized in ANOVA form in Table 10.38. Since
the data were hypothetical, no expected mean squares are given. Neither
are any F tests performed. The reader is referred to the problems at the

TABLE 10.34–Totals T_{ijk} from the Data of Table 10.33

	a_1				a_2				a_3			
	b_1	b_2	b_3	b_4	b_1	b_2	b_3	b_4	b_1	b_2	b_3	b_4
c_1	29	45	47	56	115	71	53	39	40	69	38	47
c_2	46	51	37	43	133	75	67	36	53	44	29	49
c_3	47	52	55	64	117	68	81	45	58	58	33	73

TABLE 10.35–Totals T_{ij} from the Data of Table 10.33

	a_1	a_2	a_3
b_1	122	365	151
b_2	148	214	171
b_3	139	201	100
b_4	163	120	169

TABLE 10.36–Totals T_{ik} from the Data of Table 10.33

	a_1	a_2	a_3
c_1	177	278	194
c_2	177	311	175
c_3	218	311	222

TABLE 10.37–Totals T_{jk} from the Data of Table 10.33

	b_1	b_2	b_3	b_4
c_1	184	185	138	142
c_2	232	170	133	128
c_3	222	178	169	182

TABLE 10.38–ANOVA for Data of Table 10.33

Source of Variation	Degrees of Freedom	Sum of Squares	Mean Square
Mean	1	19,703.56	19,703.56
Treatments			
A	2	941.79	470.90
B	3	463.79	154.60
C	2	84.93	42.46
AB	6	1,507.69	251.28
AC	4	38.60	9.65
BC	6	122.11	20.35
ABC	12	124.36	10.36
Experimental error	180	4,994.17	27.75
Total	216	27,981.00	. . .

end of the chapter for illustrations of various tests and the resulting inferences.

Having outlined the methods of calculation associated with two- and three-factor factorials in a completely randomized design, we are now ready to discuss the expected mean squares given in Tables 10.26 and 10.31 and the F ratios given in Tables 10.27 and 10.32. Perhaps the best way to approach this topic is to talk about the types of inferences that the researcher wishes to make. You will recall that the various models (I, II, and III) reflect the researcher's desire to make inferences about (1) only the levels of the factors present in the experiment; (2) populations of levels of factors, of which only a random sample of levels from each population is present in the experiment; and (3) a mixture of the two preceding situations respectively. In each of these situations the researcher may reason as follows:

1. When dealing with a situation in which Model I applies, the conclusions reached about any particular effect will be uncontaminated by any other effect since, by proper definition of the terms in the statistical model, the average contribution of every other effect can be made equal to zero. Consequently, all F values will be calculated by forming the ratio of the mean square for the effect under scrutiny and the experimental error mean square. That is, all effects are tested against experimental error.

2. When dealing with a situation in which Model II applies, the conclusions reached about any particular effect will be contaminated by all those effects that represent interactions between the effect under scrutiny and other effects present in the experiment. This reflects the researcher's realization that his conclusions (inferences) about the effect under scrutiny are uncertain not only because of the ϵ's but also because of the chance contributions of the randomly selected levels of any factor. That is, a different random sample of levels of any factor might lead to different conclusions, and the researcher attempts to incorporate this uncertainty into his conclusions by testing a particular effect against an "error" that includes an estimate of this additional variability. Thus the expected mean squares will be as shown in Tables 10.26 and 10.31, where it is observed that each expected mean square contains all the components of variance whose subscripts contain all the letters representing the effect under scrutiny. This is the mathematical way of expressing the "contamination" discussed above. Consequently, the F tests are as specified in Tables 10.27 and 10.32. [**Note:** This illustrates the remark made in Chapter 9, namely, "...the (proper) experimental error for testing a particular effect."]

3. When dealing with a situation in which Model III applies, the conclusions reached about any particular effect may or may not be contaminated by other effects. That is, we have a mixture of cases 1 and 2. To summarize what could be a rather involved discussion, let us state the following rule: The expected mean square for any effect will contain, in addition to its own special term, all components of variance that represent interactions between the effect under scrutiny and other effects whose levels were randomly selected. It will not contain components of variance representing interactions between the effect under scrutiny and other effects whose levels comprise the entire population of levels to be investigated.

The expected mean squares specified in Tables 10.26 and 10.31 are simply results of the above reasoning, and as a consequence the F tests are as shown in Tables 10.27 and 10.32. [**Note:** Again, this illustrates the remark made in Chapter 9, namely, "...the (proper) experimental error for testing a particular effect."]

To aid the researcher in writing out expected mean squares for different situations, the following sequence of steps is recommended:

1. Include when applicable a component of variance for each sub-sampling stage.
2. Include a component of variance representing experimental error.
3. Include every component of variance whose subscripts include *all* the letters specifying the effect with which the expected mean square is associated.
4. Insert coefficients in front of each component of variance in accordance with the approach discussed in Section 10.5.
5. Delete from the set specified in step 3 all terms representing interactions between the effect associated with the expected mean square and other effects whose levels were *not* randomly selected.
6. For a main effect replace the component of variance for that effect by a "sum of squares divided by the appropriate degrees of freedom" if the effect is a "fixed effect."

Before presenting illustrations of the methods discussed in this section, some additional remarks need to be made. These are presented here in the briefest form possible:

1. It will have been noted that the degrees of freedom for interaction effects were specified without any explanation. The general rule is: For an interaction effect denoted by $ABCD\ldots$, the degrees of freedom are $\nu = (a - 1)(b - 1)(c - 1)(d - 1)\ldots$.
2. When, as in Table 10.32, no exact tests of certain hypotheses are available, approximate tests can be made following Satterthwaite's procedure. (See Section 10.7.) For example, when Model II was assumed in Table 10.31, an approximate test of $H:\sigma_\alpha^2 = 0$ is $F \simeq A/(AB + AC - ABC)$.
3. Conclusions (inferences) about one factor in a factorial must take due cognizance of all interactions of this factor with other factors. That is, recommendations about one factor must give consideration to the way in which its effect is influenced by other factors.

Example 10.21

Consider an agronomic experiment to assess the effects of date of planting (early or late) and type of fertilizer (none, Aero, Na, or K) on the yield of soybeans. Thirty-two homogeneous experimental plots were available. The treatments were assigned to the plots at random, subject only to the restriction that 4 plots be associated with each of the 8 treatment combinations. The data are given in Table 10.39, and the ANOVA (assuming Model I) is given in Table 10.40.

TABLE 10.39–Yields of Soybeans (bushels/acre) at the
Agronomy Farm, Ames, Iowa, 1949

Date of Planting	Fertilizer	Experimental Unit within Treatments			
		1	2	3	4
Early	Check	28.6	36.8	32.7	32.6
	Aero	29.1	29.2	30.6	29.1
	Na	28.4	27.4	26.0	29.3
	K	29.2	28.2	27.7	32.0
Late	Check	30.3	32.3	31.6	30.9
	Aero	32.7	30.8	31.0	33.8
	Na	30.3	32.7	33.0	33.9
	K	32.7	31.7	31.8	29.4

TABLE 10.40–ANOVA for Experiment Described in Example 10.21
(data given in Table 10.39)

Source of Variation	Degrees of Freedom	Sum of Squares	Mean Square	Expected Mean Square	F Ratio
Mean	1	30,368.80	30,368.80	$\cdots$	$\cdots$
Treatments Dates of planting	1	32.00	32.00	$\sigma^2 + (16/1)\sum_{i=1}^{2} \alpha_i^2$	10.42
Fertilizers	3	16.40	5.47	$\sigma^2 + (8/3)\sum_{j=1}^{4} \beta_j^2$	1.78
Fertilizers × dates of planting	3	38.40	12.80	$\sigma^2 + (4/3)\sum_{i=1}^{2}\sum_{j=1}^{4} (\alpha\beta)_{ij}^2$	4.17
Experimental error	24	73.74	3.07	σ^2	$\cdots$
Total	32	30,529.34	$\cdots$	$\cdots$	$\cdots$

Assuming $\alpha = 0.01$, it is seen that the hypothesis "date of planting has no effect" must be rejected. Examination of the mean yields indicates that the later date of planting is better (i.e., is associated with higher yields). More information is needed concerning the distinction between "early" and "late" before explicit recommendations can be made. No statistically significant effects were noted either for fertilizers or for the interaction between fertilizers and date of planting. (**Note:** Had α been chosen as 0.05, the interaction effect would have been significant. This illustrates the dependence of the inferences upon the choice of significance level, a fact that is sometimes overlooked or forgotten by the analyst. That is, we must always remember that a statement about significance or nonsignificance is a direct function of the selected value of α.)

Several items need mentioning before we leave the subject of cross-classification factorial models. These are: (1) general computational procedures for factorials involving four or more factors, (2) special computational methods for 2^n and 3^n factorials, (3) subsampling in completely randomized designs involving factorial treatment combinations, and (4) analysis of response curves associated with the various main effects and interactions. A brief discussion of each of these will be given in the following paragraphs.

The general computational procedure for factorials proceeds as follows. First compute $\sum Y^2$, M_{yy}, T_{yy}, E_{yy}, and any sums of squares required because of subsampling. Then, to subdivide T_{yy} in, say, a four-factor factorial, form in succession the four-way table, all three-way tables, and all two-way tables. As each table is formed, compute the border totals as a check on the entries you have made in the cells of the tables. Then, starting with the two-way tables, calculate the sums of squares for each of the main effects and for each of the two-factor interactions. Then, proceeding to the three-way tables, calculate the sums of squares associated with each of the three-factor interactions. Finally, utilizing the four-way table, the sum of squares associated with the four-factor interaction may be obtained. The extension to $5, 6, \ldots, N$ factors is easy. After obtaining the basic sums of squares, form the N-way table, all possible $(N - 1)$-way tables, all possible $(N - 2)$-way tables, $\ldots$, all possible three-way tables, and all possible two-way tables in the order mentioned. Then calculate, in the following order, all main effect sums of squares, all two-factor interaction sums of squares, $\ldots$, all $(N - 1)$-factor interaction sums of squares, and the N-factor interaction sum of squares.

Whenever all the factors are at p levels and there are n factors, the statistician refers to such an arrangement as a p^n factorial. Of particular interest are those cases where $p = 2$ or 3. When such cases arise, certain special computational techniques are available to the research worker. These are explained in considerable detail in such references as Yates (34) and Kempthorne (19) and may be pursued by readers whose primary interest is in computation. Since the methods outlined earlier in this section are valid for *all* cases, we shall merely point out the existence of the methods and give pertinent references for the use of interested persons.

When subsampling occurs in a completely randomized design involving factorial treatment combinations, the methods of analysis are simply a combination of those given in this section and Section 10.4. Thus no detailed discussion of computational techniques will be presented at this time. However, to illustrate the nature of the ANOVA's, two cases will be mentioned. The first of these will involve only one subsampling stage, while the second will involve two stages. If only one stage of subsampling is involved, the appropriate statistical model for a two-factor factorial is

$$Y_{ijkl} = \mu + \alpha_i + \beta_j + (\alpha\beta)_{ij} + \epsilon_{ijk} + \eta_{ijkl}$$
$$i = 1, \ldots, a; j = 1, \ldots, b; k = 1, \ldots, n; l = 1, \ldots, p \qquad (10.110)$$

TABLE 10.41–Abbreviated ANOVA for a Two-Factor Factorial in a Completely Randomized Design Involving One Stage of Subsampling (Model I)

Source of Variation	Degrees of Freedom	Expected Mean Square
Mean	1	. . .
Treatments		
A	$a - 1$	$\sigma_\eta^2 + p\sigma^2 + pnb \sum_{i=1}^{a} \alpha_i^2/(a - 1)$
B	$b - 1$	$\sigma_\eta^2 + p\sigma^2 + pna \sum_{j=1}^{b} \beta_j^2/(b - 1)$
AB	$(a - 1)(b - 1)$	$\sigma_\eta^2 + p\sigma^2 + pn \sum_{i=1}^{a} \sum_{j=1}^{b} (\alpha\beta)_{ij}^2/(a - 1)(b - 1)$
Experimental error	$ab(n - 1)$	$\sigma_\eta^2 + p\sigma^2$
Sampling error	$abn(p - 1)$	σ_η^2
Total	$abnp$	. . .

and the ANOVA would appear (in abbreviated form) as in Table 10.41. (**Note:** If only *one* sample were obtained from each experimental unit; e.g., if one small sample is taken from a field plot to estimate the yield of the entire plot, p in Table 10.41 is set equal to 1 and the line for "sampling error" is deleted. However, if the whole plot is harvested, the sampling error is 0 and the ANOVA would be as shown in Table 10.25.) In the second case to be examined (i.e., a case involving two stages of subsampling) the appropriate statistical model (for a two-factor factorial) is

$$Y_{ijklm} = \mu + \alpha_i + \beta_j + (\alpha\beta)_{ij} + \epsilon_{ijk} + \eta_{ijkl} + \delta_{ijklm}$$
$$i = 1, \ldots, a; j = 1, \ldots, b; k = 1, \ldots, n$$
$$l = 1, \ldots, p; m = 1, \ldots, d \qquad (10.111)$$

and the ANOVA would appear (in abbreviated form) as in Table 10.42. The extension to cases involving more than two stages of subsampling should be obvious.

As indicated in Section 10.11, it is often advisable to examine the response curve that summarizes the effects of the various levels of a factor upon the characteristic being measured. When our data fit a factorial arrangement, we may find it possible to examine response curves associated with the levels of two or more factors. For example, if we have two factors, a and b, we may subdivide the two sums of squares A_{yy} and B_{yy} into parts designated as $(A_L)_{yy}$, $(A_Q)_{yy}$, $\ldots$, and $(B_L)_{yy}$, $(B_Q)_{yy}$, $\ldots$ respectively. That is, we may obtain the linear, quadratic, $\ldots$, sums of squares associated with each of the factors a and b. However, since we are now dealing with factorials, it is also

TABLE 10.42–Abbreviated ANOVA for a Two-Factor Factorial in a Completely Randomized Design Involving Two Stages of Subsampling (Model I)

Source of Variation	Degrees of Freedom	Expected Mean Square
Mean Treatments	1	$\cdots$
A	$a - 1$	$\sigma_\delta^2 + d\sigma_\eta^2 + dp\sigma^2 + dpnb \sum_{i=1}^{a} \alpha_i^2/(a - 1)$
B	$b - 1$	$\sigma_\delta^2 + d\sigma_\eta^2 + dp\sigma^2 + dpna \sum_{j=1}^{b} \beta_j^2/(b - 1)$
AB	$(a - 1)(b - 1)$	$\sigma_\delta^2 + d\sigma_\eta^2 + dp\sigma^2 + dpn \sum_{i=1}^{a} \sum_{j=1}^{b} (\alpha\beta)_{ij}^2/(a - 1)(b - 1)$
Experimental error	$ab(n - 1)$	$\sigma_\delta^2 + d\sigma_\eta^2 + dp\sigma^2$
First-stage sampling error	$abn(p - 1)$	$\sigma_\delta^2 + d\sigma_\eta^2$
Second-stage sampling error	$abnp(d - 1)$	σ_δ^2
Total	$abnpd$	$\cdots$

possible to subdivide the interaction sum of squares $(AB)_{yy}$. The parts into which $(AB)_{yy}$ may be subdivided will be designated as $(A_L B_L)_{yy}$, $(A_L B_Q)_{yy}$, $(A_Q B_L)_{yy}$, $(A_Q B_Q)_{yy}$, ... If a third factor c were present, we would then have such quantities as $(C_L)_{yy}$, $(C_Q)_{yy}$, $(A_L C_L)_{yy}$, $(A_Q C_L)_{yy}$, $(B_L C_Q)_{yy}$, $(A_L B_L C_L)_{yy}$, $(A_L B_L C_Q)_{yy}$, etc. The number of possible subdivisions is limited by the number of levels of the various factors involved. Because we have already devoted so much time to the discussion of factorials in a completely randomized design, the details of this technique (i.e., response curve analyses for the various main effects and interactions) will not be discussed here. The technique will be discussed in the following chapter in connection with a randomized complete block design. Since the method is the same regardless of the design (as long as the completely randomized design has equal numbers of observations in each category), the person desiring the details now can jump ahead and read Section 11.11.

The reader should realize that the foregoing discussion has only scratched the surface of the subject of analyzing factorials. However, we believe that sufficient material has been given to enable the researcher to handle the most commonly occurring situations. Should more complex situations arise, reference to one or more of the books listed at the end of the chapter should prove helpful. If not, a professional statistician should be consulted.

10.12.2 Nested (hierarchal) Classification Model

A nested classification model arises in a two-factor factorial experiment when a different set of levels of factor b is associated with each level of factor a. This differs from the crossed classification model in which the same set of levels of factor b is associated with all levels of factor a.

Example 10.22

Suppose a factory has several (5) machines that perform the same operation even though the machines are of different types; e.g., they are produced by 5 different manufacturers. An experiment is planned to compare the volume of output among the different machines. Also, several operators will be used so a two-factor factorial experiment involving machines and operators is planned. The experiment can be conducted in two ways:

1. The same operators (3) can be used at all 5 machines; the model here is a 5 × 3 crossed classified model.
2. Five different sets of three operators (a total of 15) are used; the model here is a nested classification model with factor a at 5 levels and factor b at 3 levels.

Let Y_{ijk} denote the observation from the kth experimental unit subjected to the treatment formed by combining the ith level of factor a with the jth level of factor b *within* the ith level of factor a. The linear statistical model for the observation is

$$Y_{ijk} = \mu + \alpha_i + \beta_{ij} + \epsilon_{ijk} \qquad i = 1, \ldots, a; j = 1, \ldots, b; k = 1, \ldots, n$$

$$(10.112)$$

where μ is the true mean effect, α_i is the true effect of the ith level of factor a, β_{ij} is the true effect of the jth level of factor b within the ith level of factor a, and ϵ_{ijk} is the error associated with the kth experimental unit subjected to the ijth treatment combination. As usual, μ is assumed to be a constant and $\epsilon_{ijk} \sim NI(0, \sigma^2)$. We have also assumed equal numbers of levels of factor b for each level of factor a and the same number of experimental units per treatment. The analysis for unequal numbers is considered in Chapter 12. The ANOVA is summarized in Table 10.43. Formulas for computing the sums of squares included in Table 10.43 are

$$\sum Y^2 = \sum_{i=1}^{a} \sum_{j=1}^{b} \sum_{k=1}^{n} Y_{ijk}^2 \qquad (10.113)$$

$$M_{yy} = abn\,(\overline{Y}_{\ldots})^2 = T^2/abn \qquad (10.114)$$

$$A_{yy} = bn \sum_{i=1}^{a} (\overline{Y}_{i\ldots} - \overline{Y}_{\ldots})^2 = \sum_{i=1}^{a} A_i^2/bn - M_{yy} \qquad (10.115)$$

$$(B \mid A)_{yy} = \sum_{i=1}^{a} \sum_{j=1}^{b} (\overline{Y}_{ij\cdot} - \overline{Y}_{i\cdot\cdot})^2 = \sum_{i=1}^{a} \sum_{j=1}^{b} B_{ij}^2/n - \sum_{i=1}^{a} A_i^2/bn$$

$$(10.116)$$

TABLE 10.43—ANOVA for a Two-Factor Factorial (nested classification) in a Completely Randomized Design

Source of Variation	Degrees of Freedom	Sum of Squares	Mean Square
Mean	1	M_{yy}	**M**
Treatments			
A	$a-1$	A_{yy}	**A**
B within A	$a(b-1)$	$(B\mid A)_{yy}$	$(\mathbf{B}\mid\mathbf{A})$
Experimental error	$ab(n-1)$	E_{yy}	**E**
Total	abn	$\sum Y^2$	...

$$E_{yy} = \sum_{i=1}^{a}\sum_{j=1}^{b}\sum_{k=1}^{n} (Y_{ijk} - \bar{Y}_{ij.})^2 = \sum Y^2 - \sum_{i=1}^{a}\sum_{j=1}^{b} B_{ij}^2/n$$

(10.117)

where $T = Y_{...}$ is the total of all observations and

$$A_i = Y_{i..} = \sum_{j=1}^{b}\sum_{k=1}^{n} Y_{ijk}$$

(10.118)

$$B_{ij} = Y_{ij.} = \sum_{k=1}^{n} Y_{ijk}$$

(10.119)

Just as with the cross-classification model, several assumptions can be made about the true treatment effects. These are:

1. Fixed effects model: $\sum_{i=1}^{a}\alpha_i = 0$; $\sum_{j=1}^{b}\beta_{ij} = 0$ for each i.
2. Random effects model: $\alpha_i \sim NI(0, \sigma_\alpha^2)$; $\beta_{ij} \sim NI(0, \sigma_\beta^2)$ for each i.
3. Mixed model (a fixed, b random): $\sum_{i=1}^{a}\alpha_i = 0$; $\beta_{ij} \sim NI(0, \sigma_\beta^2)$ for

TABLE 10.44–Expected Mean Squares for a Two-Factor Factorial (nested classification) in a Completely Randomized Design

Source of Variation	Expected Mean Square		
	Fixed	Random	Mixed (a fixed, b random)
Mean	...	...	...
Treatments			
A	$\sigma^2 + nb\sum_{i=1}^{a}\alpha_i^2/(a-1)$	$\sigma^2 + n\sigma_\beta^2 + nb\sigma_\alpha^2$	$\sigma^2 + n\sigma_\beta^2 + nb\sum_{i=1}^{a}\alpha_i^2/(a-1)$
B within A	$\sigma^2 + n\sum_{i=1}^{a}\sum_{j=1}^{b}\beta_{ij}^2/a(b-1)$	$\sigma^2 + n\sigma_\beta^2$	$\sigma^2 + n\sigma_\beta^2$
Experimental Error	σ^2	σ^2	σ^2

each i. [**Note:** The other mixed model (a random, b fixed) is rarely appropriate and hence is not included.]

The expected mean squares for these three models are given in Table 10.44 and the corresponding F ratios are given in Table 10.45.

TABLE 10.45–F Ratios for Testing the Appropriate Hypotheses for a Two-Factor Factorial (nested classification) in a Completely Randomized Design

Source of Variation	F Ratios		
	Fixed	Random	Mixed
Mean	...	...	...
Treatments			
A	A/E	A/(B|A)	A/(B|A)
B within A	(B|A)/E	(B|A)/E	(B|A)/E
Experimental error	...	...	...

The analysis of nested classification models can readily be extended to more than two factors. Further analyses such as treatment comparisons, estimation of treatment means and variance components, etc., follow along the same lines as in the cross-classification model and hence are not further elaborated on here.

With regard to setting up the nested classification model as a regression model, the following multiple regression model can be used in a regression analysis to evaluate the appropriate sum of squares:

$$
\begin{aligned}
Y_{ijk} = \mu &+ \alpha_1 x_{1i} + \cdots + \alpha_{a-1} x_{a-1\,i} + \beta_{11} w_{1i,1j} \\
&+ \cdots + \beta_{1\,b-1} w_{1i,b-1j} + \cdots + \beta_{a\,1} w_{ai,1j} \\
&+ \cdots + \beta_{a\,b-1} w_{ai,b-1j} + \epsilon_{ijk} \\
& \qquad i = 1, \ldots, a; j = 1, \ldots, b; k = 1, \ldots, n \qquad (10.120)
\end{aligned}
$$

where the independent variables are defined as

$$
\begin{aligned}
x_{mi} &= 1 && \text{if } i = m; m = 1, \ldots, a - 1 \\
&= -1 && \text{if } i = a \\
&= 0 && \text{otherwise}
\end{aligned}
$$

and for $m = 1, \ldots, a$ and $p = 1, \ldots, b - 1$

$$
\begin{aligned}
w_{mi,pj} &= 1 && \text{if } i = m \text{ and } j = p \\
&= -1 && \text{if } i = m \text{ and } j = b \\
&= 0 && \text{otherwise}
\end{aligned}
$$

To illustrate the regression model, consider the case where $a = 2, b = 3$, and $n = 1$; Eq. (10.120) becomes

$$Y_{ijk} = \mu + \alpha_1 x_{1i} + \beta_{11} w_{1i,1j} + \beta_{12} w_{1i,2j}$$
$$+ \beta_{21} w_{2i,1j} + \beta_{22} w_{2i,2j} + \epsilon_{ijk}$$

$$i = 1, 2; j = 1, 2, 3; n = 1 \qquad (10.121)$$

The regression equations written in the matrix notation

$$\mathbf{Y} = \mathbf{X}\beta + \epsilon \qquad (10.122)$$

are

$$
\begin{bmatrix} Y_{111} \\ Y_{121} \\ Y_{131} \\ Y_{211} \\ Y_{221} \\ Y_{231} \end{bmatrix}
=
\begin{bmatrix}
1 & 1 & 1 & 0 & 0 & 0 \\
1 & 1 & 0 & 1 & 0 & 0 \\
1 & 1 & -1 & -1 & 0 & 0 \\
1 & -1 & 0 & 0 & 1 & 0 \\
1 & -1 & 0 & 0 & 0 & 1 \\
1 & -1 & 0 & 0 & -1 & -1
\end{bmatrix}
\begin{bmatrix} \mu \\ \alpha_1 \\ \beta_{11} \\ \beta_{12} \\ \beta_{21} \\ \beta_{22} \end{bmatrix}
+
\begin{bmatrix} \epsilon_{111} \\ \epsilon_{121} \\ \epsilon_{131} \\ \epsilon_{211} \\ \epsilon_{221} \\ \epsilon_{231} \end{bmatrix} \qquad (10.123)
$$

The restrictions $\sum_{i=1}^{a} \alpha_i = 0$ and $\sum_{j=1}^{b} \beta_{ij} = 0$ for $i = 1, \ldots, a$ have been included in the model to ensure a unique solution to the regression analysis. Since the μ, α's, and β's are orthogonal, the sum of squares included in the ANOVA in Table 10.43 can easily be accumulated from a regression analysis of variance output. We also mention again that many generalized computer factorial programs exist that will analyze a nested classification model as well as the cross-classification model.

10.12.3 Mixed Classification Models

Before concluding this section on factorial experiments, we want to indicate some mixed models that can exist when the experiment includes three or more factors. Such models will have some of the factors nested and other factors crossed. Examples of some of the possible three-factor factorial models are:

1. a crossed with b; c nested in $a \times b$

$$Y_{ijkl} = \mu + \alpha_i + \beta_j + (\alpha\beta)_{ij} + \gamma_{ijk} + \epsilon_{ijkl} \qquad (10.124)$$

2. b and c nested in a; c crossed with b

$$Y_{ijkl} = \mu + \alpha_i + \beta_{ij} + \gamma_{ik} + (\beta\gamma)_{ijk} + \epsilon_{ijkl} \qquad (10.125)$$

3. b nested in a; c crossed with a and b

$$Y_{ijkl} = \mu + \alpha_i + \beta_{ij} + \gamma_k + (\alpha\gamma)_{ik} + (\beta\gamma)_{ijk} + \epsilon_{ijkl} \qquad (10.126)$$

For further details on the ANOVA for these models, the reader is referred to the references given at the end of the chapter.

10.13 NONCOMFORMITY TO ASSUMED STATISTICAL MODELS

By now the reader is well aware that the usual assumptions in analysis of variance involve the concepts of additivity, normality, homogeneity of variances, and independence of the errors. However, up to this point, little has been said about (1) tests to assess the validity of the assumptions, (2) the consequences if the assumptions are not satisfied, and (3) transformations which, if applied to the original data, may justify the use of the assumptions in connection with the transformed data (i.e., the data as they appear after the transformation has been applied). In this section each of these topics will be discussed briefly. For those who wish more details, several references are given. In particular, three excellent expository articles are those by Bartlett (4), Cochran (7), and Eisenhart (15).

First, let us consider various statistical tests that have been proposed to check on the validity of the several assumptions.

Homogeneity of Variances. In Section 6.8, Bartlett's test was given for testing the hypothesis $H: \sigma_1^2 = \sigma_2^2 = \cdots = \sigma_k^2$ where a random sample of n_i observations had been taken from the ith normal population $(i = 1, \ldots, k)$. Clearly, this test is appropriate for checking on the homogeneity of variances. However, Bartlett's test has been shown to be quite sensitive to nonnormality. Thus, if nonnormality is suspected or has been demonstrated, the test should be modified as suggested by Box and Andersen (6). For a discussion of other tests, the reader is referred to Anscombe and Tukey (2), Box and Andersen (6), David (9), and Dixon and Massey (11).

Normality. To check on the assumption of normality, one can use the chi-square test of goodness of fit given in Section 6.14. An alternative, and perhaps preferred method is the Kolmogorov-Smirnov test discussed in Chapter 14. For those who are satisfied with a less objective approach, the data (or the residuals) may be plotted on normal probability paper and a subjective judgment rendered.

Additivity. When the assumption of additivity is questioned, the problem is somewhat more involved because there are three major causes of nonadditivity; namely, (1) the true effects may be multiplicative, (2) interactions may exist but terms representing such effects have not been included in the assumed model, and (3) aberrant observations may be present. If the experimental design is such that interaction effects may be isolated, the methods of the preceding section may be used to check on (2). However, if this is not possible, the researcher may use the more general tests suggested by Tukey (27, 32) and by Ward and Dick (33). Rather than give the details of these tests, we refer the reader to the original publications. If access to these publications is not possible, perhaps the illustrations in Snedecor and Cochran (26) and Hamaker (17) will suffice.

Independence. The assumption of independence or, granting normality, of uncorrelated errors is crucial and its importance should not be overlooked. By utilization of the device of randomization, the researcher can do his best to see that the correlation between errors will not continually favor (or hinder) any particular treatment. If one wishes to test for randomness, methods are available. However, since these will be discussed in Chapter 14, no details will be given at this time. The interested reader may jump ahead to the appropriate sections. (**Note:** The procedure discussed in Section 10.6 may also be helpful in this situation.)

In general, the consequences are not serious when the assumptions made in connection with analyses of variance are not strictly satisfied. That is, moderate departures from the conditions specified by the assumptions need not alarm us. For example, minor deviations from normality and/or some degree of heteroscedasticity (lack of homogeneity of variances) will have little effect on the usual tests and the resulting inferences. In summary, the ANOVA technique is quite robust, and thus the researcher can rely on its doing a good job under most circumstances. However, since trouble can arise because of failure of the data to conform to the assumptions, ways of handling such situations must be examined.

When some action is needed to make the data conform to the usual assumptions, the customary approach is to transform the original data in such a way that the transformed data will meet the conditions specified by the assumptions. For example, if the true effects are multiplicative instead of additive, it is customary to take logarithms and thus change, for instance,

$$Y = \mu \alpha_i \beta_j \epsilon_{ij} \tag{10.127}$$

into

$$Y' = \log Y = \log \mu + \log \alpha_i + \log \beta_j + \log \epsilon_{ij} \tag{10.128}$$

TABLE 10.46–Some Common Transformations

Transformation		Conditions Leading to Its Application
Name	Equation	
Logarithmic	$Y' = \log Y$	1. The true effects are multiplicative (or proportional). *or* 2. The standard deviation is proportional to the mean.
Square root	$Y' = \sqrt{Y}$ *or* $Y' = \sqrt{Y + 1}$	The variance is proportional to the mean (e.g., when the original data are samples from a Poisson distribution).
Arcsine	$Y' = \text{arcsine } \sqrt{p}$	The variance is proportional to μ $(1 - \mu)$ as, for example, when the original data are samples (expressed as proportions or relative frequencies) from binomial populations.
Reciprocal	$Y' = 1/Y$	The standard deviation is proportional to the square of the mean.

Fortunately, in most cases one transformation will suffice. That is, it is usually not necessary to make a series of transformations, each to correct a separate "deficiency" in the original data. The reason for this fortunate state of affairs is that, in general, the utilization of a transformation to correct one particular deficiency (say nonadditivity) will also help with respect to another deficiency (say nonnormality). With this in mind the commoner transformations are summarized in Table 10.46. Further details may be found in Bartlett (4) and Tukey (27).

One other technique for handling heterogeneous variances should be mentioned. This technique is to partition the experimental error sum of squares in correspondence with any partitioning of the treatment sum of squares. This does have the disadvantage that each portion of E_{yy} will usually possess a small number of degrees of freedom so that the subsequent F tests will not be very powerful, i.e., not very discriminating. Rather than discuss the technique at this time, an example of subdividing the experimental error sum of squares will be presented in Chapter 11.

10.14 PRESENTATION OF RESULTS

Even though an ANOVA table is very convenient for summarizing certain aspects of the analysis of a set of data, it suffers from a rather serious deficiency, namely, that it tends to overemphasize tests of hypotheses and underemphasize estimation. Since estimation is the more important of these two aspects of statistical inference, this can be serious if steps are not taken to remedy the situation. Two steps that can be taken to improve matters are: (1) Always accompany an ANOVA table with tables of means, together with their standard errors, and (2) whenever possible, portray the results in graphical form. If these two steps are taken and a *readable* report is prepared, the results of your research will be more easily understood and appreciated.

Example 10.23

Reexamination of Example 10.3 will show that the means were given in Table 10.3, the ANOVA in Table 10.4, and the standard error of the mean in the discussion. Actually, the standard error, which was the same for each mean because of the equal sample sizes, might better have been included in Table 10.3.

Example 10.24

Reexamination of Example 10.4 will show that the suggestion made in Example 10.23 was adopted in that case. That is, the standard errors were presented along with the means to which they applied.

Example 10.25

Reexamination of Example 10.8 will show that the ANOVA was given in Table 10.13 and the standard error of a treatment mean was included in the discussion. However, the treatment means were not explicitly exhibited, although they could easily have been obtained. Had a complete report of the research been prepared, this deficiency would have been noted and removed.

Example 10.26

Reexamination of Examples 10.13 and 10.14 will show that standard errors were (implicitly) found for each of the selected contrasts. As noted in the discussion of Example 10.14, the point and interval estimates of the true effects of the contrasts could then be calculated. These would be included in the research report.

Example 10.27

Reexamination of Example 10.17 will indicate that any research report concerning this experiment would have benefited by a graph showing the treatment means (average yields) as a function of the amount of fertilizer applied to the experimental plots. It is suggested that the reader plot these means and examine the graph in connection with the recommendations made in Example 10.17.

Example 10.28

Reexamination of Example 10.21 will reveal that the treatment means were not given. Since they are pertinent to the conclusions, we give them in Table 10.47. The standard errors of the treatment means shown in

TABLE 10.47–Treatment Means for the Experiment Discussed in Example 10.21 (data in Table 10.39; ANOVA in Table 10.40)

| Fertilizer | Date of Planting | | Average |
	Early	Late	
Check	32.68 (0.88)	31.28 (0.88)	31.98 (0.62)
Aero	29.50 (0.88)	32.08 (0.88)	30.79 (0.62)
Na	27.78 (0.88)	32.48 (0.88)	30.12 (0.62)
K	29.28 (0.88)	31.40 (0.88)	30.34 (0.62)
Average	29.81 (0.44)	31.81 (0.44)	30.81

Note: The figures in parentheses in the table are the standard errors of the means to which they are appended.

Table 10.47 were calculated by taking the square roots of the following estimated variances:

$$\hat{V}(\overline{Y}_{i.}) = \hat{V}(\text{date of planting mean}) = 3.07/16 = 0.1919$$
$$\hat{V}(\overline{Y}_{.j}) = \hat{V}(\text{fertilizer mean}) = 3.07/8 = 0.3838$$
$$\hat{V}(\overline{Y}_{ij}) = \hat{V}(\text{date of planting} \times \text{fertilizer mean}) = 3.07/4 = 0.7675$$

A graphical presentation of the means is given in Figure 10.1 where the reader must realize that the slopes of the lines are a direct reflection of the scales adopted. However, since our main use of the graph will be in the interpretation of the interaction, this will not matter, for we shall be con-

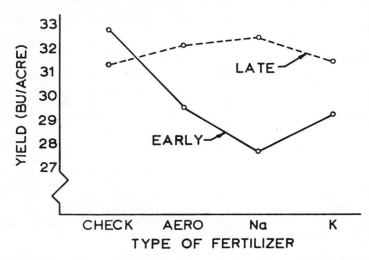

Fig. 10.1–Graphical representation of the mean yields given in Table 10.47.

cerned only with the slopes of the lines relative to one another. A study of Figure 10.1 will confirm the conclusions reached in Example 10.21; namely, (1) the late date of planting is apparently better than the early date of planting, (2) there is little difference among the main effects of the four fertilizers, and (3) there is some indication of a possible interaction. (**Note:** This last conclusion is suggested by the lack of "parallelism" of the plotted lines.)

In addition to the remarks made in the first paragraph of this section and illustrated in Examples 10.23–10.28, the reader should realize that many experiments are conducted and analyses of variance performed merely to estimate components of variance. Important as this topic is, it is felt that the discussion given earlier in the chapter will prove sufficient for most applications. Should further details be desired, it is suggested that a professional statistician be consulted.

One other topic should be mentioned in connection with the presentation of results and is concerned with the general way in which ANOVA's are commonly presented. Two customs have become quite firmly established over the years, and they are as follows:

1. (a) If an F ratio exceeds the 95 percent point but does not exceed the 99 percent point, the F ratio (or the mean square for the effect being tested) is tagged with a single asterisk (*).
   (b) If an F ratio exceeds the 99 percent point, the F ratio (or the mean square for the effect being tested) is tagged with a double asterisk (**).
2. If space is at a premium, only an abbreviated ANOVA will be presented. When this is done, it is customary to include only the columns for (1) sources of variations, (2) degrees of freedom, and (3) mean squares.

Incidentally, when the asterisk convention is used, it is good practice to define the symbols at the bottom of every ANOVA table by use of the following footnotes: *Significant at $\alpha = 0.05$. **Significant at $\alpha = 0.01$. The use of these customs will be illustrated in succeeding chapters.

If any words can be put together to summarize the implications of this section, they are as follows: *Do not forget the reader. Remember, you are not writing for yourself but for others. Anything you can do to make your assumptions, procedures, results, analyses, and conclusions more understandable will add to the value of your research.*

PROBLEMS

10.1   What are the proper objectives of analyses of variance (using experimental or survey data); i.e., for what purposes may we properly use analyses of variance?

10.2   Forty technicians were available to investigate 5 methods of determining the iron content of a certain chemical mixture. Eight of the technicians used method 1, 8 used method 2, etc. The assignment of technicians to methods was performed in a random manner. Each technician made only one determination. Given that (1) the total of the 40 observations was 80, (2) the among methods mean square was 6, and (3) the pooled variance among technicians within methods was 8, fill in the following ANOVA table. (**Note:** Omit the spaces marked X.)

Source of Variation	Degrees of Freedom	Sum of Squares	Mean Square	Expected Mean Square
Mean				X
Among methods				
Among technicians within methods				
Total			X	X

10.3   Given the following abbreviated ANOVA:

Source of Variation	Degrees of Freedom	Sum of Squares	Mean Square	Expected Mean Square
Among treatments	4	244	61	$\sigma^2 + 7\sum_{i=1}^{5} \tau_i^2/4$
Among experimental units within treatments	30	270	9	σ^2

(a)   Write out the appropriate model.
(b)   State the null hypothesis, both in words and symbolically, that the experiment was probably designed to test.
(c)   Test the hypothesis given in the answer to (b) using a probability of Type I error equal to 0.05.

10.4   A process is designed to produce a fishline that will have a "15-lb-test" rating. The braided line may be treated with 4 different waterproofings. The hypothesis is that the 4 treatments have the same, if any, effect on the test rating of the cord. Twenty samples of each type of treated cord are tested for breaking strength. Assuming that ANOVA is a valid technique to use in this case, set up the appropriate table showing the proper subdivision of the degrees of freedom. Discuss any further analyses that might be useful in investigating the treatments.

10.5   It is desired to test 10 different baking temperatures when we use a standard cake mix. Fifty sample batches of mix are prepared, and 5 are assigned at random to each of the 10 temperatures. Six judges score the cakes, and the average score is recorded for each. Give the proper subdivision of the degrees of freedom, and write out the mathematical model assumed. State the hypothesis to be tested. Discuss and evaluate the method of analysis.

10.6   Four methods of performing a certain operation have been tried and we have 10 observations for each method. The mean productivities under each method are 60, 70, 80, and 90 respectively. Not having the original data from which to calculate the sums of squares, we assume that the coefficient of variation (square root of the pooled estimate of σ^2 divided by the average of all observations) is 0.1. On this assumption, test the hypothesis that the "method population means" are equal.

10.7   Given that the means of 10 individuals in each of 5 groups are 30, 32, 34, 36, and 38 and that the variance of a group mean is 8, compute the ANOVA.

10.8   An investigation to study the variation in average daily gains made by pigs among and within litters when fed the same ration gave the following results:

Source of Variation	Degrees of Freedom	Mean Square
Among litters	29	0.0576
Among pigs in the same litter	180	0.0144

How would you use this information to design experiments to test the effects of different rations on average daily gains?

10.9   Community X and community Y are two neighboring small towns. Community X is supplied with electricity by a private power company, while community Y operates a municipally owned but inefficient high-cost power plant. As a result, cost of electricity to home users is higher in community Y than in community X; e.g., the charge for the first 50 watts is $3.00 in X and $4.50 in Y. A random sample of household meter readings for the same month was taken in each community. The following values were obtained:

X Sample (kw-hr used)		Y Sample (kw-hr used)	
28	16	6	12
12	22	36	18
14	28	24	18
4	4	58	16
16	22	60	22
28	4	6	14
30	34	14	16
76	30	54	26
		22	44
		16	58
			18

Analyze these data in two ways:

(a) Compare home consumption of electricity in the two communities by means of the comparison of two groups using "Student's" t test.

(b) Prepare an ANOVA of these two samples.

10.10 It is suspected that five filling machines in a certain plant are filling cans to different levels. Random samples of the production from each machine were taken with the following results. Analyze the data and state your conclusions.

		Machine		
A	B	C	D	E
11.95	12.18	12.16	12.25	12.10
12.00	12.11	12.15	12.30	12.04
12.25		12.08	12.10	12.02
12.10				12.02

10.11 The amount of carbon used in the manufacture of steel is assumed to have an effect on the tensile strength of the steel. Given the following data, perform the appropriate analysis and interpret your results. The tensile strengths of 6 specimens of steel for each of 3 different percentages of carbon are shown. (The data have been coded for easy calculation.)

Percentage of Carbon		
0.10	0.20	0.30
23	42	47
36	26	43
31	47	43
33	34	39
31	37	42
31	31	35

10.12   A public utility company has a stock of voltmeters that are used interchangeably by the employees. The question arises as to whether all the
voltmeters are homogeneous. Since it would be too expensive to check
all meters, a random sample of 6 meters is obtained and all are read
3 times while being subjected to a constant voltage. The following data,
expressed as deviations from the test voltage, were recorded. Analyze
and interpret.

			Meter		
1	2	3	4	5	6
0.95	0.33	−2.15	−1.20	1.80	−1.05
1.06	−1.46	1.70	0.62	0.88	−0.65
1.96	0.20	0.48	1.50	0.20	0.80

10.13   An experiment had as its objective the evaluation of variance components for the variation in ascorbic acid concentration (mg/100 g)
in turnip greens. Two leaves were taken from near the center of each
of 5 plants. Ascorbic acid concentration was determined for each leaf.
This was repeated on each of 6 days, a new selection of plants being
obtained each day. The following data were collected:

Day	Leaf	Plant				
		1	2	3	4	5
1	A	9.1	7.3	7.3	10.7	7.7
	B	7.3	9.0	8.9	12.7	9.4
2	A	12.6	9.1	10.9	8.0	8.9
	B	14.5	10.8	12.8	9.8	10.7
3	A	7.3	6.6	5.2	5.3	6.7
	B	9.0	8.4	6.9	6.8	8.3
4	A	6.0	8.0	6.8	9.1	8.4
	B	7.4	9.7	8.6	11.2	10.3
5	A	10.8	9.3	7.3	9.3	10.4
	B	12.5	11.0	8.9	11.2	12.0
6	A	10.6	10.9	10.4	13.1	7.7
	B	12.3	12.8	12.1	14.6	9.4

Plants, days, and leaves are to be considered as random variables
(there might be some question about days). Calculate the ANOVA
and evaluate the variance components for leaves of the same plant,
plants of the same day, and days.

10.14   Suppose we have the mathematical model

$$Y_{ijk} = \mu + \tau_i + \epsilon_{ij} + \delta_{ijk} \qquad i = 1, 2, 3, 4; j = 1, 2; k = 1, 2$$

where τ_i is the true effect of the ith treatment, ϵ_{ij} is the effect of the jth
experimental unit subjected to the ith treatment, and δ_{ijk} is the kth
determination on the ijth experimental unit. We wish to test the hy-

pothesis $H:\tau_i = 0$ for all i. The following values are known:

$$
\begin{array}{lll}
T_1 = 8 & E_{11} = 3 & E_{12} = 5 \\
T_2 = 7 & E_{21} = 3 & E_{22} = 4 \\
T_3 = 10 & E_{31} = 2 & E_{32} = 8 \\
T_4 = 7 & E_{41} = 5 & E_{42} = 2
\end{array}
$$

and $\sum(Y - \overline{Y})^2 = 18$. Complete the appropriate ANOVA, test the hypothesis, and interpret your results.

10.15　Given the following abbreviated ANOVA of data collected from an experiment involving 6 treatments, 10 experimental units per treatment, and 3 determinations per experimental unit:

Source of Variation	Degrees of Freedom	Mean Square	Expected Mean Square
Treatments	5	12,489	$\sigma_\delta^2 + 3\sigma^2 + \dfrac{30}{5}\sum_{i=1}^{6}\tau_i^2$
Experimental units within treatments	54	3,339	$\sigma_\delta^2 + 3\sigma^2$
Determinations per experimental unit	120	627	σ_δ^2

(a)　Write out the model assumed, stating explicitly what each term represents.
(b)　Test the hypothesis that the 6 treatments have the same population mean.
(c)　Estimate the variance of a treatment mean.
(d)　Given that the sample mean for treatment 3 is 193.7, compute and interpret the 95 percent confidence interval for estimating the true population mean of treatment 3.
(e)　Assuming that the estimates of the components of variance would remain unchanged, would it be more or less efficient to use 9 experimental units per treatment and 4 determinations per experimental unit? Show all calculations necessary to support your answer. What is the gain or loss in information?

10.16　We conducted a completely randomized experiment to study some chemical characteristics of 5 varieties of oats. We assigned each variety at random to 6 plots, making a total of 30 plots. Instead of harvesting the entire plot, we selected at random eight $3' \times 3'$ samples from each. For each sample we made 3 chemical determinations. Indicate the proper complete subdivision of the total degrees of freedom. Give the expected mean square for each source of variation. Indicate the proper F test to test the hypothesis that the population means for the 5 varieties are equal.

10.17　Given the following abbreviated ANOVA:

Source of Variation	Degrees of Freedom	Mean Square
Groups	3	600
Experimental units within groups	36	120
Determinations per experimental unit	80	12

(a) Give the expected mean squares, assuming that we are interested in just these groups but that experimental units and determinations are random variables.

(b) Test the hypothesis that the group population means are equal. Interpret your result.

(c) Estimate the variance of a group mean (per determination).

10.18   Given the following abbreviated ANOVA:

Source of Variation	Degrees of Freedom	Mean Square
Among treatments	4	20
Experimental units within treatments	15	15
Determinations per experimental unit	20	4

Obtain estimates of *all* the components of variance, and interpret each in terms of the model $Y_{ijk} = \mu + \tau_i + \epsilon_{ij} + \delta_{ijk}$, stating explicitly all assumptions that you make.

10.19   Given the following abbreviated ANOVA:

Source of Variation	Degrees of Freedom	Sum of Squares	Mean Square	Expected Mean Square
Treatments	3	1800	600	$\sigma_\delta^2 + 3\sigma^2 + 30\sigma_\tau^2$
Experimental units within treatments	36	3600	100	$\sigma_\delta^2 + 3\sigma^2$
Determinations per experimental unit	80	960	12	σ_δ^2

(a) Estimate the variance of a treatment mean.

(b) Test the null hypotheses $H: \sigma_\tau^2 = 0$ and interpret your answer.

(c) The sample mean of treatment 1 is given to be 80. Compute a 95 percent confidence interval for estimating the true population mean of treatment 1.

10.20   Given the following abbreviated ANOVA:

Source of Variation	Degrees of Freedom	Mean Square
Treatments	4	960
Experimental units within treatments	35	320
Determinations per experimental unit	40	20

(a) Test the hypothesis that the population treatment means are all equal. Interpret your answer.

(b) Give the expected mean squares in the above ANOVA.

(c) Estimate the variance of a treatment mean.

(d) Estimate the gain or loss in information if the above experiment were to be repeated with 10 experimental units per treatment and a single determination per experimental unit. State all your assumptions.

10.21 Given the following abbreviated ANOVA:

Source of Variation	Degrees of Freedom	Mean Square
Among treatments	9	570
Experimental units within treatments	90	190
Among determinations on same experimental unit	200	10

(a) Compute F to test the hypothesis that the 10 treatments have the same true effect.

(b) Estimate the variance of a treatment mean per determination in the above experiment.

(c) Assuming that the estimates of the components of variance would not change, estimate the gain or loss in information in estimating the treatment means if 20 experimental units per treatment were selected with a single determination on each experimental unit in repeating the experiment.

10.22 The cities and towns of Arizona have been allocated to 5 strata (or groups) according to population. In each stratum we select at random 10 cities (or towns), in each of these cities we select at random 4 blocks, and in each of these blocks we select at random 2 households. Indicate the proper subdivision of the degrees of freedom for a complete ANOVA of some item such as the average income of the head of each household.

10.23 Set up the ANOVA table and show the degrees of freedom for the following experiment: Six spray treatments are applied completely at random in an orchard of 100 trees (all being used). Each treatment is applied to sets of 2 trees; then the yield of each tree is estimated by obtaining 4 samples around the perimeter. Note that all but 2 treatments contain 8 sets; the remaining 2 treatments contain 9 sets. Show the expected mean squares.

10.24 The following abbreviated ANOVA was prepared from chemical determinations made on samples of a legume hay. The hay samples were obtained from a completely randomized experiment involving 16 treatments.

Source of Variation	Degrees of Freedom	Mean Square
Treatments	15	550
Plots with same treatment		220
Samples from plots treated alike	256	20

(a) State a suitable hypothesis about these treatments and make the proper test. The experimenter wished to reject the hypothesis only under the condition of a 1 percent chance of a Type I error. What is your conclusion?

(b) Indicate the function of replication in this experiment for studying the effects of various treatments on a legume hay.

(c) How might the replication and sampling procedure for this experiment be changed in order to increase replication without changing the total number of samples to be analyzed?

(d) Estimate the possible gain in relative efficiency for your proposed change in the experiment.

(e) What assumption is required for making this calculation?

10.25 In an effort to develop objective methods of estimating the yield of corn, an experimental survey was conducted in a district of central Iowa. A random selection of fields was made, and within those fields 2 sampling units (consisting of 10 hills each) were selected at random and the grain yield was determined by harvesting and weighing. The ANOVA (on a 10-hill sampling unit basis) is as follows:

Source of Variation	Degrees of Freedom	Sum of Squares	Mean Square
Fields	47	2098.7	44.65
Sampling units within fields	48	554.5	11.55

How much information would have been lost if only *one* sampling unit per field had been taken?

10.26 Data from a sample survey of farms in the Midwest were to be summarized by means of ANOVA. Eight types of farming areas were included in the study, and within each area 5 counties were selected at random. Within each of the chosen counties, 20 farms were selected at random and farm management records taken for each. A partial list of the summary calculations was as follows for the item "farm income":

Total corrected sum of squares = 8,183,000
Sum of squares for among counties within areas = 352,000
Mean square for type of farming areas = 33,000

(a) Prepare and complete an ANOVA for "farm income" from the above information.

(b) What is the variance of a type of farming area mean as determined from your ANOVA?

(c) Suppose it had been decided to select only 2 counties in each area and sample 50 farms in each county. What is the relative efficiency of the plan used to the procedure suggested here?

(d) The type of farming areas included in the study were arbitrarily selected upon the basis of some known differences. Are the areas different with respect to "farm income?"

(e) Write out the expected values of the mean squares used in answering (d) above.

10.27 Given the following abbreviated ANOVA of the data from a completely randomized experiment with 4 treatments, 8 experimental units per treatment, 3 samples per experimental unit, and 2 determinations (of some chemical or physical characteristic) per sample:

Source of Variation	Degrees of Freedom	Mean Square
Treatments	3	19,200
Among experimental units treated alike	28	4,800
Among samples per experimental unit	64	2,400
Between determinations per sample	96	1,200

Estimate the gain or loss in efficiency in estimating the treatment effects if we had used 12 experimental units per treatment, 2 samples per experimental unit, and 1 determination per sample.

10.28 Describe the assumptions underlying the application of the ANOVA technique.
(a) Which of these assumptions can the research worker check for any particular analysis?
(b) Which assumption can be fulfilled by the research worker in an experimental situation by appropriate procedures?
(c) For what purposes do we employ the ANOVA technique?
(d) What criterion should be applied for judging the validity of an F ratio obtained from an analysis of variance?

10.29 (a) Explain in your own words the meaning in the ANOVA of (1) a variance component and (2) a fixed effect?
(b) Consider the following abbreviated ANOVA of calories consumed in one day for a sample of Iowa women over the age of 30:

Source of Variation	Degrees of Freedom	Sum of Squares	Mean Square
Among zones	2	16,960,000	8,480,000
Among counties in zones	97	41,128,000	424,000
Between segments in counties in zones	100	40,000,000	400,000
Among individuals in segments in counties in zones	600	180,000,000	300,000

For this problem we shall assume that 4 individuals (women over 30) were interviewed in each segment and that 2 segments were selected at random in each county. The zones are open country, rural community, and urban. Counties appearing in the sample for the zones were 50, 25, and 25 respectively, yielding the 97 degrees of freedom for counties in zones.
(a) Estimate the variance components for individuals, segments, and counties from this ANOVA.
(b) Test the hypothesis: Calories consumed on this day are the same for all zones. Show the mean squares used in forming the F ratio and indicate in general why these are the proper mean squares to use for the test.

10.30 With reference to Example 10.11 and Table 10.18 show that the expected mean squares for the three comparisons are:

$$\sigma^2 + \frac{[(n_3 + n_4)(n_1\tau_1 + n_2\tau_2) - (n_1 + n_2)(n_3\tau_3 + n_4\tau_4)]^2}{(n_1 + n_2)(n_3 + n_4)\sum_{i=1}^{4} n_i}$$

$$\sigma^2 + \frac{n_1 n_2}{n_1 + n_2}(\tau_1 - \tau_2)^2$$

$$\sigma^2 + \frac{n_3 n_4}{n_3 + n_4}(\tau_3 - \tau_4)^2$$

10.31 With reference to Example 10.13 and Table 10.20, show that the expected mean squares for the four comparisons are:

$$\sigma^2 + (1/5)(\tau_1 + \tau_3 + \tau_4 + \tau_5 - 4\tau_2)^2$$
$$\sigma^2 + (\tau_1 + \tau_3 - \tau_4 - \tau_5)^2$$
$$\sigma^2 + 2(\tau_1 - \tau_3)^2$$
$$\sigma^2 + 2(\tau_4 - \tau_5)^2$$

10.32 Given the additional information that in Problem 10.10 machine A is a standard machine and machines B, C, D, and E are experimental models, modify the original analysis to assess the relative performance of the five machines.

10.33 Apply the technique of Section 10.10 to the following problems:
(a) 10.10   (c) 10.12
(b) 10.11   (d) 10.14

10.34 If you did not use the technique of Section 10.11 in the analysis of Problem 10.11, please do so now.

10.35 It is suspected that the age of a furnace used in curing silicon wafers influences the percentage of defective items produced. An experiment was conducted using four different furnaces and the data given below were obtained for the percentage of good wafers in 8 experimental trials per furnace (the same number of wafers were used in each furnace in each trial). Analyze and interpret the data.

Furnace			
A (age 1 year)	B (age 2 years)	C (age 3 years)	D (age 4 years)
95	95	80	70
92	85	80	65
92	92	82	70
90	83	78	72
92	83	77	72
94	88	75	66
92	89	78	50
91	90	78	66

10.36 It is suspected that the environmental temperature in which batteries are activated affects their activated life. Thirty homogeneous batteries were tested, 6 at each of 5 temperatures, and the data shown below were obtained for activated life, in seconds. Analyze and interpret the data.

Temperature (C)				
0	25	50	75	100
55	60	70	72	65
55	61	72	72	66
57	60	73	72	60
54	60	68	70	64
54	60	77	68	65
56	60	77	69	65

10.37 It is suspected that both the machine on which bearings are produced and the operator of the machine influence the critical dimension, namely, the inside diameter. To check on this, the data given below for inside diameters of bearings in inches were obtained under normal production conditions. Analyze and interpret the data.

Machine:	1		2	3	
Operator:	A	B	C	D	E
	1.02	1.03	1.05	1.03	1.02
	1.03	1.03	1.06	1.03	1.03
	1.02	1.03	1.04	1.02	1.04
		1.03	1.06		1.02
			1.07		1.02
			1.06		
			1.05		

10.38 An experiment was conducted to assess the effects of temperature and humidity on the effective resistance of a standard type of resistor. The following data on coded resistance values were obtained. Analyze and interpret the data.

Temperature:	− 20 F		70 F		160 F	
Humidity:	10%	50%	10%	50%	10%	50%
	23	24	26	24	25	27
	24	24	25	25	26	26
	25	25	26	26	26	28
	24	26	26	26	28	28

10.39 Given the following abbreviated ANOVA of net income per crop acre:

Source of Variation	Degrees of Freedom	Mean Square
Among soil areas	4	625
Soil conservation programs	3	400
SA × SCP	12	225
Among farms in subclasses	80	100

(a) Assuming both of the main classifications are fixed effects, indicate the appropriate F ratios for tests of the hypotheses: (1) soil areas do not differ in income; (2) soil conservation programs have no effect on income per crop acre.

(b) Assuming that soil areas and soil conservation programs were selected at random from a larger number (not entirely realistic, but possible), indicate the F ratios for the tests listed in (a) above.

10.40 In the accompanying table, pasture acres per farm for a sample of 36 farms in Audubon County, Iowa, for the year 1934 are presented. The sample consists of 3 farms from each soil-tenure grouping; there are 12 such groups. It is expected that there may be some interaction effects among the soil and tenure classes. We shall consider the tenure grouping as a fixed effect and the soil groupings as random sampling from a larger population. Perform the necessary calculations, set up the ANOVA table, and discuss the results. Also examine the homogeneity of variance in the tenure groups by Bartlett's test.

Tenure Group	Soil Group			
	I	II	III	IV
Owners	(pasture acres per farm)			
Farm 1	37.0	50.0	49.0	56.0
Farm 2	40.1	28.6	43.7	69.0
Farm 3	57.0	37.2	27.0	54.7
Tenants				
Farm 1	36.0	42.0	50.9	55.0
Farm 2	52.0	54.5	34.0	41.0
Farm 3	38.0	58.0	43.8	54.6
Mixed				
Farm 1	72.6	54.0	67.4	63.0
Farm 2	65.2	58.0	32.5	45.0
Farm 3	71.0	29.0	43.8	60.0

10.41 Five varieties and 4 fertilizers were tested. From each experimental plot 3 quadrats were selected at random and their yields recorded as follows:

Fertilizer	Variety				
	1	2	3	4	5
1	57	26	39	23	48
	46	38	39	36	35
	28	20	43	18	48
2	67	44	57	74	61
	72	68	61	47	60
	66	64	61	69	75
3	95	92	91	98	78
	90	89	82	85	89
	89	99	98	85	95
4	92	96	98	99	99
	88	95	93	90	98
	99	99	98	98	99

(a) Construct an ANOVA table.

(b) On the basis of the appropriate model, write the expected mean squares conforming to the following assumptions: (1) varieties and fertilizers, random selections; (2) varieties and fertilizers, both given sets; (3) varieties a random selection, fertilizers a given set.

(c) Test the hypothesis of equal variety means. Test the hypothesis of equal fertilizer means.

(d) Construct a table showing the means and their standard errors.

(e) What conclusions do you reach as a result of this experiment?

10.42 A building superintendent wishes to compare the relative performance ratings of various combinations of floor wax and length of polishing time. Three waxes are to be investigated along with 3 polishing times. Eighteen homogeneous floor areas are selected, and 2 are assigned at random to each of the 9 treatment combinations. Analyze and evaluate the following performance data (high ratings are better than low).

Wax:	A			B			C		
Polishing time (min):	15	30	45	15	30	45	15	30	45
	7	7.5	8.2	7	7.2	7.1	8	9.2	9.6
	8	7.4	8.6	7	7.6	7	8	9.4	9.5

10.43 An experiment was performed to assess the effects of type of material and heat treatment on the abrasive wear of bearings. Two bearings were tested at each of 10 treatment combinations. Analyze and interpret the following coded data.

Material:	A		B		C		D		E	
Heat treatment*:	O	M	O	M	O	M	O	M	O	M
	23	30	42	45	37	39	41	44	20	24
	25	31	44	50	38	39	42	49	25	30

*O = oven dried; M = moisture saturated.

10.44   From each of 5 lots of insulating material, 10 lengthwise specimens and 10 crosswise specimens are cut. The following table gives the impact strength in foot-pounds from tests on the specimens. Analyze and interpret the data.

Type of Cut	Lot Number				
	I	II	III	IV	V
Lengthwise specimens	1.15	1.16	0.79	0.96	0.49
	0.84	0.85	0.68	0.82	0.61
	0.88	1.00	0.64	0.98	0.59
	0.91	1.08	0.72	0.93	0.51
	0.86	0.80	0.63	0.81	0.53
	0.88	1.01	0.59	0.79	0.72
	0.92	1.14	0.81	0.79	0.67
	0.87	0.87	0.65	0.86	0.47
	0.93	0.97	0.64	0.84	0.44
	0.95	1.09	0.75	0.92	0.48
Crosswise specimens	0.89	0.86	0.52	0.86	0.52
	0.69	1.17	0.52	1.06	0.53
	0.46	1.18	0.80	0.81	0.47
	0.85	1.32	0.64	0.97	0.47
	0.73	1.03	0.63	0.90	0.57
	0.67	0.84	0.58	0.93	0.54
	0.78	0.89	0.65	0.87	0.56
	0.77	0.84	0.60	0.88	0.55
	0.80	1.03	0.71	0.89	0.45
	0.79	1.06	0.59	0.82	0.60

10.45   Five batches of ground meat are charged consecutively into a rotary filling machine for packing into cans. The machine has 6 filling cylinders. Three filled cans are taken from each cylinder at random while each batch is being run. The coded weights of the filled cans are given below. Analyze and interpret the data.

Cylinder	Batch 1	Batch 2	Batch 3	Batch 4	Batch 5
1	1	4	6	3	1
	1	3	3	1	3
	2	5	7	3	3
2	−1	−2	3	2	1
	3	1	1	0	0
	−1	0	5	1	1
3	1	2	2	1	3
	1	0	4	3	3
	1	1	3	3	3
4	−2	−2	3	0	0
	3	0	3	0	1
	0	1	4	2	1
5	1	2	0	1	−2
	1	1	1	0	3
	−1	5	2	−1	1
6	0	0	3	3	3
	1	0	3	0	1
	1	3	4	2	2

10.46 The following data on the density of small bricks resulted from an experiment involving 3 different sizes of powder particles, 3 pressures, and 3 temperatures of firing. The 27 combinations of this $3 \times 3 \times 3$ factorial were run in duplicate. Analyze and interpret the following coded data.

Size	Pressure	Temperature 1900		Temperature 2000		Temperature 2300	
5–10	5.0	340	375	316	386	374	350
	12.5	388	370	338	214	334	366
	20.0	378	378	348	378	380	398
10–15	5.0	260	244	388	304	266	234
	12.5	322	342	300	420	234	258
	20.0	330	298	260	366	350	284
15–20	5.0	134	140	146	194	152	212
	12.5	186	30	412	428	194	208
	20.0	40	210	436	490	230	254

10.47 During the manufacture of sheets of building material the permeability was determined for 3 sheets from each of 3 machines on each day. The table below gives the logarithms of the permeability in seconds for sheets selected from the 3 machines during a production period of 9 days. The 3 machines received their raw materials from a common store. Analyze and interpret the data.

Day	Machine	Log of Permeability		
1	1	1.404	1.346	1.618
	2	1.306	1.628	1.410
	3	1.932	1.674	1.399
2	1	1.447	1.569	1.820
	2	1.241	1.185	1.516
	3	1.426	1.768	1.859
3	1	1.914	1.477	1.894
	2	1.506	1.575	1.649
	3	1.382	1.690	1.361
4	1	1.887	1.485	1.392
	2	1.673	1.372	1.114
	3	1.721	1.528	1.371
5	1	1.772	1.728	1.545
	2	1.227	1.397	1.531
	3	1.320	1.489	1.336
6	1	1.665	1.539	1.690
	2	1.404	1.452	1.627
	3	1.633	1.612	1.359
7	1	1.918	1.931	2.129
	2	1.229	1.508	1.436
	3	1.328	1.802	1.385
8	1	1.845	1.790	2.042
	2	1.583	1.627	1.282
	3	1.689	2.248	1.795
9	1	1.540	1.428	1.704
	2	1.636	1.067	1.384
	3	1.703	1.370	1.839

REFERENCES

1. Anderson, R. L.; and Houseman, E. E. 1942. Tables of orthogonal polynomial values extended to $N = 104$. Agr. Exp. Sta. Res. Bull. 297, Iowa State University, Ames.
2. Anscombe, F. J.; and Tukey, J. W. 1957. The analysis of residuals. Unpubl. handout, Gordon Conference on Statistics in Chemistry and Chemical Engineering. New Hampton, N.H.
3. Barr, A. J.; and Goodnight, J. H. 1971. Statistical analysis system. Student Supply Store, North Carolina State University, Raleigh.
4. Bartlett, M. S. 1947. The use of transformations. *Biometrics* 3:39.
5. Bechhofer, R. E. 1958. A sequential multiple-decision procedure for selecting the best one of several normal populations with a common unknown variance, and its use with various experimental designs. *Biometrics* 14:408–29.
6. Box, G. E. P.; and Andersen, S. L. 1955. Permutation theory in the derivation of robust criteria and the study of departures from assumption. *J. Roy. Stat. Soc.* Series B 17:1–34.
7. Cochran, W. G. 1947. Some consequences when the assumptions for the analysis of variance are not satisfied. *Biometrics* 3:22.
8. _____. 1951. Testing a linear relation among variances. *Biometrics* 7:17.
9. David, H. A. 1952. Upper 5 percent and 1 percent points of the maximum *F*-ratio. *Biometrika* 39:422–24.

10. Dixon, W. J. (ed.). 1970. Biomedical computer programs. X-Series Supplement, University of California Press, Berkeley.
11. Dixon, W. J.; and Massey, F. J. 1957. *Introduction to Statistical Analysis*, 2nd ed. McGraw-Hill, New York.
12. Duncan, D. B. 1955. Multiple range and multiple *F*-tests. *Biometrics* 11:1–42.
13. ———. 1957. Multiple range tests for correlated and heteroschedastic means. *Biometrics* 13:164–76.
14. Dunnett, C. W. 1955. A multiple comparison procedure for comparing several treatments with a control. *J. Am. Stat. Assoc.* 50:1096–1121.
15. Eisenhart, C. 1947. The assumptions underlying the analysis of variance. *Biometrics* 3:1–21.
16. Federer, W. T. 1955. *Experimental Design*. Macmillan, New York.
17. Hamaker, H. C. 1955. Experimental design in industry. *Biometrics* 11:257–86.
18. Hartley, H. O. 1955. Some recent developments in analysis of variance. *Commun. Pure Appl. Math.* 8:47–72.
19. Kempthorne, O. 1952. *The Design and Analysis of Experiments*. Wiley, New York.
20. Keuls, M. 1952. The use of the "Studentized range" in connection with an analysis of variance. *Euphytica* 1:112–22.
21. Kramer, C. Y. 1956. Extension of multiple range tests to group means with unequal numbers of replications. *Biometrics* 12:307–10.
22. ———. 1957. Extension of multiple range tests to group correlated adjusted means. *Biometrics* 13:13–18.
23. Newman, D. 1939. The distribution of the range in samples from a normal population expressed in terms of an independent estimate of standard deviation. *Biometrika* 31:20–30.
24. Satterthwaite, F. E. 1946. An approximate distribution of estimates of variance components. *Biometrics* 2:110.
25. Scheffé, H. 1953. A method for judging all contrasts in the analysis of variance. *Biometrika* 40:87–104.
26. Snedecor, G. W.; and Cochran, W. G. 1967. *Statistical Methods*, 6th ed. Iowa State University Press, Ames.
27. Tukey, J. W. 1949. One degree of freedom for nonadditivity. *Biometrics* 5:232–42.
28. ———. 1949. Comparing individual means in the analysis of variance. *Biometrics* 5:99.
29. ———. 1951. Quick and dirty methods in statistics. Part II: Simple analyses for standard designs. Proc. 5th. Ann. Conv. ASQC, p. 189.
30. ———. 1952. Allowances for various types of error rates. Unpubl. material presented before the Institute of Mathematical Statistics and the Eastern North American Region of the Biometric Society, Blacksburg, Va.
31. ———. 1953. The problem of multiple comparisons. Unpubl. dittoed notes, Princeton University, Princeton, N.J.
32. ———. 1955. Reply to query number 113. *Biometrics* 11:111–13.
33. Ward, G. C.; and Dick, I. D. 1952. Nonadditivity in randomized block designs and balanced incomplete block designs. *N. Z. J. Sci. Tech.* 33:430–35.
34. Yates, F. 1937. The design and analysis of factorial experiments. Tech. Comm. No. 35, Imperial Bureau of Soil Science.

FURTHER READING

Anderson, V. L.; and McLean R. A. 1974. *Design of Experiments: A Realistic Approach.* Marcel Dekker, New York.

Bancroft, T. A. 1968. *Topics in Intermediate Statistical Methods*, Vol. 1. Iowa State University Press, Ames.

Bennett, C. A.; and Franklin, N. L. 1954. *Statistical Analysis in Chemistry and the Chemical Industry.* Wiley, New York.

Bowker, A. H.; and Lieberman, G. J. 1972. *Engineering Statistics*, 2nd ed. Prentice-Hall, Englewood Cliffs, N.J.

Brownlee, K. A. 1965. *Statistical Theory and Methodology in Science and Engineering*, 2nd ed. Wiley, New York.

Chakravarti, J. M.; Laha, R. G.; and Roy, J. 1967. *Handbook of Methods of Applied Statistics*, Vols. I and II. Wiley, New York.

Chew, V. (ed.). 1958. *Experimental Designs in Industry*. Wiley, New York.

Cochran, W. G.; and Cox, G. M. 1957. *Experimental Designs*, 2nd ed. Wiley, New York.

Cox, D. R. 1958. *Planning of Experiments*. Wiley, New York.

Davies, O. L. (ed.). 1971. *The Design and Analysis of Industrial Experiments*, 2nd ed. Hafner, New York.

Dixon, W. J.; and Massey, F. J., Jr. 1969. *Introduction to Statistical Analysis*, 3rd ed. McGraw-Hill, New York.

Dunn, O. J.; and Clark, V. A. 1974. *Applied Statistics: Analysis of Variance and Regression*. Wiley, New York.

Edwards, A. L. 1962. *Experimental Design in Psychological Research*. Holt, Rinehart and Winston, New York.

Federer, W. T. 1955. *Experimental Design*. Macmillan, New York.

Fisher, R. A. 1966. *The Design of Experiments*, 8th ed. Hafner, New York.

_____. 1946. *Statistical Methods for Research Workers*, 10th ed. Oliver and Boyd, Edinburgh.

Hicks, C. R. 1973. *Fundamental Concepts in the Design of Experiments*, 2nd ed. Holt, Rinehart and Winston, New York.

John, P. W. 1971. *Statistical Design and Analysis of Experiments*. Macmillan, New York.

Johnson, N. L.; and Leone, F. C. 1964. *Statistics and Experimental Design: In Engineering and the Physical Sciences*, Vols. I and II. Wiley, New York.

Kempthorne, O. 1952. *The Design and Analysis of Experiments*. Wiley, New York.

Mendenhall, W. 1968. *Introduction to Linear Models and the Design and Analysis of Experiments*. Wadsworth, Belmont, Calif.

Neter, J.; and Wasserman, W. 1974. *Applied Linear Statistical Models: Regression, Analysis of Variance and Experimental Design*. Irwin, Homewood, Ill.

Peng, K. E. 1967. *The Design and Analysis of Scientific Experiments*. Addison-Wesley, Reading, Mass.

Quenouille, M. H. 1953. *The Design and Analysis of Experiments*. Griffin, London.

Snedecor, G. W.; and Cochran, W. G. 1967. *Statistical Methods*, 6th ed. Iowa State University Press, Ames.

Steel, R. G. D.; and Torrie, J. H. 1960. *Principles and Procedures of Statistics*. McGraw-Hill, New York.

Wine, R. L. 1964. *Statistics for Scientists and Engineers*. Prentice-Hall, Englewood Cliffs, N.J.

Winer, B. J. 1971. *Statistical Principles in Experimental Design*, 2nd ed. McGraw-Hill, New York.

CHAPTER 11

RANDOMIZED COMPLETE BLOCK DESIGN

In this chapter the most widely used of all experimental designs, the randomized complete block design, will be discussed. The discussion will follow closely the pattern adopted in Chapter 10 with, once again, the greatest attention being given to methods of analysis.

11.1 DEFINITION OF A RANDOMIZED COMPLETE BLOCK DESIGN

A *randomized complete block* (RCB) *design* is a design in which (1) the experimental units are allocated to groups or *blocks* in such a way that the experimental units within a block are relatively homogeneous and that the number of experimental units within a block is equal to the number of treatments being investigated and (2) the treatments are assigned *at random* to the experimental units *within each block*. In the foregoing the formation of the blocks reflects the researcher's judgment as to potential differential responses from the various experimental units, while the randomization procedure acts as a justification for the assumption of independence. (See Chapters 9 and 10.)

Example 11.1

Six varieties of oats are to be compared with reference to their yields, and 30 experimental plots are available for experimentation. However, evidence is on file that indicates a fertility trend running from north to south, the northernmost plots of ground being the most fertile. Thus it seems reasonable to group the plots into 5 blocks of 6 plots each so that one block contains the most fertile plots, the next block contains the next most fertile group of plots, and so on down to the fifth (southernmost) block which contains the least fertile plots. The six varieties would then be assigned at random to the plots within each block, a new randomization being made in each block.

Example 11.2

An experiment is to be designed to study the effect of environmental temperature on the transfer time of a certain type of electrical gap. Twelve different temperatures are to be investigated. A check of the stockroom indicates that gaps are available from 6 different production lots. Since it has previously been established that gaps from different lots exhibit differerent characteristics, even when subjected to the same conditions, some blocking is desirable. Accordingly, 12 gaps are selected at random from each of the 6 production lots, and each such set of 12 gaps is hereafter referred to as a block. Then the 12 temperatures are assigned at random to the gaps within each block, a new randomization being made in each block.

Example 11.3

Ten rations are to be tested for differences in producing a gain in weight for steers. Forty steers are available for experimentation, and they are allocated to 4 blocks (10 steers per block) on the basis of their weights at the beginning of the feeding trial, with the heaviest steers being in one block, the next heaviest steers being in the second block, and so on. The 10 treatments (rations) were assigned at random to the steers within each block, as shown in Figure 11.1.

Block 1	H	B	F	A	C	I	E	J	D	G
Block 2	A	I	G	H	J	D	F	E	C	B
Block 3	E	A	C	I	B	H	D	G	J	F
Block 4	J	F	D	B	H	I	A	C	G	E

Fig. 11.1—Random arrangement of treatments as described in Example 11.3.

11.2 RANDOMIZED COMPLETE BLOCK DESIGN WITH ONE OBSERVATION PER EXPERIMENTAL UNIT

If an experiment is conducted using b blocks each consisting of t experimental units and if a single response is observed from each unit, the data can be represented as outlined in Table 11.1. As we did for the completely randomized design, we begin by introducing a model for the data derived from a randomized complete block experiment.

TABLE 11.1–Symbolic Representation of the Data in a Randomized Complete Block Design with One Observation per Experimental Unit

Block	Treatment 1	...	j	...	t	Block Total	Block Mean
1	Y_{11}	$\cdots$	Y_{1j}	$\cdots$	Y_{1t}	$B_1 = Y_{1.}$	$\overline{Y}_{1.}$
$\vdots$							
i	Y_{i1}	$\cdots$	Y_{ij}	$\cdots$	Y_{it}	$B_i = Y_{i.}$	$\overline{Y}_{i.}$
$\vdots$							
b	Y_{b1}	$\cdots$	Y_{bj}	$\cdots$	Y_{bt}	$B_b = Y_{b.}$	$\overline{Y}_{b.}$
Treatment Total Treatment Mean	$T_1 = Y_{.1}$ $\overline{Y}_{.1}$		$T_j = Y_{.j}$ $\overline{Y}_{.j}$		$T_t = Y_{.t}$ $\overline{Y}_{.t}$	$T = Y_{..}$ $\cdots$	$\overline{Y}_{..}$

11.2.1 Model

For data from an experiment involving a randomized complete block design with one observation per experimental unit, we assume that the observations may be represented by the linear statistical model

$$Y_{ij} = \mu + \rho_i + \tau_j + \epsilon_{ij} \qquad i = 1, \ldots, b; j = 1, \ldots, t \qquad (11.1)$$

where μ is the true mean effect, ρ_i is the true effect of the ith block, τ_j is the true effect of the jth treatment, and ϵ_{ij} is the true effect of the experimental unit in the ith block subjected to the jth treatment. Also, we assume $\sum_{i-1}^{b} \rho_i = 0$ and $\epsilon_{ij} \sim NI(0, \sigma^2)$. In addition, with regard to the treatment effect τ_j we may assume either (1) Model I (fixed effects), $\sum_{j-1}^{t} \tau_j = 0$, or (2) Model II (random effects), $\tau_j \sim NI(0, \sigma_\tau^2)$.

11.2.2 Analysis of Variance

The ANOVA associated with a randomized complete block design is based on the sum of squares identity

$$\sum_{i=1}^{b} \sum_{j=1}^{t} Y_{ij}^2 = bt(\overline{Y}_{..})^2 + t \sum_{i=1}^{b} (\overline{Y}_{i.} - \overline{Y}_{..})^2$$

$$+ b \sum_{j=1}^{t} (\overline{Y}_{.j} - \overline{Y}_{..})^2 + \sum_{i=1}^{b} \sum_{j=1}^{t} (Y_{ij} - \overline{Y}_{i.} - \overline{Y}_{.j} + \overline{Y}_{..})^2 \qquad (11.2)$$

which is a partition of the total sum of squares into the sum of squares attributable to the mean, the block effects, the treatment effects, and the random error respectively. This partition of the sum of squares is summarized in the generalized ANOVA in Table 11.2. The sums of

TABLE 11.2–Generalized ANOVA for a Randomized Complete Block Design with One Observation per Experimental Unit

Source of Variation	Degrees of Freedom	Sum of Squares	Mean Square	Expected Mean Square
Mean	1	M_{yy}	M	$\cdots$
Blocks	$b - 1$	B_{yy}	B	$\sigma^2 + t \sum_{i=1}^{b} \rho_i^2/(b - 1)$
Treatments	$t - 1$	T_{yy}	T	$\begin{cases} \sigma^2 + b \sum_{j=1}^{t} \tau_j^2/(t - 1) \\ \sigma^2 + b\sigma_\tau^2 \end{cases}$
Experimental error	$(b - 1)(t - 1)$	E_{yy}	E	σ^2
Total	bt	$\sum Y^2$	$\cdots$	$\cdots$

squares in Table 11.2 can be computed using the following equations:

$$\sum Y^2 = \sum_{i=1}^{b} \sum_{j=1}^{t} Y_{ij}^2 \qquad (11.3)$$

$$M_{yy} = bt(\bar{Y}_{..})^2 = T^2/bt \qquad (11.4)$$

$$B_{yy} = t \sum_{i=1}^{b} (\bar{Y}_{i.} - \bar{Y}_{..})^2 = \sum_{i=1}^{b} B_i^2/t - M_{yy} \qquad (11.5)$$

$$T_{yy} = b \sum_{j=1}^{t} (\bar{Y}_{.j} - \bar{Y}_{..})^2 = \sum_{j=1}^{t} T_j^2/b - M_{yy} \qquad (11.6)$$

$$E_{yy} = \sum_{i=1}^{b} \sum_{j=1}^{t} (Y_{ij} - \bar{Y}_{i.} - \bar{Y}_{.j} + \bar{Y}_{..})^2$$

$$= \sum Y^2 - B_{yy} - T_{yy} - M_{yy} \qquad (11.7)$$

where $B_i = Y_{i.}$, $T_j = Y_{.j}$, and $T = Y_{..}$ are the block, treatment, and overall totals respectively.

Following the line of reasoning developed in Section 10.2.2, it is easily verified that the hypothesis

$$H:\tau_j = 0(j = 1,\ldots,t) \qquad \text{(Model I)}$$
$$H:\sigma_\tau^2 = 0 \qquad \text{(Model II)}$$

may be tested using the statistic

$$F = T/E = \frac{\text{treatment mean square}}{\text{experimental error mean square}} \qquad (11.8)$$

which, if H is true, is distributed as F with $\nu_1 = t - 1$ and $\nu_2 = (b - 1)(t - 1)$ degrees of freedom. If the value of F specified by Eq. (11.8) exceeds $F_{(1-\alpha)(\nu_1,\nu_2)}$, where 100α percent is the chosen significance level, H will be rejected and the conclusion reached that there are significant differences among the t treatments.

11.2.3 Estimation of the Model Parameters

As before, we now indicate some estimates of the parameters in the model for a randomized complete block design. An estimate of the experimental error variance σ^2 is

$$s^2 = E \qquad (11.9)$$

If Model II is assumed, the variance of the treatment effects σ_τ^2 would be estimated by

$$s_\tau^2 = (T - E)/b \qquad (11.10)$$

In either case (i.e., Model I or Model II) the estimated variance of a treatment mean is given by

$$\hat{V}(\bar{Y}_{.j}) = E/b \qquad (11.11)$$

and the standard error of a treatment mean is given by

$$\sqrt{\hat{V}(\bar{Y}_{.j})} = \sqrt{E/b} = \sqrt{s^2/b} \qquad (11.12)$$

A 100γ percent confidence interval for estimating $\mu_j = \mu + \tau_j$ is then found by calculating

$$\left.\begin{array}{c} L \\ U \end{array}\right\} = \bar{Y}_{.j} \mp t_{[(1+\gamma)/2](\nu)} \sqrt{E/b} \qquad (11.13)$$

where $\nu = (b - 1)(t - 1)$.

Example 11.4

The experiment described in Example 11.3 was performed and the data in Table 11.3 were obtained. Use of Eqs. (11.3)–(11.7) yields the ANOVA shown in Table 11.4. Because the F ratio is significant, we reject $H{:}\tau_j = 0$ ($j = 1, \ldots, 10$) and decide that in all likelihood the 10 treatments (rations) are not equally effective in producing a weight gain on steers. The treatment means and their standard error, $\sqrt{3.43/4} = 0.93$, may then be used to determine the best treatment or treatments and to indicate the direction future research should take.

TABLE 11.3–Gains in Weight (lb) of Forty Steers
Fed Different Rations

Treat- ment	Block				Treatment Total	Treatment Mean
	1	2	3	4		
A	2	3	3	5	13	3.25
B	5	4	5	5	19	4.25
C	8	7	10	9	34	8.50
D	6	5	5	2	18	4.50
E	1	2	1	2	6	1.50
F	3	5	7	8	23	5.75
G	8	8	7	8	31	7.75
H	6	12	2	5	25	6.25
I	4	5	6	3	18	4.50
J	4	4	2	3	13	3.25
Block Total	47	55	48	50	200	. . .
Block Mean	4.7	5.5	4.8	5.0	. . .	5.0

Note: Data coded for easy calculation.

To indicate further estimates, suppose

$$\theta_k = \sum_{j=1}^{t} c_{jk}\mu_j \qquad k = 1, \ldots, m \qquad (11.14)$$

is a set of m linear combinations of the treatment means. An individual

TABLE 11.4–ANOVA for Experiment Described in Example 11.3 and Discussed in Example 11.4 (data in Table 11.3)

Source of Variation	Degrees of Freedom	Sum of Squares	Mean Square	Expected Mean Square	F Ratio
Mean	1	1000.0	1000.00	...	...
Blocks	3	3.8	1.26	$\sigma^2 + (10/3) \sum_{i=1}^{4} \rho_i^2$	...
Treatments	9	163.5	18.17	$\sigma^2 + (4/9) \sum_{j=1}^{10} \tau_j^2$	5.29**
Experimental error	27	92.7	3.43	σ^2	...
Total	40	1260.0	...	...	...

**Significant at $\alpha = 0.01$.

100γ percent confidence interval for θ_k is given by the limits

$$\left. \begin{matrix} L \\ U \end{matrix} \right\} = \hat{\theta}_k \mp t_{[(1+\gamma)/2](\nu)} \left(\mathrm{E} \sum_{j=1}^{t} c_{jk}^2 / b \right)^{1/2} \tag{11.15}$$

where $\hat{\theta}_k = \sum_{j=1}^{t} c_{jk} \overline{Y}_{.j}$ and $\nu = (b-1)(t-1)$. Simultaneous 100γ percent confidence limits for the set of θ_k's are given by

$$\left. \begin{matrix} L_k \\ U_k \end{matrix} \right\} = \hat{\theta}_k \mp \left[(t-1) F_{\gamma(\nu_1,\nu_2)} \mathrm{E} \sum_{j=1}^{t} c_{jk}^2 / b \right]^{1/2} \tag{11.16}$$

where $\nu_1 = t - 1$ and $\nu_2 = (b-1)(t-1)$.

Before setting up the analysis of a randomized complete block design as a regression analysis, one point needs discussion. That is, why do we not test $H':\rho_i = 0 (i = 1, \ldots, b)$? Examination of Table 11.2 will show that the expected mean square for blocks is of the same form as the expected mean square for treatments, and this suggests that a logical procedure would be to test H' by calculating $F = \mathbf{B}/\mathbf{E}$. Why is it then that the statistician says this should not be done? The answer may be found by noting the manner in which the randomization was performed. You will recall that the treatments were assigned at random to the experimental units within each block *but that the blocks were formed in a decidedly nonrandom fashion.* Because of this feature of the randomized complete block design, *a statistical test of the block effect should not be performed.* [Note: In some cases where "blocks" are replaced by "replications" and where the replications may be considered as random samples of all possible replications (i.e., Model II with respect to replications), an F test for replications may be appropriate. However, even in such a case the F test would be of less importance than the estimation of the component of variance for replications σ_ρ^2.]

11.2.4 Randomized Complete Block Model as a Regression Model

We now set up the linear statistical model in Eq. (11.1) as a multiple linear regression model in the same way as we did for the completely randomized model in Section 10.2.5. For the completely randomized model we had to introduce $t - 1$ independent variables associated with the first $t - 1$ treatment effects. For the randomized complete block design we must introduce a set of independent variables for the block effects as well as the treatment effects. We distinguish between these two sets of variables by using different symbols, namely, u and x respectively. The model in Eq. (11.1), written as a regression model, is

$$Y_{ij} = \mu + \rho_1 u_{1i} + \cdots + \rho_{b-1} u_{b-1i} + \tau_1 x_{1j} + \cdots + \tau_{t-1} x_{t-1j} + \epsilon_{ij}$$

$$i = 1, \ldots, b; j = 1, \ldots, t \qquad (11.17)$$

where
$$
\begin{aligned}
u_{ki} &= 1 &&\text{if } i = k; k = 1, \ldots, b - 1 \\
&= -1 &&\text{if } i = b \\
&= 0 &&\text{otherwise} \\
x_{kj} &= 1 &&\text{if } j = k; k = 1, \ldots, t - 1 \\
&= -1 &&\text{if } j = t \\
&= 0 &&\text{otherwise}
\end{aligned}
$$

As discussed in Section 10.2.5, to ensure that a unique solution for the unknown parameters exists, we must put restrictions on the parameters. The restrictions introduced in arriving at Eq. (11.17) were $\sum_{i=1}^{b} \rho_i = \sum_{j=1}^{t} \tau_j = 0$.

To illustrate Eq. (11.17), suppose $b = 3$ and $t = 3$; i.e., we have three treatments in three blocks. The equations for the nine observations, written in the matrix notation,

$$\mathbf{Y} = \mathbf{X}\mathbf{B} + \epsilon \qquad (11.18)$$

are

$$
\begin{bmatrix}
Y_{11} \\
Y_{12} \\
Y_{13} \\
Y_{21} \\
Y_{22} \\
Y_{23} \\
Y_{31} \\
Y_{32} \\
Y_{33}
\end{bmatrix}
=
\begin{bmatrix}
1 & 1 & 0 & 1 & 0 \\
1 & 1 & 0 & 0 & 1 \\
1 & 1 & 0 & -1 & -1 \\
1 & 0 & 1 & 1 & 0 \\
1 & 0 & 1 & 0 & 1 \\
1 & 0 & 1 & -1 & -1 \\
1 & -1 & -1 & 1 & 0 \\
1 & -1 & -1 & 0 & 1 \\
1 & -1 & -1 & -1 & -1
\end{bmatrix}
\begin{bmatrix}
\mu \\
\rho_1 \\
\rho_2 \\
\tau_1 \\
\tau_2
\end{bmatrix}
+
\begin{bmatrix}
\epsilon_{11} \\
\epsilon_{12} \\
\epsilon_{13} \\
\epsilon_{21} \\
\epsilon_{22} \\
\epsilon_{23} \\
\epsilon_{31} \\
\epsilon_{32} \\
\epsilon_{33}
\end{bmatrix}
\qquad (11.19)
$$

The sums of squares attributable to the block and treatment effects are part of the output of a regression analysis. Using the standard notation of Chapter 7, it can be shown that

$$M_{yy} = SS(\mu) \tag{11.20}$$

$$B_{yy} = SS(\rho_1, \ldots, \rho_{b-1} \mid \mu) = SS(\rho_1, \ldots, \rho_{b-1} \mid \mu, \tau_1, \ldots, \tau_{t-1}) \tag{11.21}$$

$$T_{yy} = SS(\tau_1, \ldots, \tau_{t-1} \mid \mu) = SS(\tau_1, \ldots, \tau_{t-1} \mid \mu, \rho_1, \ldots, \rho_{b-1}) \tag{11.22}$$

$$E_{yy} = E_{yy} \quad \text{(residual sum of squares)} \tag{11.23}$$

Thus all the entries in the ANOVA in Table 11.2 are easily retrievable from regression analysis. The identities in Eqs. (11.21) and (11.22) are a consequence of the orthogonality of the columns corresponding to the ρ's and the τ's respectively.

A method used more frequently to evaluate the sum of squares is a generalized ANOVA program or a factorial program. Both these methods were discussed in Chapter 10. With regard to the use of a factorial program to evaluate the block sum of squares B_{yy} for a randomized complete block design, the block effect ρ must be treated as a factor. In the randomized complete block model there would be no interaction between the two factors, i.e., treatments and blocks.

11.3 TUKEY'S ONE DEGREE OF FREEDOM TEST FOR NONADDITIVITY

Inherent in the analysis of the data from a randomized complete block experiment has been the important assumption that there is no interaction between the treatment and block effects. That is, we have assumed that the differences among treatments are the same for all blocks. In terms of the statistical model in Eq. (11.1) the assumption is that the block and treatment effects are *additive.*

Although it was necessary to make this assumption to analyze the data as outlined in Section 11.2, this assumption may not always be valid. There may be some experimental situations in which the additivity of the block and treatment effects does not hold. A method called Tukey's test for nonadditivity is available for testing for the existence of nonadditivity. Although the test applies to any cross-classified data when there is no additivity, we introduce it in the context of the randomized complete block design.

The method is based on partitioning the experimental error sum of squares into a term attributable to nonadditivity and a remainder sum of squares. The analysis is summarized in Table 11.5. The formulas for computing N_{yy} and R_{yy} are

$$N_{yy} = \left(\sum_{i=1}^{b} \sum_{j=1}^{t} W_{ij} Y_{ij} \right)^2 \bigg/ \sum_{i=1}^{b} \sum_{j=1}^{t} W_{ij}^2 \tag{11.24}$$

where

$$W_{ij} = (\bar{Y}_{i.} - \bar{Y}_{..})(\bar{Y}_{.j} - \bar{Y}_{..}) \tag{11.25}$$

$$R_{yy} = E_{yy} - N_{yy} \tag{11.26}$$

TABLE 11.5–ANOVA for a Randomized Complete Block Design
Including the Test for Nonadditivity

Source of Variation	Degrees of Freedom	Sum of Squares	Mean Square
Mean	1	M_{yy}	...
Blocks	$b - 1$	B_{yy}	B
Treatments	$t - 1$	T_{yy}	T
Nonadditivity	1	N_{yy}	N
Residual	$(b - 1)(t - 1) - 1$	R_{yy}	R
Total	bt	$\sum Y^2$	...

The test for nonadditivity is based on the statistic

$$F = N/R = \frac{\text{nonadditivity mean square}}{\text{residual mean square}} \qquad (11.27)$$

which, if the block and treatment effects are additive, has an F distribution with $\nu_1 = 1$ and $\nu_2 = (b - 1)(t - 1) - 1$ degrees of freedom. If the value of F in Eq. (11.27) exceeds $F_{(1-\alpha)(\nu_1, \nu_2)}$ for a 100α percent significance level, we reject the hypothesis of no difference in the treatment effects in the different blocks. This suggests that we should look at the treatment differences within each block separately.

Example 11.5

Consider the data in Example 11.4 and suppose the experimenter is interested in testing for nonadditivity of the block and treatment effects. Incorporating Tukey's test for nonadditivity into the analysis results in the ANOVA in Table 11.6. We see that $F = N/R = 2.10/3.48 < 1$, so there is no reason to believe that the treatment differences are not constant over blocks.

In terms of modeling the randomized complete block data by the regression model in Eq. (11.14), the test for nonadditivity can be in-

TABLE 11.6–ANOVA for Experiment Described in Example 11.4
Including the Test for Nonadditivity

Source of Variation	Degrees of Freedom	Sum of Squares	Mean Square
Mean	1	1000.0	...
Blocks	3	3.8	1.26
Treatments	9	163.5	18.17
Nonadditivity	1	2.1	2.10
Residual	26	90.6	3.48
Total	40	1260.0	...

corporated into the analysis by adding an additional independent variable.

$$Z_{ij} = (\overline{Y}_{i.} - \overline{Y}_{..})(\overline{Y}_{.j} - \overline{Y}_{..}) \qquad (11.28)$$

in the model. Thus the regression model becomes

$$Y_{ij} = \mu + \rho_1 u_{1i} + \cdots + \rho_{b-1} u_{b-1i} + \tau_1 x_{1j}$$
$$+ \cdots + \tau_{t-1} x_{t-1j} + \eta Z_{ij} + \epsilon_{ij} \qquad i = 1, \ldots, b; j = 1, \ldots, t$$

$$(11.29)$$

Since Z_{ij} is orthogonal to all the u's and all the x's, the ANOVA is easily partitioned as given in Table 11.5.

11.4 RELATION BETWEEN A RANDOMIZED COMPLETE BLOCK DESIGN AND "STUDENT'S" t TEST OF $H:\mu_D = 0$ WHEN PAIRED OBSERVATIONS ARE AVAILABLE

In Section 6.4, procedures were given for testing the hypothesis $H:\mu_1 = \mu_2$ for three different cases. In Section 10.3 it was stated that the analysis of a completely randomized design was equivalent to one of these, namely, Case I(1). In this section we show the equivalence of the analysis of a randomized complete block design with two treatments and "Student's" t test of $H:\mu_1 = \mu_2$ when paired observations are available, i.e., to Case II. The crucial step is to note the equivalence of "pairs" and "blocks." Once this association of terms is made, the equivalence of the techniques may easily be demonstrated. Rather than burden the reader with the details of the algebraic proof of the equivalence, we will rely on the "power of an example" to convince him of the truth of our claim. (**Note:** The reader should also reflect on the obvious connection between the material of this section and the contents of Section 8.13.)

Example 11.6

Consider again the experiment described in Example 6.13 and the data presented in Table 6.6. Denoting pairs (samples) by blocks and utilizing Eqs. (11.3)–(11.7), the ANOVA of Table 11.7 is obtained. It is seen that the F value is significant at $\alpha = 0.05$, and this permits us to reject the hypothesis that the two treatments (i.e., two different steel balls) are doing an equivalent job. (**Note:** $F = 7.89 = t^2 = (2.81)^2$; see Example 6.13.) It may be verified that the two treatment means are 54.2 and 46.2 respectively. Also, the standard error of a treatment mean is determined to be $\sqrt{60.8/15} = 2.01$.

11.5 SUBSAMPLING

When subsampling is employed in a randomized complete block design, the appropriate statistical model is

$$Y_{ijk} = \mu + \rho_i + \tau_j + \epsilon_{ij} + \eta_{ijk}$$
$$i = 1, \ldots, b; j = 1, \ldots, t; k = 1, \ldots, n \qquad (11.30)$$

TABLE 11.7–ANOVA for Experiment Described in Examples 6.13 and 11.6 (data in Table 6.6)

Source of Variation	Degrees of Freedom	Sum of Squares	Mean Square	Expected Mean Square	F Ratio
Mean	1	75,601.2	75,601.2	...	...
Blocks	14	2,649.8	189.3	$\sigma^2 + (2/14) \sum_{i=1}^{15} \rho_i^2$	...
Treatments	1	480.0	480.0	$\sigma^2 + (15/1) \sum_{j=1}^{2} \tau_j^2$	7.89*
Experimental error	14	851.0	60.8	σ^2	...
Total	30	79,582.0	...	...	...

*Significant at $\alpha = 0.05$.

where η_{ijk} is the error associated with the kth subsample taken from the experimental unit in the ith block which was subjected to the jth treatment. We assume that the experimental errors $\epsilon_{ij} \sim \text{NI}(0, \sigma^2)$ and the sampling errors $\eta_{ijk} \sim \text{NI}(0, \sigma_\eta^2)$. We also have assumed that an equal number n of subsamples are taken from each unit. The ANOVA is summarized in Table 11.8.

TABLE 11.8–Generalized ANOVA for a Randomized Complete Block Design with n Samples per Experimental Unit

Source of Variation	Degrees of Freedom	Sum of Squares	Mean Square	Expected Mean Square
Mean	1	M_{yy}	M	...
Blocks	$b - 1$	B_{yy}	B	$\sigma_\eta^2 + n\sigma^2 + tn \sum_{i=1}^{b} \rho_i^2/(b - 1)$
Treatments	$t - 1$	T_{yy}	T	$\begin{cases} \sigma_\eta^2 + n\sigma^2 + bn \sum_{j=1}^{t} \tau_j^2/(t - 1) \\ \sigma_\eta^2 + n\sigma^2 + bn\sigma_\tau^2 \end{cases}$
Experimental error	$(b - 1)(t - 1)$	E_{yy}	E	$\sigma_\eta^2 + n\sigma^2$
Sampling error	$bt(n - 1)$	S_{yy}	S	σ_η^2
Total	btn	$\sum Y^2$	...	...

The sums of squares in Table 11.8 can be computed using the formulas:

$$\sum Y^2 = \sum_{i=1}^{b} \sum_{j=1}^{t} \sum_{k=1}^{n} Y_{ijk}^2 \tag{11.31}$$

$$M_{yy} = btn(\overline{Y}_{...})^2 = T^2/btn \tag{11.32}$$

$$B_{yy} = tn \sum_{i=1}^{b} (\overline{Y}_{i..} - \overline{Y}_{...})^2 = \sum_{i=1}^{b} B_i^2/tn - M_{yy} \tag{11.33}$$

$$T_{yy} = bn \sum_{j=1}^{t} (\overline{Y}_{.j.} - \overline{Y}_{...})^2 = \sum_{j=1}^{t} T_j^2/bn - M_{yy} \tag{11.34}$$

$$E_{yy} = n \sum_{i=1}^{b} \sum_{j=1}^{t} (\overline{Y}_{ij.} - \overline{Y}_{i..} - \overline{Y}_{.j.} + \overline{Y}_{...})^2$$

$$= \sum_{i=1}^{b} \sum_{j=1}^{t} T_{ij}^2/n - B_{yy} - T_{yy} - M_{yy} \tag{11.35}$$

$$S_{yy} = \sum_{i=1}^{b} \sum_{j=1}^{t} \sum_{k=1}^{n} (Y_{ijk} - \overline{Y}_{ij.})^2 = \sum Y^2 - \sum_{i=1}^{b} \sum_{j=1}^{t} T_{ij}^2/n \tag{11.36}$$

where $B_i = Y_{i..}$ and $T_j = Y_{.j.}$ are the block and treatment totals respectively. Also, $T_{ij} = Y_{ij.}$ is the sum of the observations in the ith block that were subjected to the jth treatment and $T = Y_{...}$ is the total of all observations.

To test for the significance of the treatment effects, i.e., to test the hypothesis,

$$H:\tau_j = 0(j = 1, \ldots, t) \qquad \text{(Model I)}$$
$$H:\sigma_\tau^2 = 0 \qquad\qquad \text{(Model II)}$$

the appropriate test statistic is

$$F = T/E = \frac{\text{treatment mean square}}{\text{experimental error mean square}} \tag{11.37}$$

which under H has an F distribution with $\nu_1 = t - 1$ and $\nu_2 = (b - 1)(t - 1)$ degrees of freedom.

Example 11.7

An experiment was performed to assess the relative effects of 5 fertilizers on the yield of a certain variety of oats. The location of the 30 experimental plots available for use in the experiment was such that it seemed advisable to group the plots into 6 blocks of 5 plots each. The treatments were then randomly assigned to the plots within each block. At the end of the growing season the researcher decided to harvest (for purposes of analysis) only 3 sample quadrats from each plot. The data of Table 11.9 were obtained and these led to the ANOVA shown in Table 11.10. It is noted that the 5 fertilizer means are significantly different, and thus it is most important that the proper fertilizer be recommended to the farmer. A tabulation of the fertilizer means together with the appropriate standard error would be of great help in reaching the correct decision.

The estimates of the variance components are similar to the com-

TABLE 11.9–Coded Values of Yields from Ninety Sample Quadrats

Block	Fertilizer Treatment				
	1	2	3	4	5
1	57	67	95	102	123
	46	72	90	88	101
	28	66	89	109	113
2	26	44	92	96	93
	38	68	89	89	110
	20	64	106	106	115
3	39	57	91	102	112
	39	61	82	93	104
	43	61	98	98	112
4	23	74	105	103	120
	36	47	85	90	101
	18	69	85	105	111
5	48	61	78	99	113
	35	60	89	87	109
	48	75	95	113	111
6	50	68	85	117	124
	37	65	74	93	102
	19	61	80	107	118

TABLE 11.10–ANOVA for Data of Table 11.9 (discussion in Example 11.7)

Source of Variation	Degrees of Freedom	Sum of Squares	Mean Square	Expected Mean Square	F Ratio
Mean	1	573,921.88	573,921.88	...	...
Blocks	5	354.19	70.84	$\sigma_\eta^2 + 3\sigma^2 + (15/5) \sum_{i=1}^{6} \rho_i^2$	...
Fertilizers	4	65,246.84	16,311.71	$\sigma_\eta^2 + 3\sigma^2 + (18/4) \sum_{j=1}^{5} \tau_j^2$	220.61**
Experimental error	20	1,478.76	73.94	$\sigma_\eta^2 + 3\sigma^2$	...
Sampling error	60	5,283.33	88.06	σ_η^2	...
Total	90	646,285.00	...	...	...

**Significant at $\alpha = 0.01$.

pletely randomized design model with subsampling. These are

$$s_\eta^2 = S \tag{11.38}$$

$$s^2 = (E - S)/n \tag{11.39}$$

and for Model II,

$$s_\tau^2 = (T - E)/bn \tag{11.40}$$

where s^2 or s_τ^2 are taken to be zero if the differences in Eqs. (11.39) or (11.40) are negative. The estimated variance of a treatment mean is

$$\hat{V}(\bar{Y}_{.j.}) = E/bn \tag{11.41}$$

11.6 PRELIMINARY TESTS OF SIGNIFICANCE

The use of preliminary tests of significance could have been discussed in Chapter 10 or could be deferred until later. However, Example 11.7 brought it to our attention, and thus we shall consider it here.

Some practitioners suggest that when (as in Example 11.7) the experimental error mean square is less than the sampling error mean square, the two sums of squares and their degrees of freedom should be pooled. (The same suggestion, i.e., to pool, is also frequently made when the experimental error mean square exceeds, but not significantly, the sampling error mean square.) That is, a pooled sum of squares $(E_{yy} + S_{yy})$ is divided by the pooled degrees of freedom $[(b - 1)(t - 1) + bt(n - 1)]$, and this new mean square is then used as the denominator in the F ratio for testing $H:\tau_j = 0$ $(j = 1, \ldots, t)$. If such a procedure is followed, the statistical test of $H:\tau_j = 0$ $(j = 1, \ldots, t)$ will be based on a preceding or preliminary test of significance, the hypothesis $H':\sigma^2 = 0$ being tested by the preliminary test. Because such procedures are sometimes followed, we must make certain that we understand their advantages and disadvantages.

Problems of the above type have been investigated by Paull (2), and it will be profitable to review his conclusions and recommendations. To make the exposition easier to follow, we shall tie it in with Table 11.8. Suppose the experimenter decides he will *always pool* the two mean squares as indicated in the preceding paragraph; i.e., he will never perform a preliminary test of significance concerning $H':\sigma^2 = 0$. If σ^2 does equal 0, this procedure is fine. But suppose $\sigma^2 > 0$; then the denominator in the final F test (of $H:\tau_j = 0$ for all j) tends to be too small. Thus in such a situation the final F test tends to produce too many significant results when the null hypothesis H is really true. This is bad, for it implies that, quoting Paull (2, p. 541), "a test which the research worker thinks is being made at the 5 per cent level might actually be at, say, the 47 per cent level."

The use of a preliminary test of significance is clearly an attempt to guard against such a possibility. It will not, of course, eliminate such occurrences entirely. To be useful, however, it should keep the actual (effective) significance level achieved by the final (or dependent) F test close to the value at which the research worker desires to operate. Another property that should be required of a preliminary test is that it should increase the *power* of the final F test relative to the power of a "never pool" test. The recommendations for the performance of preliminary tests of significance for pooling mean squares in the ANOVA as formulated by Paull may be found in the reference quoted. However, if the research worker follows the rule of "never pooling," he will not go far wrong; and that is the rule we shall adopt in this text.

11.7 RELATIVE EFFICIENCY

The problem of predicting relative efficiency (RE) is no different for a randomized complete block design than it was for a completely randomized design. Thus the relative efficiencies of various allocations within a randomized complete block design are, as in the case for completely randomized designs, determined by evaluating the variances of a treatment mean.

It can be shown that for a design using b blocks and n samples per experimental unit the estimated variance of a treatment mean is

$$\hat{V}(\overline{Y}_{.j.}) = \frac{\text{experimental error mean square}}{\text{number of observations per treatment}}$$

$$= \mathbf{E}/bn = (s_\eta^2 + ns^2)/bn \tag{11.42}$$

where s^2 and s_η^2 are the estimates of the variance components σ^2 and σ_η^2 respectively. If the estimates of the components of variance remain unchanged, the efficiency of this design relative to one in which we might use b' blocks and n' samples per experimental unit would be predicted by

$$\text{RE of old to new} = 100[\,\hat{V}'(\overline{Y}_{.j.})/\hat{V}(\overline{Y}_{.j.})\,]\,\text{percent} \tag{11.43}$$

where

$$\hat{V}'(\overline{Y}_{.j.}) = (s_\eta^2 + n's^2)/b'n' \tag{11.44}$$

Example 11.8

Consider Table 11.11. It is easily seen that $s_\eta^2 = 10$ and $s^2 = 12$, yielding $\hat{V}(\overline{Y}_{.j.}) = 58/24$. To determine the efficiency of the design used relative to one involving 4 blocks and 6 samples per experimental unit, we first calculate $\hat{V}'(\overline{Y}_{.j.}) = [10 + 6(12)]/4(6) = 82/24$. Thus the estimated relative efficiency is $100(82/24)/(58/24) = 141$ percent.

The ideas of relative efficiency may easily be extended to cases involving many stages of subsampling. No examples will be given, but several of the problems at the end of the chapter will provide the necessary practice in the manipulations.

In some instances, the investigator wishes to estimate the efficiency of his use of a randomized complete block design relative to what might have happened if the treatments had been completely randomized over all the experimental units. That is, he wishes to know if he gained or

TABLE 11.11–Abbreviated ANOVA on Yields of Ten Varieties of Soybeans

Source of Variation	Degrees of Freedom	Sum of Squares	Mean Square	Expected Mean Square
Blocks	5	3000	600	$\sigma_\eta^2 + 4\sigma^2 + (40/5)\sum_{i=1}^{6} \rho_i^2$
Varieties	9	4500	500	$\sigma_\eta^2 + 4\sigma^2 + (24/4)\sum_{j=1}^{10} \tau_j^2$
Experimental error	45	2610	58	$\sigma_\eta^2 + 4\sigma^2$
Sampling error	180	1800	10	σ_η^2

lost in efficiency by grouping the experimental units into homogeneous groups (blocks). One method of comparing the efficiency of different designs is by use of *uniformity* data. Cochran (1) has discussed this particular approach, and the reader is referred to his article for further details. A second method of comparing efficiencies is to consider algebraically what might have happened to the experimental error mean square under complete randomization. To accomplish this, it is convenient to proceed as though *dummy* treatments had been applied to the experimental units. That is, we suppose that all experimental units were subjected to the same treatment (i.e., none) and then proceed to estimate what the experimental error mean square would have been under complete randomization. Since there is no treatment effect and any block effect is being attributed to experimental error, the estimated experimental error mean square for a completely randomized design is $[(b - 1)\mathbf{B} + b(t - 1)\mathbf{E}]/(bt - 1)$, where $\mathbf{B}$ and $\mathbf{E}$ refer to the mean squares (in the randomized complete block design) for blocks and experimental error respectively. Defining the efficiency of a randomized complete block design relative to a completely randomized design by

$$\text{RE} = \frac{\textit{estimated} \text{ experimental error mean square for a CR design}}{\text{experimental error mean square from the RCB design}}$$
(11.45)

it can be shown that

$$\text{RE} = \frac{(b - 1)\mathbf{B} + b(t - 1)\mathbf{E}}{(bt - 1)\mathbf{E}}$$
(11.46)

Example 11.9

Consider the ANOVA presented in Table 11.4. In this case, the efficiency of the randomized complete block design relative to a completely randomized design is estimated to be

$$\text{RE} = [3(1.26) + 4(9)(3.43)]/39(3.43) = 0.95$$

It is seen that, because of the small magnitude of $\mathbf{B}$ relative to $\mathbf{E}$, no appreciable gain in efficiency resulted from the formulation of the blocks and the use of a randomized complete block analysis. That is, apart from the "insurance" feature of the RCB design, the added effort was not worthwhile.

Example 11.10

The data in Table 11.12 resulted from a particular manufacturing operation, the operation being performed by one of 4 different machines. The data were collected on 5 different days, hereafter referred to as blocks. Calculations yielded the abbreviated ANOVA shown in Table 11.13. Proceeding according to Eq. (11.46), the randomized complete block design is estimated to be 131 percent as efficient as a completely randomized design would have been.

TABLE 11.12–Output from Four Machines Producing Part No. Z-15

Day	Machine			
---	A	B	C	D
1	293	308	323	333
2	298	353	343	363
3	280	323	350	368
4	288	358	365	345
5	260	343	340	330

Note: Output = number of units produced in one day.

TABLE 11.13–Abbreviated ANOVA for Data of Table 11.12

Source of Variation	Degrees of Freedom	Sum of Squares	Mean Square
Blocks	4	2,146.2	536.55
Treatments	3	13,444.8	4481.60**
Experimental error	12	2,626.2	218.85

**Significant at $\alpha = 0.01$.

11.8 TREATMENT COMPARISONS AND CONTRASTS

It is often desirable to make certain specific comparisons involving a selected number of the treatments. For a completely randomized design, such comparisons were discussed in Sections 10.8 and 10.9. In this section the same general topic will be examined in conjunction with a randomized complete block design.

A moment's thought should be sufficient to convince the reader that the calculations will be performed in the same manner as indicated earlier. For example, if a randomized complete block design with one observation per experimental unit is involved, the sum of squares for a particular *contrast* C_k would be given by

$$(C_k)_{yy} = \left(\sum_{j=1}^{t} c_{jk} T_j \right)^2 \bigg/ b \sum_{j=1}^{t} c_{jk}^2 \qquad (11.47)$$

where the c_{jk} are the coefficients specifying the contrast. As before, if $t - 1$ orthogonal contrasts are studied, the sum of the $t - 1$ individual sums of squares will equal the treatment sum of squares.

Example 11.11

Upon reading the complete description of the project referred to in Example 11.10, certain additional information about the four machines is brought to light. For example, machine A is the standard type of machine now in use in the industry, while machines B, C, and D are new designs that may be considered as possible substitutes. Further, it is known that B and C contain moving parts made of some aluminum alloy, while D does not have this feature. Also known from the manufacturers' specifications is the fact that B is self-lubricating, while C is not. Therefore, the comparisons represented symbolically in Table 11.14 seem to be indicated. Partitioning of the treatment sum of squares is then carried out using Eq. (11.47), and the abbreviated ANOVA of Table 11.15 is obtained.

The same comments hold true for the problem of making all possible comparisons among treatment means. These are handled in exactly the same manner as similar comparisons in a completely randomized de-

TABLE 11.14–Symbolic Representation of the Selected Treatment Comparisons Described in Example 11.11 (data of Table 11.12)

	Machine			
Comparison	A	B	C	D
1	3	-1	-1	-1
2	0	1	1	-2
3	0	-1	1	0

TABLE 11.15–Abbreviated ANOVA for Data of Table 11.12 Showing the Subdivision of the Treatment Sum of Squares

Source of Variation	Degrees of Freedom	Sum of Squares	Mean Square
Blocks	4	2,146.2	536.55
Treatments			
A vs. rest	1	13,142.4	13,142.4**
B and C vs. D	1	172.8	172.8
B vs. C	1	129.6	129.6
Experimental error	12	2,626.2	218.85

**Significant at $\alpha = 0.01$.

sign. The reader therefore is referred to Section 10.10 for the appropriate details.

11.9 SUBDIVISION OF THE EXPERIMENTAL ERROR SUM OF SQUARES WHEN CONSIDERING SELECTED TREATMENT COMPARISONS

Before proceeding to the next general topic connected with randomized complete block designs, we will be concerned with the possibility of partitioning the experimental error sum of squares. (**Note:** You will recall that this topic was mentioned in the last paragraph of Section 10.13.)

When is such a procedure in order? That is, when should the experimental error sum of squares be subdivided? The reason for subdividing E_{yy} (if such a procedure is adopted) is that we are not satisfied with our assumption of homogeneous variances of the ϵ's. If such an assumption is questioned (its validity may be investigated using Bartlett's test) and if selected treatment comparisons are being examined, it is desirable to subdivide E_{yy} in a manner similar to the subdivision of T_{yy}. Such a procedure insures that any particular treatment comparison will be tested against the appropriate error. That is, the expected value of the "error mean square for testing C_k" will contain the same components of variance (other than the treatment effects) as the expected value of the mean square associated with C_k. In other words, if we are faced with different variances σ_{ij}^2 ($i = 1, \ldots, b$; $j = 1, \ldots, t$), the procedure of subdividing E_{yy} will insure that the expected mean squares for a particular comparison and its associated error will each contain the *same* linear combination of the σ_{ij}^2. This provides us with unbiased tests for the comparisons under investigation.

Since the decision to implement the procedure (yet to be described) for subdividing E_{yy} is one everyone will have to make at some time or other, some guiding rule is needed. The following appears to be a reasonable rule: If there is any serious doubt as to the homogeneity of the variances, subdivide the experimental error sum of squares as a precautionary measure. It should be noted, though, that if the degrees of freedom associated with the various parts of E_{yy} are small, the resulting tests may be relatively insensitive (i.e., of low discriminatory power). In practice both the following conditions are usually true:

1. The degree of heterogeneity among the error variances is, as a rule, not too great. Therefore, for most practical purposes the variances may be considered homogeneous.
2. The numbers of degrees of freedom associated with the parts into which the experimental error sum of squares is subdivided are generally quite small.

Consequently, the rule stated above should be modified to read: *Because of the truth of statements (1) and (2) above, it is generally not wise to subdivide the experimental error sum of squares. However, if the heterogeneity of variances is such that a subdivision is necessary (regard-*

less of the fact that small numbers of degrees of freedom will result), the subdivision should be carried out in accordance with the procedure to be explained in the next paragraph.

The method of subdividing the experimental error sum of squares in agreement with a particular subdivision of the treatment sum of squares is as follows:

1. Set up a table showing the values of the contrasts *within each block*.
2. Calculate the portion of the experimental error sum of squares for a particular contrast using

$$(E_k)_{yy} = \text{experimental error sum of squares for } C_k$$

$$= \left[\sum_{i=1}^{b} C_{ki}^2 - C_k^2/b \right] \bigg/ \sum_{j=1}^{t} c_{jk}^2 \tag{11.48}$$

where

$$C_{ki} = \sum_{j=1}^{t} c_{jk} Y_{ij} \tag{11.49}$$

and

$$C_k = \sum_{j=1}^{t} c_{jk} T_j = \sum_{i=1}^{b} C_{ki} \tag{11.50}$$

Example 11.12

Consider the experiment discussed in Examples 11.10 and 11.11. Using Eqs. (11.49) and (11.50) in conjunction with Tables 11.12 and 11.14, we obtain Table 11.16. Then, using Eq. (11.48), we get, for example,

$(E_3)_{yy}$ = experimental error sum of squares associated with C_3

= experimental error sum of squares associated with the comparison "*B* vs. *C*"

$= [(15)^2 + (-10)^2 + (27)^2 + (7)^2 + (-3)^2 - (36)^2/5]/2 = 426.4$

TABLE 11.16–Sums for the Selected Treatment Comparisons in Each Block (data of Table 11.12)

Block	Comparison		
	C_1	C_2	C_3
1	− 85	−35	15
2	−165	−30	− 10
3	−201	−63	27
4	−204	33	7
5	−233	23	− 3
Total	−888	−72	36

TABLE 11.17–Abbreviated ANOVA for Data of Table 11.12 Showing
the Subdivision of the Experimental Error Sum of Squares

Source of Variation	Degrees of Freedom	Sum of Squares	Mean Square	Expected Mean Square*
Blocks	4	2,146.20	536.55	$\sigma^2 + (4/4)\sum_{i=1}^{5} \rho_i^2$
Treatments				
A vs. rest	1	13,142.40	13,142.40	$\sigma^2 + (25/60)(3\tau_1 - \tau_2 - \tau_3 - \tau_4)^2$
B and C vs. D	1	172.80	172.80	$\sigma^2 + (25/30)(\tau_2 + \tau_3 - 2\tau_4)^2$
B vs. C	1	129.60	129.60	$\sigma^2 + (25/10)(\tau_3 - \tau_2)^2$
Experimental error				
A vs. rest	4	1,087.27	271.82 ⎫	
B and C vs. D	4	1,112.53	278.13 ⎬ σ^2	
B vs. C	4	426.40	106.60 ⎭	

*The symbol σ^2 was used in each expected mean square as a matter of convenience. If the variances are homogeneous, it is correct; if the variances are not homogeneous, the symbol σ^2 would be replaced by various linear combinations of the σ_{ij}^2.

Similarly, $(E_1)_{yy} = 1087.27$ and $(E_2)_{yy} = 1112.53$. Thus we finally obtain the abbreviated ANOVA shown in Table 11.17.

The procedure explained and illustrated in this section can sometimes be used to advantage when analyzing a particular set of data. However, the reader should realize that it is a special technique and will therefore be used only rarely.

11.10 RESPONSE CURVES

Once again it is sufficient to state that the techniques explained in the preceding chapter are directly applicable to the present situation. Thus the reader is referred to Section 10.11 for the computational details. However, to emphasize the "sameness," an illustrative example will be presented. (**Note:** The reader will find it rewarding to compare the following example with Example 10.17.)

Example 11.13

Considering the data of Table 11.18 and using the methods described in Section 10.11, the abbreviated ANOVA shown in Table 11.19 is obtained.

11.11 FACTORIAL TREATMENT COMBINATIONS

Because of the detail with which the analysis of factorial treatment combinations was discussed in connection with the completely randomized design, only a brief discussion seems appropriate here. Accordingly, we concentrate on some of the details for the two-factor case and then summarize the three-factor case by giving the linear model, the associated ANOVA, and a numerical example.

Two-factor factorial (crossed-classification model). The linear sta-

TABLE 11.18–Yields (converted to bushels/acre) of a Certain Grain Crop in a Fertilizer Trial

	Level of Fertilizer				
Block	No treat-ment	10 lb per plot	20 lb per plot	30 lb per plot	40 lb per plot
1	20	25	36	35	43
2	25	29	37	39	40
3	23	31	29	31	36
4	27	30	40	42	48
5	19	27	33	44	47
Treatment Total	114	142	175	191	214

TABLE 11.19–Abbreviated ANOVA for Data of Table 11.18 Showing the Isolation of the Linear and Quadratic Portions of the Treatment Sum of Squares

Source of Variation	Degrees of Freedom	Sum of Squares	Mean Square
Blocks	4	154.16	38.54
Treatments	4	1256.56	314.14
Linear	1	1240.02	1240.02**
Quadratic	1	10.41	10.41
Deviations from regression	2	6.13	3.07
Experimental error	16	193.44	12.09

**Significant at $\alpha = 0.01$.

tistical model for a two-factor crossed-classification model with r blocks (replicates) and one observation per experimental unit is

$$Y_{ijk} = \mu + \rho_i + \alpha_j + \beta_k + (\alpha\beta)_{jk} + \epsilon_{ijk}$$
$$i = 1,\ldots,r; j = 1,\ldots,a; k = 1,\ldots,b \qquad (11.51)$$

where μ is the true mean effect, ρ_i is the true effect of the ith replicate (or block) and α, β, and $\alpha\beta$ are the effects of factor a, of factor b, and the interaction of the two factors respectively. We assume $\epsilon_{ijk} \sim NI(0, \sigma^2)$.

The ANOVA associated with the two-factor model in Eq. (11.51) is given in Table 11.20. The sums of squares can be computed using the formulas:

$$\sum Y^2 = \sum_{i=1}^{r} \sum_{j=1}^{a} \sum_{k=1}^{b} Y_{ijk}^2 \qquad (11.52)$$

TABLE 11.20—Generalized ANOVA for a Two-Factor Factorial in a Randomized Complete Block Design

Source of Variation	Degrees of Freedom	Sum of Squares	Mean Square	Expected Mean Square			
				Model I	Model II	Model III (a fixed, b random)	Model III (a random, b fixed)
Mean	1	M_{yy}	$\mathbf{M}$	$\dots$	$\dots$	$\dots$	$\dots$
Replicates	$r-1$	R_{yy}	$\mathbf{R}$	$\dots$	$\dots$	$\dots$	$\dots$
Treatments							
A	$a-1$	A_{yy}	$\mathbf{A}$	$\sigma^2 + rb \sum_{j=1}^{a} \alpha_j^2/(a-1)$	$\sigma^2 + r\sigma_{\alpha\beta}^2 + rb\sigma_\alpha^2$	$\sigma^2 + r\sigma_{\alpha\beta}^2 + rb \sum_{j=1}^{a} \alpha_j^2/(a-1)$	$\sigma^2 + rb\sigma_\alpha^2$
B	$b-1$	B_{yy}	$\mathbf{B}$	$\sigma^2 + ra \sum_{k=1}^{b} \beta_k^2/(b-1)$	$\sigma^2 + r\sigma_{\alpha\beta}^2 + ra\sigma_\beta^2$	$\sigma^2 + r\sigma_{\alpha\beta}^2$	$\sigma^2 + r\sigma_{\alpha\beta}^2 + ra \sum_{k=1}^{b} \beta_k^2/(b-1)$
AB	$(a-1)(b-1)$	$(AB)_{yy}$	$\mathbf{AB}$	$\sigma^2 + r \sum_{j=1}^{a}\sum_{k=1}^{b} (\alpha\beta)_{jk}^2/(a-1)(b-1)$	$\sigma^2 + r\sigma_{\alpha\beta}^2$	$\sigma^2 + r\sigma_{\alpha\beta}^2$	$\sigma^2 + r\sigma_{\alpha\beta}^2$
Experimental error	$(r-1)(ab-1)$	E_{yy}	$\mathbf{E}$	σ^2	σ^2	σ^2	σ^2
Total	rab	$\sum Y^2$	$\dots$				$\dots$

$$M_{yy} = abr(\bar{Y}_{...})^2 = T^2/abr \tag{11.53}$$

$$R_{yy} = ab \sum_{i=1}^{r} (\bar{Y}_{i..} - \bar{Y}_{...})^2 = \sum_{i=1}^{r} R_i^2/ab - M_{yy} \tag{11.54}$$

$$A_{yy} = br \sum_{j=1}^{a} (\bar{Y}_{.j.} - \bar{Y}_{...})^2 = \sum_{j=1}^{a} A_j^2/br - M_{yy} \tag{11.55}$$

$$B_{yy} = ar \sum_{k=1}^{b} (\bar{Y}_{..k} - \bar{Y}_{...})^2 = \sum_{k=1}^{b} B_k^2/ar - M_{yy} \tag{11.56}$$

$$(AB)_{yy} = r \sum_{j=1}^{a} \sum_{k=1}^{b} (\bar{Y}_{.jk} - \bar{Y}_{.j.} - \bar{Y}_{..k} + \bar{Y}_{...})^2$$

$$= \sum_{j=1}^{a} \sum_{k=1}^{b} T_{jk}^2/r - A_{yy} - B_{yy} - M_{yy} \tag{11.57}$$

$$E_{yy} = \sum_{i=1}^{r} \sum_{j=1}^{a} \sum_{k=1}^{b} (Y_{ijk} - \bar{Y}_{.jk} - \bar{Y}_{i..} + \bar{Y}_{...})^2$$

$$= \sum Y^2 - \sum_{j=1}^{a} \sum_{k=1}^{b} T_{jk}^2/r - R_{yy} \tag{11.58}$$

where $R_i = Y_{i..}$ is the ith replicate (block) total, $A_j = Y_{.j.}$ is the total for the jth level of factor a, $B_k = Y_{..k}$ is the total for the kth level of factor b, $T_{jk} = Y_{.jk}$ is the total for the jkth treatment combination, and $T = Y_{...}$ is the total of all observations.

Example 11.14

Consider the data in Table 11.21. Performing the usual calculations, we obtain the abbreviated ANOVA shown in Table 11.22. Testing $H_1:\alpha_j = 0$ ($j = 1, 2$), we calculate $F = 32.00/3.16 = 10.2$, and this leads to the rejection of H_1. Testing $H_2:\beta_k = 0$ ($k = 1, 2, 3, 4$), we calculate $F = $

TABLE 11.21–Yields of Soybeans at the Agronomy Farm, Ames, Iowa, 1949 (bu/A)

Date of Planting	Fertilizer	Replicate			
		1	2	3	4
Early	check	28.6	36.8	32.7	32.6
	aero	29.1	29.2	30.6	29.1
	Na	28.4	27.4	26.0	29.3
	K	29.2	28.2	27.7	32.0
Late	check	30.3	32.3	31.6	30.9
	areo	32.7	30.8	31.0	33.8
	Na	30.3	32.7	33.0	33.9
	K	32.7	31.7	31.8	29.4

TABLE 11.22–Abbreviated ANOVA for Data of Table 11.21

Source of Variation	Degrees of Freedom	Sum of Squares	Mean Square	Expected Mean Square
Replicates	3	7.31	2.44	$\sigma^2 + (8/3) \sum_{i=1}^{4} \rho_i^2$
Dates of planting	1	32.00	32.00	$\sigma^2 + (16/1) \sum_{j=1}^{2} \alpha_j^2$
Fertilizers	3	16.40	5.47	$\sigma^2 + (8/3) \sum_{k=1}^{4} \beta_k^2$
Fertilizers × dates of planting	3	38.40	12.80	$\sigma^2 + (4/3) \sum_{j=1}^{2} \sum_{k=1}^{4} (\alpha\beta)_{jk}^2$
Experimental error	21	66.43	3.16	σ^2

$5.47/3.16 = 1.73$, and this does *not* permit H_2 to be rejected. To test $H_3 : (\alpha\beta)_{jk} = 0 (j = 1, 2; k = 1, 2, 3, 4)$, we calculate $F = 12.80/3.16 = 4.05$, and this leads to rejection of H_3 at the 5 percent significance level but not at the 1 percent significance level. (**Note:** Depending on whether we use $\alpha = 0.05$ or $\alpha = 0.01$, the recommendations will differ. If $\alpha = 0.05$, different fertilizers would probably be suggested for each date of planting; if $\alpha = 0.01$, it is possible that the same recommendation concerning fertilizers would be made for each date of planting.)

As with all linear statistical models, we can write the model in Eq. (11.51) in the form of a linear regression model as follows:

$$Y_{ijk} = \mu + \rho_1 u_{1i} + \cdots + \rho_{r-1} u_{r-1 i} + \alpha_1 x_{1j} + \cdots + \alpha_{a-1} x_{a-1 j}$$
$$+ \beta_1 w_{1k} + \cdots + \beta_{b-1} w_{b-1 k} + (\alpha\beta)_{11} x_{1i} w_{1j}$$
$$+ \cdots + (\alpha\beta)_{a-1 b-1} x_{a-1 i} w_{b-1 j} + \epsilon_{ijk}$$
$$i = 1, \ldots, r; j = 1, \ldots, a; k = 1, \ldots, b \qquad (11.59)$$

where
$$u_{ki} = 1 \quad \text{if } i = k; k = 1, \ldots, r - 1$$
$$= -1 \quad \text{if } i = r$$
$$= 0 \quad \text{otherwise}$$

and x and w are defined in an analogous way.

Writing the model in the latter form allows us to use standard regression techniques to estimate the unknown parameters as well as to evaluate the sum of squares in the associated ANOVA. As mentioned earlier, a factorial program or a generalized ANOVA program more frequently is used to evaluate the sum of squares associated with this model. In the former case the block or replicate effect must be thought of as an additional factor in running the factorial program. The interaction of the block factor with the treatments is not included in the model.

TABLE 11.23–Abbreviated ANOVA for a Two-Factor Factorial in a Randomized Complete Block Design with n Samples per Experimental Unit

Source of Variation	Degrees of Freedom	Expected Mean Square (Model I)
Replicates	$r - 1$	$\sigma_\eta^2 + n\sigma^2 + nab \sum_{i=1}^{r} \rho_i^2/(r - 1)$
Treatments		
A	$a - 1$	$\sigma_\eta^2 + n\sigma^2 + rnb \sum_{j=1}^{a} \alpha_j^2/(a - 1)$
B	$b - 1$	$\sigma_\eta^2 + n\sigma^2 + rna \sum_{k=1}^{b} \beta_k^2/(b - 1)$
AB	$(a - 1)(b - 1)$	$\sigma_\eta^2 + n\sigma^2 + rn \sum_{j=1}^{a} \sum_{k=1}^{b} (\alpha\beta)_{jk}^2/(a - 1)(b - 1)$
Experimental error	$(r - 1)(ab - 1)$	$\sigma_\eta^2 + n\sigma^2$
Sampling error	$rab(n - 1)$	σ_η^2

Subsampling in a randomized complete block design that incorporates factorial treatment combinations leads to analyses such as shown in Tables 11.23 and 11.24. Since no new techniques are involved, numerical examples will not be given. The reader is referred to Sections 10.12 and 11.5 for further details.

TABLE 11.24–Abbreviated ANOVA for a Two-Factor Factorial in a Randomized Complete Block Design with n Samples per Experimental Unit and d Determinations per Sampling Unit

Source of Variation	Degrees of Freedom	Expected Mean Square (Model I)
Replicates	$r - 1$	$\sigma_\delta^2 + d\sigma_\eta^2 + dn\sigma^2 + dnab \sum_{i=1}^{r} \rho_i^2/(r - 1)$
Treatments		
A	$a - 1$	$\sigma_\delta^2 + d\sigma_\eta^2 + dn\sigma^2 + rdnb \sum_{j=1}^{a} \alpha_j^2/(a - 1)$
B	$b - 1$	$\sigma_\delta^2 + d\sigma_\eta^2 + dn\sigma^2 + rdna \sum_{k=1}^{b} \beta_k^2/(b - 1)$
AB	$(a - 1)(b - 1)$	$\sigma_\delta^2 + d\sigma_\eta^2 + dn\sigma^2 + rdn \sum_{j=1}^{a} \sum_{k=1}^{b} (\alpha\beta)_{jk}^2/(a - 1)(b - 1)$
Experimental error	$(r - 1)(ab - 1)$	$\sigma_\delta^2 + d\sigma_\eta^2 + dn\sigma^2$
Sampling error	$rab(n - 1)$	$\sigma_\delta^2 + d\sigma_\eta^2$
Determinations	$rabn(d - 1)$	σ_δ^2

As mentioned in Section 10.12, when factorial treatment combinations are involved, it is possible to subdivide the treatment sum of squares into several parts such as $(A_L)_{yy}$, $(A_Q)_{yy}$, ...; $(B_L)_{yy}$, $(B_Q)_{yy}$, ...; $(A_L B_L)_{yy}$, $(A_Q B_L)_{yy}$, $(A_L B_Q)_{yy}$, etc. The procedure to be followed will parallel that presented in Section 10.11, the only difference being the refinements introduced to subdivide the interaction sum of squares. Because of this, the technique will be presented in terms of a numerical example.

Example 11.15

Consider the data in Table 11.25 involving one treatment a with 4 levels and a second treatment b with 3 levels. The experiment was replicated twice ($r = 2$). The treatment sums are given in Table 11.26 to facilitate the computations. Since the number of levels for factors a and b are 4 and 3 respectively, it is possible to evaluate the linear, quadratic, and cubic effects of treatment a as well as the linear and quadratic effects of b. The joint effects are measured by subdividing the interaction sum of squares into $(A_L B_L)_{yy}$, ..., $(A_C B_Q)_{yy}$.

TABLE 11.25–Coded Data for Use in Illustrating the Calculation of the Linear, Quadratic, ... Effects in a Two-Factor Factorial Experiment Conducted in a Randomized Complete Block Design

Replicate		a_1	a_2	a_3	a_4
1	b_1	7	8	9	7
	b_2	5	6	11	10
	b_3	4	6	10	12
2	b_1	7	9	9	8
	b_2	6	6	10	11
	b_3	6	7	10	12

In terms of a linear regression model, the estimation and evaluation of these effects is possible by introducing the linear, quadratic, etc., variables for each factor. The model becomes

$$Y_{ijk} = \mu + \rho_i + \alpha_1 \xi_j^1 + \alpha_2 \xi_j^2 + \alpha_3 \xi_j^3 + \beta_1 \delta_k^1 + \beta_2 \delta_k^2 + \gamma_1 \eta_{jk}^{11}$$
$$+ \gamma_2 \eta_{jk}^{21} + \gamma_3 \eta_{jk}^{31} + \gamma_4 \eta_{jk}^{12} + \gamma_5 \eta_{jk}^{22} + \gamma_6 \eta_{jk}^{32} + \epsilon_{ijk}$$
$$i = 1, \ldots, r; j = 1, \ldots, a; k = 1, \ldots, b \qquad (11.60)$$

TABLE 11.26–Treatment Sums Formed from the Data of Table 11.25

	a_1	a_2	a_3	a_4	Total
b_1	14	17	18	15	64
b_2	11	12	21	21	65
b_3	10	13	20	24	67
Total	35	42	59	60	196

where $\eta_{jk}^{mn} = \xi_j^m \delta_k^n$. The values of the variables ξ, δ are given in Table 10.21, and the η's denote the interaction terms, i.e., the $(\alpha\beta)$'s. Application of standard regression analysis would provide the estimates and sums of squares $(A_L)_{yy}$, $(A_Q)_{yy}$, etc. These are summarized in Table 11.27.

Alternatively, we can evaluate the appropriate sum of squares by applying Eq. (10.75) with minor modifications necessary since we are dealing with a different design. For example, if we want to evaluate the sum of squares due to the linear effect of treatment a, Eq. (10.75), rewritten in the notation of the present design, becomes

$$(A_L)_{yy} = \left(\sum_{j=1}^{a} \xi_j^1 T_j\right)^2 \Bigg/ br \sum_{j=1}^{a} (\xi_j^1)^2 \qquad (11.61)$$

where for $a = 4$

$$\begin{aligned}
\xi_j^1 &= -3 & j &= 1 \\
&= -1 & j &= 2 \\
&= 1 & j &= 3 \\
&= 3 & j &= 4
\end{aligned}$$

are the values of the linear orthogonal polynomial coefficients for four treatment levels. Applying Eq. (11.61) to the example, we have

$$(A_L)_{yy} = \frac{[(-3)(35) + (-1)(42) + (1)(59) + (3)(60)]^2}{(2)(3)[(-3)^2 + (-1)^2 + (1)^2 + (3)^2]} = 70.53$$

Similarly,

$$(A_Q)_{yy} = \frac{[(1)(35) + (-1)(42) + (-1)(59) + (1)(60)]^2}{(2)(3)[(1)^2 + (-1)^2 + (-1)^2 + (1)^2]} = 1.50$$

$$(A_C)_{yy} = \frac{[(-1)(35) + (3)(42) + (-3)(59) + (1)(60)]^2}{(2)(3)[(-1)^2 + (3)^2 + (-3)^2 + (1)^2]} = 5.63$$

$$(B_L)_{yy} = \frac{[(-1)(64) + (0)(65) + (1)(67)]^2}{(2)(4)[(-1)^2 + (0)^2 + (1)^2]} = 0.56$$

$$(B_Q)_{yy} = \frac{[(1)(64) + (-2)(65) + (1)(67)]^2}{(2)(4)[(1)^2 + (-2)^2 + (1)^2]} = 0.02$$

where the divisors are:

1. For $(A_L)_{yy}$, $(A_Q)_{yy}$, and $(A_C)_{yy}$: $(rb) \times$ (sum of the squares of the coefficients).
2. For $(B_L)_{yy}$ and $(B_Q)_{yy}$: $(ra) \times$ (sum of the squares of the coefficients).

We illustrate the evaluation of the sum of squares for the interaction terms by looking at the quadratic by linear effect. The quadratic by linear sum of squares, denoted $(A_Q B_L)_{yy}$, is

$$(A_Q B_L)_{yy} = \frac{\left(\sum_{k=1}^{b} \sum_{j=1}^{a} \eta_{jk}^{21} T_{jk}\right)^2}{r \sum_{k=1}^{b} \sum_{j=1}^{a} (\eta_{jk}^{21})^2} = \frac{\left[\sum_{k=1}^{b} \delta_k^1 \left(\sum_{j=1}^{a} \xi_j^2 T_{jk}\right)\right]^2}{r \left[\sum_{j=1}^{a} (\xi_j^2)^2\right]\left[\sum_{k=1}^{b} (\delta_k^1)^2\right]} \qquad (11.62)$$

For our example the sum $\sum_{j=1}^{4} \xi_j^2 T_{jk}$ is

for $k = 1$: $(1)(14) + (-1)(17) + (-1)(18) + (1)(15) = -6$

for $k = 2$: $(1)(11) + (-1)(12) + (-1)(21) + (1)(21) = -1$

for $k = 3$: $(1)(10) + (-1)(13) + (-1)(20) + (1)(24) = 1$

and hence

$$(A_Q B_L)_{yy} = \frac{[(-1)(-6) + (0)(-1) + (1)(1)]^2}{2[(1)^2 + (-1)^2 + (-1)^2 + (1)^2][(-1)^2 + (0)^2 + (1)^2]}$$

$$= 3.06$$

The reader should verify that the remaining sum of squares $(A_L B_L)_{yy}$, $(A_L B_Q)_{yy}$, etc., can be evaluated in a like manner. The ANOVA with the treatment main effects and interaction partitioned is summarized in Table 11.27.

TABLE 11.27–Abbreviated ANOVA for Data of Table 11.25

Source of Variation	Degrees of Freedom	Sum of Squares	Mean Square
Replicates	1	1.50	1.50
Treatments			
A_L	1	70.53	70.53
A_Q	1	1.50	1.50
A_C	1	5.63	5.63
B_L	1	0.56	0.56
B_Q	1	0.02	0.02
$A_L B_L$	1	25.31	25.31
$A_L B_Q$	1	2.60	2.60
$A_Q B_L$	1	3.06	3.06
$A_Q B_Q$	1	0.20	0.20
$A_C B_L$	1	0.32	0.32
$A_C B_Q$	1	2.60	2.60
Experimental error	11	3.50	0.32

Three-factor factorial (crossed-classification model). The statistical model for a three-factor factorial in a randomized complete block design is

$$Y_{ijkl} = \mu + \rho_i + \alpha_j + \beta_k + (\alpha\beta)_{jk} + \gamma_l + (\alpha\gamma)_{jl} + (\beta\gamma)_{kl}$$

$$+ (\alpha\beta\gamma)_{jkl} + \epsilon_{ijkl}$$

$$i = 1,\ldots,r; j = 1,\ldots,a; k = 1,\ldots,b; l = 1,\ldots,c \qquad (11.63)$$

The associated generalized ANOVA is given in Table 11.28.

Example 11.16

An experiment such as described in Example 9.18 was performed. The resulting data are given in Table 11.29. The associated ANOVA is pre-

TABLE 11.28–Generalized ANOVA for a Three-Factor Factorial in a

Source of Variation	Degrees of Freedom	Sum of Squares	Mean Square	Expected Mean Square
				Model I
Mean	1	M_{yy}	M	$\cdots$
Replicates	$r - 1$	R_{yy}	R	$\cdots$
Treatments				
A	$a - 1$	A_{yy}	A	$\sigma^2 + rbc \sum_{j=1}^{a} \alpha_j^2/(a - 1)$
B	$b - 1$	B_{yy}	B	$\sigma^2 + rac \sum_{k=1}^{b} \beta_k^2/(b - 1)$
C	$c - 1$	C_{yy}	C	$\sigma^2 + rab \sum_{l=1}^{c} \gamma_l^2/(c - 1)$
AB	$(a - 1)(b - 1)$	$(AB)_{yy}$	AB	$\sigma^2 + rc \sum_{j=1}^{a} \sum_{k=1}^{b} (\alpha\beta)_{jk}^2$ $/(a - 1)(b - 1)$
AC	$(a - 1)(c - 1)$	$(AC)_{yy}$	AC	$\sigma^2 + rb \sum_{j=1}^{a} \sum_{l=1}^{c} (\alpha\gamma)_{jl}^2$ $/(a - 1)(c - 1)$
BC	$(b - 1)(c - 1)$	$(BC)_{yy}$	BC	$\sigma^2 + ra \sum_{k=1}^{b} \sum_{l=1}^{c} (\beta\gamma)_{kl}^2$ $/(b - 1)(c - 1)$
ABC	$(a - 1)(b - 1)(c - 1)$	$(ABC)_{yy}$	ABC	$\sigma^2 + r \sum_{j=1}^{a} \sum_{k=1}^{b} \sum_{l=1}^{c} (\alpha\beta\gamma)_{jkl}^2$ $/(a - 1)(b-1)(c - 1)$
Experimental error	$(r - 1)(abc - 1)$	E_{yy}	E	σ^2
Total	$rabc$	$\sum Y^2$	$\cdots$	$\cdots$

Randomized Complete Block Design

	Expected Mean Square	
Model II	Model III (*a* and *b* fixed, *c* random)	Model III (*a* fixed, *b* and *c* random)
$\cdots$ $\cdots$	$\cdots$ $\cdots$	$\cdots$ $\cdots$
$\sigma^2 + r\sigma^2_{\alpha\beta\gamma} + rc\sigma^2_{\alpha\beta} + rb\sigma^2_{\alpha\gamma} + rbc\sigma^2_{\alpha}$	$\sigma^2 + rb\sigma^2_{\alpha\gamma} + rbc\sum_{j=1}^{a} \alpha_j^2/(a-1)$	$\sigma^2 + r\sigma^2_{\alpha\beta\gamma} + rb\sigma^2_{\alpha\gamma} + rc\sigma^2_{\alpha\beta}$ $+ rbc\sum_{j=1}^{a} \alpha_j^2/(a-1)$
$\sigma^2 + r\sigma^2_{\alpha\beta\gamma} + rc\sigma^2_{\alpha\beta} + ra\sigma^2_{\beta\gamma} + rac\sigma^2_{\beta}$	$\sigma^2 + ra\sigma^2_{\beta\gamma} + rac\sum_{k=1}^{b} \beta_k^2/(b-1)$	$\sigma^2 + ra\sigma^2_{\beta\gamma} + rac\sigma^2_{\beta}$
$\sigma^2 + r\sigma^2_{\alpha\beta\gamma} + rb\sigma^2_{\alpha\gamma} + ra\sigma^2_{\beta\gamma} + rab\sigma^2_{\gamma}$	$\sigma^2 + rab\sigma^2_{\gamma}$	$\sigma^2 + ra\sigma^2_{\beta\gamma} + rab\sigma^2_{\gamma}$
$\sigma^2 + r\sigma^2_{\alpha\beta\gamma} + rc\sigma^2_{\alpha\beta}$	$\sigma^2 + r\sigma^2_{\alpha\beta\gamma} + rc\sum_{j=1}^{a}\sum_{k=1}^{b} (\alpha\beta)_{jk}^2$ $/(a-1)(b-1)$	$\sigma^2 + r\sigma^2_{\alpha\beta\gamma} + rc\sigma^2_{\alpha\beta}$
$\sigma^2 + r\sigma^2_{\alpha\beta\gamma} + rb\sigma^2_{\alpha\gamma}$	$\sigma^2 + rb\sigma^2_{\alpha\gamma}$	$\sigma^2 + r\sigma^2_{\alpha\beta\gamma} + rb\sigma^2_{\alpha\gamma}$
$\sigma^2 + r\sigma^2_{\alpha\beta\gamma} + ra\sigma^2_{\beta\gamma}$	$\sigma^2 + ra\sigma^2_{\beta\gamma}$	$\sigma^2 + ra\sigma^2_{\beta\gamma}$
$\sigma^2 + r\sigma^2_{\alpha\beta\gamma}$	$\sigma^2 + r\sigma^2_{\alpha\beta\gamma}$	$\sigma^2 + r\sigma^2_{\alpha\beta\gamma}$
σ^2	σ^2	σ^2
$\cdots$	$\cdots$	$\cdots$

TABLE 11.29–Surge Voltages Resulting from the Experiment
Described in Example 9.18

Electrolyte (a)	Heat Paper (b)	Temperature (c)	Replicate	
			I	II
0	0	0	6.08	6.79
0	0	1	6.31	6.77
0	1	0	6.53	6.73
1	0	0	6.04	6.68
0	1	1	6.12	6.49
1	0	1	6.09	6.38
1	1	0	6.43	6.08
1	1	1	6.36	6.23

TABLE 11.30–ANOVA for Data of Table 11.29

Source of Variation	Degrees of Freedom	Sum of Squares	Mean Square	F Ratio
Mean	1	651.6533	651.6533	...
Replicates	1	0.2997	0.2997	...
Treatments				
A	1	0.1462	0.1462	2.21
B	1	0.0018	0.0018	0.03
C	1	0.0232	0.0232	0.35
AB	1	0.0001	0.0001	0.00
AC	1	0.0047	0.0047	0.07
BC	1	0.0176	0.0176	0.27
ABC	1	0.0883	0.0883	1.33
Experimental error	7	0.4632	0.0662	...
Total	16	652.6981	...	...

sented in Table 11.30. It will be noted that none of the factors led to significant results.

To further illustrate the use of response curve techniques in a randomized complete block design, consider the following example involving three factors:

Example 11.17

Consider the data of Table 11.31. To save time, the calculation of the
sums of squares will be illustrated for only three effects: A_L, $A_Q B_L$, and
$A_Q B_C C_L$. These are

$$(A_L)_{yy} = \frac{[(-1)(747) + (0)(1404) + (1)(925)]^2}{(2)(4)(6)[(-1)^2 + (0)^2 + (1)^2]} = 330.04$$

$$(A_Q B_L)_{yy} = \frac{[(1)(325) + (-2)(432) + (1)(-659)]^2}{(2)(6)[(1)^2 + (-2)^2 + (1)^2][(-3)^2 + (-1)^2 + (1)^2 + (3)^2]}$$

$$= 996.67$$

$$(A_Q B_C C_L)_{yy} = (-4232)^2/D = 1066.06$$

where

$$D = 2[(1)^2 + (-2)^2 + (1)^2][(-1)^2 + (3)^2 + (-3)^2 + (1)^2][(-5)^2$$
$$+ (-3)^2 + (-1)^2 + (1)^2 + (3)^2 + (5)^2]$$

and the divisors are:

1. For $(A_L)_{yy}$: $(rbc) \times$ (sum of the squares of the coefficients).
2. For $(A_Q B_L)_{yy}$: $(rc) \times$ (product of the sums of the squares of the coefficients).
3. $(A_Q B_C C_L)_{yy}$: $(r) \times$ (product of the sums of the squares of the coefficients).

The numerators for $(A_L)_{yy}$ and $(A_Q B_L)_{yy}$ were found by the same procedures as similar quantities in Example 11.15. The numerator for $(A_Q B_C C_L)_{yy}$ was obtained by a simple extension of the same principles. In this case the extension may be explained as follows: Operate with the C_L coefficients on the c entries in Table 11.31 for each level of b within each level of a and obtain a set of sums of products, one for each level of b within each level of a. Next, use the B_C coefficients and operate on the sums just obtained. This will provide us with some "$B_C C_L$ totals" for each level of a. Then use the A_Q coefficients to give us the numerator for $(A_Q B_C C_L)_{yy}$. The procedure should now be clear, and the extension to any number of factors will be a simple even if time-consuming, job. A more

TABLE 11.31–Hypothetical Data for Use in Illustrating the Computation of Certain Sums of Squares in a Three-Factor Factorial, the Basic Design Being a Randomized Complete Block

Repli-cate		a_1				a_2				a_3			
		b_1	b_2	b_3	b_4	b_1	b_2	b_3	b_4	b_1	b_2	b_3	b_4
1	c_1	7	7	9	7	15	36	60	15	24	29	17	19
	c_2	23	18	25	15	13	35	61	18	30	26	11	8
	c_3	9	18	24	23	12	43	62	14	31	24	15	23
	c_4	7	13	25	36	11	12	63	26	32	15	12	5
	c_5	6	8	20	7	15	46	18	28	15	32	13	6
	c_6	10	12	30	11	10	42	27	12	17	29	8	7
2	c_1	7	6	11	7	15	35	60	20	25	30	20	20
	c_2	20	19	25	16	13	30	64	20	30	25	15	10
	c_3	9	22	26	24	13	40	66	15	32	25	15	22
	c_4	8	15	26	30	13	10	66	25	34	15	15	4
	c_5	8	10	20	8	17	40	20	30	18	35	15	5
	c_6	9	12	28	11	8	45	30	15	19	30	10	8

efficient way of partitioning the sum of squares into linear, quadratic, etc., components is to set up a regression model similar to Eq. (11.60) and retrieve the sums of squares from the output of the regression analysis.

11.12 MISSING DATA

Many times, even after considerable effort has been expended and due diligence exercised in planning an experiment, things occur to the disadvantage of the research worker. One of the most common of these "disturbances" is the problem of missing observations. Missing observations arise for many reasons: an animal may die, an experimental plot may be flooded out, a worker may be ill and not turn up on the job, a jar of jelly may be dropped on the floor, or the recorded data may be lost. What effect does this have on our methods of analysis? Since most experiments are designed with at least some degree of balance or symmetry, any missing observations will usually destroy this balance. Thus we now expect our original planned analysis to be complicated and some modifications in procedure to be required. We could in many instances treat the data as a case of disproportionate subclass numbers and use methods of analysis appropriate to such situations (see Chapter 12). However, other approaches are sometimes open to the statistician, and we shall examine these in this section, pointing out the difficulties that arise and indicating the computational procedures to be followed in each case.

First, let us mention two cases of missing data in a randomized complete block design that present no difficulties as to computational procedures; i.e., (1) a complete block is missing or (2) a treatment is completely missing. When one, or more, complete blocks are missing, we simply proceed with the standard type of analysis, provided we still have at least two blocks remaining; i.e., we analyze the data as though we had planned only on the number of blocks that are actually available. For the case in which no data are available on one or more treatments (assuming we still have at least two treatments remaining), we may again proceed in the regular manner. However, in this instance the research worker should certainly inquire into the reasons for the lack of data on certain treatments. It is apparent that many things might have caused such a happening, each of which could possibly lead to different decisions or recommendations on the part of the experimenter. Without a specific example, further discussion on such points can only be of a vague nature.

A more commonly occurring situation is the one in which one observation is missing. Here we run into difficulty in our analysis. Either we must treat the data by methods appropriate to unequal frequencies, or we must find some other scheme, which we hope will be simpler to apply. One such device is to treat the missing observation as an unknown value to be estimated from the observed data. The estimate of the missing observation most frequently used by statisticians is the value that minimizes the experimental error sum of squares when the regular analysis is performed. This is a familiar calculus problem in mathematics. The analysis may be summarized by the following formula, which will provide an estimate of the missing observation in accordance with the above principle:

$$M = (tT + bB - S)/(t - 1)(b - 1) \qquad (11.64)$$

where t = number of treatments
b = number of blocks
T = sum of observations with the same treatment as the missing observation
B = sum of observations in the same block as the missing observation
S = sum of all the actual observations

This value, i.e., M, is then entered in the appropriate place in the table of data and the augmented data are analyzed in the customary manner.

Before constructing the ANOVA table and testing the hypothesis $H{:}\tau_j = 0 \, (j = 1, \ldots, t)$, certain changes must be made to avoid biased results. First, the experimental error and total degrees of freedom must each be reduced by 1. Second, it can be shown that under the null hypothesis the expected value of **T**, the treatment mean square calculated using the augmented data, is greater than σ^2, the expected value of the experimental error mean square. Thus any test of hypothesis that does not correct for this fact will be a biased test and can only be considered approximate. The correction for this bias, the second change mentioned above, is to decrease the treatment sum of squares T_{yy} by the amount

$$\text{Correction for bias} = Z = [B - (t - 1)M]^2/t(t - 1) \qquad (11.65)$$

which gives us a new treatment sum of squares

$$T'_{yy} = T_{yy} - Z \qquad (11.66)$$

and the ANOVA indicated in Table 11.32 is finally obtained. An alternative procedure is through the use of covariance analysis. This is discussed in Chapter 13.

TABLE 11.32–Generalized ANOVA for a Randomized Complete Block Design with One Missing Observation

Source of Variation	Degrees of Freedom	Sum of Squares	Mean Square	F Ratio
Mean	1	M_{yy}	**M**	...
Blocks	$b - 1$	B_{yy}	**B**	...
Treatments	$t - 1$	T'_{yy}	**T'**	**T'/E**
Experimental error	$(b - 1)(t - 1) - 1$	E_{yy}	**E**	...
Total	$bt - 1$	$\sum Y^2 - Z$	...	...

Example 11.18

An experiment was conducted by Tinker (3) to investigate the consistency of blink rates during reading. Data were recorded for 6 successive 5-minute periods of reading. As we have extracted only part of the available data for our example, care should be exercised in drawing conclusions from the analysis that follows. The original paper should be consulted by those desiring further information on the subject matter. We will assume that the experiment was performed on 4 individuals (they will be our blocks), and the 6 periods will represent the treatments. Moreover, to illustrate the techniques of this section, we will assume that the observation on Subject A for the fourth period is missing. Our observed data are given in Table 11.33.

TABLE 11.33–Number of Blinks for Successive
Five-Minute Periods of Reading

Subject	Period					
	1	2	3	4	5	6
A	24	23	28	...	30	41
B	18	17	17	19	19	18
C	41	41	49	39	19	27
D	46	69	74	58	54	50

Source: Adapted from Tinker (3), p. 421.

Substituting in our formula, we find our estimate of the missing value to be

$$M = \frac{tT + bB - S}{(t-1)(b-1)} = \frac{6(116) + 4(146) - 821}{5(3)} = 30.6$$

The correction for bias in the treatment sum of squares is found to be

$$Z = \frac{[B - (t-1)M]^2}{t(t-1)} = \frac{[146 - 5(30.6)]^2}{6(5)} = 1.63$$

and so we arrive at the ANOVA presented in Table 11.34. The test of the null hypothesis $H: \tau_j = 0 (j = 1, \ldots, 6)$ gives rise to an F value less than unity. Noting that $F' = 1/F$ is not significant, we conclude that the blink rate is consistent during reading when measured over 6 successive 5-minute periods. It is evident there are wide differences among individuals, a fact that is not surprising and confirms our judgment in performing the experiment as we did, i.e., by removing the interindividual differences that otherwise would have appeared as part of the experimental error sum of squares. (**Note:** Actually, the value we have assumed to be missing was recorded as 27 in the original source of data. It will pay the reader to do the analysis with the true value entered in the table for comparison with the approximate solution presented above. This should give him an indication, *but only an indication,* of how reliable the estimation procedure is.)

TABLE 11.34–Abbreviated ANOVA of Number of Blinks during Reading

Source of Variation	Degrees of Freedom	Sum of Squares	Mean Square
Blocks	3	5233.82	1744.61
Treatments	5	339.90	67.98
Experimental error	14	1068.40	76.29

If two or more values are missing, the same general procedure (using the calculus) may be followed to provide estimates. For the person not familiar with the requisite mathematical techniques, equivalent results may be obtained by use of the following iterative method. Suppose that two values are missing. For one of these substitute the mean of all recorded observations and then estimate the second missing value using Eq. (11.64); next, place this estimate in its proper place in the table, remove the general mean from its position in place of the first missing observation, and then estimate a value for the first missing observation using Eq. (11.64). After about two cycles you will find very little or no change in successive estimates of the same missing value. When this point is reached, you have the estimated values. This procedure may easily be extended to cases where three or more observations are missing.

What changes are necessary before one proceeds with the usual F tests if we have been forced to estimate several missing values? First, we must reduce the degrees of freedom associated with both experimental error and total by the number of observations estimated. Second, the treatment sum of squares must be reduced by a specified quantity to avoid a biased test procedure. If we have only two missing observations (not in the same block), the necessary correction for bias is given by

$$Z = \{[B' - (t - 1)M']^2 + [B'' - (t - 1)M'']^2\}/t(t - 1) \qquad (11.67)$$

where t = number of treatments
B' = total of all the observations in the same block as the first missing observation
B'' = total of all the observations in the same block as the second missing observation
M' = estimate of the first missing observation.
M'' = estimate of the second missing observation

If more than two observations are missing or if two observations are missing in the *same* block, a formula giving the correction for the bias in the treatment sum of squares may be found in Yates (4, 5).

PROBLEMS

11.1   The following data are from an experiment involving a randomized complete block design. Complete the appropriate ANOVA and test the hypothesis that the true effects of the four treatments are equal. State all your assumptions.

	Treatment			
Block	1	2	3	4
1	20	18	16	17
2	18	18	16	20
3	20	18	17	18
4	20	16	20	17
5	19	16	16	20

11.2   In a randomized complete block experiment with 5 treatments in 10 replications, the variance among the 5 treatment means was 100. Complete the following abbreviated ANOVA and test the hypothesis that the 5 treatment effects are the same.

Source of Variation	Degrees of Freedom	Sum of Squares	Mean Square
Replicates		90	
Treatments			
Experimental error			5

11.3   Upon calculating the ANOVA of the yields of 6 varieties planted in 8 randomized complete blocks, the 3 sums of squares—for varieties, for blocks, and for experimental error (or remainder)—were each 245. Complete, as far as possible, the appropriate ANOVA and compute a value of F for testing the significance of the differences among varieties. Interpret your result in terms of the appropriate model and give your conclusions.

11.4   Examine the results given below to learn about the effectiveness of chalk and lime applications in neutralizing soil acidity and thus increasing the stand of beets.

	Number of Beets per Plot		
Block	Control	Chalk	Lime
1	49	135	147
2	37	151	131
3	114	143	103
4	140	146	147

11.5   Analyze the data on the ratio of dry to wet wheat in the following table and interpret the results.

Block	Nitrogen Applied			
	None	Early	Middle	Late
1	0.718	0.732	0.734	0.792
2	0.725	0.781	0.725	0.716
3	0.704	1.035	0.763	0.758
4	0.726	0.765	0.738	0.781

11.6   To study the relative efficiencies of 5 different types of filter, an experiment is to be performed using a certain brand of oil. Fifteen quarts of oil (in 1-qt tins) are purchased and the same amount of foreign material is added to each quart. Since only 5 tests can be performed in any one day, we proceed as follows: (1) Allocate, at random, the 15 quarts into three groups of 5 each; (2) allocating the groups to the days, assign the treatments at random to the quarts within groups; (3) perform the experiment to determine the amount of foreign material caught by the filter; and (4) collect, analyze, and interpret the data.

Block (day)	Type of Filter				
	A	*B*	*C*	*D*	*E*
1	16.9	18.2	17.0	15.1	18.3
2	16.5	19.2	18.1	16.0	18.3
3	17.5	17.1	17.3	17.8	19.8

11.7   In a paired experiment there were 10 pairs with the sum of the squares of the deviations of the differences from their mean being $\sum d^2 = 360$. The totals for the two treatments were $T_1 = 160$ and $T_2 = 120$. Complete the following abbreviated ANOVA.

Source of Variation	Degrees of Freedom	Sum of Squares	Mean Square
Pairs or replications			100
Treatments			
Experimental error			

11.8   Discuss the following statement: "If only *one* sample is obtained from each experimental unit; e.g., if one small sample is taken from a field plot to estimate the effect on the whole plot, *n* is set equal to 1 in Table 11.8, and the line for sampling error is omitted. However, if the whole plot in our field plot example is harvested, then the sampling error is reduced to 0, and we have an analysis as in Table 11.2."

11.9   In a randomized complete block experiment on the accuracy of determination of ascorbic acid concentration in turnip greens (Heinze-Kanapaugh method), 4 weights of sample were tried in 5 replications. Two determinations, A and B, were made on each sample. The results (in micrograms per milliliter of filtrate) follow. Complete the ANOVA for these data.

Sample Weight (g)	Replicate									
	1		2		3		4		5	
	A	B	A	B	A	B	A	B	A	B
5	34.2	37.2	47.0	52.5	48.5	46.5	44.2	44.2	42.5	43.5
2	12.8	12.8	21.5	22.0	24.5	23.0	17.8	17.8	17.0	17.5
1	5.8	8.2	10.2	13.0	16.5	11.0	9.5	15.2	11.0	10.5
0.5	3.5	3.5	5.0	6.0	9.8	6.8	5.2	3.5	3.8	4.7

11.10 Given the following abbreviated ANOVA:

Source of Variation	Degrees of Freedom	Mean Square	Expected Mean Square
Replicates	3	176	
Treatments	7	352	
Experimental error	21	88	
Sampling error	96	40	
Determinations	256	10	

(a) Give the experimental error mean square in the above analysis for the following:
   (1) If 10 has been added to each determination.
   (2) If each determination had been multiplied by 10.
(b) Fill in the expected mean squares in the above table, assuming we are interested in just these 8 treatments but that replicates, samples, and determinations may be considered as random variables.

11.11 Given the following abbreviated ANOVA:

Source of Variation	Degrees of Freedom	Mean Square
Blocks	3	
Treatments	8	
Experimental error	24	1084
Samples within plots	144	381

(a) What is the construction of the experiment?
(b) What is the variance of a treatment mean?
(c) Give the answer to (b) if we have only 1 sample per plot.

(d) What is the maximum precision obtainable by sampling, i.e., if we take k samples, $k \to \infty$?

11.12 We conducted a field experiment to estimate the effect of 9 fertilizers on the yield of oats. Instead of harvesting each plot completely, we took 12 samples, $3' \times 3'$, from each plot. The abbreviated ANOVA is as follows:

Source of Variation	Degrees of Freedom	Mean Square
Replicates	3	384
Treatments	8	960
Experimental error	24	192
Samples within plots	396	24

(a) Assuming that the components of variance do not change, estimate the gain or loss in information in the above experiment if 6 replicates had been used with 8 samples per plot.

(b) What would the above mean squares be if the ANOVA had been computed using the totals of the 12 samples in each plot?

11.13 Given the following abbreviated ANOVA:

Source of Variation	Degrees of Freedom	Mean Square	Expected Mean Square
Replicates	3	288	
Treatments	7	432	
Experimental error	21	144	
Samples within experimental units	96	72	
Determinations per sample	256	6	

(a) Compute the variance of a treatment mean.

(b) Give the expected mean squares.

(c) Compute the gain or loss in efficiency or information if 6 replicates had been used with 8 samples from each experimental unit and 1 determination per sample.

(d) Give the experimental error mean square for the following:
    (1) If 10 had been added to each determination.
    (2) If each determination had been multiplied by 10.

(e) Test the hypothesis that there are no differences among the true effects of the 8 treatments.

11.14 A chemist is confronted with the problem of just where he should expend his efforts in the following situation: A series of 8 soil treatments are applied in a randomized complete block design with 2 replications, 3 soil samples from each plot are taken in the field, each sample is divided into 2 portions in the laboratory, and duplicate determinations for each portion are analyzed for nitrate nitrogen. The following mean squares are given:

Source of Variation	Degrees of Freedom	Mean Square
Treatments	7	11,700
Experimental error	7	1,300
Samples within plots	32	100
Portions within samples	48	20
Determinations within portions	96	16

Find the expected mean squares and estimate the variance components. What might be his gain or loss in efficiency in future experiments if he used 6 replicates but still continued to run only 24 analyses per treatment, e.g., 2 samples per plot, 2 portions per sample, and 1 determination per portion?

11.15   In an experiment to test the effect of 6 treatments on some soil characteristic, we obtained the following abbreviated ANOVA. A total of 6 soil samples was selected at random from each plot, and 2 chemical determinations were made of each sample.

Source of Variation	Degrees of Freedom	Mean Square
Replicates	4	240
Treatments	5	360
Experimental error	20	120
Samples within plots	150	60
Determinations per sample	180	4

(a)   Compute the variance of a treatment mean (per determination).
(b)   Estimate the gain or loss in efficiency in the above experiment if we had taken 8 samples per plot and had made only 1 determination per sample.

11.16   Given the following abbreviated ANOVA for a randomized complete block design:

Source of Variation	Degrees of Freedom	Sum of Squares	Expected Mean Square
Blocks	9	0.4074	
Treatments	3	1.1986	
Experimental error	27	0.6249	

(a)   Complete the analysis; fill in expected mean squares.
(b)   Estimate the efficiency of this design relative to a completely randomized design.
(c)   Compute the standard error for a treatment mean and for the difference between two treatment means.
(d)   The treatment means are 1.464, 1.195, 1.325, and 1.662. What mean or means do you suspect might represent different populations?

11.17 The following data give the gains in weight of pigs in a comparative feeding trial. Analyze and interpret the data, paying attention to the comparison of rations I, II, and III with rations IV and V.

Replicate	Ration I	Ration II	Ration III	Ration IV	Ration V
1	165	168	164	185	201
2	156	180	156	195	189
3	159	180	189	186	173
4	167	166	138	201	193
5	170	170	153	165	164
6	146	161	190	175	160
7	130	171	160	187	200
8	151	169	172	177	142
9	164	179	142	166	184
10	158	191	155	165	149

11.18 The following data are extracted from a larger experiment concerned with oat-seed treatment. The following yields in grams were obtained with 2 rates of the same compound over 7 replicates. What conclusion do you draw? (The yields for the untreated seed are in a separate column.) Is seed treatment worth the added expense in this instance?

Replicate	Rate 1	2	Check
1	360	391	408
2	436	382	409
3	413	414	340
4	353	416	324
5	328	375	304
6	269	422	268
7	220	227	290

11.19 Results similar to those in Problem 11.18 are also available for flax. What advice would you give about the use of Ceresan M as against 224, and about the advisability of seed treatment?

Replicate	Ceresan M	224	Check
1	19.2	14.4	13.2
2	14.8	24.6	19.2
3	26.7	22.9	17.4
4	17.6	22.7	16.4
5	22.1	22.0	15.8
6	21.7	22.0	14.6
7	23.9	20.4	12.5
8	19.1	16.0	13.0

11.20   A project studying farm structures was concerned with the insulation of poultry houses. The data obtained from a study of a set of model structures (total number of eggs over 4 replicates of each treatment) were as follows:

Standard house + laying mash	250
3″ wall insulation + laying mash	280
3″ wall insulation + laying mash + cod liver oil	350
6″ wall insulation + laying mash	310
6″ wall insulation + laying mash + cod liver oil	400

Construct a *reasonable* set of 4 orthogonal comparisons based on the above treatments. Calculate the sum of squares for *one* of your comparisons and test for significance. The following is part of the original analysis:

Source of Variation	Degrees of Freedom	Sum of Squares
Treatments	4	3470
Experimental error	12	1728

11.21   Assume a randomized complete block experiment with 4 treatments and 8 replicates. One of the treatments is a control or check and the other 3 are different methods of treatment. Assume that the mean effect of all 32 experimental units is 40, the mean effect for the control is 34, and the mean effect for method B is 42. Also, the following abbreviated ANOVA is given:

Source of Variation	Degrees of Freedom	Mean Square
Replicates	7	32
Treatments	3	64
Experimental error	21	16

(a)   What is the experimental error variance per experimental unit?
(b)   Compute the coefficient of variation per experimental unit.
(c)   Compute the variance of a treatment mean.
(d)   Is the difference between the mean effects of control and method B significant at the 1 percent level?
(e)   Compute and interpret the 95 percent confidence interval estimate of the mean difference between the control and method B.

11.22   An experiment was conducted to assess the relative merits of 5 different gasolines. Since vehicle-to-vehicle variations in performance are inevitable, the test was run using 5 cars, hereafter called blocks. The following descriptions of the 5 gasolines are available:

A: control
B: control + additive X manufactured by company I
C: control + additive Y manufactured by company I
D: control + additive U manufactured by company II
E: control + additive V manufactured by company II

The data, in miles per gallon, are given below. Please analyze and interpret the data.

Treatment (gasoline)	Block (car)				
	1	2	3	4	5
A	22	20	18	17	19
B	28	24	23	19	25
C	21	23	25	25	27
D	26	21	21	22	20
E	27	25	22	20	24

11.23 Subdivide the experimental error sum of squares in each of the following problems in accordance with the principles given in Section 11.9:
  (a) 11.4   (c) 11.18   (e) 11.22
  (b) 11.17   (d) 11.19

11.24 Using the technique presented in Section 10.10, analyze further the data given in the following problems:
  (a) 11.1   (d) 11.6   (g) 11.19
  (b) 11.4   (e) 11.17   (h) 11.20
  (c) 11.5   (f) 11.18   (i) 11.22

11.25 In Table 11.33 we presented some data on blinking rates in successive 5-minute periods of reading. After substituting for a missing observation, these data were analyzed using a randomized complete block design. Ignoring the fact that we had to estimate a missing observation, discuss critically the use of a randomized complete block design in analyzing data of this type. If you feel that the use of a randomized complete block design was unjustified, state reasons to support your contention and give what you believe to be an appropriate method of handling such data. Examine all your assumptions carefully.

11.26 Five levels of fertilizer—0, 10, 20, 30, and 40—were applied to corn in a randomized complete block design. A preliminary ANOVA gave the following results:

Source of Variation	Degrees of Freedom	Mean Square
Replicates	4	2500
Fertilizers	4	2800
Experimental error	16	1500

The sums of the yields in the 5 plots of each level were:

Level	0	10	20	30	40
Total yield	20	140	260	300	280

What additional computations would you make to interpret the effect of treatments? Make these computations, and interpret the results.

11.27 The strength index of cotton fibers was thought to be affected by the application of potash to the soil. A randomized complete block experiment was conducted to get evidence. Here is a summary of the plot strength indexes:

Treatment (lb K_2O/A)	Replicate		
	1	2	3
36	7.62	8.00	7.93
54	8.14	8.15	7.87
72	7.76	7.73	7.74
108	7.17	7.57	7.80
144	7.46	7.68	7.21

Analyze the data. Plot the mean strength index for each treatment Y against the pounds of fertilizer per acre X. The sum of squares attributable to regression is 0.5662 with 1 degree of freedom (verify this). Subtract this from your sum of squares for treatments (4 degrees of freedom). The remainder (3 degrees of freedom) is the sum of squares of deviations from regression. Complete the analysis of variance. Test the hypothesis of 0 regression. What conclusions do you draw?

11.28   The following are the yields (tons per acre) of sugar beets on plots which had been treated with lime two years earlier:

Treatment (tons/acre)	Replicate				
	1	2	3	4	5
1	13.7	13.3	12.6	14.7	10.8
2	16.9	17.1	14.7	15.7	15.4
3	17.3	17.1	16.9	16.2	14.6
4	17.8	16.5	17.9	15.7	16.3

Analyze the data. Test the hypothesis that there is no effect of treatment. Plot the treatment means Y against the rate of application of lime X. Do you think the regression is linear? As evidence divide the sum of squares for treatments into two parts: attributable to regression (35.52) and remainder. Test the null hypothesis that there is no deviation from linear regression. Instead of thinking about regression, you might have divided the treatment sum of squares into these two parts: (1) due to difference between mean of first treatment and mean of the other 3 combined (43.02) and (2) differences among means of the last 3 treatments. What conclusions do you reach?

11.29   Consider an experiment to assess the relative effects of 4 different treatments (i.e., packing pressures) on the function time of a certain explosive actuator. Casings are available from 4 different production lots. Four casings were randomly selected from each of the lots and the treatments were assigned at random within each lot. Given the data shown below (operation time in milliseconds), analyze and interpret the results.

Block	Packing Pressure (psi)			
(lot)	10,000	20,000	30,000	40,000
1	12	17	10	12
2	11	16	9	11
3	10	15	8	11
4	9	15	8	10

11.30 Analyze and interpret the following data on yields of sweet potatoes obtained with various combinations of fertilizer (n = N, p = P_2O_5, k = K_2O).

		Replicate 1						Replicate 2				
npk	Yield	npk	Yield	npk	Yield	npk	Yield	npk	Yield	npk	Yield
133	45	211	39	333	70	212	83	211	56	133	65
111	34	313	62	311	40	221	52	321	49	112	48
221	42	222	65	212	45	322	65	333	92	311	56
323	69	233	92	132	53	313	101	122	75	332	79
213	58	123	56	121	54	111	50	312	86	213	95
331	51	332	91	223	69	331	61	232	74	222	81
232	72	131	73	322	85	132	89	223	109	231	84
122	56	112	55	113	60	123	90	113	68	323	103
312	82	321	75	231	78	233	122	131	98	121	64
Total	509	...	608	...	554	...	713	...	707	...	675
Grand Total	...	...	...	...	1671	...	...	...	...	...	2095

11.31 Given the following abbreviated ANOVA:

Source of Variation	Degrees of Freedom	Mean Square
Replicates	4	70
Treatments		
A	3	50
B	3	160
AB	9	40
Experimental error	60	10

Interpret the effects of a and b assuming that:
(a) The various levels of both a and b are fixed or selected.
(b) The various levels of both a and b are random variables.
(c) The levels of a are fixed, but the levels of b are random.
(d) The levels of a are random, but the levels of b are selected.

11.32 Mr. X sprayed apple leaves with different concentrations of a nitrogen compound, then determined the amounts of nitrogen (mg/dm^2) remaining on the leaves immediately and at two subsequent times. The object was to learn the rate at which the nitrogen was absorbed by the leaves. There were two replications of each treatment. The first entry in each cell of the table is for the first replication.

		Level of Nitrogen		
Time		n_1	n_2	n_3
t_0		2.29	6.50	8.75
		2.24	5.94	9.52
t_1		0.46	3.03	2.49
		0.19	1.00	2.04
t_2		0	0.75	1.40
		0.26	1.16	1.81

Obtain the ANOVA that subdivides the 8 degrees of freedom for treatments into individual comparisons: N_L, N_Q, T_L, T_Q, $N_L T_L$, $N_L T_Q$, $N_Q T_L$, and $N_Q T_Q$.

11.33 Given the following 2^3 factorial field plan with yields:

Replicate 1				Replicate 2				Replicate 3				Replicate 4			
(1)	7	b	24	ab	36	bc	31	a	28	ac	31	abc	66	(1)	11
abc	39	ac	31	(1)	19	ac	36	c	24	b	19	a	31	bc	29
a	30	c	21	abc	41	b	30	ab	35	(1)	13	c	21	ac	33
bc	27	ab	39	c	30	a	33	bc	26	abc	36	b	25	ab	43

Complete the ANOVA, computing the treatment sum of squares for each of the individual treatment effects and subdividing the experimental error corresponding to the subdivision of the treatment sum of squares.

11.34 Given the following ANOVA and table of treatment totals:

Source of Variation	Degrees of Freedom	Mean Square
Replicates	3	192
Treatments		
A	1	100
B	1	2500
AB	1	900
Experimental error	9	32

	a_0	a_1	Total
b_0	120	80	200
b_1	160	240	400
Total	280	320	600

(a) Interpret the effects of a and b assuming both are fixed variates.

(b) Compute and interpret the 95 percent confidence interval estimate of the true mean difference between treatments $a_1 b_1$ and $a_0 b_1$.

11.35 The following yields of grass were reported for one year (in dry matter per 1/57-acre plot). This was a randomized complete block design. Discuss the complete 2×3 factorial experiment, displaying the pertinent estimates. Outline tentative conclusions before making the ANOVA and tests of hypotheses.

Block	Elephant Grass (harvests per year)			Guatemala Grass (harvests per year)		
	2	3	4	2	3	4
1	109	222	187	277	246	252
2	97	125	163	293	263	181
3	133	134	143	260	194	224
4	113	173	179	325	190	248

11.36 Analyze and interpret the following set of experimental data: crop—oats; location—Flathus, Correctionville; year—1944; comment—yield in bushels per acre.

Treatment	Replicate			Treatment Total
	1	2	3	
$n_1 p_1 k_1$	32.2	33.9	34.6	100.7
$n_2 p_1 k_1$	37.4	40.9	38.9	117.2
$n_1 p_2 k_1$	30.6	39.4	33.8	103.8
$n_2 p_2 k_1$	52.4	48.0	43.9	144.3
$n_1 p_1 k_2$	29.9	34.5	36.5	100.9
$n_2 p_1 k_2$	42.2	29.9	34.1	106.2
$n_1 p_2 k_2$	31.8	32.5	34.2	98.5
$n_2 p_2 k_2$	46.6	49.5	46.7	142.8
Total	303.1	308.6	302.7	914.4

11.37 An experiment was conducted to assess the effects of 3 raw material sources (i.e., suppliers) and 4 mixtures (i.e., compositions) on the crushing strength of concrete blocks. Twenty-four blocks were selected, 2 at random from those manufactured by each of the 12 treatments, and the experiment was conducted as a randomized complete block with 2 replicates. The resulting data are given below. Analyze and interpret.

Supplier	Replicate	Mixture			
		A	B	C	D
1	1	57	65	93	102
	2	46	73	92	108
2	1	26	44	81	96
	2	38	67	90	99
3	1	39	57	96	105
	2	40	60	100	116

11.38   The following is a randomized complete block design with two missing plots. Fill in estimates for the missing values and complete the analysis of the data.

Block	Treatment				Block Total
	1	2	3	4	
1	43	35	37	42	157
2	45	39	40	47	171
3	42	30	M''	43	$115 + M''$
4	M'	43	48	49	$140 + M'$
5	41	34	36	44	155
Treatment Total	$171 + M'$	181	$161 + M''$	225	$738 + M' + M''$

REFERENCES

1. Cochran, W. G. 1937. Catalogue of uniformity trial data. *J. Roy. Stat. Soc.* (Suppl.), Vol. 4, No. 2.
2. Paull, A. E. 1950. On a preliminary test for pooling mean squares in the analysis of variance. *Ann. Math. Stat.* 21:539.
3. Tinker, M. A. 1945. Reliability of blinking frequency employed as a measure of readability. *J. Exp. Psych.* 35:418.
4. Yates, F. 1933. The analysis of replicated experiments when the field results are incomplete. *Emp. J. Exp. Agr.* 1:129.
5. ———. 1936. Incomplete randomized blocks. *Ann. Eugen.* 7:121.

FURTHER READING

See Chapter 10.

CHAPTER 12

ADDITIONAL DESIGNS AND TECHNIQUES RELATED TO EXPERIMENTAL DESIGNS

The completely randomized and randomized complete block designs are only two of many useful statistical designs that have been developed for special situations. Although we cannot discuss all those available, we present several additional designs and analysis techniques in this chapter. The Latin square, Graeco-Latin square, and split plot design are outlined. Also, some material on fractional factorial experiments, two-way analysis of variance with unequal frequencies, and response surface technology is presented.

12.1 LATIN AND GRAECO-LATIN SQUARES

The *Latin square* (LS) *design* is frequently used in agricultural and industrial experimentation. It is a special design that permits the researcher to assess the relative effects of various treatments when a double type of blocking restriction is imposed on the experimental units. Viewed in this way, the Latin square design is a logical extension of the randomized complete block design, and two examples should be sufficient to illustrate the ideas involved.

Example 12.1

Suppose we have 5 fertilizer treatments to be investigated and 25 plots available for experimentation. If the soil shows a fertility trend in *two* directions (say N $\rightarrow$ S and E $\rightarrow$ W), it would seem reasonable to set up blocks of 5 plots in *both* directions. This is precisely what is done under the names *rows* and *columns*. The treatments are then applied *at random*, subject to the restriction that each treatment appear but once in each row and each column.

Example 12.2

Consider the problem of testing 4 machines to see if they differ significantly in their ability to produce a certain manufactured part. It is well known that different operators and different time periods in the work day will have an effect on production. Thus we set up 4 operators as "columns" and 4 time periods as "rows" and then assign *at random* the machines to the various cells in the square, subject to the restriction that each machine be used only once by each operator and in each time period.

These two examples should acquaint the reader with the basic concepts involved in a Latin square design. The idea of a square is evident since if *m* treatments are to be investigated, we need m^2 experimental units.

The basic assumption for a Latin square design with one observation per experimental unit is that the observations may be represented by the linear statistical model

$$Y_{ij(k)} = \mu + \rho_i + \gamma_j + \tau_k + \epsilon_{ij(k)} \qquad i = 1, \ldots, m$$
$$j = 1, \ldots, m; k = 1, \ldots, m \qquad (12.1)$$

where

$$\sum_{i=1}^{m} \rho_i = \sum_{j=1}^{m} \gamma_j = \sum_{k=1}^{m} \tau_k = 0$$

and the $\epsilon_{ij(k)}$ are independently and normally distributed with mean 0 and common variance σ^2. The subscript k is placed in parentheses to indicate that it is not independent of i and j. The constants ρ_i, γ_j, and τ_k are the true effects associated with the ith row, jth column, and kth treatment respectively.

Because of the possible economies due to reduced sample sizes, the Latin square design has great appeal to researchers in all fields. Unfortunately, the design has not always been used wisely. An examination of the postulated statistical model will show that the interactions among rows, columns, and treatments have been assumed to be zero. In many experiments involving a Latin square design (e.g., in engineering applications where rows and columns usually refer to real chemical, physical, or other factors), it is precisely this assumption that appears to have been overlooked by the researcher. When information about interactions is lacking or when the assumption of zero interaction is of doubtful validity, one should consider running a full factorial experiment.

Having pointed out some of the advantages and limitations of a Latin square design, let us now summarize the appropriate analyses. The analysis of variance (ANOVA) is given in Table 12.1.

Equations useful for computing the sums of squares in Table 12.1 are:

$$\sum Y^2 = \sum_{i=1}^{m} \sum_{j=1}^{m} Y_{ij(.)}^2 \qquad (12.2)$$

$$M_{yy} = m^2(\overline{Y}_{..(.)})^2 = \left[\sum_{i=1}^{m} \sum_{j=1}^{m} Y_{ij(.)} \right]^2 \Big/ m^2 \qquad (12.3)$$

$$R_{yy} = m \sum_{i=1}^{m} [\overline{Y}_{i.(.)} - \overline{Y}_{..(.)}]^2 = \sum_{i=1}^{m} R_i^2/m - M_{yy} \qquad (12.4)$$

$$C_{yy} = m \sum_{j=1}^{m} [\overline{Y}_{.j(.)} - \overline{Y}_{..(.)}]^2 = \sum_{j=1}^{m} C_j^2/m - M_{yy} \qquad (12.5)$$

$$T_{yy.} = m \sum_{k=1}^{m} [\overline{Y}_{..(k)} - \overline{Y}_{..(.)}]^2 = \sum_{k=1}^{m} T_k^2/m - M_{yy} \qquad (12.6)$$

TABLE 12.1–Generalized ANOVA for an $m \times m$ Latin Square
Design with One Observation per Experimental Unit

Source of Variation	Degrees of Freedom	Sum of Squares	Mean Square	Expected Mean Square	F Ratio
Mean	1	M_{yy}	**M**		
Rows	$m - 1$	R_{yy}	**R**	$\sigma^2 + [m/(m-1)]\sum_{i=1}^{m} \rho_i^2$	...
Columns	$m - 1$	C_{yy}	**C**	$\sigma^2 + [m/(m-1)]\sum_{j=1}^{m} \gamma_j^2$	...
Treatments	$m - 1$	T_{yy}	**T**	$\sigma^2 + [m/(m-1)]\sum_{k=1}^{m} \tau_k^2$	**T/E**
Experimental error	$(m-1)(m-2)$	E_{yy}	**E**	σ^2	...
Total	m^2	$\sum Y^2$	...	...	...

$$E_{yy} = \sum_{i=1}^{m} \sum_{j=1}^{m} \sum_{k=1}^{m} [Y_{ij(k)} - \overline{Y}_{i.(.)} - \overline{Y}_{.j(.)} - \overline{Y}_{..(k)} + 2\overline{Y}_{..(.)}]^2$$
$$= \sum Y^2 - M_{yy} - R_{yy} - C_{yy} - T_{yy} \tag{12.7}$$

where

$$R_i = \sum_{j=1}^{m} Y_{ij(.)} \qquad C_j = \sum_{i=1}^{m} Y_{ij(.)} \qquad T_k = \sum_{i=1}^{m} \sum_{j=1}^{m} Y_{ij(k)}$$

are the row, column, and treatment totals ($\overline{Y}_{i.(.)}$, $\overline{Y}_{.j(.)}$, $\overline{Y}_{..(k)}$, and $\overline{Y}_{..(.)}$ also are the row, column, treatment, and overall mean respectively).

To write the model in Eq. (12.1) as a multiple regression model, we must introduce independent variables for the row, column, and treatment effects. These we label u, v, and x respectively. The model for the Latin square design is then

$$Y_{ij(k)} = \mu + \rho_1 u_{1i} + \cdots + \rho_{m-1} u_{m-1 i} + \gamma_1 v_{1j} + \cdots + \gamma_{m-1} v_{m-1 j}$$
$$+ \tau_1 x_{1(k)} + \cdots + \tau_{m-1} x_{m-1(k)} + \epsilon_{ij(k)}$$
$$i = 1, \ldots, m; j = 1, \ldots, m; k = 1, \ldots, m \tag{12.8}$$

where all three independent variables are of the form

$$w_{np} = 1 \quad \text{if } p = n; n = 1, \ldots, m - 1$$
$$= -1 \quad \text{if } p = m$$
$$= 0 \quad \text{otherwise}$$

The restrictions $\sum_{i=1}^{m} \rho_i = \sum_{j=1}^{m} \gamma_j = \sum_{k=1}^{m} \tau_k = 0$ have been included in the definitions of the independent variables to ensure a unique solution to the normal equations.

Example 12.3

Suppose an experiment involves three treatments and the Latin square

A	C	B
B	A	C
C	B	A

When the model in Eq. (12.8) is used, the regression equations, in matrix notation, are

$$
\begin{bmatrix}
Y_{11(1)} \\
Y_{12(3)} \\
Y_{13(2)} \\
Y_{21(2)} \\
Y_{22(1)} \\
Y_{23(3)} \\
Y_{31(3)} \\
Y_{32(2)} \\
Y_{33(1)}
\end{bmatrix}
=
\begin{bmatrix}
1 & 1 & 0 & 1 & 0 & 1 & 0 \\
1 & 1 & 0 & 0 & 1 & -1 & -1 \\
1 & 1 & 0 & -1 & -1 & 0 & 1 \\
1 & 0 & 1 & 1 & 0 & 0 & 1 \\
1 & 0 & 1 & 0 & 1 & 1 & 0 \\
1 & 0 & 1 & -1 & -1 & -1 & -1 \\
1 & -1 & -1 & 1 & 0 & -1 & -1 \\
1 & -1 & -1 & 0 & 1 & 0 & 1 \\
1 & -1 & -1 & -1 & -1 & 1 & 0
\end{bmatrix}
\begin{bmatrix}
\mu \\
\rho_1 \\
\rho_2 \\
\gamma_1 \\
\gamma_2 \\
\tau_1 \\
\tau_2
\end{bmatrix}
+
\begin{bmatrix}
\epsilon_{11(1)} \\
\epsilon_{12(3)} \\
\epsilon_{13(2)} \\
\epsilon_{21(2)} \\
\epsilon_{22(1)} \\
\epsilon_{23(3)} \\
\epsilon_{31(3)} \\
\epsilon_{32(2)} \\
\epsilon_{33(1)}
\end{bmatrix}
$$

Since the row, column, and treatment effects are orthogonal (i.e., the respective columns in the X matrix are orthogonal) the row, column, and treatment sums of squares are easily retrievable from the regression analysis. In particular, using the notation of Table 12.1, $SS(\rho_1, \ldots, \rho_{m-1} \mid \mu) = R_{yy}$; $SS(\gamma_1, \ldots, \gamma_{m-1} \mid \mu) = C_{yy}$; and $SS(\tau_1, \ldots, \tau_{m-1} \mid \mu) = T_{yy}$.

Example 12.4

The data shown in Table 12.2 resulted from an experiment such as described in Example 12.2. Assuming that time periods, operators, and machines do not interact (either pairwise or as a complete set), the ANOVA of Table 12.3 is obtained. This leads to the conclusion that there are significant differences among the outputs of the 4 machines. Further examination of the data should permit selection of the most productive machine or machines.

If a single observation should be missing in an experiment conducted according to an $m \times m$ Latin square design, its value may be estimated using

$$M = [m(R + C + T) - 2S]/(m - 1)(m - 2) \qquad (12.9)$$

TABLE 12.2–Number of Units Produced by Four Machines
in a Latin Square Design

Time Period	Operator			
	1	2	3	4
1	31 (C)	43 (D)	67 (A)	36 (B)
2	39 (D)	96 (A)	40 (B)	48 (C)
3	57 (B)	33 (C)	40 (D)	84 (A)
4	85 (A)	46 (B)	48 (C)	50 (D)

Note: The random assignment of the machines is shown by the letters in parentheses.

TABLE 12.3–Abbreviated ANOVA for Data of Table 12.2

Source of Variation	Degrees of Freedom	Sum of Squares	Mean Square	Expected Mean Square
Time periods	3	408.188	136.06	$\sigma^2 + (4/3)\sum_{i=1}^{4} \rho_i^2$
Operators	3	88.688	29.56	$\sigma^2 + (4/3)\sum_{j=1}^{4} \gamma_j^2$
Machines	3	4946.688	1648.90**	$\sigma^2 + (4/3)\sum_{k=1}^{4} \tau_k^2$
Experimental error	6	515.874	85.98	σ^2

**Significant at $\alpha = 0.01$.

where R = sum of observations in the same row as the missing observation

C = sum of observations in the same column as the missing observation

T = sum of observations with the same treatment as the missing observation

S = sum of all the actual observations

After substituting the value of M in the table, the various sums of squares are calculated as indicated above. However, it must be remembered that the treatment sum of squares T_{yy} will be biased upwards, and a correction must be applied before we test the hypothesis $H:\tau_k = 0$ $(k = 1, \ldots, m)$. This correction is made by computing a new treatment sum of squares T'_{yy} defined as

$$T'_{yy} = T_{yy} - Z \tag{12.10}$$

where

$$Z = [S - R - C - (m - 1)T]^2/(m - 1)^2(m - 2)^2 \tag{12.11}$$

Remember that the degrees of freedom associated with experimental error and total are *each reduced by one* (in view of the single missing observation); i.e., the degrees of freedom for experimental error are now $(m - 1)(m - 2) - 1$, and the degrees of freedom for total are now $m^2 - 1$. No example will be given for the above technique, but one or two of the problems at the end of this chapter will illustrate the principles involved.

By now the reader should be sufficiently adept at the calculations involved in analyses of variance so that lengthy discussions of such topics as subsampling, selected treatment comparisons, factorials, analysis of response functions, estimation of components of variance, and predictions of the relative efficiencies of various allocations of the observations in terms of experimental and sampling units would be a waste of time. Accordingly, we will do no more than state that the techniques introduced in Chapters 10 and 11 may easily be extended and adapted for use with Latin square designs. However, to make certain that the previously mentioned extensions and adaptations are made properly, a few problems requiring their use have been included at the end of this chapter.

Before terminating our discussion of the Latin square design, mention must be made of its efficiency relative to completely randomized and randomized complete block designs. (**Note:** This discussion will be closely related to that of Section 11.7.) If we designate the mean squares in the Latin square for rows, columns, and experimental error by **R**, **C**, and **E** respectively, we may readily evaluate the efficiency of a Latin square design relative to either a completely randomized or randomized complete block design. For the efficiency of a Latin square design relative to a completely randomized design, we calculate

$$\text{RE} = [\mathbf{R} + \mathbf{C} + (m - 1)\mathbf{E}]/(m + 1)\mathbf{E} \tag{12.12}$$

If, however, we wish to compare a Latin square design with what might have happened had a randomized complete block design been utilized (assuming the rows were used as blocks), the following formula is appropriate:

$$\text{RE} = [\mathbf{C} + (m - 1)\mathbf{E}]/m\mathbf{E} \tag{12.13}$$

If columns were used as the blocks, we put **R** in place of **C** in Eq. (12.13).

The concept of a Latin square design can be extended rather easily to that of a *Graeco-Latin square* (G-LS) *design*. Rather than go into the details of a Graeco-Latin square, we shall only indicate by example the nature of the design. Persons interested in using such a design are advised to consult a professional statistician.

Example 12.5

Chew (6) describes an experiment that could be used to compare 5 formulations $(\alpha, \beta, \gamma, \delta, \epsilon)$ for making concrete bricks, using material from 5 batches, prepared on each of 5 days, and tested on 5 different machines (A, B, C, D, E). One possible randomization of a Graeco-Latin

TABLE 12.4–Symbolic Representation of the Graeco-Latin Square
Used in Example 12.5

Row (batches)	Column (days)				
	1	2	3	4	5
1	$A\alpha$	$B\gamma$	$C\epsilon$	$D\beta$	$E\delta$
2	$B\beta$	$C\delta$	$D\alpha$	$E\gamma$	$A\epsilon$
3	$C\gamma$	$D\epsilon$	$E\beta$	$A\delta$	$B\alpha$
4	$D\delta$	$E\alpha$	$A\gamma$	$B\epsilon$	$C\beta$
5	$E\epsilon$	$A\beta$	$B\delta$	$C\alpha$	$D\gamma$

square design for this situation is shown in Table 12.4. It will be noted that (1) each Latin letter appears exactly once in each row and each column, (2) each Greek letter appears exactly once in each row and each column, and (3) each Latin letter appears exactly once with each Greek letter.

As tempting as Graeco-Latin squares are to an experimenter (because of the potential savings in numbers of observations), they should be used with caution. This recommendation stems from the same type of limitation that was emphasized for Latin square designs; namely, no interactions are tolerated.

12.2 SPLIT PLOT DESIGNS

A fairly simple design that is frequently used in experimental work is the *split plot* (SP) *design.* In this design we are concerned with two factors, but we wish more precise information on one of them than on the other. Let us assume that we have factors *a* and *b* and desire more accurate information on *b* than on *a*. The usual scheme is to assign the various levels of factor *a* at random to the whole plots (main plots) in each replicate as in a randomized complete block design. Following this, the levels of *b* are assigned at random to the split plots (subplots) *within each whole plot.* Such a scheme of randomization may arise not only from the desire for more precise information on one factor than on another but also because of the nature of the factors and the way in which they must be applied to the experimental units.

Example 12.6

An experiment similar to that described in Example 9.19 was performed except that the weight of the heat paper was not a factor. As before, we assume that the test engineer wishes to place 4 batteries in the temperature chamber at the same time. Thus for each replicate the test temperature is maintained and the 4 levels of electrolyte are tested simultaneously at the same temperature. The way the two factors (test temperature and electrolyte) are applied to the experimental units (batteries) leads one to a split plot analysis with temperature as the whole plot factor and electrolyte as the split plot factor.

The linear statistical model for a split plot design with one observation per experimental unit is

$$Y_{ijk} = \mu + \rho_i + \alpha_j + \eta_{ij} + \beta_k + (\alpha\beta)_{jk} + \epsilon_{ijk}$$

$$i = 1,\ldots,r; j = 1,\ldots,a; k = 1,\ldots,b \qquad (12.14)$$

where η_{ij} and ϵ_{ijk} are independent normal random variables with mean 0 and variances σ_w^2 and σ_s^2 respectively and are the whole plot and split plot experimental errors. The constants ρ_i, α_j, and β_k are the true effects associated with the ith replicate, jth level of a, and kth level of b, respectively.

TABLE 12.5–Generalized ANOVA for a Split Plot Design

Source of Variation	Degrees of Freedom	Sum of Squares	Mean Square	Expected Mean Square	F Ratio
Mean	1	M_{yy}	M	$\ldots$	$\ldots$
Whole plots					
Replicates	$r - 1$	R_{yy}	R	$\ldots$	$\ldots$
A	$a - 1$	A_{yy}	A	$\sigma_s^2 + b\sigma_w^2 + rb \sum_{j=1}^{a} \alpha_j^2/(a-1)$	A/E_w
Whole plot error	$(r-1)(a-1)$	$(E_w)_{yy}$	E_w	$\sigma_s^2 + b\sigma_w^2$	$\ldots$
Split plots					
B	$b - 1$	B_{yy}	B	$\sigma_s^2 + ra \sum_{k=1}^{b} \beta_k^2/(b-1)$	B/E_s
AB	$(a-1)(b-1)$	$(AB)_{yy}$	AB	$\sigma_s^2 + r \sum_{j=1}^{a} \sum_{k=1}^{b} (\alpha\beta)_{jk}^2/(a-1)(b-1)$	AB/E_s
Split plot error	$(r-1)a(b-1)$	$(E_s)_{yy}$	E_s	σ_s^2	$\ldots$
Total	rab	$\sum Y^2$	$\ldots$	$\ldots$	$\ldots$

The ANOVA associated with the split plot design is given in Table 12.5. Formulas useful for computing the sums of squares are:

$$\sum Y^2 = \sum_{i=1}^{r} \sum_{j=1}^{a} \sum_{k=1}^{b} Y_{ijk}^2 \qquad (12.15)$$

$$M_{yy} = rab(\overline{Y}_{...})^2 = T^2/rab \qquad (12.16)$$

$$R_{yy} = ab \sum_{i=1}^{r} (\overline{Y}_{i..} - \overline{Y}_{...})^2 = \sum_{i=1}^{r} R_i^2/ab - M_{yy} \qquad (12.17)$$

$$A_{yy} = rb \sum_{j=1}^{a} (\overline{Y}_{.j.} - \overline{Y}_{...})^2 = \sum_{j=1}^{a} A_j^2/rb - M_{yy} \qquad (12.18)$$

$$(E_w)_{yy} = b \sum_{i=1}^{r} \sum_{j=1}^{a} (\overline{Y}_{ij.} - \overline{Y}_{i..} - \overline{Y}_{.j.} + \overline{Y}_{...})^2$$

$$= \sum_{i=1}^{r} \sum_{j=1}^{a} W_{ij}^2/b - A_{yy} - R_{yy} - M_{yy} \qquad (12.19)$$

$$B_{yy} = ra \sum_{k=1}^{b} (\overline{Y}_{..k} - \overline{Y}_{...})^2 = \sum_{k=1}^{b} B_k^2/ra - M_{yy} \tag{12.20}$$

$$(AB)_{yy} = r \sum_{j=1}^{a} \sum_{k=1}^{b} (\overline{Y}_{.jk} - \overline{Y}_{.j.} - \overline{Y}_{..k} + \overline{Y}_{...})^2$$

$$= \sum_{j=1}^{a} \sum_{k=1}^{b} T_{jk}^2/r - A_{yy} - B_{yy} - M_{yy} \tag{12.21}$$

$$(E_s)_{yy} = \sum_{i=1}^{r} \sum_{j=1}^{a} \sum_{k=1}^{b} (Y_{ijk} - \overline{Y}_{.jk} - \overline{Y}_{ij.} + \overline{Y}_{.j.})^2$$

$$= \sum Y^2 - M_{yy} - R_{yy} - A_{yy} - (E_w)_{yy} - B_{yy} - (AB)_{yy} \tag{12.22}$$

where $R_i = Y_{i..}$, $A_j = Y_{.j.}$, and $B_k = Y_{..k}$ are the replicate, whole plot, and split plot treatment totals respectively; $W_{ij} = Y_{ij.}$ is the sum of the observations in the jth whole plot in the ith replicate; $T_{jk} = Y_{.jk}$ is the jkth treatment total; and $T = Y_{...}$ is the total of all the observations.

With regard to the use of regression analysis and a computer to do the necessary computations for the ANOVA, the regression model is similar to the models set up for the other designs. One difference is the whole plot error η_{ij} in the model in Eq. (12.14). A convenient way to evaluate the sum of squares $(E_w)_{yy}$ for the whole plot error is to treat η_{ij} as the interaction between the replicates and the whole plot treatment. With that change in the model, the analysis is straightforward.

For the variances of a treatment mean we must distinguish between the whole plot and split plot treatments. The estimated variance of a split plot mean is

$$\hat{V}(\overline{Y}_{..k}) = E_s/ra \tag{12.23}$$

while the estimated variance of a whole plot mean is

$$\hat{V}(\overline{Y}_{.j.}) = E_w/rb \tag{12.24}$$

The estimates of the variance components σ_s^2 and σ_w^2 are

$$s_s^2 = E_s \tag{12.25}$$

and

$$s_w^2 = (E_w - E_s)/b \tag{12.26}$$

unless $E_s > E_w$, when the estimate of σ_w^2 would be zero.

Example 12.7

Data for the experiment in Example 12.6 are given in Table 12.6. The resulting ANOVA is given in Table 12.7. No further interpretation of the data is possible because of the lack of information regarding the exact nature of the treatments.

TABLE 12.6–Activated Lives (hr) of 72 Thermal Batteries Tested in a Split Plot Design Which Used Temperatures as Whole Plots and Electrolytes as Split Plots

Temperature	Electrolyte	Replicate					
		1	2	3	4	5	6
Low	A	2.17	1.88	1.62	2.34	1.58	1.66
	B	1.58	1.26	1.22	1.59	1.25	0.94
	C	2.29	1.60	1.67	1.91	1.39	1.12
	D	2.23	2.01	1.82	2.10	1.66	1.10
Medium	A	2.33	2.01	1.70	1.78	1.42	1.35
	B	1.38	1.30	1.85	1.09	1.13	1.06
	C	1.86	1.70	1.81	1.54	1.67	0.88
	D	2.27	1.81	2.01	1.40	1.31	1.06
High[*]	A	1.75	1.95	2.13	1.78	1.31	1.30
	B	1.52	1.47	1.80	1.37	1.01	1.31
	C	1.55	1.61	1.82	1.56	1.23	1.13
	D	1.56	1.72	1.99	1.55	1.51	1.33

TABLE 12.7–Abbreviated ANOVA for Data of Table 12.6

Source of Variation	Degrees of Freedom	Sum of Squares	Mean Square
Whole plots			
Replicates	5	4.1499	0.8300
Temperatures	2	0.1781	0.0890
Whole plot error	10	1.3622	0.1362
Split plots			
Electrolytes	3	1.9625	0.6542**
Temperature × electrolyte	6	0.2105	0.0351
Split plot error	45	1.2586	0.0280

**Significant at $\alpha = 0.01$.

Before leaving the subject of split plot designs, we note that the principle of "splitting" may be carried on for several stages; i.e., we may employ split–split plot designs, etc. For more detailed discussion of such designs and for some illustrative examples, the reader should consult the list of further readings at the end of Chapter 10.

12.3 COMPLETE FACTORIALS WITHOUT REPLICATION, FRACTIONAL FACTORIALS, AND INCOMPLETE BLOCKS

Because most experimenters are interested in investigating the effects on a response variable of the simultaneous variation of many factors, a large number of designs incorporate factorial treatment combinations. However, as the number of factors increases, the size of the experiment becomes prohibitive. In addition, it becomes difficult to control the magnitude of the experimental error within reasonable bounds.

In an attempt to reduce the experimental error to a reasonable magnitude, the principle of confounding (see Chapter 9) was utilized to create a group of designs known as *incomplete block designs*. These designs are so named because not all the treatment combinations are present in each block; i.e., the blocks are incomplete. With adequate replication, these designs have proved very useful in agricultural experimentation.

Since incomplete block designs are not usually included in a first course in statistical methods, we will not discuss them in this book. However, references (7) and (12) as well as several of the books listed at the end of Chapter 10 discuss at length the methods of analyses appropriate to such designs.

When engineers and physical scientists became interested in statistically designed, multifactor experiments, they decided that both replicated complete factorials and incomplete block designs were unsatisfactory, since they required too many experimental units. Further, it was evident that as a general rule the experimental errors in industrial experiments were much smaller than those encountered in agricultural experiments. Because of the small experimental errors, one common approach has been to avoid replication (i.e., subject *only one* experimental unit to each treatment combination) and to estimate the experimental error by pooling the mean squares associated with the higher order interactions. (**Note:** This is equivalent to assuming that the true high-order interaction effects are 0.) This technique, referred to in the title of this section as *complete factorials without replication,* is used quite often. (**Note:** Actually, it is a completely randomized design involving factorial treatment combinations and utilizing only one experimental unit per treatment combination.)

Rather than devote a lot of space to the discussion of the models and assumptions for the many possible situations, let us consider an example. It is hoped that this will prove sufficient for a reasonable understanding of the principles involved. For persons wishing to consider the matter more thoroughly, we recommend reference (1) as well as the books listed at the end of Chapter 10.

Example 12.8

Davies (9) considered a laboratory experiment to investigate the yield of an isatin derivative as a function of acid strength a, time of reaction b, amount of acid c, and temperature of reaction d. Two levels of each factor were used, namely:

a: 87 percent, 93 percent
b: 15 min, 30 min
c: 35 ml, 45 ml
d: 60 C, 70 C

The data shown in Table 12.8 led to the ANOVA of Table 12.9.

TABLE 12.8–Yield of Isatin Derivative (g/10 g base material)

Acid Strength (a)	Reaction Time (b)	Temperature of Reaction (d)			
		60 ± 1		70 ± 1	
		Amount of acid (c)		Amount of acid (c)	
		35 ml	45 ml	35 ml	45 ml
87	15 min	6.08 (1)	6.31 (c)	6.79 (d)	6.77 (cd)
	30 min	6.53 (b)	6.12 (bc)	6.73 (bd)	6.49 (bcd)
93	15 min	6.04 (a)	6.09 (ac)	6.68 (ad)	6.38 (acd)
	30 min	6.43 (ab)	6.36 (abc)	6.08 (abd)	6.23 (abcd)

Source: Davies (9), p. 275, Table 7.7. By permission of the author and the publishers.

TABLE 12.9–Abbreviated ANOVA for Data of Table 12.8

Source of Variation	Degrees of Freedom	Mean Square
Main effects		
A	1	0.1463
B	1	0.0018
C	1	0.0233
D	1	0.2998
Two-factor interactions		
AB	1	0.0000
AC	1	0.0046
AD	1	0.1040
BC	1	0.0176
BD	1	0.2525
CD	1	0.0028
Experimental error (pooled high-order interactions)	5	0.0385

Source: Davies (9), p. 277, Table 7.72. By permission of the author and the publishers.

As helpful as it was, the approach taken in the two preceding paragraphs and illustrated in Example 12.8 (i.e., the utilization of a complete factorial without replication) was not enough. Experiments were still too large to suit the researcher. Some other way had to be found to reduce the size and cost. One such attempt was the development of *fractional factorials* in which only some (a fraction) of the treatment combinations are actually investigated.

Once an experimenter has decided that some form of fractional factorial is appropriate for his needs, the question naturally arises, "Which treatment combinations should be included in the experiment?" The answer to this question depends on what assumptions the experimenter is willing to make or, to phrase it differently, what information he is willing to forego. As we all know, you can seldom get something for nothing, and the desired smaller experiment with its associated savings can only be achieved at a cost, namely, the cost of giving up part of the information usually derived from complete factorials.

TABLE 12.10–Treatment Combinations to Be Used in a One-Half Replicate of a 2^6 Factorial in Which the Defining Relation Is $I = ABCDEF$

Experimental Unit	Treatment Combination	Experimental Unit	Treatment Combination
1	(1)	17	ad
2	de	18	ae
3	ef	19	adef
4	df	20	af
5	ab	21	bd
6	abde	22	be
7	abef	23	bdef
8	abdf	24	bf
9	ac	25	cd
10	acde	26	ce
11	acef	27	cdef
12	acdf	28	cf
13	bc	29	abcd
14	bcde	30	abce
15	bcef	31	abcdef
16	bcdf	32	abcf

Note: The use of the symbol I rather than M (as in Chapter 9) is to agree with convention. The equality sign is used as an abbreviation for "is completely confounded with."

To illustrate the nature of a fractional factorial, let us consider a one-half replicate of a 2^6 factorial. If we have 32 experimental units and subject them to the treatment combinations shown in Table 12.10, the principles introduced in Section 9.14 may be invoked to show the equivalences (i.e., confoundings) of effects listed in Table 12.11. If the experimenter is willing to assume that all interaction effects involving three or more factors are 0, this fractional factorial is adequate to estimate all main effects, all two-factor interactions, and experimental error. Under such an assumption, the appropriate ANOVA is as given in Table 12.12.

Fractional factorials have wide application in industrial experimentation. Thus it will pay research workers in both engineering and the physical sciences to become better acquainted with these valuable aids to efficient experimentation. As with other topics mentioned in this section, it is felt that a detailed discussion is beyond the scope of this book. [Persons desiring further details should consult references (5), (8), and (10) as well as the books listed at the end of Chapter 10.]

TABLE 12.11–Confounded Effects in a One-Half
Replicate of a 2^6 Factorial in Which the
Defining Relation Is $I = ABCDEF$

$I = ABCDEF$	$E = ABCDF$
$A = BCDEF$	$AE = BCDF$
$B = ACDEF$	$BE = ACDF$
$AB = CDEF$	$ABE = CDF$
$C = ABDEF$	$CE = ABDF$
$AC = BDEF$	$ACE = BDF$
$BC = ADEF$	$BCE = ADF$
$ABC = DEF$	$ABCE = DF$
$D = ABCEF$	$DE = ABCF$
$AD = BCEF$	$ADE = BCF$
$BD = ACEF$	$BDE = ACF$
$ABD = CEF$	$ABDE = CF$
$CD = ABEF$	$CDE = ABF$
$ACD = BEF$	$ACDE = BF$
$BCD = AEF$	$BCDE = AF$
$ABCD = EF$	$ABCDE = F$

TABLE 12.12–Abbreviated ANOVA for the Experiment
of Table 12.10

Source of Variation	Degrees of Freedom
Mean	1
A	1
B	1
C	1
D	1
E	1
F	1
AB	1
AC	1
AD	1
AE	1
AF	1
BC	1
BD	1
BE	1
BF	1
CD	1
CE	1
CF	1
DE	1
DF	1
EF	1
Experimental error (higher order interactions)	10
Total	32

12.4 UNEQUAL BUT PROPORTIONATE SUBCLASS NUMBERS

The reader will have noticed that practically all the recommended statistical designs require a balanced configuration, i.e., an equal number of observations in each group. The one exception was the completely randomized design. However, even in that case it was noticed that unequal numbers of observations in the subgroups could lead to difficulties in interpretation. (See Section 10.4.)

In this section we propose to examine one other case of unequal frequencies that presents little difficulty in the way of calculation. This case involves a factorial set of two treatments in which the cells of the two-way table contain different numbers of observations, *but these numbers happen to be proportional.* That is, the number of observations in the *ij*th cell are such that $n_{ij} = u_i v_j$, where $u_1:u_2 \cdots :u_a$ are the proportions in the rows and $v_1:v_2: \cdots :v_b$ are the proportions in the columns. Rather than including details, a numerical example will be given and it is hoped this will be sufficient to illustrate the ideas involved. Persons desiring further details should consult some of the books listed at the end of Chapter 10.

Example 12.9

Suppose we have 3 varieties of oats to be tested for yield differences and we also wish to investigate the effects of 3 fertilizers. There are 28 experimental plots available to the researcher. Further, we will assume that from previous experiments we already know considerably more about varieties B and C than about variety A; thus we shall plant variety A on twice as many plots as varieties B and C. It is also considered desirable

TABLE 12.13–Yields of Three Varieties of Oats Subjected to Three Different Fertilizer Treatments (bushels/acre)

Oat Variety	Fertilizer		
	1	2	3
A	50, 51, 52, 56, 60, 55	42, 40, 38, 38	55, 56, 56, 58
B	65, 69, 67	50, 50	62, 62
C	67, 67, 69	48, 50	65, 67

TABLE 12.14–Abbreviated ANOVA for Data of Table 12.13

Source of Variation	Degrees of Freedom	Sum of Squares	Mean Square
Treatments			
Varieties	2	818.9	409.45
Fertilizers	2	1455.0	727.50
Varieties × fertilizers	4	52.3	13.075
Plots treated alike	19	100.5	5.289

to assign the 3 fertilizers to the plots in the ratio 3:2:2, i.e., we shall apply fertilizer 1 to 12 plots and each of fertilizers 2 and 3 to 8 plots. The assignment of the treatment combinations to the plots was made completely at random, and the resulting yields (bushels/acre) are recorded in Table 12.13. Calculating the various sums of squares in the usual manner, we arrive at the abbreviated ANOVA of Table 12.14.

12.5 UNEQUAL AND DISPROPORTIONATE SUBCLASS NUMBERS

Let us now examine the case where our data may be represented by the model

$$Y_{ijk} = \mu + \alpha_i + \beta_j + (\alpha\beta)_{ij} + \epsilon_{ijk}$$
$$i = 1, \ldots, a; j = 1, \ldots, b; k = 1, \ldots, n_{ij} \qquad (12.27)$$

where the various terms are defined as before but the n_{ij} are not equal for the various cells of the two-way table. Further, the n_{ij} are not proportionate as they were in Section 12.4. What difficulties in analysis result from this fact? Why is it that we refer to the case of "unequal and disproportionate subclass numbers" as an undesirable situation? The answer is because we encounter complications in analyzing such data. Let us now take note of some of the problems that arise.

Suppose we went ahead, ignoring the fact that the subclass numbers are disproportionate, and calculated the various sums of squares in the usual fashion. If this procedure were followed, we would find that the sums of squares so calculated (assuming that each sum of squares was calculated directly; i.e., none was obtained by subtraction) would *not* sum up to agree with the total sum of squares. In other words, because of the disproportionality of the subclass numbers, the different comparisons with which the sums of squares are associated are *nonorthogonal*. This would lead to biased test procedures unless some adjustment were made. The other major difficulty that arises when dealing with cases involving disproportionate subclass numbers is that the simple (unweighted) treatment means obtained from the data are *biased* estimates of the true treatment effects. This could lead to serious errors if inferences were made without attempting to correct for this bias.

What then should be the method of analysis for such situations? The usual approach is to utilize regression techniques and to obtain a general least squares solution. However, because of the many variations that may be employed (e.g., different models and/or different orders of estimating the unknown parameters), neither detailed explanations nor numerical illustrations of such solutions will be included in this book. If you should encounter a situation in which a general least squares solution is required, we suggest that you do three things: (1) review the contents of Chapter 7, (2) study the appropriate sections of some of the books listed at the end of Chapter 10, and (3) consult a professional statistician.

12.6 RANDOM BALANCE

Another interesting contribution to experimental design is the concept of *random balance* investigations in multi-multifactor experiments. As proposed by Satterthwaite (13), the random balance technique permits the researcher to screen a large number of possible contributing factors in an experiment involving a limited number of test runs. That is, random balance is a device for considering (simultaneously) the many factors involved and (as is always important) keeping the size of the experiment within reasonable bounds. When the experiment has been performed, examination of the results should permit isolation of the more important factors for further investigation.

In a random balance experiment, all factors and levels are considered by choosing *at random* the level of each factor to be used in forming a particular treatment combination. (**Note:** From a practical point of view, the following restriction on complete randomization has been found desirable: Each level of a particular factor should be used an equal or nearly equal number of times.) Since random balance experimentation was first proposed, there has been much discussion, both pro and con, as to its worth. We believe that random balance has much to recommend it and that its use will increase, especially in industrial experimentation. However, the theory on which it is based has not been fully explored, and thus the controversy over its merits continues. For those interested in the possibilities and/or wisdom of using random balance in their own experimentation, we suggest a careful reading of references (3), (4), (13), and (14).

12.7 RESPONSE SURFACE TECHNIQUES

A significant contribution to statistical methodology has been the development of systematic procedures for experimentally determining levels of the factors under investigation that produce an optimum response. These procedures, frequently referred to as *response surface techniques,* can be of value to researchers in almost every field of specialization. Unfortunately, a satisfactory description of the many ramifications of these techniques is more than can be accomplished in this text. Therefore, we shall be content with a few general observations on the topic and then refer the reader to other sources where these ideas are discussed in greater detail.

Response surface techniques are in essence a blending of regression analysis (Chapter 7) and experimental design (Chapters 9–12) to provide an economical means of locating a set of experimental conditions (i.e., a combination of factor levels) that will yield a maximum (or minimum) response. However, one very important feature has been added, i.e., the *sequential* nature of the exploration of the response surface. While it is true that most research is of the continuing variety (and therefore sequential), the majority of the techniques discussed heretofore in this book have been of the nonsequential type. Thus the insertion of the sequential element into the pattern of the investigation is, from one point of view, a long overdue step.

In capsule form the steps involved in the application of response surface techniques are as follows:

1. Choose base levels of the factors to be investigated. (Depending on the judgment of the experimenter, these levels may be close to or far removed from the optimum levels.)
2. Since at this stage linear effects are thought to be dominant over nonlinear effects, select one other level of each factor.
3. Utilizing either a complete or fractional 2^n factorial, estimate (by examining the effects, i.e., the linear regression coefficients) the direction in which the greatest gain may be expected.
4. Moving in this direction (i.e., along the *path of steepest ascent*) to the extent that the experimenter deems reasonable, a second experiment (again utilizing a complete or fractional 2^n factorial) is performed.
5. Repeat steps 3 and 4 until a nearly stationary region is found.
6. Then, utilizing a complete or fractional 3^n factorial or a *composite design* to estimate the second-order effects, the nature of the response surface may be explored in the nearly stationary region and the optimum conditions located. (**Note:** A *composite design* is essentially a complete 2^n factorial with sufficient points added to permit estimation of the second-order effects.)

It will be realized that the preceding steps are only an indication of the procedure. Depending on the problem and the assumptions that the experimenter is willing to make, the "rules" may be modified. (For example, 3^n factorials or composite designs might be used in step 3.) However, regardless of the details, the philosophy of sequential experimentation has much to recommend it. In fact, the concept of exploring a response surface in a sequential manner with the objective of locating a maximum (or minimum) point on the surface is one with which all experimenters should be familiar. For those interested in pursuing this topic further we suggest reading Davies (9) or Myers (11). Both the theory and the application of response surface techniques are also discussed in a number of the books listed at the end of Chapter 10.

12.8 EVOLUTIONARY OPERATION

In the previous section the reader was introduced to the subject known as "response surface techniques." In this section a related technique will be introduced. This technique, known as *evolutionary operation* (EVOP), is an application of the concepts of response surface methodology to the problem of improving the performance of industrial processes. Consequently, this technique should be of great interest to those concerned with production processes.

Because no detailed discussion of response surface methodology has been undertaken, a full description of EVOP cannot be given here. However, in capsule form the important elements of the technique are:

1. It is a method of process operation that includes a built-in procedure for improving productivity.
2. It uses some relatively simple statistical concepts.
3. It is run during normal routine production by plant personnel.
4. The basic philosophy of EVOP is: A process should be run not only to produce a product but also to provide information on how to improve the process and/or product.
5. Through the planned introduction of minor variants into the process, the customary "static" operating conditions are made "dynamic" in nature.
6. Utilizing the elementary principles of response surface methodology and making small changes (in a prescribed fashion) in the "controlled" variables of the process, the effects of the forced changes can be assessed.
7. If a simple pattern of operating conditions is employed [e.g., a 2^2 factorial plus a center point (usually that nominal set of operating conditions specified in the engineering drawings or production manual)] and the operation of the process under each of these conditions is termed a "cycle," the running of several cycles will yield sufficient information to permit a judgment to be made as to what is a better nominal set of operating conditions.
8. Constant repetition of this program will lead to continual improvement of the process.
9. Two important items in an EVOP program are:
   (a) All data should be prominently displayed on an information board.
   (b) An evolutionary operation committee (composed of research, development, and production personnel) should make periodic reviews of the EVOP program if the maximum benefit is to be derived from this approach.

Inasmuch as the preceding remarks are only a skeletal description of the EVOP technique, those desiring more details must consult other sources. The book by Box and Draper (2) is especially recommended.

12.9 OTHER DESIGNS AND TECHNIQUES

As stated in the opening paragraph of this chapter, the number of designs and analysis techniques developed for special purposes are many. Thus it has been possible to mention only a few in this book. The two commonest designs, the completely randomized design and the randomized complete block design, were discussed in detail in Chapters 10 and 11 respectively. In this chapter a few of the more specialized designs and techniques have been described or mentioned. An examination of some of the books listed at the end of Chapter 10 will bring many other special designs to your attention. It is our hope that the presentation thus far will have whetted your appetite and that you will continue your readings and studies in the related areas of experimental design and research techniques.

PROBLEMS

12.1   A 5 × 5 Latin square was laid out to test the effects of 5 fertilizers on the yield of potatoes. Perform a complete analysis of the data.

Row	Column 1	Column 2	Column 3	Column 4	Column 5	Row Total
1	A(449)	B(444)	C(401)	D(299)	E(292)	1885
2	B(463)	C(375)	D(323)	E(264)	A(415)	1840
3	C(393)	D(353)	E(278)	A(404)	B(425)	1853
4	D(371)	E(241)	A(441)	B(410)	C(392)	1855
5	E(258)	A(430)	B(450)	C(385)	D(347)	1870
Column Total	1934	1843	1893	1762	1871	9303

Treatment Total	A: 2139	B: 2192	C: 1946	D: 1693	E: 1333

12.2   Shown below are the yields (cwt/1/40-acre plots) of sugarcane in a Latin square experiment comparing fertilizers.

A 14	E 22	B 20	C 18	D 25
B 19	D 21	A 16	E 23	C 18
D 23	A 15	C 20	B 18	E 23
C 21	B 46	E 24	D 21	A 18
E 23	C 16	D 23	A 17	B 19

A: no fertilizer
B: complete inorganic fertilizer
C: 10 tons manure/acre
D: 20 tons manure/acre
E: 30 tons manure/acre

What conclusions do you draw from this experiment?

12.3   Analyze the following data from a cacao experiment consisting of 3 separately randomized Latin squares. The 3 treatments were:

A: no fertilizer (check)
B: 1.5 lb superphosphate/tree
C: 3 lb superphosphate/tree

The field plans of the squares, together with plot yields in average pods per tree, follow. (**Note:** Do not consider a transformation because these are averages (rounded) from the trees on 1/15-acre plots. The total numbers of pods were large enough to approximate a continuous distribution.)

B	C	A
41	25	15
A	B	C
20	32	24
C	A	B
22	12	21

C	B	A
27	28	3
A	C	B
4	17	9
B	A	C
22	4	17

A	C	B
11	15	17
B	A	C
24	14	33
C	B	A
22	20	15

12.4 Five levels of a fertilizer were tried in a 5 × 5 Latin square. This is the analysis;

Source of Variation	Degrees of Freedom	Mean Square
Rows	4	25
Columns	4	20
Treatments	4	28
Error	12	15

The sums of the yields in the 5 plots of each level were:

Level	1	2	3	4	5
Sum of yields	2	14	26	30	28

Subdivide the 4 degrees of freedom for treatments into

Source of Variation	Degrees of Freedom
Linear regression	1
Second-degree term	1
Remainder	2

Is any comparison significant?

12.5 Crop—wheat; location—R. W. Gt. Harpenden (175); year—1935; type—6 × 6 Latin square; comment—yield in pounds of grain per 1/40-acre plot

4	0	2	1	3	5	Total
77.2	88.0	89.7	92.6	72.1	76.2	495.8
3	4	0	5	1	2	
93.2	95.8	94.1	93.9	91.6	67.3	535.9
5	2	3	4	0	1	
90.2	87.0	86.1	85.5	93.4	68.5	510.7
2	3	1	0	5	4	
72.5	76.7	96.3	95.3	95.9	78.2	514.9
0	1	5	2	4	3	
84.2	96.5	98.5	81.6	90.1	81.8	532.7
1	5	4	3	2	0	
77.0	91.9	95.1	86.3	82.8	60.5	493.6
Total 494.3	535.9	559.8	535.2	525.9	432.5	3083.6

Treatment	Treatment Total
0—No $(NH_4)_2SO_4$	515.5
1—$(NH_4)_2SO_4$ applied Oct. 26 at 0.4 cwt N/acre	522.5
2—$(NH_4)_2SO_4$ applied Jan. 19 at 0.4 cwt N/acre	480.9
3—$(NH_4)_2SO_4$ applied Mar. 18 at 0.4 cwt N/acre	496.2
4—$(NH_4)_2SO_4$ applied Apr. 27 at 0.4 cwt N/acre	521.9
5—$(NH_4)_2SO_4$ applied May 24 at 0.4 cwt N/acre	546.6

Analyze and interpret the above data.

12.6   We wish to conduct a field experiment to test the yielding ability of 6 varieties of soybeans and have available an area of land sufficient for 36 plots. Indicate the proper subdivision of the total degrees of freedom for the following experimental designs:
(a)   Completely randomized.
(b)   Randomized complete block.
(c)   Latin square.
Indicate, by means of arrows, the proper F tests for testing variety differences in each design.

12.7   Given that the coded value of crushing strengths shown below resulted from an experiment such as described in Example 12.5, perform the analysis and give your interpretations of the results.

Row (batches)	Column (days)				
	1	2	3	4	5
1	257	230	279	287	202
2	245	283	245	280	260
3	182	252	280	246	250
4	203	204	227	193	259
5	231	271	266	334	338

Note: The treatments are assumed to have been imposed exactly as shown in Example 12.5.

12.8   An experiment was conducted to assess the relative resistances to abrasion of 4 grades of leather (A, B, C, D). A machine was used in which the samples could be tested in any one of 4 positions. Since different runs (replications) are known to yield variable results, it was decided to make 4 runs. A Latin square design was utilized and the following results obtained. Analyze and interpret the data.

Run	Position			
	1	2	3	4
1	118(B)	136(D)	168(A)	135(C)
2	127(D)	141(B)	129(C)	151(A)
3	174(A)	173(C)	126(B)	134(D)
4	130(C)	170(A)	125(D)	95(B)

12.9 Another experiment such as described in Problem 12.8 was conducted at a second laboratory. In this case, the data shown below were obtained. Analyze and interpret. (**Note:** M represents a missing observation.)

Run	Position 4	2	1	3
2	$A(150)$	$B(145)$	$D(130)$	$C(133)$
3	$D(130)$	$C(172)$	$A(170)$	$B(127)$
1	$C(M)$	$D(132)$	$B(115)$	$A(170)$
4	$B(98)$	$A(171)$	$C(132)$	$D(120)$

12.10 The experiment described in Problem 12.8 was conducted once more, this time at a third laboratory. Analyze and interpret the data that follow. [**Hint:** Use Eq. (12.9) and the iterative technique discussed in Section 11.12.]

Run	Position 1	2	3	4
1	$C(131)$	$D(M')$	$A(167)$	$B(136)$
2	$D(139)$	$A(196)$	$B(140)$	$C(148)$
3	$B(157)$	$C(133)$	$D(140)$	$A(184)$
4	$A(185)$	$B(146)$	$C(M'')$	$D(150)$

12.11 On checking the original data sheets, it was discovered that the technician took two independent abrasion readings on the samples tested in the experiment described in Problem 12.8. The second set of readings is reproduced below. Pooling these data with those given in Problem 12.8, analyze and interpret the complete results.

Run	Position 1	2	3	4
1	$120(B)$	$130(D)$	$165(A)$	$140(C)$
2	$125(D)$	$142(B)$	$120(C)$	$140(A)$
3	$175(A)$	$180(C)$	$120(B)$	$140(D)$
4	$132(C)$	$170(A)$	$130(D)$	$102(B)$

12.12 An experiment was performed to compare the effects of 3 catalysts on the yield of a chemical process. Three runs were started, one using catalyst A, another using B, and the third C. After 3 days a sample was drawn from each run and an analysis performed. A similar operation (i.e., taking samples and performing the analyses) was performed after 5 days. The whole experiment was repeated 4 times. Analyze and interpret the resulting coded yields given below.

	Catalyst					
	A		*B*		*C*	
Replicate	3 days	5 days	3 days	5 days	3 days	5 days
1	68	82	90	96	82	88
2	83	79	68	80	71	78
3	66	75	70	91	68	78
4	66	76	84	92	74	80

12.13 A split–split plot design was used in an experiment concerned with the yield of cotton. Four replications (or blocks) were involved. Each main plot was subjected to 1 of 2 levels of irrigation, each subplot was subjected to 1 of 3 rates of planting, and each sub-subplot was subjected to 1 of 3 levels of fertilizer application. Analyze and interpret the following experimental coded yields.

Irrigation	Rate of Planting (density of plants)	Fertilizer Rate	Block			
			1	2	3	4
Light	Thin	None	9.0	8.2	8.5	8.2
		Average	9.5	8.1	8.8	7.9
		Heavy	10.6	9.4	8.8	8.6
	Medium	None	9.0	9.7	11.1	7.8
		Average	8.9	9.7	10.3	8.5
		Heavy	9.3	10.4	9.1	8.6
	Thick	None	8.1	7.4	8.2	8.5
		Average	9.0	8.1	7.6	8.8
		Heavy	9.6	7.5	9.4	8.4
Heavy	Thin	None	8.1	10.3	6.0	7.2
		Average	8.6	10.8	10.4	11.6
		Heavy	10.2	10.4	11.5	11.6
	Medium	None	12.2	9.8	9.1	11.0
		Average	11.0	9.5	11.7	13.2
		Heavy	12.0	12.4	11.6	13.0
	Thick	None	7.9	13.4	12.0	11.7
		Average	10.0	14.2	12.2	13.8
		Heavy	12.5	14.0	13.8	13.4

12.14 The following coded yields of a chemical process resulted from an un-replicated complete factorial. Analyze and interpret. State all your assumptions.

Temperature (C)	Concentration of Solvent		
	Low	Medium	High
100	44	46	42
200	51	55	55
300	50	50	48

12.15 In a manufacturing company the micrometers used in checking quality are themselves checked by use of gauge blocks. However, there are 5 departments and each has its own micrometers and gauge blocks. Because of a suspicion that there is too much variation among micrometers and/or gauge blocks, the quality control engineer ran a test utilizing a random sample of instruments. Analyze and interpret the following data.

Gauge Block	Micrometer				
	1	2	3	4	5
A	0.0110	0.0115	0.0130	0.0151	0.0121
B	0.0135	0.0127	0.0132	0.0155	0.0128
C	0.0127	0.0124	0.0132	0.0152	0.0130

12.16 An experiment was run to investigate the effect of temperature, type of powder, amount of powder, and packing pressure on the function time (msec) of an explosive actuator. An unreplicated complete factorial yielded the following data. Analyze and interpret.

Temperature (F)	Packing Pressure (psi)	Powder A (mg)			Powder B (mg)			Powder C (mg)		
		5	10	15	5	10	15	5	10	15
		(msec)			(msec)			(msec)		
−50	10,000	7.4	7.0	6.8	5.4	5.0	4.8	7.2	6.9	6.6
	15,000	7.5	7.2	6.7	5.5	5.2	4.7	7.2	6.6	6.5
	20,000	7.4	7.4	6.0	5.4	5.4	4.0	7.2	6.7	6.2
75	10,000	6.6	6.6	5.8	4.6	4.6	3.8	6.8	7.2	4.9
	15,000	6.8	6.6	6.6	4.8	4.6	4.6	6.9	7.0	5.0
	20,000	6.8	6.2	5.9	4.8	4.2	3.9	7.0	7.1	5.0
200	10,000	5.1	5.1	5.1	3.1	3.1	3.1	6.0	4.9	4.8
	15,000	5.1	4.8	4.9	3.1	2.8	2.9	6.4	4.8	4.1
	20,000	5.2	4.7	5.0	3.2	2.7	3.0	5.9	4.9	2.0

12.17 A complete but unreplicated factorial was used to investigate the effects of type of metal (a qualitative factor), amount of primary initiator (a quantitative factor), and packing pressure (a quantitative factor) on the firing time (msec) of explosive switches. Analyze and interpret the following data:

Metal	Primary Initiator (mg)	Packing Pressure (psi)		
		12,000	20,000	28,000
2 s al	5	12.3	10.6	15.2
	10	10.4	9.5	15.0
	15	8.8	9.1	14.5
Teflon	5	12.4	11.7	15.0
	10	11.0	11.0	14.6
	15	11.0	9.8	14.6

12.18   A certain type of capacitor was to be tested to assess its performance as a function of a number of specified factors. The 4 factors considered were:

a = potted (+) or not potted (−)
b = wedged (+) or not wedged (−)
c = impregnated (+) or not impregnated (−)
d = high temperature (+) or low temperature (−)

The performance characteristic measured was the high voltage breakdown when the capacitors were subjected to a voltage rise of 250 V/sec. Some hypothetical data that could have resulted from such an experiment are:

Capacitor	Level of Factor				High Voltage Breakdown (kV)
	a	b	c	d	
1	−	−	−	−	10.7
2	+	−	−	+	11.4
3	−	+	−	+	12.2
4	+	+	−	−	13.0
5	−	−	+	+	10.6
6	+	−	+	−	12.1
7	−	+	+	−	12.0
8	+	+	+	+	13.2

Analyze and interpret the one-half replicate of a 2^4 factorial described above.

12.19   An experiment was to be performed to assess the effects of the following factors on the surge voltage of a specific model of thermal battery: temperature, humidity, amount of electrolyte, amount of heat paper, and type of electrolyte. These 5 factors were denoted as a, b, c, d, and e respectively. Since this was only a preliminary experiment (in the development phase) and since all 3-, 4-, and 5-factor interactions could be assumed to be negligible, a one-half replicate of the 2^5 factorial was performed. Analyze and interpret the following data.

Treatment Combination	Surge Voltage (V)
(1)	14.0
ae	14.6
be	11.7
ab	16.3
ce	11.2
ac	16.6
bc	15.6
abce	10.2
de	13.9
ad	13.8
bd	15.1
abde	13.2
cd	14.6
acde	14.3
bcde	12.6
abcd	15.4

REFERENCES

1. Anderson, R. L. 1958. Complete factorials, fractional factorials, and confounding. *Experimental Designs in Industry* (ed., V. Chew). Wiley, New York.
2. Box, G. E. P.; and Draper, N. 1969. *Evolutionary Operation: A Statistical Method for Process Improvement.* Wiley, New York.
3. Budne, T. A. 1959. Random balance. Part I. The missing statistical link in fact-finding techniques. Part II. The techniques of analysis. Part III. Case histories. *Ind. Qual. Control* 15(Apr.):5–10, 15(May):11–16, 15(June):16–19.
4. _____. 1959. The application of random balance designs. *Technometrics* 1(May): 139–55.
5. Carroll, M. B.; and Dykstra, O., Jr. 1958. Application of fractional factorials in a food research laboratory. *Experimental Designs in Industry* (ed., V. Chew). Wiley, New York.
6. Chew, V. 1958. Basic experimental designs. *Experimental Designs in Industry* (ed., V. Chew). Wiley, New York.
7. Connor, W. S. 1958. Experiences with incomplete block designs. *Experimental Designs in Industry* (ed., V. Chew). Wiley, New York.
8. Connor, W. S.; and Young, S. 1959. Fractional factorial designs for experiments with factors at two and three levels. Nat. Bur. Stand. Appl. Math. Ser. 54, USGPO, Washington, D.C.
9. Davies, O.L. (ed.) 1956. *The Design and Analysis of Industrial Experiments,* 2nd ed. Oliver and Boyd, Edinburgh.
10. Horton, W. H. 1958. Experiences with fractional factorials. *Experimental Designs in Industry* (ed., V. Chew). Wiley, New York.
11. Myers, R. H. 1971. *Response Surface Methodology.* Allyn and Bacon, Boston.
12. Rao, C. R. 1947. General methods of analysis for incomplete block designs. *J. Am. Stat. Assoc.* 42:541.
13. Satterthwaite, F. E. 1959. Random balance experimentation. *Technometrics* 1(May): 111–37.
14. Youden, W. J.; Kempthorne, O.; Tukey, J. W.; Box, G. E. P.; and Hunter, J. S. 1959. Discussion of the papers of Satterthwaite and Budne (including authors' responses to discussion). *Technometrics* 1(May):157–93.

FURTHER READING

See Chapter 10.

CHAPTER 13

ANALYSIS OF COVARIANCE

In preceding chapters great emphasis has been placed on two very important techniques, namely, regression analysis and analysis of variance. Further, in Sections 10.11, 11.10, and 12.7 these two techniques were combined to handle particular problems associated with the exploration of response curves or surfaces. In this chapter we shall investigate another blending of these two fundamental tools. This technique, known as *analysis of covariance*, is one that has proved very useful in many areas of research.

13.1 INTRODUCTION

One reason for using the randomized complete block design in some experimental situations is to eliminate certain types of experimental effects that might influence comparisons between treatments. For example, in Example 11.1 the experimental plots were blocked into groups of fairly equal fertility levels. In this way the effect due to different levels of fertility was removed when making comparisons among the six varieties of oats. In some experimental situations it is not possible to design the experiment so as to control for differences in the experimental units. Even if some experimental effects are controlled by blocking, certain differences still may exist among the units within a block. In such experimental situations it may still be possible to control certain sources of variation by taking additional observations. In the field experiment mentioned above, it may be possible to observe some measure of fertility for each unit. This measure can be used to adjust the yields according to the fertility of the experimental unit and then compare the yields for the six varieties based on the adjusted observations. Whenever it is possible to take additional observations that may be used to adjust for certain sources of variation, the analysis of covariance is a useful tool for analyzing the data.

To further indicate the ideas of covariance analysis and to illustrate the type of experimental situations in which the analysis of covariance has been profitably employed, let us give several examples. These will indicate to the reader the nature of the combination of the ideas of regression and the analysis of variance in the analysis of covariance.

Example 13.1

Consider a case where the researcher is interested in the effects of various rations on the weights of hogs. If a randomized complete block design is utilized and the final weights Y of the animals after a specified number of days of feeding are analyzed, the differences among the effects of the various rations may or may not be significant. In either case, however, the good researcher will think more about the conduct of the experiment before drawing any conclusions from the ANOVA implied in the preceding sentence. He might say to himself, "If the experimental animals varied greatly with respect to their initial weights at the time the experiment was started, how do we know that differences among final weights reflect ration effects rather than just varying initial weights?" Calling the initial weights X, he might *adjust* the Y values according to the associated X values and then analyze and interpret the experimental data. The method by which this is carried out is known as covariance analysis.

Example 13.2

When dealing with an experiment to compare several methods of teaching statistics in which the criterion is to be the final score Y obtained by the students, all of whom take the same examination, final judgment concerning the various methods of teaching should not be rendered until the IQ ratings X of the individual students have been examined and the necessary corrections (adjustments) made.

Example 13.3

Another example of the use of covariance analysis involves an observational study undertaken in a large city in Iowa to study the effect of the type of traffic signal (traffic lights, turn lanes, etc.) on the accident rates at an intersection. In accumulating the data, it was found that intersections with sophisticated signals usually also had a high volume of traffic, while intersections with low volume usually had simple traffic signals. Thus to compare the accident rates for different types of signals, it was necessary to adjust the observed data for the volume of traffic.

In all the examples given above the response variable Y is a function of a classification variable (treatment) as well as some independent variable X called the covariate. Of primary interest in each case was a comparison of the response for the different treatments. The analysis of covariance is a technique for adjusting the observations according to the value of the covariate and then analyzing the adjusted experimental data.

13.2 UNDERLYING ASSUMPTIONS

As would be expected, the assumptions we make when performing a covariance analysis are similar to those required for linear regression and analysis of variance. Thus we find the usual assumptions of independence, normality, homogeneous variances, fixed X's, etc. To be more specific, let us look at some of the mathematical models associated with the completely randomized design when a covariance analysis is contemplated. Models for other designs are similar and are given in later sections.

Let Y be the response variable and let $X_1, X_2, \ldots$ be the independent variables or covariates. If the experiment involves a completely randomized design, some models are

1. Simple linear covariance model (one covariate),

$$Y_{ij} = \mu + \tau_i + \beta X_{ij} + \epsilon_{ij} \qquad i = 1, \ldots, t; j = 1, \ldots, n_i \qquad (13.1)$$

2. Multiple linear covariance model (p covariates),

$$Y_{ij} = \mu + \tau_i + \beta_1 X_{1ij} + \beta_2 X_{2ij} + \cdots + \beta_p X_{pij} + \epsilon_{ij}$$
$$i = 1, \ldots, t; j = 1, \ldots, n_i \qquad (13.2)$$

3. Two-factor factorial with a simple linear covariance model,

$$Y_{ijk} = \mu + \alpha_i + \gamma_j + (\alpha\gamma)_{ij} + \beta X_{ijk} + \epsilon_{ijk}$$
$$i = 1, \ldots, a; j = 1, \ldots, c; k = 1, \ldots, n_{ij} \qquad (13.3)$$

The notation is consistent with our previous use of the symbols in Chapter 10. In particular, μ is the overall mean effect, τ_i is the true effect of the ith treatment, and α_i is the true effect of the ith level of factor a, γ_j is the true effect of the jth level of factor c, and $(\alpha\gamma)_{ij}$ is the true interaction of the ith level of factor a and jth level of factor c. The assumptions regarding the treatments (Model I or II) are still applicable. As we have in the previous chapters, we assume that the $\epsilon_{ijk} \sim NI(0, \sigma_E^2)$. Also, as we did in Chapter 7, we assume the X's are fixed, i.e., measured without error.

Although we have not mentioned it, another item is commonly considered as a necessary assumption for a valid analysis of covariance, namely, that the concomitant variable X should not be affected by the treatments. That is, the treatments that have been applied to the experimental units so we may observe and judge their effects on the Y variable should not influence the observed values of X. However, this is too restrictive an assumption. Even though the treatments do affect the X values, a covariance analysis may be profitably employed if proper care is exercised in the interpretation of the experimental results. It is clear, then, that the inferences that may be made are different in the two cases, depending upon whether or not the X variable is affected by the treatments. The researcher is therefore cautioned to be extremely careful when dealing with the interpretation of covariance analyses.

We now consider some special cases in more detail so that the reader may become more familiar with the interpretation of data amenable to an analysis of covariance. The computational procedure for an analysis of covariance involves computing sums of squares as for an ANOVA. As we did for the analysis of variance in Chapters 10 through 12, we also outline the analysis using regression analysis on a computer.

13.3 COMPLETELY RANDOMIZED DESIGN

If the treatments can be thought to be applied at random to a set of homogeneous (except for the treatments and the covariate) experimental units and the functional relationship between the response Y, and a

TABLE 13.1–General Analysis of Covariance for Data Conforming to a Completely Randomized Design

Source of Variation	Degrees of Freedom	Sum of Squares and Products*			Deviation about Regression		
		$\sum x^2$	$\sum xy$	$\sum y^2$	$\sum y^2 - \left(\sum xy\right)^2 / \sum x^2$	Degrees of Freedom	Mean Square
Among treatments	$t-1$	T_{xx}	T_{xy}	T_{yy}	...	...	...
Among experimental units treated alike (within treatments)	$\displaystyle\sum_{i=1}^{t} n_i - t$	E_{xx}	E_{xy}	E_{yy}	$S_E = E_{yy} - E_{xy}^2/E_{xx}$	$\displaystyle\sum_{i=1}^{t} n_i - t - 1$	$s_E^2 = S_E \Big/ \left(\displaystyle\sum_{i=1}^{t} n_i - t - 1\right)$
Among treatments + within treatments (= total)	$\displaystyle\sum_{i=1}^{t} n_i - 1$	$S_{xx} = T_{xx} + E_{xx}$	$S_{xy} = T_{xy} + E_{xy}$	$S_{yy} = T_{yy} + E_{yy}$	$S_{T+E} = S_{yy} - S_{xy}^2/S_{xx}$	$\displaystyle\sum_{i=1}^{t} n_i - 2$	...
			Difference for testing among adjusted treatment means		$S_{T+E} - S_E$ $= T_{yy} - S_{xy}^2/S_{xx}$ $+ E_{xy}^2/E_{xx}$	$t-1$	$(S_{T+E} - S_E)/(t-1)$

*The symbols S_{xx}, S_{xy}, and S_{yy} have been used in place of $\sum x^2$, $\sum xy$, and $\sum y^2$ respectively in an attempt to standardize the notation in this and other tables that follow.

single covariate X is simple linear, the linear statistical model is given in Eq. (13.1). For convenience we restate the model,

$$Y_{ij} = \mu + \tau_i + \beta X_{ij} + \epsilon_{ij} \qquad i = 1, \ldots, t; j = 1, \ldots, n_i \qquad (13.4)$$

where n_i is the number of experimental units associated with the ith treatment and τ_i is the true effect of the ith treatment. Also, we assume $\epsilon_{ij} \sim \mathrm{NI}(0, \sigma_E^2)$.

The computations for an analysis of covariance based on the model in Eq. (13.4) are summarized in Table 13.1. Formulas for computing the sums of squares included in Table 13.1 are:

T_{xx} = treatment sum of squares for X

$$= \sum_{i=1}^{t} X_{i.}^2 / n_i - (X_{..})^2 / \sum_{i=1}^{t} n_i \qquad (13.5)$$

T_{xy} = treatment sum of products for X and Y

$$= \sum_{i=1}^{t} X_{i.} Y_{i.} / n_i - (X_{..})(Y_{..}) / \sum_{i=1}^{t} n_i \qquad (13.6)$$

T_{yy} = treatment sum of squares for Y

$$= \sum_{i=1}^{t} Y_{i.}^2 / n_i - (Y_{..})^2 / \sum_{i=1}^{t} n_i \qquad (13.7)$$

E_{xx} = experimental error sum of squares for X

$$= \sum_{i=1}^{t} \sum_{j=1}^{n_i} X_{ij}^2 - \sum_{i=1}^{t} X_{i.}^2 / n_i \qquad (13.8)$$

E_{xy} = experimental error sum of products for X and Y

$$= \sum_{i=1}^{t} \sum_{j=1}^{n_i} X_{ij} Y_{ij} - \sum_{i=1}^{t} X_{i.} Y_{i.} / n_i \qquad (13.9)$$

E_{yy} = experimental error sum of squares for Y

$$= \sum_{i=1}^{t} \sum_{j=1}^{n_i} Y_{ij}^2 - \sum_{i=1}^{t} Y_{i.}^2 / n_i \qquad (13.10)$$

where the dot notation is consistent with our previous use.

The purpose of the analysis is to make comparisons among the treatments after adjusting for the covariate. Thus we begin by testing the hypothesis that there are no differences among the true effects of the t treatments on the Y variable after adjusting for the effect of the X variable. The appropriate test statistic is

$$F = \frac{(S_{T+E} - S_E)/(t - 1)}{S_E \Big/ \left(\sum_{i=1}^{t} n_i - t - 1 \right)} = \frac{(S_{T+E} - S_E)/(t - 1)}{s_E^2} \qquad (13.11)$$

which has $\nu_1 = t - 1$ and $\nu_2 = \sum_{i=1}^{t} n_i - t - 1$ degrees of freedom.

Example 13.4

Consider an experiment in which the gains in weight of pigs for 4 different feeds were compared. The concomitant variable X was the initial weight of the pig. Pigs were assigned to feeds completely at random. The data are recorded in Table 13.2. The computations based on Eqs. (13.5)–(13.10) are summarized in Table 13.3.

Carrying out the F test for comparing the treatments, we obtain $F = 536.53/276.58 = 1.94$, which has $\nu_1 = 3$ and $\nu_2 = 19$ degrees of freedom. This is not significant at the 5 percent level, and thus we are unable to reject the hypothesis of no differences among the true effects of the 4 treatments on the gain in weight of pigs after adjusting for the varying initial weights of the experimental animals. Incidentally, in this case the same decision would have been reached had no adjustment been made for the concomitant variable. However, in many instances the conclusions may change considerably depending on whether or not the covariance technique is used, and thus the researcher should always see if it is applicable to the problem at hand.

TABLE 13.2–Gains in Weight Y and Initial Weights X of Pigs in a Feeding Trial

Treatment							
1		2		3		4	
X	Y	X	Y	X	Y	X	Y
30	165	24	180	34	156	41	201
27	170	31	169	32	189	32	173
20	130	20	171	35	138	30	200
21	156	26	161	35	190	35	193
33	167	20	180	30	160	28	142
29	151	25	170	29	172	36	189
Total 160	939	146	1031	195	1005	202	1098

TABLE 13.3–Analysis of Covariance for Data in Table 13.2

Source of Variation	Degrees of Freedom	Sum of Squares and Products			Deviation about Regression		
		$\sum x^2$	$\sum xy$	$\sum y^2$	$\sum y^2 - \dfrac{(\sum xy)^2}{\sum x^2}$	Degrees of Freedom	Mean Square
Among treatments	3	365.46	451.21	2163.13	...	...	...
Among animals treated alike	20	361.50	496.83	5937.83	5255.01	19	276.58
Total	23	726.96	948.04	8100.96	6864.61	22	...
Difference for testing among adjusted treatment means					1609.60	3	536.53

Just as we did for the analysis of variance in the previous chapters, we now outline the analysis of covariance as a special case of a regression analysis. The linear statistical model in Eq. (13.4) can be restated as the regression model

$$Y_{ij} = \mu + \tau_1 z_{1i} + \cdots + \tau_{t-1} z_{t-1\,i} + \beta X_{ij} + \epsilon_{ij}$$
$$i = 1, \ldots, t; j = 1, \ldots, n_i \quad (13.12)$$

where

$$z_{km} = 1 \quad \text{if } m = k; k = 1, \ldots, t - 1$$
$$= -1 \quad \text{if } m = t$$
$$= 0 \quad \text{otherwise}$$

Note that for computational convenience and to ensure that a unique solution to the corresponding normal equations exists, we have incorporated the restriction $\sum_{i=1}^{t} \tau_i = 0$ as we did in Chapter 10.

Since the purpose of the analysis is to compare the t treatments after adjusting for the covariate, the analysis is based on adjusting for the covariate and then testing to determine if there is any difference in the true effects of the treatments. In terms of the regression model in Eq. (13.12) the analysis involves comparing the model with the treatment effects included to the model with the treatment effects not included. The latter model is based on the assumption that all the treatment effects are the same and hence the τ_j's are all zero. The general procedure is as follows:

1. Perform a regression analysis using the regression model in Eq. (13.12). From this analysis we evaluate the regression sum of squares $SS(\beta, \tau_1, \ldots, \tau_{t-1} \mid \mu)$ and the residual sum of squares E_{yy}. These are respectively the sums of squares due to the independent variables, in this case the true treatment effects and the concomitant variable X and the remainder sum of squares attributed to experimental error.

2. Perform a regression analysis using the regression model

$$Y_{ij} = \mu + \beta X_{ij} + \epsilon_{ij} \qquad i = 1, \ldots, t; j = 1, \ldots, n_i \quad (13.13)$$

This model assumes there is no difference in the treatment effects on Y after adjusting for the effect of the X variable. From this analysis we evaluate the regression sum of squares $SS(\beta \mid \mu)$, which is the sum of squares due to the X variable ignoring the different treatments.

3. The difference in the regression sum of squares using the models in Eqs. (13.12) and (13.13) respectively; i.e., the difference

$$SS(\tau_1, \ldots, \tau_{t-1} \mid \mu, \beta) = SS(\beta, \tau_1, \ldots, \tau_{t-1} \mid \mu) - SS(\beta \mid \mu)$$

is the sum of squares due to the treatment effects after adjusting for the effect of the X variable. Thus, to test the hypothesis of no difference among the true effects of the t treatments on Y after adjusting for the effect of X, we use the test statistic

$$F = \frac{[SS(\beta, \tau_1, \ldots, \tau_{t-1} \mid \mu) - SS(\beta \mid \mu)]/(t - 1)}{E_{yy} \Big/ \Big(\sum_{i=1}^{t} n_i - t - 1\Big)}$$

$$= \frac{SS(\tau_1, \ldots, \tau_{t-1} \mid \mu, \beta)/(t - 1)}{E_{yy} \Big/ \Big(\sum_{i=1}^{t} n_i - t - 1\Big)} \qquad (13.14)$$

When the hypothesis is true, the test statistic has an F distribution with $\nu_1 = t - 1$ and $\nu_2 = \sum_{i=1}^{t} n_i - t - 1$ degrees of freedom. A summary of the appropriate sums of squares and mean squares associated with this test is usually given in an analysis of covariance table such as Table 13.4. In terms of the entries in Table 13.4, the test statistic is

$$F = SS(\tau_1, \ldots, \tau_{t-1} \mid \mu, \beta)/(t - 1)\mathbf{E} \qquad (13.15)$$

TABLE 13.4–Analysis of Covariance for a Completely Randomized Design

Source of Variation	Degrees of Freedom	Sum of Squares	Mean Square
Mean	1	$SS(\mu) = (Y_{..})^2 \Big/ \sum_{i=1}^{t} n_i$	...
Regression	t	$SS(\tau_1, \ldots, \tau_{t-1}, \beta \mid \mu)$	...
Due to β	1	$SS(\beta \mid \mu)$	...
Treatments $\mid \beta$	$t - 1$	$SS(\tau_1, \ldots, \tau_{t-1} \mid \mu, \beta)$	$SS(\tau_1, \ldots, \tau_{t-1} \mid \mu, \beta)/(t - 1)$
Experimental error	$\sum_{i=1}^{t} n_i - t - 1$	E_{yy}	$\mathbf{E}$
Total	$\sum_{i=1}^{t} n_i$	$\sum Y^2$	...

Example 13.5

The data used in Example 13.4 to illustrate the analysis of covariance was analyzed, using the procedure based on regression analysis. The appropriate regression sums of squares are summarized in Table 13.5. Using the test statistic in Eq. (13.15), the value of F is $536.53/276.58 = 1.94$, which is the same as the value in Example 13.4.

With the availability of computers to do the computations, the analysis of covariance is just a special type of regression analysis. In particular, referring back to Section 7.8 where we discussed methods of comparing several regressions, note that the analysis of covariance is a

TABLE 13.5–Analysis of Covariance for Data Given in Table 13.2

Source of Variation	Degrees of Freedom	Sum of Squares	Mean Square
Mean	1	691,222.04	. . .
Regression	4	2,845.95	. . .
Due to β	1	1,236.35	. . .
Treatments $\mid \beta$	3	1,609.60	536.53
Experimental error	19	5,255.01	276.58
Total	24	699,323.00	. . .

special case of that analysis, assuming that the slopes β are all identical. If any doubt exists as to the correctness of this assumption, a test of the hypothesis of equal slopes should be entertained. We refer the reader to Section 7.8 for a discussion of those methods.

It should be clear that the regression coefficient β in the model has been assumed to be nonzero. If such were not the case, the introduction of the concomitant variable X into the model has only complicated the analysis. Sometimes the researcher will want to check this assumption. To make this check, set up the hypothesis $H:\beta = 0$. The test statistic, in terms of the entries in Table 13.1, is

$$F = E_{xy}^2/s_E^2 E_{xx} \qquad (13.16)$$

with $\nu_1 = 1$ and $\nu_2 = \sum_{i=1}^t n_i - t - 1$ degrees of freedom.

To use regression analysis to test the hypothesis $H:\beta = 0$, we must fit the regression model in Eq. (13.12), assuming the hypothesis is true. That is, we perform a regression analysis using the model

$$Y_{ij} = \mu + \tau_1 z_{1i} + \cdots + \tau_{t-1} z_{t-1 i} + \epsilon_{ij}$$
$$i = 1, \ldots, t; j = 1, \ldots, n_i \qquad (13.17)$$

The regression sum of squares is $SS(\tau_1, \ldots, \tau_{t-1} \mid \mu)$. The sum of squares due to β after adjusting for the different treatment effects is the difference in regression sum of squares for the full model in Eq. (13.12) and the model in Eq. (13.17), assuming $\beta = 0$. This difference is

$$SS(\beta \mid \mu, \tau_1, \ldots, \tau_{t-1}) = SS(\beta, \tau_1, \ldots, \tau_{t-1} \mid \mu) - SS(\tau_1, \ldots, \tau_{t-1} \mid \mu)$$

The test statistic for testing $H:\beta = 0$ is

$$F = SS(\beta \mid \mu, \tau_1, \ldots, \tau_{t-1})/\mathbf{E} \qquad (13.18)$$

with $\nu_1 = 1$ and $\nu_2 = \sum_{i=1}^t n_i - t - 1$ degrees of freedom. The denominator $\mathbf{E}$ is the error mean square in Table 13.4.

In addition to performing the F test of significance of the treatments, it is frequently of interest to look at the adjusted treatment means for further aid in interpreting the experimental results. The adjusted treatment means may be found using the formula

$$\text{adj } \overline{Y}_{i.} = \overline{Y}_{i.} - b(\overline{X}_{i.} - \overline{X}_{..}) \qquad i = 1, \ldots, t \qquad (13.19)$$

where b is the estimated regression coefficient. The estimate b is either part of the output of a regression fit or is calculated from the experimental error sums of squares and products, using the equation

$$b = E_{xy}/E_{xx} \qquad (13.20)$$

The estimated variance of an adjusted treatment mean is given by

$$\hat{V}(\text{adj } \overline{Y}_{i.}) = s_E^2 \left[\frac{1}{n_i} + \frac{(\overline{X}_{i.} - \overline{X}_{..})^2}{\sum\limits_{i=1}^{t} \sum\limits_{j=1}^{n_i} (X_{ij} - \overline{X}_{i.})^2} \right] \qquad (13.21)$$

where $s_E^2 = \mathbf{E}$ is the experimental error mean square. To make a comparison of two treatments based on a comparison of the corresponding adjusted treatment means requires evaluation of the variance of the difference. The estimated variance of the difference between two adjusted treatment means is given by

$$\hat{V}(\text{adj } \overline{Y}_{i.} - \text{adj } \overline{Y}_{i'.}) = s_E^2 \left[\frac{1}{n_i} + \frac{1}{n_{i'}} + \frac{(\overline{X}_{i.} - \overline{X}_{i'.})^2}{\sum\limits_{i=1}^{t} \sum\limits_{j=1}^{n_i} (X_{ij} - \overline{X}_{i.})^2} \right] \qquad (13.22)$$

Example 13.6

For the data included in Example 13.4 $\overline{X}_{..} = 29.29$, $\overline{Y}_{..} = 169.71$, $b = 1.374$, so the adjusted treatment means and their corresponding standard errors are:

Treatment	1	2	3	4
adj $\overline{Y}_{i.}$	160.10	178.65	163.09	176.98
Standard error, $s_{\text{adj } \overline{Y}_i.}$	7.17	8.06	7.35	7.80

13.4 RANDOMIZED COMPLETE BLOCK DESIGN

When the data conform to a randomized complete block design, the linear statistical model involving a single covariate is

$$Y_{ij} = \mu + \rho_i + \tau_j + \beta X_{ij} + \epsilon_{ij} \qquad i = 1, \ldots, r; j = 1, \ldots, t \qquad (13.23)$$

where ρ_i is the true effect of the ith block and τ_j is the true effect of the jth treatment. The experimental error $\epsilon_{ij} \sim \text{NI}(0, \sigma_E^2)$.

The analysis of covariance is summarized in Table 13.6. The sums of squares R_{xx}, T_{xx}, E_{xx}, R_{yy}, T_{yy}, and E_{yy} can be obtained as in any randomized complete block design (see Chapter 11). The sums of prod-

TABLE 13.6–General Analysis of Covariance for Data Conforming to a Randomized Complete Block Design

Source of Variation	Degrees of Freedom	Sum of Squares and Products			Deviation about Regression		
		$\sum x^2$	$\sum xy$	$\sum y^2$	$\sum y^2 - (\sum xy)^2/\sum x^2$	Degrees of Freedom	Mean square
Replicates (blocks)	$r - 1$	R_{xx}	R_{xy}	R_{yy}			
Treatments	$t - 1$	T_{xx}	T_{xy}	T_{yy}	$\cdots$	$\cdots$	$\cdots$
Experimental error	$(r - 1)(t - 1)$	E_{xx}	E_{xy}	E_{yy}	$S_E = E_{yy} - E_{xy}^2/E_{xx}$	$(r - 1)(t - 1) - 1$	$s_E^2 = S_E/[(r - 1)(t - 1) - 1]$
Treatments + error	$r(t - 1)$	$S_{xx} = T_{xx} + E_{xx}$	$S_{xy} = T_{xy} + E_{xy}$	$S_{yy} = T_{yy} + E_{yy}$	$S_{T+E} = S_{yy} - S_{xy}^2/S_{xx}$	$r(t - 1) - 1$	$\cdots$
Difference for testing among adjusted treatment means					$S_{T+E} - S_E = T_{yy} - S_{xy}^2/S_{xx} + E_{xy}^2/E_{xx}$	$t - 1$	$(S_{T+E} - S_E)/(t - 1)$

ucts can be calculated from the equations:

$$R_{xy} = t \sum_{i=1}^{r} (\bar{X}_{i.} - \bar{X}_{..})(\bar{Y}_{i.} - \bar{Y}_{..})$$

$$= \sum_{i=1}^{r} X_{i.} Y_{i.}/t - (X_{..})(Y_{..})/rt \qquad (13.24)$$

$$T_{xy} = r \sum_{j=1}^{t} (\bar{X}_{.j} - \bar{X}_{..})(\bar{Y}_{.j} - \bar{Y}_{..})$$

$$= \sum_{j=1}^{t} X_{.j} Y_{.j}/r - (X_{..})(Y_{..})/rt \qquad (13.25)$$

$$E_{xy} = \sum_{i=1}^{r}\sum_{j=1}^{t} (X_{ij} - \bar{X}_{i.} - \bar{X}_{.j} + \bar{X}_{..})(Y_{ij} - \bar{Y}_{i.} - \bar{Y}_{.j} + \bar{Y}_{..})$$

$$= \sum_{i=1}^{r}\sum_{j=1}^{t} X_{ij} Y_{ij} - T_{xy} - R_{xy} - (X_{..})(Y_{..})/rt \qquad (13.26)$$

In terms of the entries in Table 13.6, the F ratio for testing the hypothesis of no differences among the true effects of the treatments on the Y variable after adjusting for the effect of the X variable is

$$F = \frac{(S_{T+E} - S_E)/(t - 1)}{S_E/[(r - 1)(t - 1) - 1]} = \frac{(S_{T+E} - S_E)/(t - 1)}{s_E^2} \qquad (13.27)$$

To test the hypothesis that β equals 0, the test statistic is

$$F = \frac{E_{xy}^2/E_{xx}}{S_E/[(r - 1)(t - 1) - 1]} = \frac{E_{xy}^2/E_{xx}}{s_E^2} \qquad (13.28)$$

which has $\nu_1 = 1$ and $\nu_2 = (r - 1)(t - 1) - 1$ degrees of freedom.

Example 13.7

Consider the data of Table 13.7. These data have been examined in considerable detail by Wishart (4); we shall, however, consider them from a more limited point of view that will be sufficient for our purposes. The computations are summarized in the analysis of covariance in Table 13.8.

Before testing the hypothesis of no difference among the true effects of adjusted varieties, let us test the hypothesis that the true regression coefficient β is 0. After all, it is more reasonable to examine this point first, for unless we can conclude that $\beta \neq 0$, the decision to perform an analysis of covariance is questionable. Accordingly, using the test statistic in Eq. (13.28) to test $H:\beta = 0$, we have $F = [(46)^2/86]/4.68 = 5.26$. Since $F_{0.90(1,5)} = 4.06$ and $F_{0.95(1,5)} = 6.61$, we would reject the hypothesis that $\beta = 0$ at $\alpha = 0.10$. Although this might not be considered significant, we shall continue with the analysis as if $\beta \neq 0$.

To examine the variety differences, we use the test statistic in Eq. (13.27). We have $F = 22.25/4.68 = 4.75$. Again, we conclude there are differences in the varieties after adjusting for X at $\alpha = 0.10$ but not at $\alpha = 0.05$.

TABLE 13.7–Yields for Three Varieties of a Certain Crop in a Randomized Complete Block Design with Four Blocks

Block		Variety			Block Total
		A	B	C	
1	X	54	51	57	162
	Y	64	65	72	201
2	X	62	64	60	186
	Y	68	69	70	207
3	X	51	47	46	144
	Y	54	60	57	171
4	X	53	50	41	144
	Y	62	66	61	189
Variety	X	220	212	204	636
Total	Y	248	260	260	768

Source: Reproduced from Table 7 in Wishart (4), with permission of author and publishers.

Note: X = yield of a plot in a preliminary year under uniformity trial conditions; Y = yield on the same plot in the experimental year when the 3 varieties were used.

TABLE 13.8–Analysis of Covariance for Data of Table 13.7

Source of Variation	Degrees of Freedom	Sum of Squares and Products			Deviation about Regression		
		$\sum x^2$	$\sum xy$	$\sum y^2$	$\sum y^2 - (\sum xy)^2 / \sum x^2$	Degrees of Freedom	Mean Square
Replicates (blocks)	3	396	264	252	...	...	...
Treatments (varieties)	2	32	−24	24	...	...	...
Experimental error	6	86	46	48	23.4	5	4.68
Treatments + error	8	118	22	72	67.9	7	...
Difference for testing among adjusted variety means					44.5	2	22.25

Ordinarily, variety differences would not be called statistically significant. However, a review of the analysis indicates that the F statistic increased after adjustment ($F = 1.5$ under the ordinary ANOVA). This suggests that perhaps the fertility differences among the plots are tending to obscure the true differences among varieties.

As we did for the completely randomized model with a single covariate, we outline briefly how regression analysis can be used to evaluate the necessary sums of squares for analysis of covariance. Rewriting

Eq. (13.23) as a regression model, we have

$$Y_{ij} = \mu + \rho_1 u_{1i} + \cdots + \rho_{r-1} u_{r-1i} + \tau_1 z_{1j} + \cdots + \tau_{t-1} z_{t-1j}$$
$$+ \beta X_{ij} + \epsilon_{ij} \qquad i = 1, \ldots, r; j = 1, \ldots, t \qquad (13.29)$$

where
$$\begin{aligned} z_{kj} &= 1 \quad \text{if } j = k; k = 1, \ldots, t-1 \\ &= -1 \quad \text{if } j = t \\ &= 0 \quad \text{otherwise} \\ u_{ki} &= 1 \quad \text{if } i = k; k = 1, \ldots, r-1 \\ &= -1 \quad \text{if } i = r \\ &= 0 \quad \text{otherwise} \end{aligned}$$

The model is the same as the regression model (Eq. 11.17) for a randomized complete block design except for the added covariate.

The procedure for evaluating the necessary sums of squares to test for no differences among the true effects of the treatments on Y after adjusting for the effect of X is similar to the procedure outlined for the completely randomized design. In particular,

1. Perform a regression analysis using the (full) regression model in Eq. (13.29). From this analysis we evaluate the regression sum of squares $SS(\beta, \rho_i\text{'s}, \tau_j\text{'s} \mid \mu)$ and the residual sum of squares E_{yy}.
2. Perform a regression analysis using the (reduced) regression model

$$Y_{ij} = \mu + \rho_1 u_{1i} + \cdots + \rho_{r-1} u_{r-1i} + \beta X_{ij} + \epsilon_{ij}$$
$$i = 1, \ldots, r; j = 1, \ldots, t \qquad (13.30)$$

That is, we fit a model assuming the hypothesis that there are no differences among the true effects of the treatments on Y after adjusting for the effect of X. In terms of the model parameters, the model in Eq. (13.30) assumes the τ's are all zero. From this analysis we evaluate the regression sum of squares $SS(\beta, \rho_i\text{'s} \mid \mu)$.

The difference in the regression sum of squares for the two models $SS(\beta, \rho_i\text{'s}, \tau_j\text{'s} \mid \mu) - SS(\beta, \rho_i\text{'s} \mid \mu)$, is the sum of squares due to differences among the treatment effects after adjusting for the concomitant variable X. We denote this sum of squares by $SS(\tau_j\text{'s} \mid \mu, \beta, \rho_i\text{'s})$. Note that after adjusting for the concomitant variable X, the block and treatment effects are no longer orthogonal as was true for the randomized complete block design without the covariate in Chapter 11. Thus it is no longer true that $SS(\tau_j\text{'s} \mid \mu, \beta, \rho_i\text{'s}) = SS(\tau_j\text{'s} \mid \mu, \beta)$ and it is important that the block effects [the ρ_i's in Eq. (13.30)] are included in the reduced regression model. The analysis of covariance based on regression is summarized in Table 13.9. To test the hypothesis of no differences among the treatment effects after adjusting for the effect of X, the test statistic is

$$F = SS(\tau_j\text{'s} \mid \mu, \beta)/(t-1)E \qquad (13.31)$$

which has $\nu_1 = t - 1$ and $\nu_2 = (r-1)(t-1) - 1$ degrees of freedom.

TABLE 13.9–Analysis of Covariance for
a Randomized Complete Block Design

Source of Variation	Degrees of Freedom	Sum of Squares	Mean Square
Mean	1	$SS(\mu) = (Y_{..})^2/rt$	$\cdots$
Regression (full)	$r + t - 1$	$SS(\beta, \rho_i's, \tau_j's \mid \mu)$	$\cdots$
Regression (reduced)	r	$SS(\beta, \rho_i's \mid \mu)$	$\cdots$
Treatments $\mid \beta$	$t - 1$	$SS(\tau_j's \mid \mu, \beta, \rho_i's)$	$SS(\tau_j's \mid \mu, \beta, \rho_i's)/(t - 1)$
Experimental error	$(r - 1)(t - 1) - 1$	E_{yy}	$\mathbf{E}$
Total	rt	$\sum Y^2$	$\cdots$

We can also use regression analysis to test the hypothesis $H:\beta = 0$. To do this, we must perform a regression analysis using the model without the covariate,

$$Y_{ij} = \mu + \rho_1 u_{1i} + \cdots + \rho_{r-1} u_{r-1i} + \tau_1 z_{1j} + \cdots + \tau_{t-1} z_{t-1j} + \epsilon_{ij}$$
$$i = 1, \ldots, r;\ j = 1, \ldots, t \qquad (13.32)$$

From this model we can evaluate the regression sum of squares $SS(\rho_i's, \tau_j's \mid \mu)$. The test statistic for testing $H:\beta = 0$ is

$$F = SS(\beta \mid \mu, \rho_i's, \tau_j's)/\mathbf{E}$$
$$= [SS(\beta, \rho_i's, \tau_j's \mid \mu) - SS(\rho_i's, \tau_j's \mid \mu)]/\mathbf{E} \qquad (13.33)$$

which has $\nu_1 = 1$ and $\nu_2 = (r - 1)(t - 1) - 1$ degrees of freedom.

Example 13.8

Consider again the data given in Table 13.7. We repeat the analysis using the sums of squares derived from regression analyses. The results are summarized in Table 13.10, where we have partitioned the regression sum of

TABLE 13.10–Analysis of Covariance for the Data in Table 13.7

Source of Variation	Degrees of Freedom	Sum of Squares	Mean Square
Mean	1	49,152.00	$\cdots$
Regression	6	300.605	$\cdots$
Due to β	1	159.136	$\cdots$
Blocks $\mid \beta$	3	96.965	$\cdots$
Treatments $\mid \beta$	2	44.503	22.25
Experimental error	5	23.394	4.68
Total	12	49,476.00	$\cdots$

squares for the reduced model $SS(\beta, \rho_i\text{'s} \mid \mu)$ into the sum of squares due to β, $SS(\beta \mid \mu)$, and the sum of squares due to blocks after fitting β. This is frequently easy to do and can give us some information about the effectiveness of blocking.

To test $H:\beta = 0$, the regression model in Eq. (13.32) with $i = 4$ and $j = 3$ was used. The regression sum of squares $SS(\rho_i\text{'s}, \tau_j\text{'s} \mid \mu) = 276.0$. Thus, the value of the test statistic in Eq. (13.33) is $F = (300.605 - 276.0)/4.68 = 24.605/4.68 = 5.26$. Similarly, to test the hypothesis of no differences among the treatments after adjusting for the effect of the concomitant variable, the value of the test statistic is $F = 22.25/4.68 = 4.75$. Both statistics are the same as the corresponding statistic in Example 13.7.

We can easily calculate the adjusted treatment means. The adjusted mean for the jth treatment is

$$\text{adj } \overline{Y}_{.j} = \overline{Y}_{.j} - b(\overline{X}_{.j} - \overline{X}_{..}) \qquad j = 1, \ldots, t \qquad (13.34)$$

where b is the estimated regression coefficient. The estimated variance of an adjusted treatment mean is

$$\hat{V}(\text{adj } \overline{Y}_{.j}) = s_E^2\left[\frac{1}{r} + \frac{(\overline{X}_{.j} - \overline{X}_{..})^2}{E_{xx}}\right] \qquad (13.35)$$

where $s_E^2 = \mathbf{E}$ is the error mean square and E_{xx} is the sum of squares,

$$E_{xx} = \sum_{i=1}^{r} \sum_{j=1}^{t} (X_{ij} - \overline{X}_{i.} - \overline{X}_{.j} + \overline{X}_{..})^2 \qquad (13.36)$$

The estimated variance of the differences between two adjusted treatment means is

$$\hat{V}(\text{adj } \overline{Y}_{.j} - \text{adj } \overline{Y}_{.j'}) = s_E^2\left[\frac{2}{r} + \frac{(\overline{X}_{.j} - \overline{X}_{.j'})^2}{E_{xx}}\right] \qquad (13.37)$$

13.5 TWO-FACTOR FACTORIAL IN A RANDOMIZED COMPLETE BLOCK DESIGN

When the treatments involved in an experiment are combinations of several factors, the analysis consists of partitioning the treatment sum of squares into main effects and interactions. The same analysis holds when a covariate is included in the model. Thus we are able to test among the adjusted means for each of the factors and all the interactions. To illustrate the procedure, we consider a two-factor factorial (cross-classification) involving a randomized complete block design with a single covariate. The linear statistical model is

$$Y_{ijk} = \mu + \rho_i + \alpha_j + \gamma_k + (\alpha\gamma)_{jk} + \beta X_{ijk} + \epsilon_{ijk}$$
$$i = 1, \ldots, r; j = 1, \ldots, a; k = 1, \ldots, c \qquad (13.38)$$

The analysis of covariance is summarized in Table 13.11. The sums of squares R_{xx}, R_{yy}, A_{xx}, etc., are the same as outlined in Chapter 11. In

TABLE 13.11–General Analysis of Covariance for a Two-factor Factorial in a Randomized Complete Block Design

Source of Variation	Degrees of Freedom	Sum of Squares and Products			Deviation about Regression		
		$\sum x^2$	$\sum xy$	$\sum y^2$	$\sum y^2 - (\sum xy)^2/\sum x^2$	Degrees of Freedom	Mean Square
Replicates	$r-1$	R_{xx}	R_{xy}	R_{yy}	...	...	...
Treatments							
A	$a-1$	A_{xx}	A_{xy}	A_{yy}	...	...	...
C	$c-1$	C_{xx}	C_{xy}	C_{yy}	...	...	...
AC	$(a-1)(c-1)$	$(AC)_{xx}$	$(AC)_{xy}$	$(AC)_{yy}$	...	...	...
Experimental error	$(r-1)(ac-1)$	E_{xx}	E_{xy}	E_{yy}	$S_E = E_{yy} - E_{xy}^2/E_{xx}$	$(r-1)(ac-1)-1$	$s_E^2 = S_E/[(r-1)(ac-1)-1]$
A + error	$(a-1)+(r-1)(ac-1)$	$AS_{xx} = A_{xx}+E_{xx}$	$AS_{xy}=A_{xy}+E_{xy}$	$AS_{yy}=A_{yy}+E_{yy}$	$S_{A+E} = AS_{yy} - AS_{xy}^2/AS_{xx}$	$(a-1)+(r-1)(ac-1)-1$	...
		Difference for testing among adjusted A-means			$S_A + E - S_E = A_{yy} - AS_{xy}^2/AS_{xx} + E_{xy}^2/E_{xx}$	$a-1$	$(S_A+E-S_E)/(a-1)$
C + error	$(c-1)+(r-1)(ac-1)$	$CS_{xx} = C_{xx}+E_{xx}$	$CS_{xy}=C_{xy}+E_{xy}$	$CS_{yy}=C_{yy}+E_{yy}$	$S_{C+E} = CS_{yy} - CS_{xy}^2/CS_{xx}$	$(c-1)+(r-1)(ac-1)-1$	...
		Difference for testing among adjusted C-means			$S_C + E - S_E = C_{yy} - CS_{xy}^2/CS_{xx} + E_{xy}^2/E_{xx}$	$c-1$	$(S_C+E-S_E)/(c-1)$
AC + error	$(a-1)(c-1)$ $+ (r-1)(ac-1)$	ACS_{xx} $=(AC)_{xx}$ $+ E_{xx}$	ACS_{xy} $=(AC)_{xy}$ $+ E_{xy}$	ACS_{yy} $=(AC)_{yy}$ $+ E_{yy}$	$S_{AC+E} = ACS_{yy}$ $- ACS_{xy}^2/ACS_{xx}$	$(a-1)(c-1)$ $+ (r-1)(ac-1) - 1$	...
		Difference for testing among adjusted AC-effects			$S_{AC} + E - S_E = (AC)_{yy}$ $- ACS_{xy}^2/ACS_{xx}$ $+ E_{xy}^2/E_{xx}$	$(a-1)(c-1)$	$(S_{AC}+E-S_E)/$ $(a-1)(c-1)$

addition, the sums of products of X and Y can be calculated using the equations:

$$R_{xy} = ac \sum_{i=1}^{r} (\bar{X}_{i..} - \bar{X}_{...})(\bar{Y}_{i..} - \bar{Y}_{...})$$

$$= \sum_{i=1}^{r} (X_{i..})(Y_{i..})/ac - (X_{...})(Y_{...})/rac \qquad (13.39)$$

$$A_{xy} = rc \sum_{j=1}^{a} (\bar{X}_{.j.} - \bar{X}_{...})(\bar{Y}_{.j.} - \bar{Y}_{...})$$

$$= \sum_{j=1}^{a} (X_{.j.})(Y_{.j.})/rc - (X_{...})(Y_{...})/rac \qquad (13.40)$$

$$C_{xy} = ra \sum_{k=1}^{c} (\bar{X}_{..k} - \bar{X}_{...})(\bar{Y}_{..k} - \bar{Y}_{...})$$

$$= \sum_{k=1}^{c} (X_{..k})(Y_{..k})/ra - (X_{...})(Y_{...})/rac \qquad (13.41)$$

$$(AC)_{xy} = r \sum_{j=1}^{a} \sum_{k=1}^{c} (\bar{X}_{.jk} - \bar{X}_{.j.} - \bar{X}_{..k} + \bar{X}_{...})(\bar{Y}_{.jk} - \bar{Y}_{.j.} - \bar{Y}_{..k} + \bar{Y}_{...})$$

$$= \sum_{j=1}^{a} \sum_{k=1}^{c} (X_{.jk})(Y_{.jk})/r - A_{xy} - C_{xy} - (X_{...})(Y_{...})/rac \qquad (13.42)$$

$$E_{xy} = \sum_{i=1}^{r} \sum_{j=1}^{a} \sum_{k=1}^{c} (X_{ijk})(Y_{ijk}) - R_{xy} - A_{xy} - C_{xy}$$

$$- (AC)_{xy} - (X_{...})(Y_{...})/rac \qquad (13.43)$$

Before indicating the tests, we rewrite the model as a regression equation. Allowing for the usual restrictions on the parameters $\sum_{j=1}^{a} \alpha_j = \sum_{k=1}^{c} \gamma_k = \sum_{j=1}^{a} (\alpha\gamma)_{jk} = \sum_{k=1}^{c} (\alpha\gamma)_{jk} = \sum_{i=1}^{r} \rho_i = 0$, the model is

$$Y_{ijk} = \mu + \rho_1 u_{1i} + \cdots + \rho_{r-1} u_{r-1i} + \alpha_1 z_{1j} + \cdots + \alpha_{a-1} z_{a-1j}$$
$$+ \gamma_1 w_{1k} + \cdots + \gamma_{c-1} w_{c-1k}$$
$$+ (\alpha\gamma)_{11} z_{1j} w_{1k} + \cdots + (\alpha\gamma)_{a-1,c-1} z_{a-1j} w_{c-1k} + \beta X_{ijk} + \epsilon_{ijk}$$
$$i = 1,\ldots,r; j = 1,\ldots,a; k = 1,\ldots,c \qquad (13.44)$$

The analysis based on the regression model in Eq. (13.44) follows the pattern outlined previously. The pertinent sums of squares and mean squares are given in Table 13.12. Since the factor effects are no longer orthogonal after adjusting for X, the reduced model excludes only terms corresponding to the effect to be tested.

The hypotheses and test statistics are summarized in Table 13.13, where, for example, the hypothesis $H:(\alpha\gamma)_{11} = \cdots = (\alpha\gamma)_{ac} = 0$ denotes the hypothesis of no interaction between factor a and factor c as they affect the variable Y after adjusting for the effect of X.

TABLE 13.12–Analysis of Covariance for Two-Factor Factorial (cross-classification) with a Randomized Complete Block Design

Source of Variation	Degrees of Freedom	Sum of Squares	Mean Square
Mean	1	$SS(\mu) = (Y_{...})^2/rab$	$\cdots$
Regression (full)	$ac + r - 1$	$SS(\beta, \rho_i\text{'s}, \alpha_j\text{'s}, \gamma_k\text{'s}, (\alpha\gamma)_{jk}\text{'s} \mid \mu)$	$\cdots$
Due to β	1	$SS(\beta \mid \mu)$	$\cdots$
Blocks $\mid \beta$	$r - 1$	$SS(\rho_i\text{'s} \mid \mu, \beta)$	$\cdots$
Due to $A \mid \beta$	$a - 1$	$SS[\alpha_j\text{'s} \mid \mu, \beta, \rho_i\text{'s}, \gamma_k\text{'s}, (\alpha\gamma)_{jk}\text{'s}]$	$MS[\alpha_j\text{'s} \mid \mu, \beta, \rho_i\text{'s}, \gamma_k\text{'s}, (\alpha\gamma)_{jk}\text{'s}]$
Due to $C \mid \beta$	$c - 1$	$SS[\gamma_k\text{'s} \mid \mu, \beta, \rho_i\text{'s}, \alpha_j\text{'s}, (\alpha\gamma)_{jk}\text{'s}]$	$MS[\gamma_k\text{'s} \mid \mu, \beta, \rho_i\text{'s}, \alpha_j\text{'s}, (\alpha\gamma)_{jk}\text{'s}]$
Due to $AC \mid \beta$	$(a-1)(c-1)$	$SS[(\alpha\gamma)_{jk}\text{'s} \mid \mu, \beta, \rho_i\text{'s}, \alpha_j\text{'s}, \gamma_k\text{'s}]$	$MS[(\alpha\gamma)_{jk}\text{'s} \mid \mu, \beta, \rho_i, \alpha_j\text{'s}, \gamma_k\text{'s}]$
Experimental error	$(r-1)(ac-1) - 1$	E_{yy}	E
Total	rac	$\sum Y^2$	

TABLE 13.13–Tests for a Two-Factor Factorial in a Randomized Complete Block Design

Hypothesis	Test Statistic		Degrees of Freedom
	Table 13.11	Table 13.12	
$H: \beta = 0$	$E_{xy}^2/E_{xx}s_E^2$	$SS[\beta \mid \mu, \rho_i\text{'s}, \alpha_j\text{'s}, \gamma_k\text{'s} (\alpha\gamma)_{jk}\text{'s}]/E$	$\nu_1 = 1,$ $\nu_2 = (r-1)(ac-1) - 1$
$H:(\alpha\gamma)_{11} = \cdots$ $= (\alpha\gamma)_{ac} = 0$	$\dfrac{(S_{AC+E} - S_E)/(a-1)}{(c-1)s_E^2}$	$\dfrac{MS[(\alpha\gamma)_{jk}\text{'s} \mid \mu, \beta, \rho_i\text{'s}, \alpha_j\text{'s}, \gamma_k\text{'s}]}{E}$	$\nu_1 = (a-1)(c-1)$ $\nu_2 = (r-1)(ac-1) - 1$
$H:\alpha_1 = \cdots$ $= \alpha_a = 0$	$\dfrac{(S_{A+E} - S_E)/(a-1)s_E^2}{}$	$\dfrac{MS[\alpha_j\text{'s} \mid \mu, \beta, \rho_i\text{'s}, \gamma_k\text{'s}, (\alpha\gamma)_{jk}\text{'s}]}{E}$	$\nu_1 = (a-1)$ $\nu_2 = (r-1)(ac-1) - 1$
$H:\gamma_1 = \cdots$ $= \gamma_c = 0$	$\dfrac{(S_{C+E} - S_E)/(c-1)s_E^2}{}$	$\dfrac{MS[\gamma_k\text{'s} \mid \mu, \beta, \rho_i\text{'s}, \alpha_j\text{'s}, (\alpha\gamma)_{jk}\text{'s}]}{E}$	$\nu_1 = (c-1)$ $\nu_2 = (r-1)(ac-1) - 1$

Again the adjusted means are easily calculated. For example, the adjusted mean for the jth level of factor a is

$$\text{adj } \overline{Y}_{.j.} = \overline{Y}_{.j.} - b(\overline{X}_{.j.} - \overline{X}_{...}) \tag{13.45}$$

The appropriate standard errors for the various mean effects are found from the following estimated variances:

$$A\text{-effect} \quad \hat{V}(\text{adj } \overline{Y}_{.j.}) = s_E^2\left[\frac{1}{rc} + \frac{(\overline{X}_{.j.} - \overline{X}_{...})^2}{E_{xx}}\right] \tag{13.46}$$

$$C\text{-effect} \quad \hat{V}(\text{adj } \overline{Y}_{..k}) = s_E^2\left[\frac{1}{ra} + \frac{(\overline{X}_{..k} - \overline{X}_{...})^2}{E_{xx}}\right] \tag{13.47}$$

$$AC\text{-effect} \qquad \hat{V}(\text{adj } \overline{Y}_{.jk}) = s_E^2 \left[\frac{1}{r} + \frac{(\overline{X}_{.jk} - \overline{X}_{...})^2}{E_{xx}} \right] \qquad (13.48)$$

where

$$E_{xx} = \sum_{i=1}^{r} \sum_{j=1}^{a} \sum_{k=1}^{c} X_{ijk}^2 - R_{xx} - A_{xx} - C_{xx} - (AC)_{xx} - (X_{...})^2/rac \qquad (13.49)$$

Example 13.9

As an example of a covariance analysis in a randomized complete block design where the treatments are of a factorial nature, consider the data of Table 13.14. These data were originally examined by Wishart (4), and the interested reader is referred to his study for a more detailed discussion.

In discussing the data of Table 13.14, we shall consider the 5 pens as 5 replicates, and thus we have a 3×2 factorial in a randomized complete block design. Following the calculational procedure outlined in Table 13.11, we arrive at the results presented in Table 13.15.

To test $H:\beta = 0$, we calculate $F = [(39.367)^2/442.93]/0.2534 = 13.81$, with $\nu_1 = 1$ and $\nu_2 = 19$ degrees of freedom, and this is significant at $\alpha = 0.01$. The various treatment effects may also be tested for significance, the appropriate variance ratios being

$$\text{Food:} \qquad F = 1.16825/0.2534 = 4.61$$
$$\text{Sex:} \qquad F = 1.2594/0.2534 = 4.97$$
$$\text{Food} \times \text{Sex:} \quad F = 0.0489/0.2534 = 0.19$$

where the degrees of freedom are as given in Table 13.15. These F ratios

TABLE 13.14–Initial Weights and Gains in Weight of Young Pigs in a Comparative Feeding Trial

Pen		\multicolumn{6}{c}{Feeding Treatment}	Total
		\multicolumn{2}{c}{A}	\multicolumn{2}{c}{B}	\multicolumn{2}{c}{C}	
		Male	Female	Male	Female	Male	Female	
I	X	38	48	39	48	48	48	269
	Y	9.52	9.94	8.51	10.00	9.11	9.75	56.83
II	X	35	32	38	32	37	28	202
	Y	8.21	9.48	9.95	9.24	8.50	8.66	54.04
III	X	41	35	46	41	42	33	238
	Y	9.32	9.32	8.43	9.34	8.90	7.63	52.94
IV	X	48	46	40	46	42	50	272
	Y	10.56	10.90	8.86	9.68	9.51	10.37	59.88
V	X	43	32	40	37	40	30	222
	Y	10.42	8.82	9.20	9.67	8.76	8.57	55.44
Total	X	205	193	203	204	209	189	1203
	Y	48.03	48.46	44.95	47.93	44.78	44.98	279.13

Source: Reproduced from Table 11 in Wishart (4), with permission of author and publishers.

Note: X = initial weight in pounds; Y = gain in weight in pounds.

TABLE 13.15–Analysis of Covariance for the Data of Table 13.14

Source of Variation	Degrees of Freedom	Sum of Squares and Products			Deviation about Regression		
		$\sum x^2$	$\sum xy$	$\sum y^2$	$\sum y^2 - \dfrac{(\sum xy)^2}{\sum x^2}$	Degrees of Freedom	Mean Square
Replicates (pens)	4	605.87	39.905	4.8518	...	...	...
Treatments							
Food	2	5.40	−0.147	2.2686	...	...	...
Sex	1	32.03	−3.730	0.4344	...	...	...
Food × sex	2	22.47	3.112	0.4761	...	...	...
Experimental error	20	442.93	39.367	8.3144	4.8155	19	0.2534
Food + error	22	448.33	39.220	10.5830	7.1520	21	...
Difference for testing among adjusted food means					2.3365	2	1.16825
Sex + error	21	474.96	35.637	8.7488	6.0749	20	...
Difference for testing among adjusted sex means					1.2594	1	1.2594
(Food × sex) + (error)	22	465.40	42.479	8.7905	4.9133	21	...
Difference for testing among adjusted food × sex effects					0.0978	2	0.0489

(and the corresponding inferences) should be compared with those resulting from an ANOVA on the gains in weight, taking no account of the varying initial weights. Such comparisons will aid the reader in understanding the principles of covariance analyses and, in our example, will help to explain the effect of initial weights on weight gains subject to the chosen experimental conditions. A table of adjusted treatment means, together with the appropriate standard errors, should also be presented to make the analysis complete.

13.6 LATIN SQUARE DESIGN

The performance of an analysis of covariance on data resulting from a Latin square design introduces no new concepts. Thus we shall outline briefly the computations and tests associated with the Latin square design. The linear statistical model is

$$Y_{ij(k)} = \mu + \rho_i + \gamma_j + \tau_{(k)} + \beta X_{ij(k)} + \epsilon_{ij(k)}$$
$$i = 1, \ldots, m; j = 1, \ldots, m; k = 1, \ldots, m \qquad (13.50)$$

The analysis of covariance based on sum of squares and cross-product computations is summarized in Table 13.16. Besides the usual calculations R_{yy}, C_{yy}, T_{yy}, etc., for the ordinary ANOVA, the only calculations are the sums of cross products. These are computable using the equations

$$R_{xy} = \sum_{i=1}^{m} Y_{i.(.)} X_{i.(.)}/m - Y_{..(.)} X_{..(.)}/m^2 \qquad (13.51)$$

TABLE 13.16–General Analysis of Covariance for an $m \times m$ Latin Square

Source of Variation	Degrees of Freedom	Sum of Squares and Products			Deviation about Regression		
		$\sum x^2$	$\sum xy$	$\sum y^2$	$\sum y^2 - (\sum xy)^2/\sum x^2$	Degrees of Freedom	Mean Square
Rows	$m-1$	R_{xx}	R_{xy}	R_{yy}			
Columns	$m-1$	C_{xx}	C_{xy}	C_{yy}	$\cdots$	$\cdots$	$\cdots$
Treatments	$m-1$	T_{xx}	T_{xy}	T_{yy}	$\cdots$	$\cdots$	$\cdots$
Experimental error	$(m-1)(m-2)$	E_{xx}	E_{xy}	E_{yy}	$S_E = E_{yy} - E_{xy}^2/E_{xx}$	$(m-1)(m-2)-1$	$s_E^2 = S_E/[(m-1)(m-2)-1]$
Treatment + error	$(m-1)^2$	$S_{xx} = T_{xx} + E_{xx}$	$S_{xy} = T_{xy} + E_{xy}$	$S_{yy} = T_{yy} + E_{yy}$	$S_{T+E} = S_{yy} - S_{xy}^2/S_{xx}$	$(m-1)^2 - 1$	$\cdots$
Difference for testing among adjusted treatment means					$S_{T+E} - S_E = T_{yy} - S_{xy}^2/S_{xx} + E_{xy}^2/E_{xx}$	$m-1$	$(S_{T+E} - S_E)/(m-1)$

$$C_{xy} = \sum_{j=1}^{m} Y_{.j(.)} X_{.j(.)}/m - Y_{..(.)} X_{..(.)}/m^2 \tag{13.52}$$

$$T_{xy} = \sum_{k=1}^{m} Y_{..(k)} X_{..(k)}/m - Y_{..(.)} X_{..(.)}/m^2 \tag{13.53}$$

$$E_{xy} = \sum_{i=1}^{m}\sum_{j=1}^{m}\sum_{k=1}^{m} Y_{ij(k)} X_{ij(k)} - R_{xy} - C_{xy} - T_{xy} - Y_{..(.)} X_{..(.)}/m^2 \tag{13.54}$$

where the dot notation is consistent with our previous use. The corresponding regression model is

$$Y_{ij(k)} = \mu + \rho_1 u_{1i} + \cdots + \rho_{m-1} u_{m-1i} + \gamma_1 v_{1j} + \cdots + \gamma_{m-1} v_{m-1j}$$
$$+ \tau_1 z_{1k} + \cdots + \tau_{m-1} z_{m-1k} + \beta X_{ij(k)} + \epsilon_{ij(k)}$$
$$i = 1,\ldots,m; j = 1,\ldots,m; k = 1,\ldots,m \tag{13.55}$$

where u, v, and z are of the form

$$u_{ki} = \quad 1 \quad \text{if } i = k; k = 1,\ldots,m-1$$
$$= -1 \quad \text{if } i = m$$
$$= \quad 0 \quad \text{otherwise}$$

The relevant sums of squares obtained from a regression analysis using Eq. (13.55) are summarized in Table 13.17.

TABLE 13.17–Analysis of Covariance for a Latin Square Design

Source of Variation	Degrees of Freedom	Sum of Squares	Mean Square
Mean	1	$SS(\mu) = (Y_{...})^2/m^2$	$\cdots$
Regression (full)	$3m-2$	$SS(\beta, \rho_i\text{'s}, \gamma_j\text{'s}, \tau_k\text{'s} \mid \mu)$	$\cdots$
Regression (reduced)	$2m-1$	$SS(\beta, \rho_i\text{'s}, \gamma_j\text{'s} \mid \mu)$	$\cdots$
Treatments $\mid \beta$	$m-1$	$SS(\tau_k\text{'s} \mid \mu, \beta, \rho_i\text{'s}, \gamma_j\text{'s})$	$SS(\tau_k\text{'s} \mid \mu,\beta,\rho_i\text{'s},\gamma_j\text{'s})/(m-1)$
Experimental error	$(m-1)(m-2)-1$	E_{yy}	E
Total	m^2	$\sum Y^2$	$\cdots$

The test of $H{:}\beta = 0$ uses the test statistic

$$F = SS(\tau_k\text{'s} \mid \mu, \beta, \rho_i\text{'s}, \gamma_j\text{'s})/(m-1)E, \text{ using Eq. (13.55)}$$
$$= E_{xy}^2/E_{xx}s_E^2, \text{ using Eq. (13.50)} \tag{13.56}$$

with $\nu_1 = 1$ and $\nu_2 = (m-1)(m-2) - 1$ degrees of freedom. To test among adjusted treatment means, we compute

$$F = SS(\tau_k\text{'s} \mid \mu,\beta)/(m-1)E, \text{ using Eq. (13.55)}$$
$$= (S_{T+E} - S_E)/(m-1)s_E^2, \text{ using Eq. (13.50)} \tag{13.57}$$

with $\nu_1 = m - 1$ and $\nu_2 = (m - 1)(m - 2) - 1$ degrees of freedom. The adjusted trestment means may be found using

$$\text{adj } \overline{Y}_{..(k)} = \overline{Y}_{..(k)} - b(\overline{X}_{..(k)} - \overline{X}_{..(.)}) \tag{13.58}$$

where b is the regression coefficient associated with experimental error; i.e., $b = E_{xy}/E_{xx}$. The estimated variance of an adjusted treatment mean is

$$\hat{V}(\text{adj } \overline{Y}_{..(k)}) = s_E^2 \left[\frac{1}{m} + \frac{(\overline{X}_{..(k)} - \overline{X}_{..(.)})^2}{E_{xx}} \right] \tag{13.59}$$

and the estimated variance of the difference between two adjusted treatment means is

$$\hat{V}(\text{adj } \overline{Y}_{..(k)} - \text{adj } \overline{Y}_{..(k')}) = s_E^2 \left[\frac{2}{m} + \frac{(\overline{X}_{..(k)} - \overline{X}_{..(k')})^2}{E_{xx}} \right] \tag{13.60}$$

13.7 MULTIPLE COVARIANCE

Occasionally, two or more concomitant variables are measured in addition to the response variable, and hence multiple covariance procedures must be considered. The multiple covariance model involving p concomitant variables and the completely randomized design is given in Eq. (13.2).

The analysis of covariance when there are several concomitant variables is most easily done by applying regression analysis and using a computer. The regression procedure with use of a computer is identical with that involving only one X variable. That is, first fit the model including the true treatment effects and evaluate the regression sum of squares $SS(\beta_1, \ldots, \beta_p, \tau_j\text{'s} \mid \mu)$, then fit the model under the assumption that all the treatment effects are identical, i.e., all the τ_j's are zero. From this model evaluate $SS(\beta_1, \ldots, \beta_p \mid \mu)$. The difference between these sums of squares,

$$SS(\tau_j\text{'s} \mid \mu, \beta_1, \ldots, \beta_p) = SS(\beta_1, \ldots, \beta_p, \tau_j\text{'s} \mid \mu) - SS(\beta_1, \ldots, \beta_p \mid \mu)$$

is a measure of the difference in the effects of the treatments after adjusting for the effect of the p concomitant variables. To illustrate the analysis involved in multiple covariance, we consider an example.

Example 13.10

Crampton and Hopkins (3) studied the effects of initial weight and food consumption on the gaining ability of pigs when they were given different feeds. The data are presented in Table 13.18.

The response variable measured was Y, the final weight. In addition, two concomitant variables measured were X_1, the initial weight, and X_2, feed eaten. A randomized complete block design was used. The linear statistical model for the data is

$$Y_{ij} = \mu + \rho_i + \tau_j + \beta_1 X_{1ij} + \beta_2 X_{2ij} + \epsilon_{ij}$$

$$i = 1, \ldots, 10; j = 1, \ldots, 5 \tag{13.61}$$

TABLE 13.18– Weights, Gains, and Feed Consumption of Pigs in Comparative Feeding Trials

Replicate	Treatment I				Treatment II				Treatment III				Treatment IV				Treatment V			
	Initial weight	Feed eaten	Gain	Final weight	Initial weight	Feed eaten	Gain	Final weight	Initial weight	Feed eaten	Gain	Final weight	Initial weight	Feed eaten	Gain	Final weight	Initial weight	Feed eaten	Gain	Final weight
	X_1	X_2	Y_1	Y_2	X_1	X_2	Y_1	Y_2	X_1	X_2	Y_1	Y_2	X_1	X_2	Y_1	Y_2	X_1	X_2	Y_1	Y_2
1	30	674	165	195	26	699	168	194	39	708	164	203	41	716	185	226	41	831	201	242
2	21	628	156	177	24	626	180	204	34	614	156	190	35	769	195	230	36	754	189	225
3	21	661	159	180	20	668	180	200	32	733	189	221	32	733	186	218	32	722	173	205
4	33	694	167	200	35	668	166	201	35	663	138	173	34	742	201	235	35	728	193	228
5	27	713	170	197	25	707	170	195	32	607	153	185	32	624	165	197	32	646	164	196
6	24	585	146	170	26	651	161	187	35	745	190	225	35	710	175	210	36	678	160	196
7	20	575	130	150	20	672	171	191	30	637	160	190	30	742	187	217	30	763	200	230
8	29	638	151	180	31	660	169	200	29	662	172	201	28	648	177	205	28	625	142	170
9	28	632	164	192	29	769	179	208	32	609	142	174	34	628	166	200	32	710	184	216
10	26	637	158	184	27	666	191	218	25	596	155	180	26	601	165	191	26	651	149	175
Mean	25.9	643.7	156.6	182.5	26.3	678.6	173.5	199.8	32.3	657.4	161.9	194.2	32.7	691.3	180.2	212.9	32.8	710.8	175.5	208.3

Source: Reproduced from Crampton and Hopkins (3), p. 335, by permission of the authors and publishers.

The regression model is similar to the model in Eq. (13.29) for one concomitant variable. For this example the model is

$$Y_{ij} = \mu + \rho_1 u_{1i} + \cdots + \rho_9 u_{9i} + \tau_1 z_{1j} + \cdots + \tau_4 z_{4j}$$
$$+ \beta_1 X_{1ij} + \beta_2 X_{2ij} + \epsilon_{ij} \qquad i = 1, \ldots, 10; j = 1, \ldots, 5 \qquad (13.62)$$

where

$$
\begin{aligned}
z_{km} &= 1 && \text{if } m = k; k = 1, 2, 3, 4 \\
&= -1 && \text{if } m = 5 \\
&= 0 && \text{otherwise} \\
u_{km} &= 1 && \text{if } m = k; k = 1, \ldots, 9 \\
&= -1 && \text{if } m = 10 \\
&= 0 && \text{otherwise}
\end{aligned}
$$

A regression analysis program on a computer was used to analyze the data. The results are summarized in an analysis of covariance table (Table 13.19). To test the hypothesis of no differences among the true effects of the treatments after adjusting for the effect of X_1 and X_2, we evaluate the F statistic

$$F = SS(\tau_j\text{'s} \mid \beta_1, \beta_2, \rho_i\text{'s})/(4)E = 241.47/103.11 = 2.34$$

Comparing this with the F distribution with $\nu_1 = 4$ and $\nu_2 = 34$ degrees of freedom, we conclude that the feeds are not different at $\alpha = 0.05$.

Adjusted treatment means and their standard errors may be calculated following methods similar to those outlined in earlier sections of this chapter.

TABLE 13.19–Analysis of Covariance for Data in Table 13.18

Source of Variation	Degrees of Freedom	Sum of Squares	Mean Square
Mean	1	1,990,810.58	...
Regression	15	15,128.62	...
Due to β_1, β_2	2	13,688.02	...
Blocks $\mid \beta_1, \beta_2$	9	474.71	...
Treatments $\mid \beta_1, \beta_2$	4	965.88	241.47
Experimental error	34	3,505.80	103.11
Total	50	2,009,445.00	...

13.8 COVARIANCE WHEN THE X VARIABLE IS AFFECTED BY THE TREATMENTS

When the treatments being employed in the experiment are such that they have an appreciable effect on the concomitant variate X as well as on the Y variate, the researcher should proceed with caution. Computationally, each step is carried through as before, but the final inferences must take account of the effect the treatments have had on the concomitant variate. For example, if the concomitant variate in a feeding experiment had been "amount of feed consumed" rather than "initial weight," it is quite possible that the different treatments (feeds) would have a significant effect on the food consumption. Thus any covariance analysis of gain in weight should take cognizance of the weight-producing effects of the different feeds due to increased (or decreased) consumption apart from any nutritional differences among the feeds. Other examples of covariance analyses involving similar problems are available in the literature and should be studied critically by those who desire a fuller understanding of covariance techniques. The reader is especially referred to Cochran and Cox (2) and Bartlett (1) for discussions of this type of problem.

PROBLEMS

13.1 An experiment using a randomized complete block design gave the following corrected sums of squares and products:

Source of Variation	Degrees of Freedom	$\sum x^2$	$\sum xy$	$\sum y^2$	b
Replicates	5	200	600	4000	...
Treatments	5	100	200	2500	2
Experimental error	25	300	1200	7500	4

(a) Based on the experimental error sum of squares and products, is the regression of Y on X significant at $\alpha = 0.05$?
(b) Are the differences among the treatment means for Y adjusted for variation attributed to X significant at $\alpha = 0.05$?
(c) What conclusions do you draw from the above data about the effect of treatments? Make any additional computations that you consider necessary.

13.2 Given the following data:

Source of Variation	Degrees of Freedom	$\sum x^2$	$\sum xy$	$\sum y^2$
Replicates	4	100	140	400
Treatments	10	100	100	900
Experimental error	40	400	900	2500

(a) What conclusions may be drawn about the effect of treatments on Y?
(b) Test the regression coefficient based on experimental error for significance at the 5 percent level.

13.3 Ten lines of soybeans were compared in randomized complete blocks with 4 replications. The differences in yield Y were not significant, but it was observed that the incidence of an infestation X differed among the varieties. Following is the table of sums of squares and products. Test the hypothesis that the yields adjusted for infestation do not differ in the sampled populations. What fraction of $\sum y^2$ for lines is unexplained by the regression?

Source of Variation	Degrees of Freedom	$\sum x^2$	$\sum xy$	$\sum y^2$
Lines	9	4684	-532	112
Error	27	3317	-650	216

13.4   For an experiment involving 9 soil sterilization treatments, the effect on the number of seedling alfalfa plants X and on the green weight of plants at 3 weeks Y is summarized by the following sums of squares and products. Complete the analysis, making appropriate tests to indicate the reason for your conclusions. If the mean of the X's is 15 and the mean of the Y's is 25, give the regression equation for error.

Source of Variation	Degrees of Freedom	$\sum x^2$	$\sum xy$	$\sum y^2$
Replicates	5	4	16	96
Treatments	8	16	32	80
Error	40	20	40	160

13.5   A study of eastern Iowa farms included one group of tenants who were not related to their landlords and another group of tenants each of whom was related to his landlord. It was assumed that soil improvements would be more generally undertaken when landlord and tenant were related. Hence value of crops should be greater in those situations. An analysis of variance was undertaken to examine this hypothesis. Since size of farm could confuse the comparison, the size of farm was introduced as a covariate. The following table was prepared. Value of crops has been coded for this analysis.

Source of Variation	Degrees of Freedom	$\sum x^2$	$\sum xy$	$\sum y^2$
Total	59	125,000	33,000	36,600
Subareas (replicates)	4	20,000	14,010	13,600
Between groups of tenants	1	61,000	13,260	4,200
Interaction	4	4,000	$-13,270$	6,100
Within subclasses	50	40,000	19,000	12,700

(a) Is the acceptance of the hypothesis H (no difference between groups of tenants) changed by the introduction of farm size as the covariate?

(b) Is the error regression significant?

13.6   A sample of farms was taken in the eastern livestock area of Iowa for
the purpose of studying certain types of farm lease arrangements. For
this problem we are taking a portion of the data to study the difference
in "gross value of crops" produced on two groups of cash-rented farms:
(1) farms for which landlord and tenant are related and (2) farms for
which landlord and tenant are not related. The variates measured are
value of crops produced Y and size of farm X. The data presented are
given for 3 hypothetical blocks. In practice, these blocks might be
strata, e.g., different counties, different soil areas, type-of-farming
areas or groups of farms enumerated by different enumerators (i.e., 3
agricultural economics students for the 3 blocks of our example). The
data are presented in the following table. Perform an analysis of covariance on these data.

Block I			Block II			Block III		
Farm No.	Y (related)	X	Farm No.	Y (related)	X	Farm No.	Y (related)	X
22	6399	160	27	2,490	90	17	4,489	120
13	8456	320	24	5,349	154	25	10,026	245
20	8453	200	11	5,518	160	1	5,659	160
8	4891	160	34	10,417	234	26	5,475	160
21	3491	120	38	4,278	120	4	11,382	320
	(not related)			(not related)			(not related)	
31	6944	160	13	4,936	160	20	5,731	160
30	6971	160	1	7,376	200	15	6,787	173
11	4053	120	19	6,216	160	7	5,814	134
6	8767	280	32	10,313	240	5	9,607	239
16	6765	160	28	5,124	120	25	9,817	320

13.7   (a) What are the assumptions behind a covariance analysis?
        (b) In the process of analyzing data by a covariance analysis, what tests
            of significance are made?
        (c) Explain the interpretation or inferences and the course of action
            indicated when each of the above tests is significant; when each is
            nonsignificant.

13.8   The following is an experiment involving randomized complete blocks
with 4 replications. Eleven lines of soybeans were planted. The data are
as follows:

X_1 = maturity, measured in days later than the Hawkeye variety
X_2 = lodging, measured on a scale from 0 to 5
Y = infection by stem canker measured as a percentage of stalks
          infected

Line	Replicate 1			Replicate 2			Replicate 3			Replicate 4		
	X_1	X_2	Y	X_1	X_2	Y	X_1	X_2	Y	X_1	X_2	Y
Lincoln	9	3.0	19.3	10	2.0	29.2	12	3.0	1.0	9	2.5	6.4
A7-6102	10	3.0	10.1	10	2.0	34.7	9	2.0	14.0	9	3.0	5.6
A7-6323	10	2.5	13.1	9	1.5	59.3	12	2.5	1.1	10	2.5	8.1
A7-6520	8	2.0	15.6	5	2.0	49.0	8	2.0	17.4	6	2.0	11.7
A7-6905	12	2.5	4.3	11	1.0	48.2	13	3.0	6.3	10	2.5	6.7
C-739	4	2.0	25.2	2	1.5	36.5	2	2.0	23.4	1	2.0	12.9
C-776	3	1.5	67.6	4	1.0	79.3	6	2.0	13.6	2	1.5	39.4
H-6150	7	2.0	35.1	8	2.0	40.0	7	2.0	24.7	7	2.0	4.8
L6-8477	8	2.0	14.0	8	1.5	30.2	10	1.5	7.2	7	2.0	8.9
L7-1287	9	2.5	3.3	9	2.0	35.8	13	3.0	1.1	9	3.0	2.0
Bav. Sp.	10	3.5	3.1	10	3.0	9.6	11	3.0	1.0	10	3.5	0.1

The principal objective is to learn whether maturity or lodging is more closely related to infection. Determine this from the multiple regression. Test the hypothesis of no differences among adjusted mean infection for the varieties.

13.9   Discuss the use of covariance analysis. What factors must be considered in interpreting the results of the analysis?

13.10   The data for this problem consists of 54 pairs of observations on the calories consumed Y on one particular day by a respondent and her age X. The respondents were adult Iowa women over the age of 30 who were interviewed to obtain information on nutrition and health. About 1000 women were so enumerated for this survey, and our group of 54 is a subgroup from the total, which was taken so as to make numbers in the subclasses equal.

Among the items observed for each respondent in addition to caloric intake and age was place of residence (zone) and income class. These are listed below:

Zone	Income Group	
1—open country	1	$ 0– 999
2—rural place	2	1000–1499
3—urban	3	1500–1999
	4	2000–2999
	5	3000–3999
	6	over 4000

Education, height, weight, national origin, marital status, family composition, and many other factors were recorded for each respondent.

The nutritionists studying these data were interested in determining how food intake and health are related to these other observed factors. A few relevant hypotheses could be advanced. Preliminary analysis consisted of preparing tables of means for several classifications of the total sample and graphical analysis (plotting on scattergrams of a subsample of 60 stratified by age). A number of nutritive factors exhibited an apparent negative regression on age. Age thus seemed a useful covariate. Other factors such as education, height, and weight seemed to indicate no relation to nutritive intake.

With this background we shall use these data to undertake an analysis of covariance for the purpose of testing hypotheses about zone and income group effects after taking account of the regression on age. The table below gives the 54 pairs of observations with the sums, sums of squares, and sums of products. Both zone and income group may be considered as fixed effects.

(a) Prepare the analysis of covariance table.
(b) Find the regression of calories on age.
(c) Test the hypotheses that zone and income effects are equal to zero (separately, of course).
(d) It would also be of interest to test for interaction of zone and income group. What do you conclude on this point?
(e) The regression of calories on age may not be homogeneous over the zones. Indicate by a schematic analysis of variance of regression how you would examine these regressions.

	Zone 1		Zone 2		Zone 3	
Income Group	Y	X	Y	X	Y	X
1	1911	46	1318	80	1127	74
	1560	66	1541	67	1509	71
	2639	38	1350	73	1756	60
2	1034	50	1559	58	1054	83
	2096	33	1260	74	2238	47
	1356	44	1772	44	1599	71
3	2130	35	2027	32	1479	56
	1878	45	1414	51	1837	40
	1152	59	1526	34	1437	66
4	1297	68	1938	33	2136	31
	2093	43	1551	40	1765	56
	2035	59	1450	39	1056	70
5	2189	33	1183	54	1156	47
	2078	36	1967	36	2660	43
	1905	38	1452	53	1474	50
6	1156	57	2599	35	1015	63
	1809	52	2355	64	2555	34
	1997	44	1932	79	1436	54

$\sum Y^2$	166,416,926	$\sum XY$	4,573,454	$\sum X^2$	157,356
$-CT$	156,053,200	$-CT$	4,773,496	$-CT$	146,016
$\sum y^2 =$	10,363,726	$\sum xy =$	$-200,042$	$\sum x^2 =$	11,340
$\sum Y =$	91,798		$n = 54$	$\sum X =$	2,808

REFERENCES

1. Bartlett, M. S. 1936. A note on the analysis of covariance. *J. Agr. Sci.* 26:448.
2. Cochran, W. G.; and Cox, G. M. 1957. *Experimental Designs,* 2nd ed. Wiley, New York.
3. Crampton, E. W.; and Hopkins, J. W. 1934. The use of the method of partial regression in the analysis of comparative feeding trial data. Part II. *J. Nutr.* 8:329.
4. Wishart, J. 1950. Field trials II: The analysis of covariance. Tech. Comm. 15, Commonwealth Bureau of Plant Breeding and Genetics, Cambridge, England.

FURTHER READING

See Chapter 10.

CHAPTER 14

DISTRIBUTION-FREE METHODS

In preceding chapters the emphasis has been on statistical techniques that assume the sampled populations to be of a known form. This chapter includes a discussion of some methods that are designed to be used even if the form of the population is unknown.

14.1 INTRODUCTION

Most of the statistical techniques discussed so far have assumed that the sampled population is of a known form (e.g., it is a normal population or it is an exponential population). However, because the analyst is not always certain of the validity of such assumptions and/or because not all statistical techniques are robust (i.e., insensitive to departures from such assumptions), much work has been done to devise procedures that are free of these restrictions. These techniques are referred to as *distribution-free methods*. (**Note:** Many authors refer to distribution-free methods as *nonparametric methods* and, although the expressions are not strictly equivalent, they have been, and probably will continue to be, used interchangeably.)

The literature on distribution-free methods is quite extensive; thus it is possible to mention only a few of the more popular and useful methods in this book. Those who wish to delve deeper into this area of statistics are encouraged to consult the list of further readings at the end of the chapter.

Four widely used distribution-free methods have already been introduced. These are (1) Tchebycheff's inequality discussed in Section 2.19, (2) the distribution-free tolerance limits referred to in Section 5.11, (3) the chi-square goodness of fit test described in Section 6.14, and (4) the measures of rank correlation described in Section 8.11. Because these methods were discussed in the aforementioned sections, it would be superfluous to repeat their descriptions at this time. It is recommended, however, that the indicated sections be reread in the present context. Let us now proceed to the study of some additional distribution-free methods that have been found useful in a variety of situations.

14.2 SIGN TEST

In many experimental situations, the investigator wishes to compare the effects of two treatments. When the data occur in pairs, one member of the pair being associated with treatment A and the other with treatment B, one test of wide applicability is the *sign test*. Using inequality signs to denote the relationship between the members of a pair, whether the comparison be qualitative or quantitative, the sign test proceeds as follows:

1. Examine each of the pairs (X_i, Y_i).
2. If $X_i > Y_i$, assign a plus sign; if $X_i < Y_i$, assign a minus sign; if $X_i = Y_i$, discard the pair.
3. Denote the number of pairs remaining (i.e., the number of pairs resulting in either a plus or minus sign) by n.
4. Denote by r the number of times the less frequent sign occurs.
5. To test the hypothesis of no difference between the effects of the two treatments, compare r with the critical values tabulated in Appendix 13.
6. If the observed value of r is less than or equal to the tabulated value for the chosen significance level, the hypothesis is rejected; otherwise, it is not rejected.

(**Note:** If measurements are recorded, then $X_i > Y_i$ will signify that in the ith pair treatment A resulted in a higher reading than treatment B. If no measurements are available, then $X_i > Y_i$ will signify that in the ith pair treatment A resulted in something larger than (or better than or preferred over) treatment B.)

Before giving numerical illustrations, it is appropriate that attention be called to different hypotheses that can be tested in the manner indicated above. Some of these are:

1. Each difference $X_i - Y_i$ has a probability distribution (which need not be the same for all differences) with median equal to 0, i.e., $H:P(X_i > Y_i) = 0.5$ for all i.
2. If the underlying distributions are assumed to be symmetric, the sign test may be used to test the hypothesis $H:\mu_{X_i} = \mu_{Y_i}$.
3. If it can be assumed that the underlying distributions differ only in their means, then a test of $H:\mu_{X_i} = \mu_{Y_i}$ is equivalent to testing the hypothesis that the probability distributions of each pair are the same.
4. Questions such as (a) Is A better than B by P percent? and (b) Is A better than B by U units? may also be studied by applying the sign test to the differences $D = A - (1 + P/100)B$ and $D = A - (B + U)$ respectively.

Example 14.1

Consider once again the data given in Table 6.6 and discussed in Example 6.14. We note that $n = 15$ and $r = 4$. Assuming $\alpha = 0.05$, it is seen that the hypothesis of equal hardness indications by the two steel balls cannot be rejected since the critical value of r tabulated in Appendix 13 was 3. You will note that this is the opposite decision to that reached in Example 6.14 because, when normality can be assumed, the sign test is less efficient (i.e., less sensitive) than "Student's" t test.

Example 14.2

In a marketing study, 2 brands of lemonade were compared. Each of 50 judges tasted 2 samples, one of brand A and one of brand B, with the following results: 35 preferred brand A, 10 preferred brand B, and 5 could not tell the difference. Thus, $n = 45$ and $r = 10$. Assuming $\alpha = 0.01$, we reject the hypothesis of equal preference (since $r = 10 < 13 =$ critical value) and conclude that brand A is preferred.

14.3 SIGNED RANK TEST

The sign test described in Section 14.2 was simple to apply. However, when measurement data have been obtained, it is not the most efficient distribution-free test available. A better test, sometimes referred to as the *Wilcoxon signed rank test* and at other times more simply as the *signed rank test*, is one that takes account of the magnitude of the observed differences. It proceeds as follows:

1. Rank the differences without regard to sign; i.e., rank the absolute values of the differences. (The smallest difference is given rank 1, and ties are assigned average ranks.)
2. Assign to each rank the sign of the observed difference.
3. Obtain the sum of the negative ranks and the sum of the positive ranks.
4. Denote by T the absolute value of the smaller of the two sums of ranks found in the previous step.
5. To test the hypothesis of no difference between the effects of the two treatments, compare T with the critical values tabulated in Appendix 14.
6. If the observed value of T is less than or equal to the tabulated value for the chosen significance level, the hypothesis is rejected; otherwise it is not rejected.

Before giving numerical illustrations, we should note that the signed rank test is also applicable in the following situations:

1. To test the hypothesis that the median of a population is equal to some specified value, say M_0.
2. To test the hypothesis that the median of a population of differences is equal to some specified value, say M_0.

It should be clear that in Case 1 the basic variable is $|X - M_0|$, while in Case 2 it is $|(X - Y) - M_0|$. Apart from this obvious transformation, the procedure is exactly as specified above.

Example 14.3

Consider again the data of Table 6.6. These are reproduced in Table 14.1 for your convenience. Applying the procedure for the signed rank test, it is seen that $T = 18.5$. Assuming $\alpha = 0.05$ and consulting Appendix 14, it may be verified that $T = 18.5 < T_C = 25$; therefore, the hypothesis of equal treatment effects is rejected. The reader should compare this result with those reached in Examples 6.14 and 14.1.

Example 14.4

Given the data of Table 14.2, test the hypothesis that the population median equals 12. It is easily seen that the calculations, also shown in

TABLE 14.1–Data Obtained in a Brinell Hardness Test

Sample No.	Differences (D)	Rank of | D |	Signed Rank Positive	Signed Rank Negative
1	22	13	13	...
2	2	2.5	2.5	...
3	4	4.5	4.5	...
4	12	11	11	...
5	11	10	10	...
6	15	12	12	...
7	28	15	15	...
8	−5	6	...	−6
9	8	8	8	...
10	4	4.5	4.5	...
11	−1	1	...	−1
12	−10	9	...	−9
13	−2	2.5	...	−2.5
14	25	14	14	...
15	7	7	7	...
				...
		Total	101.5	−18.5

Table 14.2 for convenience, lead to $T = 6$ which, for $n = 8$ and $\alpha = 0.05$, tells us we are unable to reject the stated hypothesis.

14.4 RUN TEST

Among other things, the theory of runs may be used to test the following two hypotheses:

1. The observations have been drawn at random from a single population.

TABLE 14.2–Hypothetical Data to Illustrate the Procedure
of the Signed Rank Test

Observation (X)	$X - M_0$	Rank of $\| X - M_0 \|$	Signed Rank Positive	Signed Rank Negative
12.55	0.55	3	3	...
14.62	2.62	8	8	...
12.93	0.93	4	4	...
12.46	0.46	2	2	...
11.95	−0.05	1	...	−1
14.55	2.55	7	7	...
13.11	1.11	6	6	...
10.90	−1.10	5	...	−5
		Total	30	−6

2. Two random samples come from populations having the same distribution.

Because the mathematics of the theory of runs is quite involved, we shall do no more than sketch the approach.

Case 1:

1. List the observations in the order in which they were obtained, i.e., in their order of occurrence.
2. Determine the sample median.
3. Denote observations below the median by minus signs and observations above the median by plus signs.
4. Denote the number of minus signs by n_1 and the number of plus signs by n_2.
5. Count the number of runs and denote this number by r. (In terms of our symbols, a run is a sequence of signs of the same kind bounded by signs of the other kind.)
6. If r is less than or equal to the critical value tabulated in Appendix 15 (Table 1) or greater than or equal to the critical value tabulated in Appendix 15 (Table 2), the hypothesis is rejected at the 5 percent significance level.

Case 2:

1. List the $n_1 + n_2$ observations from the two samples in order of magnitude; i.e., arrange them in one sequence according to their values.
2. Denoting observations from one population by x's and observations from the other population by y's, count the number of runs.
3. Denote the observed number of runs by r.
4. If r is less than or equal to the critical value tabulated in Appendix 15 (Table 1), the hypothesis is rejected at the 5 percent significance level.

Example 14.5

Suppose a manufacturing process is turning out washers, and the characteristic of interest is the outside diameter. In the first 40 washers tested, there were 16 runs above and below the sample median. Noting that $n_1 = n_2 = 20$, we refer to Appendix 15 and find that $r_L = 14 < 16 < 28 = r_U$. Thus at the 5 percent significance level we are unable to reject the hypothesis that the 40 observations constitute a random sample from a single population.

Example 14.6

Consider the data of Table 14.3. Listing according to ranks, we have: $B, AAAAAA, B, A, B, AA, BBB, A, BBB, A, B, A$. That is, we have $r = 12$ runs. In addition, $n_A = 12$ and $n_B = 10$. Reference to Appendix 15 (Table 1) tells us that, since $r = 12 > r_C = 7$, we are unable to reject the hypothesis that the two random samples came from populations having the same distribution. In other words, using the run test and operating at the

TABLE 14.3–Outside Diameters of Washers Produced
by Two Different Production Lines

Line A	Line B
1.63 (6)	1.65 (8)
1.68 (9)	1.69 (10)
1.59 (4)	1.72 (13)
1.64 (7)	1.91 (21)
1.70 (11)	1.74 (14)
1.58 (3)	1.75 (15)
1.62 (5)	1.55 (1)
1.71 (12)	1.86 (17)
1.57 (2)	1.87 (18)
1.84 (16)	1.88 (19)
1.90 (20)	. . .
1.96 (22)	. . .

Note: Figures in parentheses are the ranks.

5 percent significance level, we are unable to reject the hypothesis that the two lines are producing equivalent product.

14.5 KOLMOGOROV-SMIRNOV TEST OF GOODNESS OF FIT

An alternative to the chi-square goodness of fit test described in Section 6.14 is provided by the Kolmogorov-Smirnov test to be described here. Since the Kolmogorov-Smirnov test is more powerful than the chi-square test, its use is to be encouraged. It proceeds as follows:

1. Let $F(x)$ be the completely specified theoretical cumulative distribution function under the null hypothesis.
2. Let $S_n(x)$ be the sample cumulative distribution function based on n observations. For any observed x, $S_n(x) = k/n$, where k is the number of observations less than or equal to x.
3. Determine the maximum deviation D defined by $D = \max | F(x) - S_n(x) |$
4. If, for the chosen significance level, the observed value of D is greater than or equal to the critical value tabulated in Appendix 16, the hypothesis will be rejected.

Example 14.7

To illustrate the Kolmogorov-Smirnov test of goodness of fit, we shall apply it to the data of Table 6.25. These data are reproduced here as Table 14.4. To test the hypothesis that the data constitute a random sample from a Poisson population with a mean of 10.44, calculations are carried out as shown in Table 14.4. The values of $F(x)$ were found by consulting Appendix 2 and using $\lambda = 10.5$. (**Note:** If a more precise evaluation is needed, $F(x)$ should be determined using $\lambda = 10.44$. The approximate value 10.5 was used since $\lambda = 10.44$ would require interpolation in Appendix 2.)

TABLE 14.4–Application of the Kolmogorov-Smirnov Goodness of Fit Test to the Number of Busy Senders in a Telephone Exchange

Number Busy	Observed Frequency	Observed Cumulative Frequency	Relative Cumulative Frequency $S_n(x)$	Expected Relative Cumulative Frequency $F(x)$	$\|F(x) - S_n(x)\|$
0	0	0	0	0	0
1	5	5	0.001	0	0.001
2	14	19	0.005	0.002	0.003
3	24	43	0.011	0.007	0.004
4	57	100	0.027	0.021	0.006
5	111	211	0.056	0.050	0.006
6	197	408	0.109	0.102	0.007
7	278	686	0.183	0.179	0.004
8	378	1064	0.283	0.279	0.004
9	418	1482	0.395	0.397	0.002
10	461	1943	0.518	0.521	0.003
11	433	2376	0.633	0.639	0.006
12	413	2789	0.743	0.742	0.001
13	358	3147	0.838	0.825	0.013
14	219	3366	0.897	0.888	0.009
15	145	3511	0.935	0.932	0.003
16	109	3620	0.964	0.960	0.004
17	57	3677	0.979	0.978	0.001
18	43	3720	0.991	0.988	0.003
19	16	3736	0.995	0.994	0.001
20	7	3743	0.997	0.997	0
21	8	3751	0.999	0.999	0
22	3	3754	1.000	0.999	0.001

Source: Fry (3), p. 295.

Since $D = \max |F(x) - S_n(x)| = 0.013 < 1.63/\sqrt{3754} = 0.027$, the hypothesis may not be rejected at the 1 percent significance level. The reader should compare this result with that obtained in Example 6.31.

14.6 MEDIAN TESTS

The procedures to be described in this section are of value when testing the following hypotheses:

1. That $k(k \geq 2)$ random samples were drawn from identically distributed populations.
2. That in a one-factor experiment the k levels of the factor have the same effect.
3. That in a two-factor experiment (a) the a levels of factor a have the same effect, (b) the b levels of factor b have the same effect, and (c) there is no true interaction between factors a and b.

For our purposes it is deemed sufficient to concentrate on Case 1. Those wishing to investigate Cases 2 and 3 are referred to Brown and Mood (1, 2) and Mood (4).

If k random samples consisting of $n_1, \ldots, n_k$ observations respectively are available, determine the numbers of observations in each sample that are above and below the median of the combined samples. These data may then be analyzed as a $2 \times k$ contingency table in the manner specified in Section 6.15.

Example 14.8

Consider the two samples in Table 14.3. Examination of these data leads to the 2×2 contingency table shown in Table 14.5. Using Eq. (6.42), we obtain

$$\chi^2 = 22(\,|\,(4)(3) - (8)(7)\,| - 11)^2/(12)(10)(11)(11) = 2.07$$

Since $\chi^2 = 2.07 < \chi^2_{0.95(1)} = 3.84$, we are unable to reject the hypothesis that the two random samples were drawn from identically distributed populations.

TABLE 14.5–Contingency Table Formed from the Data of Table 14.3

Position	Line A	Line B	Total
Above median	4	7	11
Below median	8	3	11
Total	12	10	22

PROBLEMS

14.1 Apply the method described in Section 14.2 to the following problems: (a) 6.29, (b) 6.32, and (c) 6.33. Discuss any differences between the method used here and that used in Chapter 6.

14.2 Apply the method described in Section 14.3 to the following problems: (a) 6.29, (b) 6.32, and (c) 6.33. Discuss your results with reference to those obtained in Problem 14.1 and in Chapter 6.

14.3 Apply the method described in Section 14.3 to Problems 6.1 and 6.2 (d). State any changes in the assumptions and the wording of the hypotheses that you make. Discuss the implications of these changes. Compare the results of using distribution-free tests with those obtained using parametric tests in Chapter 6.

14.4 Apply the method described in Section 14.4 to the data given in the following problems:

(a) 5.1	(f) 6.1	(k) 6.35
(b) 5.5	(g) 6.2	(l) 6.36
(c) 5.20	(h) 6.30	(m) 6.37
(d) 5.22	(i) 6.31	(n) 6.39
(e) 5.23	(j) 6.33	(o) 6.40

In each case, state the hypothesis being tested and specify any assumptions you make.

14.5 Apply the method described in Section 14.5 to check on the assumption of normality in Problems 5.5 and 6.1.

14.6 Apply the method described in Section 14.6 to the data given in the following problems:

(a) 6.30	(e) 6.37	(i) 10.10
(b) 6.31	(f) 6.39	(j) 10.11
(c) 6.35	(g) 6.40	(k) 10.12
(d) 6.36	(h) 10.9	

Compare the conclusions reached here with those reached in the earlier chapters. Discuss any discrepancies.

REFERENCES

1. Brown, G. W.; and Mood, A. M. 1948. Homogeneity of several samples. *Am. Stat.* 2:22.

2. ———. 1951. On median tests for linear hypotheses. Proc. 2nd Berkeley Symposium on Mathematical Statistics and Probability, pp. 159–66. University of California Press, Berkeley.

3. Fry, T. C. 1928. *Probability and Its Engineering Uses.* Van Nostrand, New York.

4. Mood, A. M. 1950. *Introduction to the Theory of Statistics.* McGraw-Hill, New York.

FURTHER READING

Bradley, J. V. 1968. *Distribution-Free Statistical Tests.* Prentice-Hall, Englewood Cliffs, N.J.

Conover, W. J. 1971. *Practical Nonparametric Statistics.* Wiley, New York.

Edgington, E. S. 1969. *Statistical Inference; The Distribution-Free Approach.* McGraw-Hill, New York.

Gibbons, J. D. 1971. *Nonparametric Statistical Inference.* McGraw-Hill, New York.

Hollander, M.; and Wolfe, D. A. 1973. *Nonparametric Statistical Methods.* Wiley, New York.

Kendall, M. G. 1970. *Rank Correlation Methods,* 4th ed. Griffin, London.

Mosteller, F.; and Rourke, R. E. K. 1973. *Sturdy Statistics: Nonparametric and Order Statistics.* Addison-Wesley, Reading, Mass.

Siegel, S. 1956. *Nonparametric Statistics for the Behavioral Sciences.* McGraw-Hill, New York.

Walsh, J. E. 1962, 1965. *Handbook of Nonparametric Statistics,* Vols. I and II. Van Nostrand, New York.

CHAPTER 15

TOPICS IN STATISTICAL QUALITY CONTROL AND RELIABILITY

Since the early 1940s the use of statistical methods in industry has been on the upswing. This increased use of statistics has been particularly noticeable in two areas: (1) *research and development* and (2) *reliability and quality control*. Because the major part of this book has been concerned with statistical methods that have proved most valuable in research and development, this chapter will be devoted to a presentation of several techniques especially useful for controlling quality and reliability. Two of these techniques useful in controlling and improving quality are *control charts* and *acceptance sampling plans*. Tolerances as they relate to engineering systems and system reliability are also discussed.

Because these topics and techniques are discussed at great length in books devoted entirely to statistical quality control and reliability, only a brief outline will be given. However, the material to be presented will be sufficient to acquaint the reader with the basic concepts. Those in need of more detailed explanations are referred to the list of further reading given at the end of the chapter.

15.1 CONTROL CHARTS

Although control charts may prove useful in many situations, they are most commonly employed in the analysis and control of production processes. For this reason, discussion of these charts will be in terms perhaps more familiar to the engineer than to the research worker.

It has long been recognized that some variation is inevitable in any repetitive process. For example, Grant (4, p. 3) states: "Measured quality of manufactured product is always subject to a certain amount of variation as a result of chance. Some stable "system of chance causes" is inherent in any particular scheme of production and inspection. Variation within this stable pattern is inevitable. The reasons for variation outside this stable pattern may be discovered and corrected." Taking Grant at his word, we seek tests for detecting unnatural patterns in the plotted data.

The control chart, as conceived and developed by Shewhart (6), is a simple pictorial device for detecting unnatural patterns of variation in data resulting from repetitive processes. That is, the control chart provides criteria for detecting lack of statistical control. (**Note:** When a process is operating under a constant system of chance causes, it is said to be in *statistical control*.) Rather than covering details concerning the theory underlying control charts, we shall (1) indicate the proper pro-

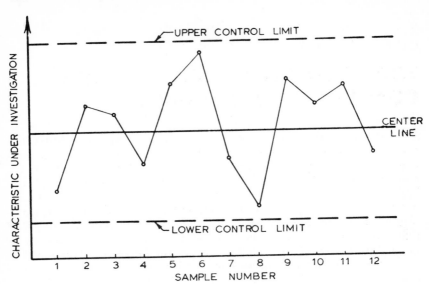

Fig. 15.1–Illustration of the general appearance of a control chart.

cedures for constructing the charts, (2) state criteria to be used for indicating unnatural patterns of variation, and (3) give numerical examples of the four most commonly used charts.

Basically, all control charts appear as in Figure 15.1. The sample points are plotted in a sequential manner, i.e., as obtained. The plotted points are joined together solely as an aid to visual interpretation.

What tests should be employed to detect unnatural patterns in the plotted data? Depending on the degree of effectiveness desired, many criteria might be used. However, since our sole purpose is to indicate the basic nature of the control chart technique, only the most common tests will be mentioned. Those wishing to investigate more sophisticated tests should consult the references at the end of the chapter.

The most common tests for unnatural patterns are tests for instability, i.e., for determining if the cause system is changing. As commonly employed, they refer to the A, B, and C zones shown in Figure 15.2. With reference to these zones, the observed pattern of variation is said to be unnatural, or the process is said to be "out of control," if any one or more of the following events occurs:

1. A single point falls outside the control limit, i.e., beyond zone A.
2. Two out of three successive points fall in zone A or beyond.
3. Four out of five successive points fall in zone B or beyond.
4. Eight points in succession fall in zone C or beyond.

It should be noted that the above tests apply to both halves of a control chart *but they are applied separately to each half,* not to the two halves in combination. [**Note:** The tests described here may be used when the two control limits are reasonably symmetrically located with respect to

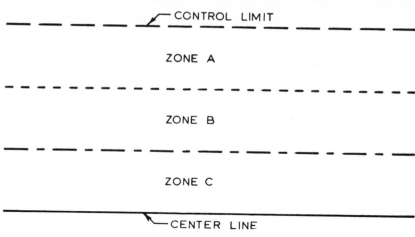

Fig. 15.2–Diagram defining the A, B, and C zones used in control chart analyses. (Each of the zones A, B, and C constitutes one-third of the area between the center line and the control limit.)

the center line. If the limits are decidedly asymmetric, the tests should be modified as described in reference (7).]

Before presenting numerical illustrations of control chart applications, it is necessary that the four most commonly used charts be introduced and that formulas be given for the calculation of the center lines and control limits. The charts most often encountered are (1) the $\overline{X}$ chart, (2) the R chart, (3) the p chart, and (4) the c chart. The first two of these deal with measurement data while the last two deal with attribute (enumeration) data. The pertinent assumptions and the formulas for the control limits are specified in Table 15.1.

Example 15.1

Consider the data of Table 15.2. It may be verified that $\overline{\overline{X}} = \sum \overline{X}/k = 213.20/20 = 10.66$ and $\overline{R} = \sum R/k = 31.8/20 = 1.59$. Then, using the

TABLE 15.1–Assumptions and Formulas for the Most Commonly Used Control Charts

Chart	Assumed Distribution	Center Line	Upper Control Limit (UCL)	Lower Control Limit (LCL)
$\overline{X}$	Normal	$\overline{\overline{X}}$	$\overline{\overline{X}} + A_2\overline{R}$	$\overline{\overline{X}} - A_2\overline{R}$
R	Normal	$\overline{R}$	$D_4\overline{R}$	$D_3\overline{R}$
p	Binomial	$\overline{p}$	$\overline{p} + 3\sqrt{\overline{p}(1 - \overline{p})/n}$	$\overline{p} - 3\sqrt{\overline{p}(1 - \overline{p})/n}$
c	Poisson	$\overline{c}$	$\overline{c} + 3\sqrt{\overline{c}}$	$\overline{c} - 3\sqrt{\overline{c}}$

Note: The constants A_2, D_3, and D_4 are given in Appendix 8, while the quantities $\overline{\overline{X}}$, $\overline{R}$, $\overline{p}$, and $\overline{c}$ are calculated from the sample data as shown in the numerical examples that follow.

TABLE 15.2–Coded Values of the Crushing Strengths of Concrete Blocks

Sample Number	Individual Value					Mean $(\overline{X})$	Range (R)
	X_1	X_2	X_3	X_4	X_5		
1	11.1	9.4	11.2	10.4	10.1	10.44	1.8
2	9.6	10.8	10.1	10.8	11.0	10.46	1.4
3	9.7	10.0	10.0	9.8	10.4	9.98	0.7
4	10.1	8.4	10.2	9.4	11.0	9.82	2.6
5	12.4	10.0	10.7	10.1	11.3	10.90	2.4
6	10.1	10.2	10.2	11.2	10.1	10.36	1.1
7	11.0	11.5	11.8	11.0	11.3	11.32	0.8
8	11.2	10.0	10.9	11.2	11.0	10.86	1.2
9	10.6	10.4	10.5	10.5	10.9	10.58	0.5
10	8.3	10.2	9.8	9.5	9.8	9.52	1.9
11	10.6	9.9	10.7	10.2	11.4	10.56	1.5
12	10.8	10.2	10.5	8.4	9.9	9.96	2.4
13	10.7	10.7	10.8	8.6	11.4	10.44	2.8
14	11.3	11.4	10.4	10.6	11.1	10.96	1.0
15	11.4	11.2	11.4	10.1	11.6	11.14	1.5
16	10.1	10.1	9.7	9.8	10.5	10.04	0.8
17	10.7	12.8	11.2	11.2	11.3	11.44	2.1
18	11.9	11.9	11.6	12.4	11.4	11.84	1.0
19	10.8	12.1	11.8	9.4	11.6	11.14	2.7
20	12.4	11.1	10.8	11.0	11.9	11.44	1.6
Average						10.66	1.59

formulas given in Table 15.1, we see that for the $\overline{X}$ chart UCL = 10.66 + (0.58)(1.59) = 11.58 and LCL = 10.66 − (0.58)(1.59) = 9.74. Similarly, for the R chart UCL = (2.11)(1.59) = 3.35 and LCL = (0)(1.59) = 0. The resulting charts are shown in Figure 15.3. The tests for unnatural patterns have been applied to the $\overline{X}$ chart and the potential trouble spots tagged in the customary manner. It is suggested that the reader consider the application of these tests to the R chart.

Example 15.2

Consider the data of Table 15.3. Here we are dealing with enumeration data, namely, the number of defective fuses in samples of size 50 taken at random times during the production process. It is easily verified that $\bar{p} = \sum p/k = 1.68/40 = 0.042$. Using the formulas specified in Table 15.1, we obtain

$$\text{UCL} = 0.042 + 3[(0.042)(0.958)/50]^{1/2} = 0.127$$

and

$$\text{LCL} = 0.042 - 3[(0.042)(0.958)/50]^{1/2} = -0.043$$

(**Note:** Since the formula leads to a negative value for the lower control limit and because the fraction defective is a nonnegative quantity, the

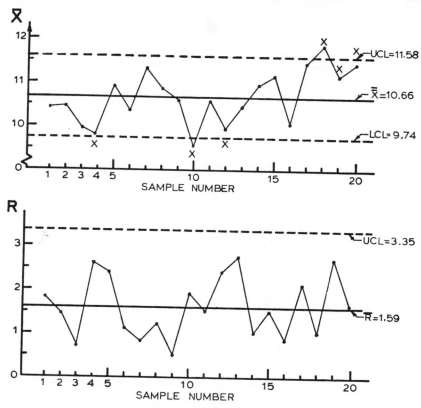

Fig. 15.3–Control charts for the data of Table 15.2.

lower control limit is arbitrarily set at 0. This makes the control limits asymmetric with respect to the center line. The tests for unnatural patterns specified earlier have therefore not been applied. The application of the modified tests is left as an exercise for the reader. The results of the preliminary analysis are presented in Figure 15.4.)

Example 15.3

Consider now a situation in which the characteristic of interest is the number of defects per unit. In such a case, a Poisson distribution would undoubtedly be assumed and a c chart would be appropriate. Given the data in Table 15.4, it is easily verified that $\bar{c} = \sum c/k = 144/24 = 6$. Using the formulas specified in Table 15.1, we obtain UCL $= 6 + 3\sqrt{6} = 13.35$ and LCL $= 6 - 3\sqrt{6} = -1.35$. (**Note:** As in Example 15.2 the negative lower control limit will arbitrarily be changed to 0 since c is by definition a nonnegative quantity.) The results of the analysis are plotted in Figure 15.5.

Examination of Figure 15.5 will reveal that some modifications have been made in the usual method of presentation, namely, the various half-days have been separated (i.e., not connected) to emphasize the breaks be-

TABLE 15.3–Number of Defective Fuses in Random Samples of Size 50

Sample No.	Number of Defectives	Fraction Defective (p)	Sample No.	Number of Defectives	Fraction Defective (p)
1	2	0.04	21	1	0.02
2	1	0.02	22	1	0.02
3	2	0.04	23	4	0.08
4	0	0.00	24	2	0.04
5	2	0.04	25	2	0.04
6	3	0.06	26	4	0.08
7	4	0.08	27	1	0.02
8	2	0.04	28	3	0.06
9	0	0.00	29	3	0.06
10	3	0.06	30	2	0.04
11	0	0.00	31	3	0.06
12	1	0.02	32	6	0.12
13	2	0.04	33	2	0.04
14	2	0.04	34	3	0.06
15	3	0.06	35	2	0.04
16	5	0.10	36	3	0.06
17	1	0.02	37	1	0.02
18	2	0.04	38	0	0.00
19	3	0.06	39	2	0.04
20	1	0.02	40	0	0.00
			Average		0.042

tween work periods. Now it will be noted that two conclusions are obvious: (1) no points plot above the upper control limit, and (2) essentially the same pattern appears in each half-day. Study of the recurring pattern suggests the existence of a fatigue factor. [**Note:** For further discussion of this example, consult Grant (4, pp. 32–35).]

Before leaving the topic of control charts, some additional remarks are necessary. These will, however, be very brief and in no particular order.

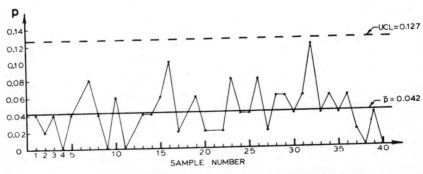

Fig. 15.4–Control chart for the data of Table 15.3.

TABLE 15.4–Number of Defects Observed in a Welded Seam

Sample No.	Date	Time of Sample	Number of Defects (c)
1	July 18	8:00 A.M.	2
2		9:05 A.M.	4
3		10:10 A.M.	7
4		11:00 A.M.	3
5		12:30 P.M.	1
6		1:35 P.M.	4
7		2:20 P.M.	8
8		3:30 P.M.	9
9	July 19	8:10 A.M.	5
10		9:00 A.M.	3
11		10:05 A.M.	7
12		11:15 A.M.	11
13		12:25 P.M.	6
14		1:30 P.M.	4
15		2:30 P.M.	9
16		3:40 P.M.	9
17	July 20	8:00 A.M.	6
18		8:55 A.M.	4
19		10:00 A.M.	3
20		11:10 A.M.	9
21		12:25 P.M.	7
22		1:30 P.M.	4
23		2:20 P.M.	7
24		3:30 P.M.	12
Total			144

Source: Grant (4), p. 33. By permission of the author and publishers.
Note: Each count is taken on a single seam; the welder produced eight seams per hour.

1. The primary reason for using control charts is to provide a signal that some *action* is desirable.
2. Before control limits may be calculated with any assurance of their being reliable, at least 20 subgroups (samples) should be available.
3. Before the control limits, calculated from past production records, are used to monitor future production, the process should be in control.
4. If a process is in control and a point plots outside the control limits, the taking of action may be viewed as committing a Type I error.
5. The control chart is not a panacea for all production problems; it is only another useful tool.

15.2 ACCEPTANCE SAMPLING PLANS

Acceptance sampling, or sampling inspection, is of two types: *lot-by-lot sampling* and *sampling of continuous production*. In this brief exposure to the concepts and procedures of acceptance sampling, only the

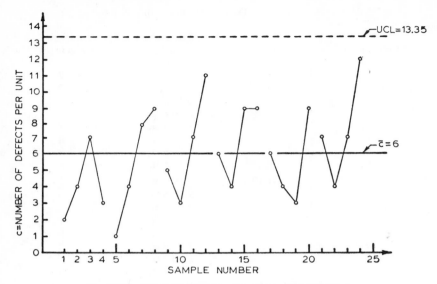

Fig. 15.5–Control chart for the data of Table 15.4.

first of these two types will be discussed. Persons desiring information on continuous sampling plans are referred to the further reading list at the end of the chapter. In addition to the distinction between lot-by-lot and continuous sampling, it is customary to classify sampling plans as either *attributes* or *variables* plans. Attributes plans refer to cases in which each item is classified simply as either defective or nondefective; variables plans refer to cases in which a measurement is taken and recorded numerically on each item inspected. In this section only attributes plans will be considered. There is one other way acceptance sampling plans may be classified, namely, as *single, double,* or *multiple* (including *sequential*) plans. These categories refer to the number of samples selected and will become clear as the exposition continues.

An *attributes single sampling plan that operates on a lot-by-lot basis* is completely defined by three numbers: the lot size N, the sample size n, and the acceptance number a. (**Note:** In many publications, the acceptance number is denoted by c. However, we prefer a for two reasons: it stands for the word acceptance and it permits an easier extension to double and multiple sampling.) Such a plan operates as follows:

1. A single sample of n items is selected by chance from a lot of items.
2. Each item in the sample is then classified as either defective or nondefective.
3. If the number of defective items in the sample does not exceed a, the lot is accepted.
4. If the number of defective items in the sample exceeds a, the lot is rejected.

As with all statistical procedures, the risks associated with decisions (inferences) resulting from sampling inspection must be assessed. The customary manner of presenting these risks is by means of a graph of the operating characteristic (OC) function of the sampling plan, i.e., by plotting the probability of accepting the lot as a function of the fraction defective of the lot. The protection afforded by various sampling plans may then be compared by examining their OC curves.

For the single sampling plan specified previously, the OC function is

$$P_{acc} = \sum_{d=0}^{a} C(D, d)\,[C(N - D, n - d)/C(N, n)] \qquad (15.1)$$

where D represents the number of defective items in the lot and d represents the number of defective items in the sample. Eq. (15.1) may be evaluated for $D = 0, 1, \ldots, N$ and the results plotted as a series of ordinates erected at the corresponding values of $p = D/N$, namely, $0, 1/N, 2/N, \ldots, 1$. It is clear, however, that these calculations may prove onerous unless a high-speed computer is available. Fortunately, helpful tables of the hypergeometric function have been provided by Lieberman and Owen (5). In addition, extensive catalogs of OC curves for lot-by-lot, single-sampling (by attributes) plans have been prepared, e.g., Wiesen (8) and Clark and Koopmans (3). If none of these publications is readily available and access to a high-speed computer is impossible, the OC function may be approximated by

$$P_{acc} \cong \sum_{d=0}^{a} C(n, d)\,p^d(1 - p)^{n-d} \qquad (15.2)$$

or

$$P_{acc} \cong \sum_{d=0}^{a} e^{-\lambda}\lambda^d/d! \qquad (15.3)$$

where $\lambda = np$. (**Note:** The reader is referred to Chapter 4 for discussion of the accuracy and relevancy of these approximations.)

Persons familiar with the presentation of OC curves for sampling plans, or those who consult the references given in the preceding paragraph, will realize that it is not customary to plot OC functions as a series of ordinates (as stated in the preceding paragraph) but to show *smooth* OC curves. That is, the functions are plotted as though p were a continuous parameter. For lot-by-lot plans this minor "tampering with the truth" is not serious and the resulting gain in ease of presentation far outweighs the inaccuracy of the graph. Consequently, OC curves usually appear as in Figure 15.6.

Rather than plot the entire OC function, the practitioner frequently contents himself with calculating two or three points on the curve. The three points most commonly determined are: p_1 = the value of p for which $P_{acc} = 0.95$, p_2 = the value of p for which $P_{acc} = 0.50$, and p_3 = the value of p for which $P_{acc} = 0.10$. These three values of p are

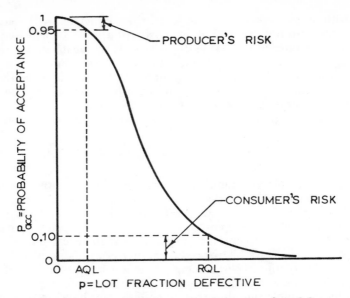

Fig. 15.6–Illustration of the general appearance of an OC curve.
(To aid in understanding the concepts, the AQL, RQL,
consumer's risk, and producer's risk are shown
in this figure.)

usually referred to as the *acceptable quality level* (AQL), the *indifference quality,* and the *rejectable quality level* (RQL) respectively. [**Note:** The *rejectable quality level* (RQL) is also known as the *lot tolerance percent defective* (LTPD).] The AQL and RQL points, as well as the associated expressions, *consumer's risk* and *producer's risk,* are shown on Figure 15.6. (**Note:** If in the definitions of p_1 and p_3 the values 0.95 and 0.10 are replaced by $1 - \alpha$ and β respectively, the reader will immediately see the close connection between the ideas of this section and those discussed in Chapter 6. Incidentally, the same remark may be made here as there; namely, the values assigned to α and β are arbitrary and the use of $\alpha = 0.05$ and $\beta = 0.10$ is only a matter of custom.)

One other common way of presenting the performance ability of acceptance sampling plans is to calculate and plot the *average outgoing quality* (AOQ) function. This function, restricted to cases in which the testing is nondestructive, depends on the assumption that all rejected lots are submitted to 100 percent inspection, with all defective items being removed and replaced by nondefective items. Under this assumption, the average outgoing quality is determined to be

$$AOQ = p \cdot P_{acc} \tag{15.4}$$

where P_{acc} is defined by Eq. (15.1). The graph of a typical AOQ function is shown in Figure 15.7. The maximum value of the average out-

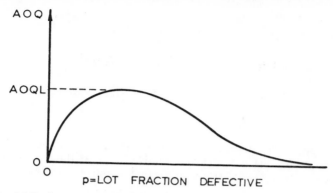

Fig. 15.7—Illustration of the general appearance of an AOQ curve.

going quality is known as the *average outgoing quality limit* (AOQL). From a practical point of view, the AOQL is perhaps the most important descriptive measure associated with any acceptance sampling plan. [**Note:** AOQ curves are also included in the catalogs of Wiesen (8) and Clark and Koopmans (3).]

Let us now consider double sampling plans. An *attributes double sampling plan that operates on a lot-by-lot basis* is completely defined by six numbers: the lot size N, the size of the first sample n_1, the acceptance number associated with the first sample a_1, the rejection number associated with the first sample r_1, the size of the second sample n_2, and the acceptance number associated with the combined samples a_2. Such a plan operates as follows:

1. A sample of n_1 items is selected at random from a lot of N items.
2. Each item in the sample is then classified as either defective or nondefective.
3. If the number of defective items in the sample does not exceed a_1, the lot is accepted.
4. If the number of defective items in the sample equals or exceeds r_1, the lot is rejected.
5. If the number of defective items in the sample exceeds a_1 but is less than r_1, a second sample (of n_2 items) is selected at random from the remainder of the lot.
6. If the number of defective items in the two samples combined does not exceed a_2, the lot is accepted.
7. If the number of defective items in the two samples combined exceeds a_2, the lot is rejected.

Rather than go into the same detail as for single sampling, only the bare essentials will be presented. The OC function is

$$P_{acc} = \sum_{d_1=0}^{a_1} C(D, d_1) \cdot C(N - D, n_1 - d_1)/C(N, n_1)$$

$$+ \sum_{d_1 = a_1 + 1}^{r_1 - 1} [C(D, d_1) \cdot C(N - D, n_1 - d_1)/C(N, n_1)] \cdot$$

$$\sum_{d_2 = 0}^{a_2 - d_1} C(D - d_1, d_2) \cdot C(N - n_1 - D + d_1, n_2 - d_2)/C(N - n_1, n_2)$$

$$(15.5)$$

where d_1 represents the number of defective items in the first sample and d_2 represents the number of defective items in the second sample. Subject to the usual restrictions, this may be approximated by

$$P_{acc} \cong \sum_{d_1 = 0}^{a_1} C(n_1, d_1) p^{d_1} (1 - p)^{n_1 - d_1}$$

$$+ \sum_{d_1 = a_1 + 1}^{r_1 - 1} C(n_1, d_1) p^{d_1} (1 - p)^{n_1 - d_1} \sum_{d_2 = 0}^{a_2 - d_1} C(n_2, d_2) p^{d_2} (1 - p)^{n_2 - d_2}$$

$$(15.6)$$

or

$$P_{acc} \cong \sum_{d_1 = 0}^{a_1} e^{-\lambda_1} \lambda_1^{d_1}/d_1! + \sum_{d_1 = a_1 + 1}^{r_1 - 1} [e^{-\lambda_1} \lambda_1^{d_1}/d_1!] \sum_{d_2 = 0}^{a_2 - d_1} e^{-\lambda_2} \lambda_2^{d_2}/d_2!$$

$$(15.7)$$

where $\lambda_1 = n_1 p$ and $\lambda_2 = n_2 p$. As before, the average outgoing quality is given by

$$AOQ = p \cdot P_{acc} \qquad (15.8)$$

where P_{acc} is defined by Eq. (15.5). In addition, since the total sample size is now a variable, it is also possible to assess the relative "costs" of sampling plans by comparing their expected sample sizes. To summarize, it is customary to calculate the *average sample number* (ASN). Proceeding on the assumption that every item in each sample is inspected, the ASN for a double sampling plan is given by

$$ASN = n_1 + n_2 \sum_{d_1 = a_1 + 1}^{r_1 - 1} C(D, d_1) \cdot C(N - D, n_1 - d_1)/C(N, n_1) \quad (15.9)$$

The graph of a typical ASN function for a double sampling plan is shown in Figure 15.8, where it is compared with the constant sample size of an equivalent (in terms of the OC function) single sampling plan.

Before proceeding to multiple sampling plans, it seems desirable to list the advantages and disadvantages of double sampling plans relative to single sampling plans.

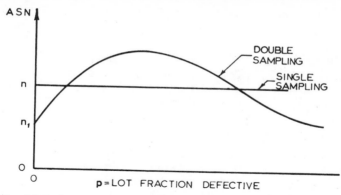

Fig. 15.8–Illustration of the general appearance of ASN functions for single and double sampling plans that exhibit essentially the same OC curve.

Advantages:

1. They have the psychological advantage of giving doubtful lots a second chance.
2. On the average they have (for the two extremes of good and bad quality) the advantage of requiring fewer inspections.

Disadvantages:

1. They are said to be more difficult to administer. (However, we do not subscribe to this point of view.)
2. The inspection load is variable.
3. The maximum number of inspections can (for intermediate quality) exceed that for comparable single sampling plans.

Since there is little if anything new in multiple sampling plans that has not been discussed in connection with single and double sampling plans, the following remarks will be very brief. If one extends the concepts and procedures of double sampling to $k > 2$ samples, it may easily be seen that such plans are completely defined by the numbers N, $n_1, \ldots, n_k, a_1, \ldots, a_k, r_1, \ldots, r_k = a_k + 1$. Proceeding in a manner analogous to that followed for double sampling plans, one may obtain OC, AOQ, and ASN functions. (**Note:** The calculations will, however, be quite involved and lengthy. If each $n_i = 1$ and if $k \to \infty$, the multiple plan is usually referred to as a sequential plan.) For those desiring more details on multiple and sequential acceptance sampling plans, the appropriate references at the end of the chapter are recommended.

Example 15.4

Suppose that lots of size 100 are submitted for acceptance. A sample of size 2 is drawn from a lot and the lot is accepted if both the sample items are nondefective; if one or both are defective, the lot is rejected. The OC function for this plan is $P_{acc} = C(D, 0) \cdot C(100 - D, 2)/C(100, 2)$, and this may be evaluated for $D = 0, 1, \ldots, 100$.

Example 15.5

With reference to Example 15.4, it is easily verified that the AOQ function is $\text{AOQ} = p \cdot P_{acc} = (D/100)[C(D, 0) \cdot C(100 - D, 2)/C(100, 2)]$, where once again this may be evaluated for $D = 0, 1, \ldots, 100$.

15.3 TOLERANCES

In Section 2.18 some remarks were made with regard to the distribution of a linear combination of random variables. At this time, it is appropriate to consider a specific application of those remarks, namely, to the subject of tolerances.

Before embarking on this discussion, a distinction must be made between *specification limits* (i.e., a nominal value plus and minus certain engineering tolerances) and *natural or statistical tolerances.* In general, specification limits are set by the designer as a statement of his requirements with respect to a certain dimension. Thus in many cases the specification limits have little connection with production capabilities. On the other hand, statistical tolerance limits reflect the capabilities of the process producing the dimensions in question. (See Sections 5.9–5.11.) Therefore, from the point of view of managerial decision making, the subject of statistical tolerances is of great importance, for it bears directly on the success or failure of the company's product.

In what follows, we are interested only in general concepts and a method of approach that may prove helpful in the design of complex equipment. Accordingly, our attention will be confined to cases in which the several variables (dimensions) are normally and independently distributed with known means and variances. Restrictive as these assumptions may be, they will not seriously limit our presentation. [**Note:** If other distributions must be used, the same concepts will apply. For example, see Breipohl (2).]

Two cases will be examined: (1) linear combinations of the variables and (2) nonlinear combinations of the variables.

Case I: Linear combinations of independent random variables

This case was covered in Section 2.18, where it was noted that if

$$U = \sum_{i=1}^{n} a_i X_i \tag{15.10}$$

then

$$\mu_U = \sum_{i=1}^{n} a_i \mu_i \tag{15.11}$$

and

$$\sigma_U^2 = \sum_{i=1}^{n} a_i^2 \sigma_i^2 \qquad (15.12)$$

where μ_i and σ_i^2 are the mean and variance respectively of $X_i(i = 1, \ldots, n)$. If, as is true in many applications, each $a_i = 1$, then Eqs. (15.11) and (15.12) reduce to

$$\mu_U = \sum_{i=1}^{n} \mu_i \qquad (15.13)$$

and

$$\sigma_U^2 = \sum_{i=1}^{n} \sigma_i^2 \qquad (15.14)$$

Because the above results are used in a variety of ways, a complete discussion of each different situation is not planned. However, some typical examples will be presented to acquaint the reader with a few of the possible applications.

Example 15.6

Consider a simple addition of components such that the dimension Y of the assembly is the sum of the dimensions of the individual components; i.e., $Y = \sum_{i=1}^{n} X_i$. If there are two components X_1 and X_2 with means 0.500 and 0.410 and with standard deviations 0.008 and 0.006 respectively, then $\mu_Y = 0.910$ and $\sigma_Y = [(0.008)^2 + (0.006)^2]^{1/2} = 0.01$. If X_1 and X_2 are assumed to be normally distributed, Y is normally distributed and the distribution of Y may be compared with the specification limits for Y to obtain the expected percentage of defective assemblies.

Example 15.7

Consider an assembly consisting of 5 components for which $Y = \sum_{i=1}^{5} X_i$ represents the simple addition of the appropriate dimensions. If the 5 dimensions are normally and independently distributed with means 0.500, 0.410, 0.200, 0.700, and 0.210 respectively and if they may be assumed to have a common variance σ^2, how large can σ be if $\mu_Y \pm 3\sigma_Y = 2.020 \pm 0.030$ units? Using Eq. (15.14), it may be verified that $\sigma_Y^2 = 5\sigma^2$. Therefore, $3\sigma_Y = 3\sigma \sqrt{5} = 0.030$, which means that $5\sigma^2 = (0.01)^2 = 0.0001$, and thus $\sigma^2 = 0.00002$. Consequently, the maximum allowable value of σ is $(0.00002)^{1/2} = 0.0045$.

Example 15.8

Another illustration of a linear combination of dimensions is the clearance between a shaft and a bearing. Let us assume that shafts are mass produced such that the outside diameters are normally and independently distributed with $\mu_S = 1.05$ inches and with standard deviation σ_S. Let us also assume that bearings are produced such that the inside diameters are normally and independently distributed with $\mu_B = 1.06$ inches and with standard deviation $\sigma_B = 0.001$ inch. If production of the shafts is to be controlled so that no more than 5 percent of randomly mated shafts and bearings will exhibit interference, what is the maximum allowable value of σ_S?

To answer this question, consider $Y = B - S$. Using Eqs. (15.11) and (15.12), it may be verified that $\mu_Y = 1.06 - 1.05 = 0.01$ inch and that $\sigma_Y = [(0.001)^2 + \sigma_S^2]^{1/2}$ inch. Now interference will occur when $Y < 0$. Thus it is appropriate to consider $P\{Y < 0\} = P\{Z < (0 - 0.01)/\sigma_Y\} = 0.05$, where $Z = (Y - \mu_Y)/\sigma_Y$. On consulting Appendix 3, it is found that $-0.01/\sigma_Y = -1.645$ and, consequently, $\sigma_Y^2 = 0.000001 + \sigma_S^2 = (1/164.5)^2 = 0.000037$. As a result it is seen that the maximum σ_S is 0.006.

It is hoped that the preceding examples will be sufficient to indicate the nature and scope of the theory of statistical tolerances when dealing with linear combinations of independent random variables. For those who wish to investigate the subject in greater depth, more details and examples are available in Bowker and Lieberman (1) and Breipohl (2).

Case II: Nonlinear combinations of independent random variables

When variables are combined in a nonlinear fashion, the tolerance problem is usually much more difficult because, in general, it is not easy to determine the distribution of a nonlinear combination of random variables. However, if a linear approximation is acceptable, the problem may be handled by expanding the function in a Taylor series about the mean values. That is, if

$$Y = \phi(X_1, \ldots, X_n) \tag{15.15}$$

it may be shown that

$$Y = \phi(\mu_1, \ldots, \mu_n) + \sum_{i=1}^{n} (X_i - \mu_i) \frac{\partial \phi}{\partial X_i}\bigg|_{\mu_1, \ldots, \mu_n}$$
$$+ \text{ terms of higher order} \tag{15.16}$$

Then if the "terms of higher order" are neglected, it may be verified that

$$\mu_Y \cong \phi(\mu_1, \ldots, \mu_n) \tag{15.17}$$

and

$$\sigma_Y^2 \cong \sum_{i=1}^{n} \sigma_i^2 \left(\frac{\partial \phi}{\partial X_i}\bigg|_{\mu_1, \ldots, \mu_n}\right)^2 \tag{15.18}$$

where μ_i and σ_i^2 are the mean and variance respectively of $X_i(i = 1, \ldots, n)$. As a consequence, Eqs. (15.17) and (15.18) may be used to obtain approximate tolerance limits for Y.

Example 15.9

Suppose that $Y = X_1 X_2 X_3$. Expanding this function in a Taylor series about μ_1, μ_2, and μ_3, we obtain $Y \cong \mu_1\mu_2\mu_3 + (X_1 - \mu_1)\mu_2\mu_3 + $

$(X_2 - \mu_2)\mu_1\mu_3 + (X_3 - \mu_3)\mu_1\mu_2$. If the X_i are normally and independently distributed with mean μ_i and variance σ_i^2, then Y is approximately normally distributed with mean $\mu_Y = \mu_1\mu_2\mu_3$ and variance $\sigma_Y^2 = (\mu_2\mu_3)^2\sigma_1^2 + (\mu_1\mu_3)^2\sigma_2^2 + (\mu_1\mu_2)^2\sigma_3^2$. These expessions may then be used to investigate the natural tolerances of Y.

Example 15.10

Consider an electrical circuit consisting of two resistors connected in parallel. For such a circuit, $R_C = R_1 R_2/(R_1 + R_2)$ where $R_i (i = 1, 2)$ signify the resistances of the two resistors connected in parallel and R_C is the circuit resistance. Using the Taylor series approximation, it is easily verified that $\mu_C \cong \mu_1\mu_2/(\mu_1 + \mu_2)$ and $\sigma_C^2 \cong [\mu_1/(\mu_1 + \mu_2)]^4\sigma_2^2 + [\mu_2/(\mu_1 + \mu_2)]^4\sigma_1^2$, where the subscripts conform with the notation used above for the R values. These equations may then be used to study the relationships among the tolerances of the resistors and the tolerances of the circuit.

15.4 SYSTEM RELIABILITY

Although many statistical techniques are applicable to the problems arising in the analysis of reliability data, we only look at one specific problem, that of estimating system reliability. Stated briefly, the problem is as follows: Given probabilities for the successful operation of the various components utilized in a system, estimate the probability that the system will operate successfully. Before attempting to present the solution of a problem such as posed in the preceding paragraph, it will be wise to adopt some standard notation. Thus in what follows we shall use:

p_i = the probability that the ith component in the system will operate successfully

q_i = the probability that the ith component in the system will fail to operate successfully = $1 - p_i$

P = the probability that the system will operate successfully

Q = the probability that the system will fail to operate successfully = $1 - P$

Given this notation and assuming that n components are utilized in the system, we may write

$$P = f(p_1, \ldots, p_n) \qquad (15.19)$$

where the form of the function will depend on the nature of the system.

If the system under consideration is either a simple series system (in which all components must work for the system to succeed) or a parallel system (in which at least one of the components must work for the system to succeed), and if the various components are statistically independent in their operation, the functional form is easy to determine and the basic equations are as follows:

Series system:

$$P = \prod_{i=1}^{n} p_i = \prod_{i=1}^{n} (1 - q_i) \qquad (15.20)$$

$$Q = 1 - \prod_{i=1}^{n} p_i = 1 - \prod_{i=1}^{n} (1 - q_i) \qquad (15.21)$$

Parallel system:

$$Q = \prod_{i=1}^{n} q_i = \prod_{i=1}^{n} (1 - p_i) \qquad (15.22)$$

$$P = 1 - \prod_{i=1}^{n} q_i = 1 - \prod_{i=1}^{n} (1 - p_i) \qquad (15.23)$$

If the system involves both series and parallel features, an equivalent system (or circuit) can always be found that will permit utilization of the preceding equations.

Example 15.11

Consider a system consisting of 4 components connected in series in which all components must operate properly if the system is to function properly. If the respective failure probabilities of the 4 components are 0.02, 0.03, 0.05, and 0.02, then $P = (0.98)(0.97)(0.95)(0.98)$ and $Q = 1 - P$.

Example 15.12

Consider a system consisting of 4 components connected in parallel in which at least one of the components must operate properly if the system is to function properly. If the respective failure probabilities of the 4 components are 0.05, 0.10, 0.02, and 0.001, then $Q = (0.05)(0.10)(0.02)(0.001)$ and $P = 1 - Q$.

We hope the brief preceding treatment has been sufficient to acquaint the reader with the nature of the problem and with the method of solution. Some of the problems at the end of the chapter will require the extension of the basic principles to more complex situations, while others will introduce certain approximations that are often used by reliability analysts.

Before leaving this topic, we must call a number of specific points to your attention.

1. The assumption of statistical independence is frequently subject to question.
2. Rarely does the analyst know the true values of the p_i and q_i. This means that he is really using $\hat{p}_i$ and $\hat{q}_i = 1 - \hat{p}_i$, and thus $\hat{P} = f(\hat{p}_1, \ldots, \hat{p}_n)$ and $\hat{Q} = 1 - \hat{P}$ are actually point estimates of the unknown parameters P and Q.
3. While confidence limits for the true p_i and q_i are easily obtained (see Chapter 5), no general methods have been developed for providing confidence limits for P and Q, although several special cases have been solved.

4. Frequently, the analyst must consider more than just *success* or *failure*. For example, in nuclear weaponry the possibility of *premature* operation must also be considered. Thus for each device and for the system as a whole, we now have three probabilities to contend with. Apart from the added complexity of the mathematics involved, this can lead to great difficulties in the logical analysis of the system.

5. In many applications, the probabilities p_i and q_i will be functions of time; i.e., the reliability analyst will be dealing with $p_i(t)$ and $q_i(t)$, where t represents operating time or age. When such is the case, the problems of analysis are complicated by such phenomena as early failures and wear-out.

The preceding are just a few of the points that plague a reliability analyst as he goes about his daily work. We are certain that you could add many other items to the above list. However, such a list would be out of place in this book. The items mentioned, however, are distinctly statistical in nature, and thus it seems appropriate to call them to your attention.

PROBLEMS

15.1   Define and discuss each of the following terms: (a) quality, (b) control, (c) quality control, (d) statistical quality control.

15.2   Why should both the $\bar{X}$ chart and the R chart be used when dealing with measurement (as opposed to attributes) data?

15.3   A control chart for $\bar{X}$ has UCL = 24.23 and LCL = 23.99. The calculations were based on samples of size 4. If the engineering specifications had been given as 24.10 ± 0.20, what percentage of the sample means would you expect to plot out of control? State any assumptions you find it necessary to make.

15.4   Given the following coded data on sample means and ranges ($n = 10$) for length of life of light bulbs, construct and interpret the appropriate control charts.

Sample No.	Mean ($\bar{X}$)	Range (R)
1	69.4	45
2	63.4	48
3	55.0	72
4	64.0	48
5	57.4	36
6	82.0	81
7	85.0	78
8	33.4	42
9	46.0	69
10	112.4	84
11	93.8	48
12	95.6	75
13	117.8	51
14	113.6	84
15	74.8	54

(continued)

Sample No.	Mean ($\overline{X}$)	Range (R)
16	80.8	45
17	71.8	57
18	53.2	75
19	74.8	48
20	59.2	63
21	65.8	129
22	109.6	42
23	44.2	51
24	73.6	51
25	51.4	27
Total	1848.0	1503

15.5 Given the following sample data, construct control charts for the sample mean and the sample range (i.e., $\overline{X}$ and R charts) and interpret the results.

Sample No.	X_1	X_2	X_3	X_4	X_5	Sample No.	X_1	X_2	X_3	X_4	X_5
1	2	−1	1	0	0	13	−2	−3	2	−1	−5
2	0	0	−3	2	−3	14	2	−1	0	−2	−1
3	3	1	4	−2	−1	15	2	−2	1	−1	0
4	1	−3	0	1	−1	16	0	4	−4	0	3
5	−1	0	1	0	1	17	0	0	−1	0	2
6	−1	0	−1	−2	−3	18	−2	−1	2	−1	0
7	0	0	3	−1	3	19	−2	0	−1	−1	0
8	0	2	−2	−3	0	20	2	−2	−1	2	−1
9	2	0	0	1	1	21	0	−1	−1	−1	1
10	0	3	2	1	1	22	0	2	1	1	1
11	1	−2	1	1	3	23	−1	−1	2	−1	1
12	0	0	−2	−2	0	24	0	1	1	−3	−2
						25	1	−1	−4	−1	1

15.6 Construct and interpret the appropriate control charts for the following data.

Sample No.	X_1	X_2	X_3	X_4	X_5	$\overline{X}$	R
1	0.831	0.829	0.836	0.840	0.826	0.8324	0.014
2	0.834	0.826	0.831	0.831	0.831	0.8306	0.008
3	0.836	0.826	0.831	0.822	0.816	0.8262	0.020
4	0.833	0.831	0.835	0.831	0.833	0.8326	0.004
5	0.830	0.831	0.831	0.833	0.820	0.8290	0.013
6	0.829	0.828	0.828	0.832	0.841	0.8316	0.013
7	0.835	0.833	0.829	0.830	0.841	0.8336	0.012

Sample No.	Individual Value					$\bar{X}$	R
	X_1	X_2	X_3	X_4	X_5		
8	0.818	0.838	0.835	0.834	0.830	0.8310	0.020
9	0.841	0.831	0.831	0.833	0.832	0.8336	0.010
10	0.832	0.828	0.836	0.832	0.825	0.8306	0.011
11	0.831	0.838	0.844	0.827	0.826	0.8332	0.018
12	0.831	0.826	0.828	0.832	0.827	0.8288	0.006
13	0.838	0.822	0.835	0.830	0.830	0.8310	0.016
14	0.815	0.832	0.831	0.831	0.838	0.8294	0.023
15	0.831	0.833	0.831	0.834	0.832	0.8322	0.003
16	0.830	0.819	0.819	0.844	0.832	0.8288	0.025
17	0.826	0.839	0.842	0.835	0.830	0.8344	0.016
18	0.813	0.833	0.819	0.834	0.836	0.8270	0.023
19	0.832	0.831	0.825	0.831	0.850	0.8338	0.025
20	0.831	0.838	0.833	0.831	0.833	0.8332	0.007
21	0.823	0.830	0.832	0.835	0.835	0.8310	0.012
22	0.835	0.829	0.834	0.826	0.828	0.8304	0.009
23	0.833	0.836	0.831	0.832	0.832	0.8328	0.005
24	0.826	0.835	0.842	0.832	0.831	0.8332	0.016
25	0.833	0.823	0.816	0.831	0.838	0.8282	0.022
26	0.829	0.830	0.830	0.833	0.831	0.8306	0.004
27	0.850	0.834	0.827	0.831	0.835	0.8354	0.023
28	0.835	0.846	0.829	0.833	0.822	0.8330	0.024
29	0.831	0.832	0.834	0.826	0.833	0.8312	0.008

15.7   Using the data below, calculate limits and plot the $\bar{X}$ and R charts. Apply the standard tests for unnatural patterns and discuss the results. (**Note:** The sample size is $n = 5$.)

Sample No.	$\bar{X}$	R	Sample No.	$\bar{X}$	R
1	1.444	0.09	18	1.424	0.05
2	1.427	0.08	19	1.434	0.05
3	1.464	0.08	20	1.414	0.09
4	1.455	0.08	21	1.406	0.07
5	1.462	0.10	22	1.418	0.14
6	1.448	0.05	23	1.438	0.09
7	1.454	0.04	24	1.416	0.07
8	1.446	0.08	25	1.419	0.06
9	1.437	0.12	26	1.406	0.08
10	1.471	0.11	27	1.428	0.06
11	1.438	0.09	28	1.430	0.06
12	1.438	0.05	29	1.421	0.07
13	1.415	0.12	30	1.434	0.07
14	1.428	0.12	31	1.408	0.05
15	1.425	0.08	32	1.414	0.08
16	1.440	0.09	33	1.410	0.03
17	1.430	0.05	34	1.406	0.06

(*continued*)

Sample No.	$\overline{X}$	R	Sample No.	$\overline{X}$	R
35	1.457	0.09	43	1.405	0.07
36	1.444	0.09	44	1.419	0.10
37	1.432	0.05	45	1.410	0.07
38	1.438	0.05	46	1.420	0.05
39	1.404	0.10	47	1.414	0.05
40	1.409	0.05	48	1.426	0.11
41	1.400	0.07	49	1.386	0.06
42	1.425	0.09	50	1.387	0.08

15.8   Following are the number of defective piston rings inspected in 23 daily samples of 100. Calculate the control limits for the type of control chart that should be used with these data. Interpret the results.

Date	Number Defective	Date	Number Defective
1	9	17	3
2	5	18	6
3	10	19	3
4	10	22	3
5	13	23	3
8	10	24	0
9	13	25	5
10	2	26	5
11	1	29	3
12	3	30	2
15	2	31	4
16	2		

15.9   Given the following data on number of defectives in samples of size 100, construct the appropriate control chart. Interpret the results.

Sample No.	Number Defective (d)	Sample No.	Number Defective (d)
1	3	14	3
2	1	15	5
3	4	16	8
4	4	17	2
5	4	18	3
6	6	19	5
7	5	20	4
8	5	21	3
9	2	22	4
10	4	23	6
11	3	24	4
12	4	25	3
13	4	26	5

Sample No.	Number Defective (d)	Sample No.	Number Defective (d)
27	4	34	9
28	7	35	6
29	6	36	4
30	5	37	3
31	5	38	1
32	6	39	3
33	4	40	1

15.10 At a certain point in the assembly process, TV sets are subjected to a critical inspection. The following data resulted from the inspection of 25 randomly selected sets. Plot and interpret the appropriate control chart.

Set No.	Number of Defects per Set	Set No.	Number of Defects per Set
1	12	14	7
2	11	15	4
3	13	16	9
4	17	17	15
5	7	18	12
6	10	19	11
7	9	20	9
8	8	21	4
9	13	22	4
10	15	23	10
11	18	24	12
12	9	25	4
13	14		

15.11 Phonograph records were selected at random times from a production line. Given the following data, construct and interpret the appropriate chart.

Record No.	Number of Defects per Record	Record No.	Number of Defects per Record
1	1	11	20
2	1	12	1
3	3	13	6
4	7	14	12
5	8	15	4
6	1	16	5
7	2	17	1
8	6	18	8
9	1	19	7
10	1	20	9

(*continued*)

Set No.	Number of Defects per Set	Set No.	Number of Defects per Set
21	10	26	2
22	5	27	3
23	0	28	14
24	19	29	6
25	16	30	8

15.12   (a) Plot the OC curve for Example 15.4.

       (b) Determine the AQL and RQL points.

15.13   (a) Plot the AOQ curve for Example 15.5.

       (b) Determine the AOQL.

15.14   Discuss the basic purpose of acceptance sampling and the general methods by which this purpose is approached.

15.15   Plot, on the same graph, the following OC curves:

       (a) $N = 500, n = 50, a = 0$

       (b) $N = 500, n = 50, a = 1$

       (c) $N = 500, n = 50, a = 2$

15.16   Plot, on the same graph, the following OC curves:

       (a) $N = 500, n = 25, a = 1$

       (b) $N = 500, n = 50, a = 1$

       (c) $N = 500, n = 100, a = 1$

15.17   For the double sampling plan specified by $N = 500, n_1 = 50, n_2 = 100$, $a_1 = 5, r_1 = 14$, and $a_2 = 13$, determine and plot the OC, AOQ, and ASN functions. Find the value of the AOQL.

15.18   The following truncated sequential sampling plan is proposed:

       (a) If the first item is defective, reject the lot.

       (b) If the first item is nondefective and there is no more than one defective up to and including the tenth item, accept the lot.

       (c) If the first item is nondefective and there are two defectives prior to or including the tenth item, reject the lot.

       Determine the OC and ASN functions.

15.19   Rework Example 15.7 assuming that $\mu_Y \pm 5\sigma_Y = 2.020 \pm 0.030$ units.

15.20   Rework Example 15.8 assuming that no more than 1 percent of randomly mated bearings and shafts may exhibit interference.

15.21   Evaluate the results of Example 15.9 if $\mu_1 = 40$, $\mu_2 = 0.5$, $\mu_3 = 3$, $3\sigma_1 = 0.5, 3\sigma_2 = 0.005$, and $3\sigma_3 = 0.06$.

15.22   If $Y = X \cos \theta + UVW \sin \theta$, where X, U, V, W, and θ are random variables, use a Taylor's expansion about the means to determine approximate expressions for the mean and variance of Y.

15.23   Assuming that each random variable is normally distributed with a mean equal to the nominal value and with a standard deviation equal to one-third of the one-sided engineering tolerance, evaluate the results determined in Problem 15.22, for the following specifications:

       (a) $X = 40 \pm 0.4$ (d) $W = 3 \pm 2$ percent

       (b) $U = 40 \pm 0.5$ (e) $\theta = 60 \pm 0.25$ degrees

       (c) $V = 0.5 \pm 1$ percent

15.24   Two resistors are assembled in series. Each is nominally rated at 20 ohms. The resistors are known to be normally distributed about the nominal value with a standard deviation of 1.5 ohms. What is the mean and standard deviation of the circuit resistance?

15.25   Rework Problem 15.24 if the resistors are uniformly distributed over the interval 18–22 ohms.

15.26   If two mating parts X and Y are each normally distributed with means $\mu_X = 2$ in. and $\mu_Y = 2.04$ in. and standard deviations $\sigma_X = 0.01$ in. and $\sigma_Y = 0.02$ in., what is the probability of interference?

15.27   With reference to Problem 15.26, what should μ_X be (assuming everything else is unchanged) if the probability of interference is to be 0.01?

15.28   For each of the following cases use a Taylor's series to find the approximate mean and variance of the dependent variable:

(a) $R = kX^b$

(b) $\Delta C = kC_{max}e^{-b/\Delta\theta}$

(c) $R = R_1 R_2 / R_3$

(d) $P_v = P_{sat} - (P_b - P_{sat})(T_{db} - T_{wb})/(2800 - 1.3T_{wb})$

(e) $C = q[1 - (A_2/A_1)^2]^{1/2}/A_2(2gh)^{1/2}$

(f) $\delta = e\{\sec[(P/EI)^{1/2}(L/2)] - 1\}$

15.29   For a system involving 4 components A, B, C, and D for which $p_A = 0.9, p_B = 0.8, p_C = 0.9$, and $p_D = 0.9$, determine P under the assumption of mutual independence and given that A and B are connected in parallel (branch I) and that this branch is then connected in series with C and D.

15.30   Consider an electrical circuit consisting of two subcircuits, the first of which involves the components X_1, X_2, X_3, and X_4 in parallel and the second of which involves the components X_5 and X_6 in parallel. If the two subcircuits are connected in series and mutual independence can be assumed, determine Q if $q_1 = 0.10$, $q_2 = 0.05$, and $q_3 = q_4 = q_5 = q_6 = 0.02$.

15.31   An equipment consists of 3 components D, E, and F connected in series. If the reliabilities of the 3 components are 0.92, 0.95, and 0.96, what is the reliability of the equipment? State all assumptions made in achieving your answer.

15.32   An equipment consists of 3 components D, E, and F connected in parallel. If the reliabilities of the 3 components are 0.92, 0.95, and 0.96, what is the reliability of the equipment? State all assumptions made in achieving your answer.

15.33   An equipment consists of 100 parts, of which 20 parts are tubes connected functionally in series (branch A). This branch is in turn connected in series to two parallel branches of 60 and 20 parts (branches B and C). The parts that comprise each of these branches are connected functionally in series. The reliability of each tube is 0.95, while the geometric mean reliability of branch B is 0.93 and the geometric mean reliability of branch C is 0.96. Draw a simplified equipment diagram and determine the reliability of the equipment. State all assumptions made in achieving your answer.

15.34   When reliability is a function of time, it is common practice to assume the validity of the exponential density function as a failure distribution. That is, it is common to use $f(t) = \lambda e^{-\lambda t}$ as the basic failure distribution. In such a case, λ is known as the *hazard rate* (or, in loose language, the *failure rate*) and $m = 1/\lambda$ is called the *mean time between failure* (MTBF). Under the preceding conditions the reliability of a component to time t is $R(t) = 1 - F(t) = e^{-\lambda t}$. Utilizing the above information, and assuming that the ith component has a constant hazard rate $\lambda_i (i = 1, \ldots, n)$, express Eqs. (15.20)–(15.23) in terms of the λ's and t.

15.35 Simplify the answer to the preceding problem on the assumption that all the λ_i are equal.

15.36 If $f(t) = \lambda e^{-\lambda t}$ and $\lambda > 0$, $t > 0$, what effect does doubling the value of λ have on (a) the MTBF, and (b) $R(t)$?

15.37 Assume an equipment consists of 2 components for which the failure rates are $\lambda_1 = 0.001$ and $\lambda_2 = 0.002$ respectively. Calculate the equipment reliability for $t = 100$ for the following two cases: (a) series connection, (b) parallel connection.

15.38 Show that if $n = 4$, Eq. (15.21) can be approximated by either $4q$ or $4q - 6q^2$. (**Note:** The reliability analyst makes frequent use of such approximations.)

15.39 The failure rate for a television receiver is 0.02 failures/hr.
   (a) Calculate the mean time between failures.
   (b) What is the probability of such a receiver failing in the first 4 hours?

REFERENCES

1. Bowker, A. H.; and Lieberman, G. J. 1972. *Engineering Statistics*, 2nd ed. Prentice-Hall, Englewood Cliffs, N.J.
2. Breipohl, A. M. 1960. A statistical approach to tolerances. Sandia Corp. Tech. Memo. SCTM 65–60 (14), Albuquerque, N. Mex.
3. Clark, C. R.; and Koopmans, L. H. 1959. Graphs of the hypergeometric OC and AOQ functions for lot sizes 10 to 225. Sandia Corp. Monogr. SCR-121, Albuquerque, N. Mex.
4. Grant, E. L. 1952. *Statistical Quality Control*, 2nd ed. McGraw-Hill, New York.
5. Lieberman, G. J.; and Owen, D. B. 1960. *Tables of the Hypergeometric Distribution*. Stanford University Press, Calif.
6. Shewhart, W. A. 1931. *Economic Control of Quality of Manufactured Product*. Van Nostrand, New York.
7. Western Electric Company, Inc. 1958. *Statistical Quality Control Handbook*, 2nd ed. New York.
8. Wiesen, J. M. 1958. Extension of existing tables of attributes sampling plans. Sandia Corp. Tech. Memo. SCTM 42-58 (12), Albuquerque, N. Mex.

FURTHER READING

American National Standards Institute. 1958. ANSI stds. Z1.1 and Z1.2. Guide for quality control and control chart method for analyzing data. New York.
———. 1958. ANSI std. Z1.3. Control chart method of controlling quality during production. New York.
American Society for Testing Materials. 1951. ASTM manual on quality control of materials. STP No. 15-C, Philadelphia.
Bazovsky, I. 1961. *Reliability Theory and Practice*. Prentice-Hall, Englewood Cliffs, N.J.
Bowker, A. H.; and Lieberman, G. J. 1972. *Engineering Statistics*, 2nd ed. Prentice-Hall, Englewood Cliffs, N.J.
Calabro, S. R. 1962. *Reliability Principles and Practices*. McGraw-Hill, New York.
Chorafas, D. N. 1960. *Statistical Processes and Reliability Engineering*. Van Nostrand, Princeton, N.J.
Dodge, H. F.; and Romig, H. G. 1959. *Sampling Inspection Tables, Single and Double Sampling*, 2nd ed. Wiley, New York.

Duncan, A. J. 1974. *Quality Control and Industrial Statistics,* 4th ed. Irwin, Homewood, Ill.

Enrick, N. L. 1972. *Quality Control and Reliability,* 6th ed. Industrial Press, New York.

Feigenbaum, A. V. 1961. *Total Quality Control: Engineering and Management.* McGraw-Hill, New York.

Grant, E. L.; and Leavenworth, R. S. 1972. *Statistical Quality Control,* 4th ed. McGraw-Hill, New York.

Hansen, B. L. 1964. *Quality Control; Theory and Applications.* Prentice-Hall, Englewood Cliffs, N.J.

Juran, J. M.; and Gryna, F. M., Jr. 1970. *Quality Planning and Analysis.* McGraw-Hill, New York.

Juran, J. M.; Seder, L. A.; and Gryna, F. M., Jr. (ed.). 1962. *Quality Control Handbook,* 2nd ed. McGraw-Hill, New York.

Kirkpatrick, E. G. 1970. *Quality Control for Managers and Engineers.* Wiley, New York.

Lloyd, D. K.; and Lipow, M. 1962. *Reliability: Management, Methods and Mathematics.* Prentice-Hall, Englewood Cliffs, N.J.

Mann, N. R.; Schafer, R. E.; and Singpurwalla, N. D. 1974. *Methods for Statistical Analysis of Reliability and Life Data.* Wiley, New York.

Peach, P. 1964. *Quality Control for Management.* Prentice-Hall, Englewood Cliffs, N.J.

Sampson, G.; Hart, P.; and Rubin, C. 1970. *Fundamentals of Statistical Quality Control.* Addison-Wesley, Reading, Mass.

Statistical Research Group, Columbia University. 1948. *Sampling Inspection.* McGraw-Hill, New York.

U.S. Government. 1963. DOD MIL-STD-105D. Sampling procedures and tables for inspection by attributes. USGPO, Washington, D.C.

———. 1957. DOD MIL-STD-414. Sampling procedures and tables for inspection by variables for percent defective. USGPO, Washington, D.C.

———. 1969. DOD MIL-STD-690-B. Failure rate sampling plans and procedures. USGPO, Washington, D.C.

———. 1967. DOD MIL-STD-781-B. Reliability tests: Exponential distribution. USGPO, Washington, D.C.

———. 1962. DOD MIL-STD-1235. Single and multi-level continuous sampling procedures and tables for inspection by attributes. USGPO, Washington, D.C.

Wetherill, G. B. 1969. *Sampling Inspection and Quality Control.* Methuen, London.

APPENDIX 1

MATHEMATICAL CONCEPTS

SET THEORY

The subject of the theory of sets is fundamental in mathematics. We are concerned here only with a few basic concepts that are useful in the theory of probability.

Definition A1.1 A *set* is a collection of elements.

Definition A1.2 The *universal set* is the set consisting of all elements under discussion. (**Note:** The universal set is sometimes referred to as a *space*.)

Definition A1.3 The *null set* is the set containing no elements at all.

Definition A1.4 Associated with each set A is another set A', called the complement of A and defined to be the set consisting of all the elements of the universal set that are not elements of A.

Definition A1.5 For any two sets A and B, the *union* of A and B is the set consisting of all elements that are either in A or in B or in both A and B. The union of A and B is commonly denoted by $A \cup B$.

Definition A1.6 For any two sets A and B, the *intersection* of A and B is the set consisting of all elements that are in both A and B. The intersection of A and B is commonly denoted by $A \cap B$ or by AB.

Theorem A1.1 If A and B are two sets having no common elements, then the set AB is the null set.

A useful device for illustrating the properties of the algebra of sets is the *Venn diagram*. In such a diagram, the points interior to a rectangle constitute the universal set. Arbitrary sets within the universal set (i.e., *subsets* of the universal set) will be represented, for convenience, by the points interior to circles within the rectangle. In Figure A1.1 the set A is shaded by vertical lines and the set B is shaded by horizontal lines. Since $AB \neq 0$ (i.e., does not equal the null set) AB appears as the crosshatched area.

Probability theory depends on the number of elements in a set. We will denote the number of elements in any arbitrary set A by $n(A)$.

Theorem A1.2 If A and B have no elements in common, $n(A \cup B) = n(A) + n(B)$.

Theorem A1.3 If A and B have no elements in common, $n(AB) = 0$.

[520]

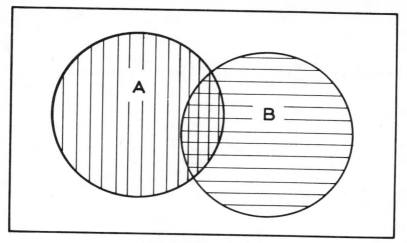

Fig. A1.1–Simple Venn diagram.

Theorem A1.4   For arbitrary sets A and B, it is true that $n(A \cup B) = n(A) + n(B) - n(AB)$.

NOTATION

As in all subjects, the system of notation employed is a matter of concern to the reader. Since statistics is so entwined with mathematics, it is no surprise that problems of notation arise; therefore, it seems appropriate to mention some of the more frequently occurring signs and symbols.

Definition A1.7   The *absolute value* of a number x, denoted by $|x|$, is its numerical value neglecting its algebraic sign. For example, $|-3| = 3$ and $|3| = 3$.

Definition A1.8 $x = y$ is read "x is equal to y."

Definition A1.9. $x \neq y$ is read "x is not equal to y."

Definition A1.10 $x \cong y$ is read "x is approximately equal to y."

Definition A1.11 $x < y$ is read "x is less than y."

Definition A1.12 $x \leq y$ is read "x is less than or equal to y."

Definition A1.13 $x > y$ is read "x is greater than y."

Definition A1.14 $x \geq y$ is read "x is greater than or equal to y"

Definition A1.15 $\displaystyle\sum_{i=1}^{n} Y_i = Y_1 + Y_2 + \cdots + Y_n$

(**Note:** The Greek capital letter sigma, $\sum$, is known as the *summation sign*. Further, i is called the *index of summation*, while 1 and n are known as the *limits of summation*.)

Theorem A1.5 $\displaystyle\sum_{i=1}^{n} Y_i = \sum_{i=1}^{k} Y_i + \sum_{i=k+1}^{n} Y_i$

Theorem A1.6 $\displaystyle\sum_{i=1}^{n} c Y_i = c \sum_{i=1}^{n} Y_i$ where c is a constant

Theorem A1.7 $\displaystyle\sum_{i=1}^{n} c = nc$

Theorem A1.8 $\displaystyle\sum_{i=1}^{n} (X_i + Y_i) = \sum_{i=1}^{n} X_i + \sum_{i=1}^{n} Y_i$

Theorem A1.9 $\displaystyle\left(\sum_{i=1}^{n} Y_i\right)^2 = \sum_{i=1}^{n} Y_i^2 + 2 \sum_{i=1}^{n-1} \sum_{j=i+1}^{n} Y_i Y_j$

$\displaystyle = \sum_{i=1}^{n} Y_i^2 + 2 \sum_{i<j} Y_i Y_j$

(**Note:** In this theorem the notation $\sum_{i<j}$ is interpreted to mean that we sum all possible products $Y_i Y_j$ letting i and j go from 1 to n, subject only to the restriction that in any particular term $i < j$.)

Definition A1.16 $\displaystyle\prod_{i=1}^{n} Y_i = (Y_1)(Y_2) \cdots (Y_n)$

(**Note:** In contrast to Definition A1.15, in which we introduced $\sum$ as the summation sign, we have here introduced the Greek capital letter pi, Π, as the *product sign.*

Theorem A1.10 $\displaystyle\prod_{i=1}^{n} i = (1)(2) \cdots (n) = n!$

(**Note:** The symbol $n!$ is called n *factorial* or *factorial n.*)

Definition A1.17 $0! = 1$

PERMUTATIONS AND COMBINATIONS

Permutations and combinations are concerned with the different subgroups and arrangements that can be formed from a given set. A *permutation* is a particular sequence (i.e., arrangement) of a given set or subset of elements, while a *combination* is the set or subset without reference to the order of the contained elements.

Definition A1.18   If an event A can occur in $n(A)$ ways and if a different event B can occur in $n(B)$ ways, then the event "either A or B" can occur in $n(A) + n(B)$ ways provided A and B cannot occur simultaneously.

(**Note:** You will notice the similarity between this definition and Theorem A1.2.)

Definition A1.19 If an event A can occur in $n(A)$ ways and a subsequent event B can occur in $n(B)$ ways, then the event "both A and B" can occur in $n(A) \cdot n(B)$ ways.

Definition A1.20 An r permutation of n things is an ordered selection or arrangement of r of them.

Definition A1.21 An r combination of n things is a selection of r of them without regard to order.

Definition A1.22 The number of different permutations that can be formed from n distinct objects taken r at a time is $P(n, r) = n(n - 1) \cdots (n - r + 1) = n!/(n - r)!$

Definition A1.23 The number of different permutations that can be formed from n objects taken n at a time, given that n_i are of type i, where $i = 1, 2, \cdots, k$, and $\sum n_i = n$, is $P(n; n_1, n_2, \ldots, n_k) = n!/n_1!n_2! \cdots n_k!$

Definition A1.24 The number of different combinations that can be formed from n distinct objects taken r at a time is $C(n, r) = P(n, r)/r! = n!/r!(n - r)!$

Definition A1.25 $C(n, r) = 0$ for $r < 0$ or $r > n$.

SOME USEFUL IDENTITIES AND SERIES

In statistical work it is often necessary to sum a series of terms or simplify a particular expression. A few of the more useful results are given here for ready reference.

Theorem A1.11 $(a + b)^n = \sum_{r=0}^{n} C(n, r) a^{n-r} b^r$

Theorem A1.12 If in Theorem A1.11 we let $a = 1$ and $b = t$, we obtain

$$(1 + t)^n = \sum_{r=0}^{n} C(n, r) t^r$$

Theorem A1.13 If in Theorem A1.11 we let $a = q$ and $b = p = 1 - q$, where $0 < p < 1$, we obtain

$$1 = \sum_{r=0}^{n} C(n, r) q^{n-r} p^r$$

This is a very useful expression in probability and statistics.

Definition A1.26 $e^x = \exp(x) = \sum_{i=0}^{\infty} x^i/i!$

Theorem A1.14 $(1 - x^n)/(1 - x) = \sum_{i=0}^{n-1} x^i$

Theorem A1.15 $1/(1 - x)^n = \sum_{i=0}^{\infty} C(n + i - 1, i)x^i$

Theorem A1.16 $\sum_{i=0}^{c} C(a, i) \cdot C(b, c - i) = C(a + b, c)$

Theorem A1.17 $\sum_{i=1}^{n} i = n(n + 1)/2$

Theorem A1.18 $\sum_{i=1}^{n} i^2 = n(n + 1)(2n + 1)/6$

Theorem A1.19 $\sum_{i=1}^{\infty} ix^i = x/(1 - x)^2$ for $-1 < x < 1$

SOME IMPORTANT FUNCTIONS

Some mathematical functions not always presented in courses in elementary mathematics are of great interest to the statistician. Two of these will be presented here for your convenience.

Definition A1.27   The *gamma function,* denoted by $\Gamma(p)$, is defined by the integral

$$\Gamma(p) = \int_{0}^{\infty} x^{p-1}e^{-x}dx$$

for $p > 0$. An alternative form for this function is

$$\Gamma(p) = 2 \int_{0}^{\infty} y^{2p-1}e^{-y^2}dy$$

where the transformation used was $x = y^2$.

Theorem A1.20   If in Definition A1.27 we let $p = n$ where n is a positive integer, we obtain $\Gamma(n) = (n - 1) \cdot \Gamma(n - 1) = (n - 1)!$

Theorem A1.21 $\Gamma(1/2) = \sqrt{\pi} = (\pi)^{1/2}$

Definition A1.28   The *beta function,* denoted by $\beta(p, q)$, is defined by the integral

$$\beta(p, q) = \int_{0}^{1} x^{p-1}(1 - x)^{q-1}dx$$

for $p > 0$ and $q > 0$. An alternative form for this

function is

$$\beta(p, q) = \int_0^{\pi/2} \sin^{2p-1}\theta \cos^{2q-1}\theta \, d\theta$$

where the transformation used was $x = \sin^2\theta$.

Theorem A1.22 $\beta(p, q) = \beta(q, p)$

Theorem A1.23 $\beta(p, q) = \Gamma(p) \cdot \Gamma(q)/\Gamma(p + q)$

MATRICES

Many of the methods discussed in this book depend on the theory of linear statistical models. This theory is most expeditiously handled in terms of matrix algebra.

Definition A1.29   A *matrix* **A** of dimension $r \times c$ is a rectangular array of elements a_{ij} arranged in r rows and c columns:

$$\mathbf{A} = [a_{ij}] = \begin{bmatrix} a_{11} & a_{12} & \cdots & a_{1c} \\ a_{21} & a_{22} & \cdots & a_{2c} \\ \vdots & & & \\ a_{r1} & a_{r2} & \cdots & a_{rc} \end{bmatrix}$$

Definition A1.30   If **A** is of dimension $n \times 1$, it is called an $n \times 1$ *vector*.

Definition A1.31 $\mathbf{A} = \mathbf{B}$ when and only when **A** and **B** are of the same dimension and $a_{ij} = b_{ij}$ for all i and j.

Definition A1.32   The product of a matrix **A** and a scalar (ordinary) number k is a matrix **B** where $b_{ij} = ka_{ij}$ for all i and j, i.e., $k\mathbf{A} = \mathbf{A}k = \mathbf{B}$.

Definition A1.33   The sum of two matrices, **A** and **B**, can be defined only when **A** and **B** are of the same dimension. Then $\mathbf{A} + \mathbf{B} = \mathbf{C}$ where $c_{ij} = a_{ij} + b_{ij}$.

Definition A1.34   The product of two matrices, say **AB**, can be defined only when the number of columns in **A** equals the number of rows in **B**. Then $\mathbf{AB} = \mathbf{C}$ where $c_{ij} = \sum_{s=1}^{c} a_{is}b_{sj}$

(**Note:** We must be very careful of the order of the factors when multiplying one matrix by another. Even if **AB** and **BA** are both defined, they are not necessarily equal.)

Definition A1.35   The transpose of a matrix **A** of dimension $r \times c$ is denoted by **A'**, where **A'** is a matrix of dimension $c \times r$ in which $a'_{ij} = a_{ji}$. That is, the rows of **A'** are the columns of **A** and the columns of **A'** are the rows of **A**.

Theorem A1.24 $(\mathbf{A}')' = \mathbf{A}$

Theorem A1.25	$(A + B)' = A' + B'$
Theorem A1.26	$(AB)' = B'A'$
Definition A1.36	If $r = c$, A is called a square matrix.
Theorem A1.27	For a square matrix A, we can write $A^2 = AA$, $A^3 = AAA$, etc.
Definition A1.37	In a square matrix of dimension $n \times n$, the elements $a_{11}, a_{22}, \ldots, a_{nn}$ form the main diagonal and are known as diagonal elements.
Definition A1.38	A square matrix that is symmetric with respect to its main diagonal is called a symmetric matrix.
Theorem A1.28	For a symmetric matrix, $A' = A$.
Definition A1.39	A symmetric matrix in which $a_{ij} = 0$ for all $i \neq j$ is called a diagonal matrix.
Definition A1.40	A diagonal matrix in which $a_{ii} = 1$ for all i is called a unit (or an identity) matrix and will be denoted by I.
Definition A1.41	A matrix having all its elements equal to zero is called the null matrix, and will be denoted by 0.
Definition A1.42	The determinant of a square matrix A of dimension $n \times n$, denoted by $\mid A \mid$, is defined by

$$\mid A \mid = \sum \pm a_{1r_1} a_{2r_2} \cdots a_{nr_n}$$

where the second subscripts $r_1, r_2, \ldots, r_n$ run through all the $n!$ possible permutations of the numbers $1, 2, \ldots, n$, and the sign of each term (either $+$ or $-$) is determined according to a well-defined rule.
Note: If A is of dimension 2×2, then

$$\mid A \mid = \begin{vmatrix} a_{11} & a_{12} \\ a_{21} & a_{22} \end{vmatrix} = a_{11}a_{22} - a_{21}a_{12}$$

Theorem A1.29	For any square matrix A, $\mid A \mid = \mid A' \mid$.
Theorem A1.30	If two rows (or columns) of a square matrix are interchanged, the determinant changes its sign.
Theorem A1.31	If two rows (or columns) of a square matrix are identical, the determinant is 0.
Theorem A1.32	If A, B, and C are square matrices such that $AB = C$, then $\mid A \mid \cdot \mid B \mid = \mid C \mid$.
Theorem A1.33	If a multiple of one row (column) is added to another row (column) of a square matrix, the determinant is unchanged.
Definition A1.43	For any arbitrary matrix A, the determinant of any square submatrix of A is called a minor of A.
Definition A1.44	For a square matrix A, the minor obtained by deleting the ith row and jth column, multiplied by $(-1)^{i+j}$, is known as the cofactor of a_{ij}. We denote the cofactor of a_{ij} by $\mathrm{cof}\, a_{ij}$.
Theorem A1.34	For a square matrix A of dimension $n \times n$, the de-

terminant $|\mathbf{A}|$ may be found by evaluating

$$|\mathbf{A}| = \sum_{j=1}^{n} a_{ij}(\text{cof } a_{ij}) = \sum_{i=1}^{n} a_{ij}(\text{cof } a_{ij})$$

Definition A1.45   If, for a square matrix, $|\mathbf{A}| \neq 0$, then $\mathbf{A}$ is of rank n and $\mathbf{A}$ is said to be nonsingular.

Definition A1.46   For a nonsingular square matrix $\mathbf{A}$, the inverse of $\mathbf{A}$ is denoted by $\mathbf{A}^{-1}$ and is defined by

$$\mathbf{A}^{-1} = \left[\frac{\text{cof } a_{ji}}{|\mathbf{A}|}\right]$$

Theorem A1.35   For a nonsingular square matrix $\mathbf{A}$, it is true that

$$\mathbf{A}\mathbf{A}^{-1} = \mathbf{A}^{-1}\mathbf{A} = \mathbf{I}$$

Theorem A1.36   For a nonsingular square matrix $\mathbf{A}$, it is true that

$$(\mathbf{A}')^{-1} = (\mathbf{A}^{-1})'$$

LINEAR EQUATIONS

Many times in statistical work we find it necessary to discuss systems of linear equations such as:

$$a_{11}x_1 + a_{12}x_2 + \cdots + a_{1n}x_n = y_1$$
$$a_{21}x_1 + a_{22}x_2 + \cdots + a_{2n}x_n = y_2$$
$$\vdots$$
$$a_{n1}x_1 + a_{n2}x_2 + \cdots + a_{nn}x_n = y_n \tag{A1.1}$$

The matrix notation introduced in the preceding section gives us an extremely concise method of representing such systems. For example, it is clear that

$$\mathbf{A}\mathbf{X} = \mathbf{Y} \tag{A1.2}$$

is the same as Eq. (A1.1) if

$$\mathbf{A} = \begin{bmatrix} a_{11} & a_{12} & \cdots & a_{1n} \\ \vdots & & & \\ a_{n1} & a_{n2} & \cdots & a_{nn} \end{bmatrix} \quad \mathbf{X} = \begin{bmatrix} x_1 \\ x_2 \\ \vdots \\ x_n \end{bmatrix} \quad \mathbf{Y} = \begin{bmatrix} y_1 \\ y_2 \\ \vdots \\ y_n \end{bmatrix}$$

Theorem A1.37   If $\mathbf{A}$ in Eq. (A1.2) is nonsingular, than $\mathbf{A}^{-1}\mathbf{A}\mathbf{X} = \mathbf{I}\mathbf{X} = \mathbf{X} = \mathbf{A}^{-1}\mathbf{Y}$. That is,

$$x_j = \frac{1}{|\mathbf{A}|} \sum_{i=1}^{n} y_i(\text{cof } a_{ij}) \qquad j = 1, 2, \ldots, n$$

Note: Another way of writing this is

$$x_j = \frac{1}{|\mathbf{A}|} \begin{vmatrix} \text{matrix } \mathbf{A} \text{ in which } a_{ij} \\ \text{has been replaced by} \\ y_i \text{ for } i = 1, 2, \ldots, n \end{vmatrix} \qquad j = 1, 2, \ldots, n$$

That is,

$$x_j = \frac{1}{|\mathbf{A}|} \begin{vmatrix} a_{11} \cdots a_{1,j-1} & y_1 & a_{1,j+1} \cdots a_{1n} \\ a_{21} \cdots a_{2,j-1} & y_2 & a_{2,j+1} \cdots a_{2n} \\ \vdots & & \\ a_{n1} \cdots a_{n,j-1} & y_n & a_{n,j+1} \cdots a_{nn} \end{vmatrix}$$

$$j = 1, \ldots, n$$

CUMULATIVE POISSON DISTRIBUTION

λ \\ x	0	1	2	3	4	5	6	7	8	9	10
0.01	0.990										
0.02	0.980										
0.03	0.970										
0.04	0.961	0.999									
0.05	0.951	0.999									
0.06	0.942	0.998									
0.07	0.932	0.998									
0.08	0.923	0.997									
0.09	0.914	0.996									
0.10	0.905	0.995									
0.15	0.861	0.990	0.999								
0.20	0.819	0.982	0.999								
0.25	0.779	0.974	0.998								
0.30	0.741	0.963	0.996								
0.35	0.705	0.951	0.994								
0.40	0.670	0.938	0.992	0.999							
0.45	0.638	0.925	0.989	0.999							
0.50	0.607	0.910	0.986	0.998							
0.55	0.577	0.894	0.982	0.998							
0.60	0.549	0.878	0.977	0.997							
0.65	0.522	0.861	0.972	0.996	0.999						

Source: Reprinted from E. C. Molina, *Poisson's Exponential Binomial Limit*, Van Nostrand, New York, 1947. By permission of the author and publishers.

Notes: (1) Entries in the table are values of $F(x)$ where

$$F(x) = P(c \leq x) = \sum_{c=0}^{x} e^{-\lambda} \lambda^c / c!$$

(2) Blank spaces to the right of the last entry in any row of the table may be read as 1; blank spaces to the left of the first entry in any row of the table may be read as 0.

λ \ x	0	1	2	3	4	5	6	7	8	9	10
0.70	0.497	0.844	0.966	0.994	0.999						
0.75	0.472	0.827	0.959	0.993	0.999						
0.80	0.449	0.809	0.953	0.991	0.999						
0.85	0.427	0.791	0.945	0.989	0.998						
0.90	0.407	0.772	0.937	0.987	0.998						
0.95	0.387	0.754	0.929	0.984	0.997						
1.00	0.368	0.736	0.920	0.981	0.996	0.999					
1.1	0.333	0.699	0.900	0.974	0.995	0.999					
1.2	0.301	0.663	0.879	0.966	0.992	0.998					
1.3	0.273	0.627	0.857	0.957	0.989	0.998					
1.4	0.247	0.592	0.833	0.946	0.986	0.997	0.999				
1.5	0.223	0.558	0.809	0.934	0.981	0.996	0.999				
1.6	0.202	0.525	0.783	0.921	0.976	0.994	0.999				
1.7	0.183	0.493	0.757	0.907	0.970	0.992	0.998				
1.8	0.165	0.463	0.731	0.891	0.964	0.990	0.997	0.999			
1.9	0.150	0.434	0.704	0.875	0.956	0.987	0.997	0.999			
2.0	0.135	0.406	0.677	0.857	0.947	0.983	0.995	0.999			
2.1	0.122	0.380	0.650	0.839	0.938	0.980	0.994	0.999			
2.2	0.111	0.355	0.623	0.819	0.928	0.975	0.993	0.998			
2.3	0.100	0.331	0.596	0.799	0.916	0.970	0.991	0.997	0.999		
2.4	0.091	0.308	0.570	0.779	0.904	0.964	0.988	0.997	0.999		
2.5	0.082	0.287	0.544	0.758	0.891	0.958	0.986	0.996	0.999		
2.6	0.074	0.267	0.518	0.736	0.877	0.951	0.983	0.995	0.999		
2.7	0.067	0.249	0.494	0.714	0.863	0.943	0.979	0.993	0.998	0.999	
2.8	0.061	0.231	0.469	0.692	0.848	0.935	0.976	0.992	0.998	0.999	
2.9	0.055	0.215	0.446	0.670	0.832	0.926	0.971	0.990	0.997	0.999	
3.0	0.050	0.199	0.423	0.647	0.815	0.916	0.966	0.988	0.996	0.999	

λ \ x	0	1	2	3	4	5	6	7	8	9	10
3.2	0.041	0.171	0.380	0.603	0.781	0.895	0.955	0.983	0.994	0.998	
3.4	0.033	0.147	0.340	0.558	0.744	0.871	0.942	0.977	0.992	0.997	0.999
3.6	0.027	0.126	0.303	0.515	0.706	0.844	0.927	0.969	0.988	0.996	0.999
3.8	0.022	0.107	0.269	0.473	0.668	0.816	0.909	0.960	0.984	0.994	0.998
4.0	0.018	0.092	0.238	0.433	0.629	0.785	0.889	0.949	0.979	0.992	0.997
4.2	0.015	0.078	0.210	0.395	0.590	0.753	0.867	0.936	0.972	0.989	0.996
4.4	0.012	0.066	0.185	0.359	0.551	0.720	0.844	0.921	0.964	0.985	0.994
4.6	0.010	0.056	0.163	0.326	0.513	0.686	0.818	0.905	0.955	0.980	0.992
4.8	0.008	0.048	0.143	0.294	0.476	0.651	0.791	0.887	0.944	0.975	0.990
5.0	0.007	0.040	0.125	0.265	0.440	0.616	0.762	0.867	0.932	0.968	0.986
5.2	0.006	0.034	0.109	0.238	0.406	0.581	0.732	0.845	0.918	0.960	0.982
5.4	0.005	0.029	0.095	0.213	0.373	0.546	0.702	0.822	0.903	0.951	0.977
5.6	0.004	0.024	0.082	0.191	0.342	0.512	0.670	0.797	0.886	0.941	0.972
5.8	0.003	0.021	0.072	0.170	0.313	0.478	0.638	0.771	0.867	0.929	0.965
6.0	0.002	0.017	0.062	0.151	0.285	0.446	0.606	0.744	0.847	0.916	0.957

λ \ x	11	12	13	14	15	16	17	18	19	20	21
3.2											
3.4											
3.6											
3.8	0.999										
4.0	0.999										
4.2	0.999										
4.4	0.998	0.999									
4.6	0.997	0.999									
4.8	0.996	0.999									
5.0	0.995	0.998	0.999								
5.2	0.993	0.997	0.999								
5.4	0.990	0.996	0.999								
5.6	0.988	0.995	0.998	0.999							
5.8	0.984	0.993	0.997	0.999							
6.0	0.980	0.991	0.996	0.999	0.999						

λ \ x	0	1	2	3	4	5	6	7	8	9	10
6.2	0.002	0.015	0.054	0.134	0.259	0.414	0.574	0.716	0.826	0.902	0.949
6.4	0.002	0.012	0.046	0.119	0.235	0.384	0.542	0.687	0.803	0.886	0.939
6.6	0.001	0.010	0.040	0.105	0.213	0.355	0.511	0.658	0.780	0.869	0.927
6.8	0.001	0.009	0.034	0.093	0.192	0.327	0.480	0.628	0.755	0.850	0.915
7.0	0.001	0.007	0.030	0.082	0.173	0.301	0.450	0.599	0.729	0.830	0.901
7.2	0.001	0.006	0.025	0.072	0.156	0.276	0.420	0.569	0.703	0.810	0.887
7.4	0.001	0.005	0.022	0.063	0.140	0.253	0.392	0.539	0.676	0.788	0.871
7.6	0.001	0.004	0.019	0.055	0.125	0.231	0.365	0.510	0.648	0.765	0.854
7.8		0.004	0.016	0.048	0.112	0.210	0.338	0.481	0.620	0.741	0.835
8.0		0.003	0.014	0.042	0.100	0.191	0.313	0.453	0.593	0.717	0.816
8.2		0.003	0.012	0.037	0.089	0.174	0.290	0.425	0.565	0.692	0.796
8.4		0.002	0.010	0.032	0.079	0.157	0.267	0.399	0.537	0.666	0.774

λ \ x	11	12	13	14	15	16	17	18	19	20	21
6.2	0.975	0.989	0.995	0.998	0.999						
6.4	0.969	0.986	0.994	0.997	0.999						
6.6	0.963	0.982	0.992	0.997	0.999	0.999					
6.8	0.955	0.978	0.990	0.996	0.998	0.999					
7.0	0.947	0.973	0.987	0.994	0.998	0.999					
7.2	0.937	0.967	0.984	0.993	0.997	0.999	0.999				
7.4	0.926	0.961	0.980	0.991	0.996	0.998	0.999				
7.6	0.915	0.954	0.976	0.989	0.995	0.998	0.999				
7.8	0.902	0.945	0.971	0.986	0.993	0.997	0.999				
8.0	0.888	0.936	0.966	0.983	0.992	0.996	0.998	0.999			
8.2	0.873	0.926	0.960	0.979	0.990	0.995	0.998	0.999			
8.4	0.857	0.915	0.952	0.975	0.987	0.994	0.997	0.999			

λ \ x	1	2	3	4	5	6	7	8	9	10	11	12	13	14	15	16	17
8.5	0.002	0.009	0.030	0.074	0.150	0.256	0.386	0.523	0.653	0.763	0.849	0.909	0.949	0.973	0.986	0.993	0.997
9.0	0.001	0.006	0.021	0.055	0.116	0.207	0.324	0.456	0.587	0.706	0.803	0.876	0.926	0.959	0.978	0.989	0.995
9.5	0.001	0.004	0.015	0.040	0.089	0.165	0.269	0.392	0.522	0.645	0.752	0.836	0.898	0.940	0.967	0.982	0.991
10.0		0.003	0.010	0.029	0.067	0.130	0.220	0.333	0.458	0.583	0.697	0.792	0.864	0.917	0.951	0.973	0.986
10.5		0.002	0.007	0.021	0.050	0.102	0.179	0.279	0.397	0.521	0.639	0.742	0.825	0.888	0.932	0.960	0.978
11.0		0.001	0.005	0.015	0.038	0.079	0.143	0.232	0.341	0.460	0.579	0.689	0.781	0.854	0.907	0.944	0.968
11.5		0.001	0.003	0.011	0.028	0.060	0.114	0.191	0.289	0.402	0.520	0.633	0.733	0.815	0.878	0.924	0.954
12.0		0.001	0.002	0.008	0.020	0.046	0.090	0.155	0.242	0.347	0.462	0.576	0.682	0.772	0.844	0.899	0.937
12.5			0.002	0.005	0.015	0.035	0.070	0.125	0.201	0.297	0.406	0.519	0.628	0.725	0.806	0.869	0.916
13.0			0.001	0.004	0.011	0.026	0.054	0.100	0.166	0.252	0.353	0.463	0.573	0.675	0.764	0.835	0.890
13.5			0.001	0.003	0.008	0.019	0.041	0.079	0.135	0.211	0.304	0.409	0.518	0.623	0.718	0.798	0.861
14.0			0.001	0.002	0.006	0.014	0.032	0.062	0.109	0.176	0.260	0.358	0.464	0.570	0.669	0.756	0.827
14.5				0.001	0.004	0.010	0.024	0.048	0.088	0.145	0.220	0.311	0.413	0.518	0.619	0.711	0.790
15.0				0.001	0.003	0.008	0.018	0.037	0.070	0.118	0.185	0.268	0.363	0.466	0.568	0.664	0.749
16.0					0.001	0.004	0.010	0.022	0.043	0.077	0.127	0.193	0.275	0.368	0.467	0.566	0.659
17.0					0.001	0.002	0.005	0.013	0.026	0.049	0.085	0.135	0.201	0.281	0.371	0.468	0.564
18.0						0.001	0.003	0.007	0.015	0.030	0.055	0.092	0.143	0.208	0.287	0.375	0.469

λ \ x	18	19	20	21	22	23	24	25	26	27	28	29	30	31	32	33	34
8.5	0.999	0.999															
9.0	0.998	0.999															
9.5	0.996	0.998	0.999														
10.0	0.993	0.997	0.998	0.999													
10.5	0.988	0.994	0.997	0.999	0.999												
11.0	0.982	0.991	0.995	0.998	0.999												
11.5	0.974	0.986	0.992	0.996	0.998	0.999											
12.0	0.963	0.979	0.988	0.994	0.997	0.999	0.999										
12.5	0.948	0.969	0.983	0.991	0.995	0.998	0.999	0.999									
13.0	0.930	0.957	0.975	0.986	0.992	0.996	0.998	0.999									
13.5	0.908	0.942	0.965	0.980	0.989	0.994	0.997	0.998	0.999								
14.0	0.883	0.923	0.952	0.971	0.983	0.991	0.995	0.997	0.999	0.999							
14.5	0.853	0.901	0.936	0.960	0.976	0.986	0.992	0.996	0.998	0.999	0.999						
15.0	0.819	0.875	0.917	0.947	0.967	0.981	0.989	0.994	0.997	0.998	0.999						
16.0	0.742	0.812	0.868	0.911	0.942	0.963	0.978	0.987	0.993	0.996	0.998	0.999	0.999				
17.0	0.655	0.736	0.805	0.861	0.905	0.937	0.959	0.975	0.985	0.991	0.995	0.997	0.999	0.999			
18.0	0.562	0.651	0.731	0.799	0.855	0.899	0.932	0.955	0.972	0.983	0.990	0.994	0.997	0.998	0.999		

λ \ x	6	7	8	9	10	11	12	13	14	15	16	17
19.0	0.001	0.002	0.004	0.009	0.018	0.035	0.061	0.098	0.150	0.215	0.292	0.378
20.0		0.001	0.002	0.005	0.011	0.021	0.039	0.066	0.105	0.157	0.221	0.297
21.0			0.001	0.003	0.006	0.013	0.025	0.043	0.072	0.111	0.163	0.227
22.0			0.001	0.002	0.004	0.008	0.015	0.028	0.048	0.077	0.117	0.169
23.0				0.001	0.002	0.004	0.009	0.017	0.031	0.052	0.082	0.123
24.0					0.001	0.003	0.005	0.011	0.020	0.034	0.056	0.087
25.0					0.001	0.001	0.003	0.006	0.012	0.022	0.038	0.060

λ \ x	18	19	20	21	22	23	24	25	26	27	28	29
19.0	0.469	0.561	0.647	0.725	0.793	0.849	0.893	0.927	0.951	0.969	0.980	0.988
20.0	0.381	0.470	0.559	0.644	0.721	0.787	0.843	0.888	0.922	0.948	0.966	0.978
21.0	0.302	0.384	0.471	0.558	0.640	0.716	0.782	0.838	0.883	0.917	0.944	0.963
22.0	0.232	0.306	0.387	0.472	0.556	0.637	0.712	0.777	0.832	0.877	0.913	0.940
23.0	0.175	0.238	0.310	0.389	0.472	0.555	0.635	0.708	0.772	0.827	0.873	0.908
24.0	0.128	0.180	0.243	0.314	0.392	0.473	0.554	0.632	0.704	0.768	0.823	0.868
25.0	0.092	0.134	0.185	0.247	0.318	0.394	0.473	0.553	0.629	0.700	0.763	0.818

λ \ x	30	31	32	33	34	35	36	37	38	39	40	41	42
19.0	0.993	0.996	0.998	0.999	0.999								
20.0	0.987	0.992	0.995	0.997	0.999	0.999							
21.0	0.976	0.985	0.991	0.994	0.997	0.998	0.999	0.999					
22.0	0.959	0.973	0.983	0.989	0.994	0.996	0.998	0.999	0.999				
23.0	0.936	0.956	0.971	0.981	0.988	0.993	0.996	0.997	0.999	0.999			
24.0	0.904	0.932	0.953	0.969	0.979	0.987	0.992	0.995	0.997	0.998	0.999	0.999	
25.0	0.863	0.900	0.929	0.950	0.966	0.978	0.985	0.991	0.994	0.997	0.998	0.999	0.999

APPENDIX 3

CUMULATIVE STANDARD NORMAL DISTRIBUTION

z	$G(z)$	z	$G(z)$	z	$G(z)$
−4.00	0.00003	−3.60	0.00016	−3.20	0.00069
−3.99	0.00003	−3.59	0.00017	−3.19	0.00071
−3.98	0.00003	−3.58	0.00017	−3.18	0.00074
−3.97	0.00004	−3.57	0.00018	−3.17	0.00076
−3.96	0.00004	−3.56	0.00019	−3.16	0.00079
−3.95	0.00004	−3.55	0.00019	−3.15	0.00082
−3.94	0.00004	−3.54	0.00020	−3.14	0.00084
−3.93	0.00004	−3.53	0.00021	−3.13	0.00087
−3.92	0.00004	−3.52	0.00022	−3.12	0.00090
−3.91	0.00005	−3.51	0.00022	−3.11	0.00094
−3.90	0.00005	−3.50	0.00023	−3.10	0.00097
−3.89	0.00005	−3.49	0.00024	−3.09	0.00100
−3.88	0.00005	−3.48	0.00025	−3.08	0.00104
−3.87	0.00005	−3.47	0.00026	−3.07	0.00107
−3.86	0.00006	−3.46	0.00027	−3.06	0.00111
−3.85	0.00006	−3.45	0.00028	−3.05	0.00114
−3.84	0.00006	−3.44	0.00029	−3.04	0.00118
−3.83	0.00006	−3.43	0.00030	−3.03	0.00122
−3.82	0.00007	−3.42	0.00031	−3.02	0.00126
−3.81	0.00007	−3.41	0.00032	−3.01	0.00131
−3.80	0.00007	−3.40	0.00034	−3.00	0.00135
−3.79	0.00008	−3.39	0.00035	−2.99	0.00139
−3.78	0.00008	−3.38	0.00036	−2.98	0.00144
−3.77	0.00008	−3.37	0.00038	−2.97	0.00149
−3.76	0.00008	−3.36	0.00039	−2.96	0.00154
−3.75	0.00009	−3.35	0.00040	−2.95	0.00159
−3.74	0.00009	−3.34	0.00042	−2.94	0.00164
−3.73	0.00010	−3.33	0.00043	−2.93	0.00169
−3.72	0.00010	−3.32	0.00045	−2.92	0.00175
−3.71	0.00010	−3.31	0.00047	−2.91	0.00181
−3.70	0.00011	−3.30	0.00048	−2.90	0.00187
−3.69	0.00011	−3.29	0.00050	−2.89	0.00193
−3.68	0.00012	−3.28	0.00052	−2.88	0.00199
−3.67	0.00012	−3.27	0.00054	−2.87	0.00205
−3.66	0.00013	−3.26	0.00056	−2.86	0.00212
−3.65	0.00013	−3.25	0.00058	−2.85	0.00219
−3.64	0.00014	−3.24	0.00060	−2.84	0.00226
−3.63	0.00014	−3.23	0.00062	−2.83	0.00233
−3.62	0.00015	−3.22	0.00064	−2.82	0.00240
−3.61	0.00015	−3.21	0.00066	−2.81	0.00248

Source: Abridged from Karl Pearson, *Tables for Statisticians and Biometricians*, Part I, Cambridge University Press, London, 1924, pp. 2–6. By permission of the author and publishers.

Note: $G(z) = P(Z \le z) = \int_{-\infty}^{z} (2\pi)^{-1/2} \exp\{-x^2/2\}\, dx$

z	$G(z)$	z	(Gz)	z	$G(z)$
−2.80	0.00256	−2.30	0.01072	−1.80	0.03593
−2.79	0.00264	−2.29	0.01101	−1.79	0.03673
−2.78	0.00272	−2.28	0.01130	−1.78	0.03754
−2.77	0.00280	−2.27	0.01160	−1.77	0.03836
−2.76	0.00289	−2.26	0.01191	−1.76	0.03920
−2.75	0.00298	−2.25	0.01222	−1.75	0.04006
−2.74	0.00307	−2.24	0.01255	−1.74	0.04093
−2.73	0.00317	−2.23	0.01287	−1.73	0.04182
−2.72	0.00326	−2.22	0.01321	−1.72	0.04272
−2.71	0.00336	−2.21	0.01355	−1.71	0.04363
−2.70	0.00347	−2.20	0.01390	−1.70	0.04457
−2.69	0.00357	−2.19	0.01426	−1.69	0.04551
−2.68	0.00368	−2.18	0.01463	−1.68	0.04648
−2.67	0.00379	−2.17	0.01500	−1.67	0.04746
−2.66	0.00391	−2.16	0.01539	−1.66	0.04846
−2.65	0.00402	−2.15	0.01578	−1.65	0.04947
−2.64	0.00415	−2.14	0.01618	−1.64	0.05050
−2.63	0.00427	−2.13	0.01659	−1.63	0.05155
−2.62	0.00440	−2.12	0.01700	−1.62	0.05262
−2.61	0.00453	−2.11	0.01743	−1.61	0.05370
−2.60	0.00466	−2.10	0.01786	−1.60	0.05480
−2.59	0.00480	−2.09	0.01831	−1.59	0.05592
−2.58	0.00494	−2.08	0.01876	−1.58	0.05705
−2.57	0.00508	−2.07	0.01923	−1.57	0.05821
−2.56	0.00523	−2.06	0.01970	−1.56	0.05938
−2.55	0.00539	−2.05	0.02018	−1.55	0.06057
−2.54	0.00554	−2.04	0.02068	−1.54	0.06178
−2.53	0.00570	−2.03	0.02118	−1.53	0.06301
−2.52	0.00587	−2.02	0.02169	−1.52	0.06426
−2.51	0.00604	−2.01	0.02222	−1.51	0.06552
−2.50	0.00621	−2.00	0.02275	−1.50	0.06681
−2.49	0.00639	−1.99	0.02330	−1.49	0.06811
−2.48	0.00657	−1.98	0.02385	−1.48	0.06944
−2.47	0.00676	−1.97	0.02442	−1.47	0.07078
−2.46	0.00695	−1.96	0.02500	−1.46	0.07215
−2.45	0.00714	−1.95	0.02559	−1.45	0.07353
−2.44	0.00734	−1.94	0.02619	−1.44	0.07493
−2.43	0.00755	−1.93	0.02680	−1.43	0.07636
−2.42	0.00776	−1.92	0.02743	−1.42	0.07780
−2.41	0.00798	−1.91	0.02807	−1.41	0.07927
−2.40	0.00820	−1.90	0.02872	−1.40	0.08076
−2.39	0.00842	−1.89	0.02938	−1.39	0.08226
−2.38	0.00866	−1.88	0.03005	−1.38	0.08379
−2.37	0.00889	−1.87	0.03074	−1.37	0.08534
−2.36	0.00914	−1.86	0.03144	−1.36	0.08691
−2.35	0.00939	−1.85	0.03216	−1.35	0.08851
−2.34	0.00964	−1.84	0.03288	−1.34	0.09012
−2.33	0.00990	−1.83	0.03362	−1.33	0.09176
−2.32	0.01017	−1.82	0.03438	−1.32	0.09342
−2.31	0.01044	−1.81	0.03515	−1.31	0.09510

z	$G(z)$	z	$G(z)$	z	$G(z)$
−1.30	0.09680	−0.85	0.19766	−0.40	0.34458
−1.29	0.09853	−0.84	0.20045	−0.39	0.34827
−1.28	0.10027	−0.83	0.20327	−0.38	0.35197
−1.27	0.10204	−0.82	0.20611	−0.37	0.35569
−1.26	0.10383	−0.81	0.20897	−0.36	0.35942
−1.25	0.10565	−0.80	0.21186	−0.35	0.36317
−1.24	0.10749	−0.79	0.21476	−0.34	0.36693
−1.23	0.10935	−0.78	0.21770	−0.33	0.37070
−1.22	0.11123	−0.77	0.22065	−0.32	0.37448
−1.21	0.11314	−0.76	0.22363	−0.31	0.37828
−1.20	0.11507	−0.75	0.22663	−0.30	0.38209
−1.19	0.11702	−0.74	0.22965	−0.29	0.38591
−1.18	0.11900	−0.73	0.23270	−0.28	0.38974
−1.17	0.12100	−0.72	0.23576	−0.27	0.39358
−1.16	0.12302	−0.71	0.23885	−0.26	0.39743
−1.15	0.12507	−0.70	0.24196	−0.25	0.40129
−1.14	0.12714	−0.69	0.24510	−0.24	0.40517
−1.13	0.12924	−0.68	0.24825	−0.23	0.40905
−1.12	0.13136	−0.67	0.25143	−0.22	0.41294
−1.11	0.13350	−0.66	0.25463	−0.21	0.41683
−1.10	0.13567	−0.65	0.25785	−0.20	0.42074
−1.09	0.13786	−0.64	0.26109	−0.19	0.42465
−1.08	0.14007	−0.63	0.26435	−0.18	0.42858
−1.07	0.14231	−0.62	0.26763	−0.17	0.43251
−1.06	0.14457	−0.61	0.27093	−0.16	0.43644
−1.05	0.14686	−0.60	0.27425	−0.15	0.44038
−1.04	0.14917	−0.59	0.27760	−0.14	0.44433
−1.03	0.15150	−0.58	0.28096	−0.13	0.44828
−1.02	0.15386	−0.57	0.28434	−0.12	0.45224
−1.01	0.15625	−0.56	0.28774	−0.11	0.45620
−1.00	0.15866	−0.55	0.29116	−0.10	0.46017
−0.99	0.16109	−0.54	0.29460	−0.09	0.46414
−0.98	0.16354	−0.53	0.29806	−0.08	0.46812
−0.97	0.16602	−0.52	0.30153	−0.07	0.47210
−0.96	0.16853	−0.51	0.30503	−0.06	0.47608
−0.95	0.17106	−0.50	0.30854	−0.05	0.48006
−0.94	0.17361	−0.49	0.31207	−0.04	0.48405
−0.93	0.17619	−0.48	0.31561	−0.03	0.48803
−0.92	0.17879	−0.47	0.31918	−0.02	0.49202
−0.91	0.18141	−0.46	0.32276	−0.01	0.49601
−0.90	0.18406	−0.45	0.32636	0.00	0.50000
−0.89	0.18673	−0.44	0.32997	0.01	0.50399
−0.88	0.18943	−0.43	0.33360	0.02	0.50798
−0.87	0.19215	−0.42	0.33724	0.03	0.51197
−0.86	0.19489	−0.41	0.34090	0.04	0.51595

z	$G(z)$	z	$G(z)$	z	$G(z)$
0.05	0.51994	0.50	0.69146	0.95	0.82894
0.06	0.52392	0.51	0.69497	0.96	0.83147
0.07	0.52790	0.52	0.69847	0.97	0.83398
0.08	0.53188	0.53	0.70194	0.98	0.83646
0.09	0.53586	0.54	0.70540	0.99	0.83891
0.10	0.53983	0.55	0.70884	1.00	0.84134
0.11	0.54380	0.56	0.71226	1.01	0.84375
0.12	0.54776	0.57	0.71566	1.02	0.84614
0.13	0.55172	0.58	0.71904	1.03	0.84850
0.14	0.55567	0.59	0.72240	1.04	0.85083
0.15	0.55962	0.60	0.72575	1.05	0.85314
0.16	0.56356	0.61	0.72907	1.06	0.85543
0.17	0.56749	0.62	0.73237	1.07	0.85769
0.18	0.57142	0.63	0.73565	1.08	0.85993
0.19	0.57535	0.64	0.73891	1.09	0.86214
0.20	0.57926	0.65	0.74215	1.10	0.86433
0.21	0.58317	0.66	0.74537	1.11	0.86650
0.22	0.58706	0.67	0.74857	1.12	0.86864
0.23	0.59095	0.68	0.75175	1.13	0.87076
0.24	0.59483	0.69	0.75490	1.14	0.87286
0.25	0.59871	0.70	0.75804	1.15	0.87493
0.26	0.60257	0.71	0.76115	1.16	0.87698
0.27	0.60642	0.72	0.76424	1.17	0.87900
0.28	0.61026	0.73	0.76730	1.18	0.88100
0.29	0.61409	0.74	0.77035	1.19	0.88298
0.30	0.61791	0.75	0.77337	1.20	0.88493
0.31	0.62172	0.76	0.77637	1.21	0.88686
0.32	0.62552	0.77	0.77935	1.22	0.88877
0.33	0.62930	0.78	0.78230	1.23	0.89065
0.34	0.63307	0.79	0.78524	1.24	0.89251
0.35	0.63683	0.80	0.78814	1.25	0.89435
0.36	0.64058	0.81	0.79103	1.26	0.89617
0.37	0.64431	0.82	0.79389	1.27	0.89796
0.38	0.64803	0.83	0.79673	1.28	0.89973
0.39	0.65173	0.84	0.79955	1.29	0.90147
0.40	0.65542	0.85	0.80234	1.30	0.90320
0.41	0.65910	0.86	0.80511	1.31	0.90490
0.42	0.66276	0.87	0.80785	1.32	0.90658
0.43	0.66640	0.88	0.81057	1.33	0.90824
0.44	0.67003	0.89	0.81327	1.34	0.90988
0.45	0.67364	0.90	0.81594	1.35	0.91149
0.46	0.67724	0.91	0.81859	1.36	0.91309
0.47	0.68082	0.92	0.82121	1.37	0.91466
0.48	0.68439	0.93	0.82381	1.38	0.91621
0.49	0.68793	0.94	0.82639	1.39	0.91774

z	$G(z)$	z	$G(z)$	z	$G(z)$
1.40	0.91924	1.85	0.96784	2.30	0.98928
1.41	0.92073	1.86	0.96856	2.31	0.98956
1.42	0.92220	1.87	0.96926	2.32	0.98983
1.43	0.92364	1.88	0.96995	2.33	0.99010
1.44	0.92507	1.89	0.97062	2.34	0.99036
1.45	0.92647	1.90	0.97128	2.35	0.99061
1.46	0.92785	1.91	0.97193	2.36	0.99086
1.47	0.92922	1.92	0.97257	2.37	0.99111
1.48	0.93056	1.93	0.97320	2.38	0.99134
1.49	0.93189	1.94	0.97381	2.39	0.99158
1.50	0.93319	1.95	0.97441	2.40	0.99180
1.51	0.93448	1.96	0.97500	2.41	0.99202
1.52	0.93574	1.97	0.97558	2.42	0.99224
1.53	0.93699	1.98	0.97615	2.43	0.99245
1.54	0.93822	1.99	0.97670	2.44	0.99266
1.55	0.93943	2.00	0.97725	2.45	0.99286
1.56	0.94062	2.01	0.97778	2.46	0.99305
1.57	0.94179	2.02	0.97831	2.47	0.99324
1.58	0.94295	2.03	0.97882	2.48	0.99343
1.59	0.94408	2.04	0.97932	2.49	0.99361
1.60	0.94520	2.05	0.97982	2.50	0.99379
1.61	0.94630	2.06	0.98030	2.51	0.99396
1.62	0.94738	2.07	0.98077	2.52	0.99413
1.63	0.94845	2.08	0.98124	2.53	0.99430
1.64	0.94950	2.09	0.98169	2.54	0.99446
1.65	0.95053	2.10	0.98214	2.55	0.99461
1.66	0.95154	2.11	0.98257	2.56	0.99477
1.67	0.95254	2.12	0.98300	2.57	0.99492
1.68	0.95352	2.13	0.98341	2.58	0.99506
1.69	0.95449	2.14	0.98382	2.59	0.99520
1.70	0.95543	2.15	0.98422	2.60	0.99534
1.71	0.95637	2.16	0.98461	2.61	0.99547
1.72	0.95728	2.17	0.98500	2.62	0.99560
1.73	0.95818	2.18	0.98537	2.63	0.99573
1.74	0.95907	2.19	0.98574	2.64	0.99585
1.75	0.95994	2.20	0.98610	2.65	0.99598
1.76	0.96080	2.21	0.98645	2.66	0.99609
1.77	0.96164	2.22	0.98679	2.67	0.99621
1.78	0.96246	2.23	0.98713	2.68	0.99632
1.79	0.96327	2.24	0.98745	2.69	0.99643
1.80	0.96407	2.25	0.98778	2.70	0.99653
1.81	0.96485	2.26	0.98809	2.71	0.99664
1.82	0.96562	2.27	0.98840	2.72	0.99674
1.83	0.96638	2.28	0.98870	2.73	0.99683
1.84	0.96712	2.29	0.98899	2.74	0.99693

z	$G(z)$	z	$G(z)$	z	$G(z)$
2.75	0.99702	3.20	0.99931	3.65	0.99987
2.76	0.99711	3.21	0.99934	3.66	0.99987
2.77	0.99720	3.22	0.99936	3.67	0.99988
2.78	0.99728	3.23	0.99938	3.68	0.99988
2.79	0.99736	3.24	0.99940	3.69	0.99989
2.80	0.99744	3.25	0.99942	3.70	0.99989
2.81	0.99752	3.26	0.99944	3.71	0.99990
2.82	0.99760	3.27	0.99946	3.72	0.99990
2.83	0.99767	3.28	0.99948	3.73	0.99990
2.84	0.99774	3.29	0.99950	3.74	0.99991
2.85	0.99781	3.30	0.99952	3.75	0.99991
2.86	0.99788	3.31	0.99953	3.76	0.99992
2.87	0.99795	3.32	0.99955	3.77	0.99992
2.88	0.99801	3.33	0.99957	3.78	0.99992
2.89	0.99807	3.34	0.99958	3.79	0.99992
2.90	0.99813	3.35	0.99960	3.80	0.99993
2.91	0.99819	3.36	0.99961	3.81	0.99993
2.92	0.99825	3.37	0.99962	3.82	0.99993
2.93	0.99831	3.38	0.99964	3.83	0.99994
2.94	0.99836	3.39	0.99965	3.84	0.99994
2.95	0.99841	3.40	0.99966	3.85	0.99994
2.96	0.99846	3.41	0.99968	3.86	0.99994
2.97	0.99851	3.42	0.99969	3.87	0.99995
2.98	0.99856	3.43	0.99970	3.88	0.99995
2.99	0.99861	3.44	0.99971	3.89	0.99995
3.00	0.99865	3.45	0.99972	3.90	0.99995
3.01	0.99869	3.46	0.99973	3.91	0.99995
3.02	0.99874	3.47	0.99974	3.92	0.99996
3.03	0.99878	3.48	0.99975	3.93	0.99996
3.04	0.99882	3.49	0.99976	3.94	0.99996
3.05	0.99886	3.50	0.99977	3.95	0.99996
3.06	0.99889	3.51	0.99978	3.96	0.99996
3.07	0.99893	3.52	0.99978	3.97	0.99996
3.08	0.99897	3.53	0.99979	3.98	0.99997
3.09	0.99900	3.54	0.99980	3.99	0.99997
3.10	0.99903	3.55	0.99981	4.00	0.99997
3.11	0.99906	3.56	0.99981		
3.12	0.99910	3.57	0.99982		
3.13	0.99913	3.58	0.99983		
3.14	0.99916	3.59	0.99983		
3.15	0.99918	3.60	0.99984		
3.16	0.99921	3.61	0.99985		
3.17	0.99924	3.62	0.99985		
3.18	0.99926	3.63	0.99986		
3.19	0.99929	3.64	0.99986		

APPENDIX 4

CUMULATIVE CHI-SQUARE DISTRIBUTION

| ν | p ||||||||||
	0.0005	0.001	0.005	0.01	0.025	0.05	0.10	0.20	0.30	0.40
1	0.0^6393	0.0^5157	0.0^4393	0.0^3157	0.0^3982	0.0^2393	0.0158	0.0642	0.148	0.275
2	0.0^2100	0.0^2200	0.0100	0.0201	0.0506	0.103	0.211	0.446	0.713	1.02
3	0.0153	0.0243	0.0717	0.115	0.216	0.352	0.584	1.00	1.42	1.87
4	0.0639	0.0908	0.207	0.297	0.484	0.711	1.06	1.65	2.19	2.75
5	0.158	0.210	0.412	0.554	0.831	1.15	1.61	2.34	3.00	3.66
6	0.299	0.381	0.676	0.872	1.24	1.64	2.20	3.07	3.83	4.57
7	0.485	0.598	0.989	1.24	1.69	2.17	2.83	3.82	4.67	5.49
8	0.710	0.857	1.34	1.65	2.18	2.73	3.49	4.59	5.53	6.42
9	0.972	1.15	1.73	2.09	2.70	3.33	4.17	5.38	6.39	7.36
10	1.26	1.48	2.16	2.56	3.25	3.94	4.87	6.18	7.27	8.30
11	1.59	1.83	2.60	3.05	3.82	4.57	5.58	6.99	8.15	9.24
12	1.93	2.21	3.07	3.57	4.40	5.23	6.30	7.81	9.03	10.2
13	2.31	2.62	3.57	4.11	5.01	5.89	7.04	8.63	9.93	11.1
14	2.70	3.04	4.07	4.66	5.63	6.57	7.79	9.47	10.8	12.1
15	3.11	3.48	4.60	5.23	6.26	7.26	8.55	10.3	11.7	13.0
16	3.54	3.94	5.14	5.81	6.91	7.96	9.31	11.2	12.6	14.0
17	3.98	4.42	5.70	6.41	7.56	8.67	10.1	12.0	13.5	14.9
18	4.44	4.90	6.26	7.01	8.23	9.39	10.9	12.9	14.4	15.9
19	4.91	5.41	6.84	7.63	8.91	10.1	11.7	13.7	15.4	16.9
20	5.40	5.92	7.43	8.26	9.59	10.9	12.4	14.6	16.3	17.8
21	5.90	6.45	8.03	8.90	10.3	11.6	13.2	15.4	17.2	18.8
22	6.40	6.98	8.64	9.54	11.0	12.3	14.0	16.3	18.1	19.7
23	6.92	7.53	9.26	10.2	11.7	13.1	14.8	17.2	19.0	20.7
24	7.45	8.08	9.89	10.9	12.4	13.8	15.7	18.1	19.9	21.7
25	7.99	8.65	10.5	11.5	13.1	14.6	16.5	18.9	20.9	22.6
26	8.54	9.22	11.2	12.2	13.8	15.4	17.3	19.8	21.8	23.6
27	9.09	9.80	11.8	12.9	14.6	16.2	18.1	20.7	22.7	24.5
28	9.66	10.4	12.5	13.6	15.3	16.9	18.9	21.6	23.6	25.5
29	10.2	11.0	13.1	14.3	16.0	17.7	19.8	22.5	24.6	26.5
30	10.8	11.6	13.8	15.0	16.8	18.5	20.6	23.4	25.5	27.4
31	11.4	12.2	14.5	15.7	17.5	19.3	21.4	24.3	26.4	28.4
32	12.0	12.8	15.1	16.4	18.3	20.1	22.3	25.1	27.4	29.4
33	12.6	13.4	15.7	17.1	19.0	20.9	23.1	26.0	28.3	30.3
34	13.2	14.1	16.5	17.8	19.8	21.7	24.0	26.9	29.2	31.3
35	13.8	14.7	17.2	18.5	20.6	22.5	24.8	27.8	30.2	32.3
36	14.4	15.3	17.9	19.2	21.3	23.3	25.6	28.7	31.1	33.3
37	15.0	16.0	18.6	20.0	22.1	24.1	26.5	29.6	32.1	34.2
38	15.6	16.6	19.3	20.7	22.9	24.9	27.3	30.5	33.0	35.2
39	16.3	17.3	20.0	21.4	23.7	25.7	28.2	31.4	33.9	36.2
40	16.9	17.9	20.7	22.2	24.4	26.5	29.1	32.3	34.9	37.1
41	17.5	18.6	21.4	22.9	25.2	27.3	29.9	33.3	35.8	38.1
42	18.2	19.2	22.1	23.7	26.0	28.1	30.8	34.2	36.8	39.1
43	18.8	19.9	22.9	24.4	26.8	29.0	31.6	35.1	37.7	40.0
44	19.5	20.6	23.6	25.1	27.6	29.8	32.5	36.0	38.6	41.0
45	20.1	21.3	24.3	25.9	28.4	30.6	33.4	36.9	39.6	42.0
46	20.8	21.9	25.0	26.7	29.2	31.4	34.2	37.8	40.5	43.0
47	21.5	22.6	25.8	27.4	30.0	32.3	35.1	38.7	41.5	43.9
48	22.1	23.3	26.5	28.2	30.8	33.1	35.9	39.6	42.4	44.9
49	22.8	24.0	27.2	28.9	31.6	33.9	36.8	40.5	43.4	45.9
50	23.5	24.7	28.0	29.7	32.4	34.8	37.7	41.4	44.3	46.9

Source: Adapted from A. Hald and S. A. Sinkbaek, A table of percentage points of the χ^2-distribution, *Skandinavisk Aktuarietidskrift*, 1950, pp. 170–75. By permission of the authors and publishers.

Notes:

(1) Entries in the table are values of χ_p^2 where $p = P(\chi^2 \leq \chi_p^2)$.

(2) For $\nu > 100$, use $\chi_p^2 \cong (z_p + \sqrt{2\nu - 1})^2/2$ where z_p is the $100\,p$th percentile of the standard normal distribution.

(3) Read 0.0^6393 as 0.000000393, etc.

ν	p 0.50	0.60	0.70	0.80	0.90	0.95	0.975	0.99	0.995	0.999	0.9995
1	0.455	0.708	1.07	1.64	2.71	3.84	5.02	6.63	7.88	10.8	12.1
2	1.39	1.83	2.41	3.22	4.61	5.99	7.38	9.21	10.6	13.8	15.2
3	2.37	2.95	3.67	4.64	6.25	7.81	9.35	11.3	12.8	16.3	17.7
4	3.36	4.04	4.88	5.99	7.78	9.49	11.1	13.3	14.9	18.5	20.0
5	4.35	5.13	6.06	7.29	9.24	11.1	12.8	15.1	16.7	20.5	22.1
6	5.35	6.21	7.23	8.56	10.6	12.6	14.4	16.8	18.5	22.5	24.1
7	6.35	7.28	8.38	9.80	12.0	14.1	16.0	18.5	20.3	24.3	26.0
8	7.34	8.35	9.52	11.0	13.4	15.5	17.5	20.1	22.0	26.1	27.9
9	8.34	9.41	10.7	12.2	14.7	16.9	19.0	21.7	23.6	27.9	29.7
10	9.34	10.5	11.8	13.4	16.0	18.3	20.5	23.2	25.2	29.6	31.4
11	10.3	11.5	12.9	14.6	17.3	19.7	21.9	24.7	26.8	31.3	33.1
12	11.3	12.6	14.0	15.8	18.5	21.0	23.3	26.2	28.3	32.9	34.8
13	12.3	13.6	15.1	17.0	19.8	22.4	24.7	27.7	29.8	34.5	36.5
14	13.3	14.7	16.2	18.2	21.1	23.7	26.1	29.1	31.3	36.1	38.1
15	14.3	15.7	17.3	19.3	22.3	25.0	27.5	30.6	32.8	37.7	39.7
16	15.3	16.8	18.4	20.5	23.5	26.3	28.8	32.0	34.3	39.3	41.3
17	16.3	17.8	19.5	21.6	24.8	27.6	30.2	33.4	35.7	40.8	42.9
18	17.3	18.9	20.6	22.8	26.0	28.9	31.5	34.8	37.2	42.3	44.4
19	18.3	19.9	21.7	23.9	27.2	30.1	32.9	36.2	38.6	43.8	46.0
20	19.3	21.0	22.8	25.0	28.4	31.4	34.2	37.6	40.0	45.3	47.5
21	20.3	22.0	23.9	26.2	29.6	32.7	35.5	38.9	41.4	46.8	49.0
22	21.3	23.0	24.9	27.3	30.8	33.9	36.8	40.3	42.8	48.3	50.5
23	22.3	24.1	26.0	28.4	32.0	35.2	38.1	41.6	44.2	49.7	52.0
24	23.3	25.1	27.1	29.6	33.2	36.4	39.4	43.0	45.6	51.2	53.5
25	24.3	26.1	28.2	30.7	34.4	37.7	40.6	44.3	46.9	52.6	54.9
26	25.3	27.2	29.2	31.8	35.6	38.9	41.9	45.6	48.3	54.1	56.4
27	26.3	28.2	30.3	32.9	36.7	40.1	43.2	47.0	49.6	55.5	57.9
28	27.3	29.2	31.4	34.0	37.9	41.3	44.5	48.3	51.0	56.9	59.3
29	28.3	30.3	32.5	35.1	39.1	42.6	45.7	49.6	52.3	58.3	60.7
30	29.3	31.3	33.5	36.3	40.3	43.8	47.0	50.9	53.7	59.7	62.2
31	30.3	32.3	34.6	37.4	41.4	45.0	48.2	52.2	55.0	61.1	63.6
32	31.3	33.4	35.7	38.5	42.6	46.2	49.5	53.5	56.3	62.5	65.0
33	32.3	34.4	36.7	39.6	43.7	47.4	50.7	54.8	57.6	63.9	66.4
34	33.3	35.4	37.8	40.7	44.9	48.6	52.0	56.1	59.0	65.2	67.8
35	34.3	36.5	38.9	41.8	46.1	49.8	53.2	57.3	60.3	66.6	69.2
36	35.3	37.5	39.9	42.9	47.2	51.0	54.4	58.6	61.6	68.0	70.6
37	36.3	38.5	41.0	44.0	48.4	52.2	55.7	59.9	62.9	69.3	72.0
38	37.3	39.6	42.0	45.1	49.5	53.4	56.9	61.2	64.2	70.7	73.4
39	38.3	40.6	43.1	46.2	50.7	54.6	58.1	62.4	65.5	72.1	74.7
40	39.3	41.6	44.2	47.3	51.8	55.8	59.3	63.7	66.8	73.4	76.1
41	40.3	42.7	45.2	48.4	52.9	56.9	60.6	65.0	68.1	74.7	77.5
42	41.3	43.7	46.3	49.5	54.1	58.1	61.8	66.2	69.3	76.1	78.8
43	42.3	44.7	47.3	50.5	55.2	59.3	63.0	67.5	70.6	77.4	80.2
44	43.3	45.7	48.4	51.6	56.4	60.5	64.2	68.7	71.9	78.7	81.5
45	44.3	46.8	49.5	52.7	57.5	61.7	65.4	70.0	73.2	80.1	82.9
46	45.3	47.8	50.5	53.8	58.6	62.8	66.6	71.2	74.4	81.4	84.2
47	46.3	48.8	51.6	54.9	59.8	64.0	67.8	72.4	75.7	82.7	85.6
48	47.3	49.8	52.6	56.0	60.9	65.2	69.0	73.7	77.0	84.0	86.9
49	48.3	50.9	53.7	57.1	62.0	66.3	70.2	74.9	78.2	85.4	88.2
50	49.3	51.9	54.7	58.2	63.2	67.5	71.4	76.2	79.5	86.7	89.6

| ν | \multicolumn{11}{c}{p} |
	0.0005	0.001	0.005	0.01	0.025	0.05	0.10	0.20	0.30	0.40	0.50
51	24.1	25.4	28.7	30.5	33.2	35.6	38.6	42.4	45.3	47.8	50.3
52	24.8	26.1	29.5	31.2	34.0	36.4	39.4	43.3	46.2	48.8	51.3
53	25.5	26.8	30.2	32.0	34.8	37.3	40.3	44.2	47.2	49.8	52.3
54	26.2	27.5	31.0	32.8	35.6	38.1	41.2	45.1	48.1	50.8	53.3
55	26.9	28.2	31.7	33.6	36.4	39.0	42.1	46.0	49.1	51.7	54.3
56	27.6	28.9	32.5	34.3	37.2	39.8	42.9	47.0	50.0	52.7	55.3
57	28.2	29.6	33.2	35.1	38.0	40.6	43.8	47.9	51.0	53.7	56.3
58	28.9	30.3	34.0	35.9	38.8	41.5	44.7	48.8	51.9	54.7	57.3
59	29.6	31.0	34.8	36.7	39.7	42.3	45.6	49.7	52.9	55.6	58.3
60	30.3	31.7	35.5	37.5	40.5	43.2	46.5	50.6	53.8	56.6	59.3
61	31.0	32.5	36.3	38.3	41.3	44.0	47.3	51.6	54.8	57.6	60.3
62	31.7	33.2	37.1	39.1	42.1	44.9	48.2	52.5	55.7	58.6	61.3
63	32.5	33.9	37.8	39.9	43.0	45.7	49.1	53.4	56.7	59.6	62.3
64	33.2	34.6	38.6	40.6	43.8	46.6	50.0	54.3	57.6	60.5	63.3
65	33.9	35.4	39.4	41.4	44.6	47.4	50.9	55.3	58.6	61.5	64.3
66	34.6	36.1	40.2	42.2	45.4	48.3	51.8	56.2	59.5	62.5	65.3
67	35.3	36.8	40.9	43.0	46.3	49.2	52.7	57.1	60.5	63.5	66.3
68	36.0	37.6	41.7	43.8	47.1	50.0	53.5	58.0	61.4	64.4	67.3
69	36.7	38.3	42.5	44.6	47.9	50.9	54.4	59.0	62.4	65.4	68.3
70	37.5	39.0	43.3	45.4	48.8	51.7	55.3	59.9	63.3	66.4	69.3
71	38.2	39.8	44.1	46.2	49.6	52.6	56.2	60.8	64.3	67.4	70.3
72	38.9	40.5	44.8	47.1	50.4	53.5	57.1	61.8	65.3	68.4	71.3
73	39.6	41.3	45.6	47.9	51.3	54.3	58.0	62.7	66.2	69.3	72.3
74	40.4	42.0	46.4	48.7	52.1	55.2	58.9	63.6	67.2	70.3	73.3
75	41.1	42.8	47.2	49.5	52.9	56.1	59.8	64.5	68.1	71.3	74.3
76	41.8	43.5	48.0	50.3	53.8	56.9	60.7	65.5	69.1	72.3	75.3
77	42.6	44.3	48.8	51.1	54.6	57.8	61.6	66.4	70.0	73.2	76.3
78	43.3	45.0	49.6	51.9	55.5	58.7	62.5	67.3	71.0	74.2	77.3
79	44.1	45.8	50.4	52.7	56.3	59.5	63.4	68.3	72.0	75.2	78.3
80	44.8	46.5	51.2	53.5	57.2	60.4	64.3	69.2	72.9	76.2	79.3
81	45.5	47.3	52.0	54.4	58.0	61.3	65.2	70.1	73.9	77.2	80.3
82	46.3	48.0	52.8	55.2	58.8	62.1	66.1	71.1	74.8	78.1	81.3
83	47.0	48.8	53.6	56.0	59.7	63.0	67.0	72.0	75.8	79.1	82.3
84	47.8	49.6	54.4	56.8	60.5	63.9	67.9	72.9	76.8	80.1	83.3
85	48.5	50.3	55.2	57.6	61.4	64.7	68.8	73.9	77.7	81.1	84.3
86	49.3	51.1	56.0	58.5	62.2	65.6	69.7	74.8	78.7	82.1	85.3
87	50.0	51.9	56.8	59.3	63.1	66.5	70.6	75.7	79.6	83.0	86.3
88	50.8	52.6	57.6	60.1	63.9	67.4	71.5	76.7	80.6	84.0	87.3
89	51.5	53.4	58.4	60.9	64.8	68.2	72.4	77.6	81.6	85.0	88.3
90	52.3	54.2	59.2	61.8	65.6	69.1	73.3	78.6	82.5	86.0	89.3
91	53.0	54.9	60.0	62.6	66.5	70.0	74.2	79.5	83.5	87.0	90.3
92	53.8	55.7	60.8	63.4	67.4	70.9	75.1	80.4	84.4	88.0	91.3
93	54.5	56.5	61.6	64.2	68.2	71.8	76.0	81.4	85.4	88.9	92.3
94	55.3	57.2	62.4	65.1	69.1	72.6	76.9	82.3	86.4	89.9	93.3
95	56.1	58.0	63.2	65.9	69.9	73.5	77.8	83.2	87.3	90.9	94.3
96	56.8	58.8	64.1	66.7	70.8	74.4	78.7	84.2	88.3	91.9	95.3
97	57.6	59.6	64.9	67.6	71.6	75.3	79.6	85.1	89.2	92.9	96.3
98	58.4	60.4	65.7	68.4	72.5	76.2	80.5	86.1	90.2	93.8	97.3
99	59.1	61.1	66.5	69.2	73.4	77.0	81.4	87.0	91.2	94.8	98.3
100	59.9	61.9	67.3	70.1	74.2	77.9	82.4	87.9	92.1	95.8	99.3

ν	0.60	0.70	0.80	0.90	0.95	0.975	0.99	0.995	0.999	0.9995
						p				
51	52.9	55.8	59.2	64.3	68.7	72.6	77.4	80.7	88.0	90.9
52	53.9	56.8	60.3	65.4	69.8	73.8	78.6	82.0	89.3	92.2
53	55.0	57.9	61.4	66.5	71.0	75.0	79.8	83.3	90.6	93.5
54	56.0	58.9	62.5	67.7	72.2	76.2	81.1	84.5	91.9	94.8
55	57.0	60.0	63.6	68.8	73.3	77.4	82.3	85.7	93.2	96.2
56	58.0	61.0	64.7	69.9	74.5	78.6	83.5	87.0	94.5	97.5
57	59.1	62.1	65.7	71.0	75.6	79.8	84.7	88.2	95.8	98.8
58	60.1	63.1	66.8	72.2	76.8	80.9	86.0	89.5	97.0	100.1
59	61.1	64.2	67.9	73.3	77.9	82.1	87.2	90.7	98.3	101.4
60	62.1	65.2	69.0	74.4	79.1	83.3	88.4	92.0	99.6	102.7
61	63.2	66.3	70.0	75.5	80.2	84.5	89.6	93.2	100.9	104.0
62	64.2	67.3	71.1	76.6	81.4	85.7	90.8	94.4	102.2	105.3
63	65.2	68.4	72.2	77.7	82.5	86.8	92.0	95.6	103.4	106.6
64	66.2	69.4	73.3	78.9	83.7	88.0	93.2	96.9	104.7	107.9
65	67.2	70.5	74.4	80.0	84.8	89.2	94.4	98.1	106.0	109.2
66	68.3	71.5	75.4	81.1	86.0	90.3	95.6	99.3	107.3	110.5
67	69.3	72.6	76.5	82.2	87.1	91.5	96.8	100.6	108.5	111.7
68	70.3	73.6	77.6	83.3	88.3	92.7	98.0	101.8	109.8	113.0
69	71.3	74.6	78.6	84.4	89.4	93.9	99.2	103.0	111.1	114.3
70	72.4	75.7	79.7	85.5	90.5	95.0	100.4	104.2	112.3	115.6
71	73.4	76.7	80.8	86.6	91.7	96.2	101.6	105.4	113.6	116.9
72	74.4	77.8	81.9	87.7	92.8	97.4	102.8	106.6	114.8	118.1
73	75.4	78.8	82.9	88.8	93.9	98.5	104.0	107.9	116.1	119.4
74	76.4	79.9	84.0	90.0	95.1	99.7	105.2	109.1	117.3	120.7
75	77.5	80.9	85.1	91.1	96.2	100.8	106.4	110.3	118.6	121.9
76	78.5	82.0	86.1	92.2	97.4	102.0	107.6	111.5	119.9	123.2
77	79.5	83.0	87.2	93.3	98.5	103.2	108.8	112.7	121.1	124.5
78	80.5	84.0	88.3	94.4	99.6	104.3	110.0	113.9	122.3	125.7
79	81.5	85.1	89.3	95.5	100.7	105.5	111.1	115.1	123.6	127.0
80	82.6	86.1	90.4	96.6	101.9	106.6	112.3	116.3	124.8	128.3
81	83.6	87.2	91.5	97.7	103.0	107.8	113.5	117.5	126.1	129.5
82	84.6	88.2	92.5	98.8	104.1	108.9	114.7	118.7	127.3	130.8
83	85.6	89.2	93.6	99.9	105.3	110.1	115.9	119.9	128.6	132.0
84	86.6	90.3	94.7	101.0	106.4	111.2	117.1	121.1	129.8	133.3
85	87.7	91.3	95.7	102.1	107.5	112.4	118.2	122.3	131.0	134.5
86	88.7	92.4	96.8	103.2	108.6	113.5	119.4	123.5	132.3	135.8
87	89.7	93.4	97.9	104.3	109.8	114.7	120.6	124.7	133.5	137.0
88	90.7	94.4	98.9	105.4	110.9	115.8	121.8	125.9	134.7	138.3
89	91.7	95.5	100.0	106.5	112.0	117.0	122.9	127.1	136.0	139.5
90	92.8	96.5	101.1	107.6	113.1	118.1	124.1	128.3	137.2	140.8
91	93.8	97.6	102.1	108.7	114.3	119.3	125.3	129.5	138.4	142.0
92	94.8	98.6	103.2	109.8	115.4	120.4	126.5	130.7	139.7	143.3
93	95.8	99.6	104.2	110.9	116.5	121.6	127.6	131.9	140.9	144.5
94	96.8	100.7	105.3	111.9	117.6	122.7	128.8	133.1	142.1	145.8
95	97.9	101.7	106.4	113.0	118.8	123.9	130.0	134.2	143.3	147.0
96	98.9	102.8	107.4	114.1	119.9	125.0	131.1	135.4	144.6	148.2
97	99.9	103.8	108.5	115.2	121.0	126.1	132.3	136.6	145.8	149.5
98	100.9	104.8	109.5	116.3	122.1	127.3	133.5	137.8	147.0	150.7
99	101.9	105.9	110.6	117.4	123.2	128.4	134.6	139.0	148.2	151.9
100	102.9	106.9	111.7	118.5	124.3	129.6	135.8	140.2	149.4	153.2

CUMULATIVE *t* DISTRIBUTION

				p				
ν	0.75	0.80	0.85	0.90	0.95	0.975	0.995	0.9995
1	1.0005	1.376	1.963	3.078	6.314	12.706	63.657	636.619
2	0.816	1.061	1.386	1.886	2.920	4.303	9.925	31.598
3	0.765	0.978	1.250	1.638	2.353	3.182	5.841	12.941
4	0.741	0.941	1.190	1.533	2.132	2.776	4.604	8.610
5	0.727	0.920	1.156	1.476	2.015	2.571	4.032	6.859
6	0.718	0.906	1.134	1.440	1.943	2.447	3.707	5.959
7	0.711	0.896	1.119	1.415	1.895	2.365	3.499	5.405
8	0.706	0.889	1.108	1.397	1.860	2.306	3.355	5.041
9	0.703	0.883	1.100	1.383	1.833	2.262	3.250	4.781
10	0.700	0.879	1.093	1.372	1.812	2.228	3.169	4.587
11	0.697	0.876	1.088	1.363	1.796	2.201	3.106	4.437
12	0.695	0.873	1.083	1.356	1.782	2.179	3.055	4.318
13	0.694	0.870	1.079	1.350	1.771	2.160	3.012	4.221
14	0.692	0.868	1.076	1.345	1.761	2.145	2.977	4.140
15	0.691	0.866	1.074	1.341	1.753	2.131	2.947	4.073
16	0.690	0.866	1.071	1.337	1.746	2.120	2.921	4.015
17	0.689	0.863	1.069	1.333	1.740	2.110	2.898	3.965
18	0.688	0.862	1.067	1.330	1.734	2.101	2.878	3.922
19	0.688	0.861	1.066	1.328	1.729	2.093	2.861	3.883
20	0.687	0.860	1.064	1.325	1.725	2.086	2.845	3.850
21	0.686	0.859	1.063	1.323	1.721	2.080	2.831	3.819
22	0.686	0.858	1.061	1.321	1.717	2.074	2.819	3.792
23	0.685	0.858	1.060	1.319	1.714	2.069	2.807	3.767
24	0.685	0.857	1.059	1.318	1.711	2.064	2.797	3.745
25	0.684	0.856	1.058	1.316	1.708	2.060	2.787	3.725
26	0.684	0.856	1.058	1.315	1.706	2.056	2.779	3.707
27	0.684	0.855	1.057	1.314	1.703	2.052	2.771	3.690
28	0.683	0.855	1.056	1.313	1.701	2.048	2.763	3.674
29	0.683	0.854	1.055	1.311	1.699	2.045	2.756	3.659
30	0.683	0.854	1.055	1.310	1.697	2.042	2.750	3.646
35	0.682	0.852	1.052	1.306	1.690	2.030	2.724	3.591
40	0.681	0.851	1.050	1.303	1.684	2.021	2.704	3.551
45	0.680	0.850	1.048	1.301	1.680	2.014	2.690	3.520
50	0.680	0.849	1.047	1.299	1.676	2.008	2.678	3.496
55	0.679	0.849	1.047	1.297	1.673	2.004	2.669	3.476
60	0.679	0.848	1.046	1.296	1.671	2.000	2.660	3.460
70	0.678	0.847	1.045	1.294	1.667	1.994	2.648	3.435
80	0.678	0.847	1.044	1.293	1.665	1.990	2.638	3.416
90	0.678	0.846	1.043	1.291	1.662	1.987	2.632	3.402
100	0.677	0.846	1.042	1.290	1.661	1.984	2.626	3.390
200	0.676	0.844	1.039	1.286	1.653	1.972	2.601	3.340
300	0.676	0.843	1.038	1.285	1.650	1.968	2.592	3.323
400	0.676	0.843	1.038	1.284	1.649	1.966	2.588	3.315
500	0.676	0.843	1.037	1.284	1.648	1.965	2.586	3.310
1000	0.675	0.842	1.037	1.283	1.647	1.962	2.581	3.301
∞	0.67449	0.84162	1.03643	1.28155	1.64485	1.95996	2.57582	3.29053

Source: Partly from Table III of R. A. Fisher and Frank Yates, *Statistical Tables for Biological, Agricultural and Medical Research*, 3rd ed., Oliver and Boyd, Edinburgh, 1948. By permission of the authors and publishers.

Note: Entries in the table are values of t_p where $p = P(t \leq t_p)$.

APPENDIX 6

CUMULATIVE *F* DISTRIBUTION

Sources: Reproduced from Table A–7c of W. J. Dixon and F. J. Massey, *Introduction to Statistical Analysis,* 2nd ed., McGraw-Hill, New York, 1957. By permission of the authors and publishers. However, since most of the values in Dixon and Massey were extracted from other publications, permission was also requested of the primary sources noted below. In each case, permission was granted to reproduce the needed material.

(a) All values for ν_1, ν_2 equal to 50, 100, 200, and 500 are from A. Hald, *Statistical Tables and Formulas,* Wiley, New York, 1952.

(b) For cumulative proportions 0.5, 0.75, 0.9, 0.95, 0.975, 0.99, and 0.995, most of the values are from M. Merrington and C. M. Thompson, Tables of percentage points of the inverted beta (*F*) distribution, *Biometrika* 33(Apr.): 74–87, 1943.

(c) For cumulative proportions 0.999, the values are from C. C. Colcord and L. S. Deming, The one-tenth percent level of Z, *Sankhyā* 2(Dec.): 423–24, 1936.

(d) As noted in Dixon and Massey, the remaining values were found by forming reciprocals or by interpolation.

Note: Entries in the table are values of F_p where $p = P(F \leq F_p)$.

ν_2 / p	ν_1 1	2	3	4	5	6	7	8	9	10	11	12
1 .0005	$.0^62$	$.0^350$	$.0^238$	$.0^294$	.016	.022	.027	.032	.036	.039	.042	.045
.001	$.0^525$	$.0^210$	$.0^260$	.013	.021	.028	.034	.039	.044	.048	.051	.054
.005	$.0^462$	$.0^251$	.018	.032	.044	.054	.062	.068	.073	.078	.082	.085
.010	$.0^325$	.010	.029	.047	.062	.073	.082	.089	.095	.100	.104	.107
.025	$.0^315$	.026	.057	.082	.100	.113	.124	.132	.139	.144	.149	.153
.05	$.0^262$	.054	.099	.130	.151	.167	.179	.188	.195	.201	.207	.211
.10	.025	.117	.181	.220	.246	.265	.279	.289	.298	.304	.310	.315
.25	.172	.389	.494	.553	.591	.617	.637	.650	.661	.670	.680	.684
.50	1.00	1.50	1.71	1.82	1.89	1.94	1.98	2.00	2.03	2.04	2.05	2.07
.75	5.83	7.50	8.20	8.58	8.82	8.98	9.10	9.19	9.26	9.32	9.36	9.41
.90	39.9	49.5	53.6	55.8	57.2	58.2	58.9	59.4	59.9	60.2	60.5	60.7
.95	161	200	216	225	230	234	237	239	241	242	243	244
.975	648	800	864	900	922	937	948	957	963	969	973	977
.99	405^1	500^1	540^1	562^1	576^1	586^1	593^1	598^1	602^1	606^1	608^1	611^1
.995	162^2	200^2	216^2	225^2	231^2	234^2	237^2	239^2	241^2	242^2	243^2	244^2
.999	406^3	500^3	540^3	562^3	576^3	586^3	593^3	598^3	602^3	606^3	609^3	611
.9995	162^4	200^4	216^4	225^4	231^4	234^4	237^4	239^4	241^4	242^4	243^4	244^4
2 .0005	$.0^650$	$.0^350$	$.0^242$	.011	.020	.029	.037	.044	.050	.056	.061	.065
.001	$.0^520$	$.0^210$	$.0^268$	.016	.027	.037	.046	.054	.061	.067	.072	.077
.005	$.0^450$	$.0^250$	.020	.038	.055	.069	.081	.091	.099	.106	.112	.118
.01	$.0^320$	.010	.032	.056	.075	.092	.105	.116	.125	.132	.139	.144
.025	$.0^213$	.026	.062	.094	.119	.138	.153	.165	.175	.183	.190	.196
.05	$.0^250$	.053	.105	.144	.173	.194	.211	.224	.235	.244	.251	.257
.10	.020	.111	.183	.231	.265	.289	.307	.321	.333	.342	.350	.356
.25	.133	.333	.439	.500	.540	.568	.588	.604	.616	.626	.633	.641
.50	.667	1.00	1.13	1.21	1.25	1.28	1.30	1.32	1.33	1.34	1.35	1.36
.75	2.57	3.00	3.15	3.23	3.28	3.31	3.34	3.35	3.37	3.38	3.39	3.39
.90	8.53	9.00	9.16	9.24	9.29	9.33	9.35	9.37	9.38	9.39	9.40	9.41
.95	18.5	19.0	19.2	19.2	19.3	19.3	19.4	19.4	19.4	19.4	19.4	19.4
.975	38.5	39.0	39.2	39.2	39.3	39.3	39.4	39.4	39.4	39.4	39.4	39.4
.99	98.5	99.0	99.2	99.2	99.3	99.3	99.4	99.4	99.4	99.4	99.4	99.4
.995	198	199	199	199	199	199	199	199	199	199	199	199
.999	998	999	999	999	999	999	999	999	999	999	999	999
.9995	200^1	200^1	200^1	200^1	200^1	200^1	200^1	200^1	200^1	200^1	200^1	200^1
3 .0005	$.0^646$	$.0^350$	$.0^244$	.012	.023	.033	.043	.052	.060	.067	.074	.079
.001	$.0^819$	$.0^210$	$.0^271$	.018	.030	.042	.053	.063	.072	.079	.086	.093
.005	$.0^446$	$.0^250$	.021	.041	.060	.077	.092	.104	.115	.124	.132	.138
.01	$.0^319$	.010	.034	.060	.083	.102	.118	.132	.143	.153	.161	.168
.025	$.0^212$	.026	.065	.100	.129	.152	.170	.185	.197	.207	.216	.224
.05	$.0^246$	.052	.108	.152	.185	.210	.230	.246	.259	.270	.279	.287
.10	.019	.109	.185	.239	.276	.304	.325	.342	.356	.367	.376	.384
.25	.122	.317	.424	.489	.531	.561	.582	.600	.613	.624	.633	.641
.50	.585	.881	1.00	1.06	1.10	1.13	1.15	1.16	1.17	1.18	1.19	1.20
.75	2.02	2.28	2.36	2.39	2.41	2.42	2.43	2.44	2.44	2.44	2.45	2.45
.90	5.54	5.46	5.39	5.34	5.31	5.28	5.27	5.25	5.24	5.23	5.22	5.22
.95	10.1	9.55	9.28	9.12	9.01	8.94	8.89	8.85	8.81	8.79	8.76	8.74
.975	17.4	16.0	15.4	15.1	14.9	14.7	14.6	14.5	14.5	14.4	14.4	14.3
.99	34.1	30.8	29.5	28.7	28.2	27.9	27.7	27.5	27.3	27.2	27.1	27.1
.995	55.6	49.8	47.5	46.2	45.4	44.8	44.4	44.1	43.9	43.7	43.5	43.4
.999	167	149	141	137	135	133	132	131	130	129	129	128
.9995	266	237	225	218	214	211	209	208	207	206	204	204

Read $.0^356$ as .00056, 200^1 as 2000, 162^4 as 1620000, etc.

ν_2	p	15	20	24	30	40	50	60	100	120	200	500	∞
1	.0005	.051	.058	.062	.066	.069	.072	.074	.077	.078	.080	.081	.083
	.001	.060	.067	.071	.075	.079	.082	.084	.087	.088	.089	.091	.092
	.005	.093	.101	.105	.109	.113	.116	.118	.121	.122	.124	.126	.127
	.01	.115	.124	.128	.132	.137	.139	.141	.145	.146	.148	.150	.151
	.025	.161	.170	.175	.180	.184	.187	.189	.193	.194	.196	.198	.199
	.05	.220	.230	.235	.240	.245	.248	.250	.254	.255	.257	.259	.261
	.10	.325	.336	.342	.347	.353	.356	.358	.362	.364	.366	.368	.370
	.25	.698	.712	.719	.727	.734	.738	.741	.747	.749	.752	.754	.756
	.50	2.09	2.12	2.13	2.15	2.16	2.17	2.17	2.18	2.18	2.19	2.19	2.20
	.75	9.49	9.58	9.63	9.67	9.71	9.74	9.76	9.78	9.80	9.82	9.84	9.85
	.90	61.2	61.7	62.0	62.3	62.5	62.7	62.8	63.0	63.1	63.2	63.3	63.3
	.95	246	248	249	250	251	252	252	253	253	254	254	254
	.975	985	993	997	100^1	101^1	101^1	101^1	101^1	101^1	102^1	102^1	102^1
	.99	616^1	621^1	623^1	626^1	629^1	630^1	631^1	633^1	634^1	635^1	636^1	637^1
	.995	246^2	248^2	249^2	250^2	251^2	252^2	253^2	253^2	254^2	254^2	254^2	255^2
	.999	616^3	621^3	623^3	626^3	629^3	630^3	631^3	633^3	634^3	635^3	636^3	637^3
	.9995	246^4	248^4	249^4	250^4	251^4	252^4	252^4	253^4	253^4	253^4	254^4	254^4
2	.0005	.076	.088	.094	.101	.108	.113	.116	.122	.124	.127	.130	.132
	.001	.088	.100	.107	.114	.121	.126	.129	.135	.137	.140	.143	.145
	.005	.130	.143	.150	.157	.165	.169	.173	.179	.181	.184	.187	.189
	.01	.157	.171	.178	.186	.193	.198	.201	.207	.209	.212	.215	.217
	.025	.210	.224	.232	.239	.247	.251	.255	.261	.263	.266	.269	.271
	.05	.272	.286	.294	.302	.309	.314	.317	.324	.326	.329	.332	.334
	.10	.371	.386	.394	.402	.410	.415	.418	.424	.426	.429	.433	.434
	.25	.657	.672	.680	.689	.697	.702	.705	.711	.713	.716	.719	.721
	.50	1.38	1.39	1.40	1.41	1.42	1.42	1.43	1.43	1.43	1.44	1.44	1.44
	.75	3.41	3.43	3.43	3.44	3.45	3.45	3.46	3.47	3.47	3.48	3.48	3.48
	.90	9.42	9.44	9.45	9.46	9.47	9.47	9.47	9.48	9.48	9.49	9.49	9.49
	.95	19.4	19.4	19.5	19.5	19.5	19.5	19.5	19.5	19.5	19.5	19.5	19.5
	.975	39.4	39.4	39.5	39.5	39.5	39.5	39.5	39.5	39.5	39.5	39.5	39.5
	.99	99.4	99.4	99.5	99.5	99.5	99.5	99.5	99.5	99.5	99.5	99.5	99.5
	.995	199	199	199	199	199	199	199	199	199	199	199	200
	.999	999	999	999	999	999	999	999	999	999	999	999	999
	.9995	200^1	200^1	200^1	200^1	200^1	200^1	200^1	200^1	200^1	200^1	200^1	200^1
3	.0005	.093	.109	.117	.127	.136	.143	.147	.156	.158	.162	.166	.169
	.001	.107	.123	.132	.142	.152	.158	.162	.171	.173	.177	.181	.184
	.005	.154	.172	.181	.191	.201	.207	.211	.220	.222	.227	.231	.234
	.01	.185	.203	.212	.222	.232	.238	.242	.251	.253	.258	.262	.264
	.025	.241	.259	.269	.279	.289	.295	.299	.308	.310	.314	.318	.321
	.05	.304	.323	.332	.342	.352	.358	.363	.370	.373	.377	.382	.384
	.10	.402	.420	.430	.439	.449	.455	.459	.467	.469	.474	.476	.480
	.25	.658	.675	.684	.693	.702	.708	.711	.719	.721	.724	.728	.730
	.50	1.21	1.23	1.23	1.24	1.25	1.25	1.25	1.26	1.26	1.26	1.27	1.27
	.75	2.46	2.46	2.46	2.47	2.47	2.47	2.47	2.47	2.47	2.47	2.47	2.47
	.90	5.20	5.18	5.18	5.17	5.16	5.15	5.15	5.14	5.14	5.14	5.14	5.13
	.95	8.70	8.66	8.63	8.62	8.59	8.58	8.57	8.55	8.55	8.54	8.53	8.53
	.975	14.3	14.2	14.1	14.1	14.0	14.0	14.0	14.0	13.9	13.9	13.9	13.9
	.99	26.9	26.7	26.6	26.5	26.4	26.4	26.3	26.2	26.2	26.2	26.1	26.1
	.995	43.1	42.8	42.6	42.5	42.3	42.2	42.1	42.0	42.0	41.9	41.9	41.8
	.999	127	126	126	125	125	125	124	124	124	124	124	123
	.9995	203	201	200	199	199	198	198	197	197	197	197	196

ν_2	ν_1 / p	1	2	3	4	5	6	7	8	9	10	11	12
4	.0005	$.0^644$	$.0^550$	$.0^246$	.013	.024	.036	.047	.057	.066	.075	.082	.089
	.001	$.0^518$	$.0^210$	$.0^273$	.019	.032	.046	.058	.069	.079	.089	.097	.104
	.005	$.0^444$	$.0^250$	.022	.043	.064	.083	.100	.114	.126	.137	.145	.153
	.01	$.0^318$	.010	.035	.063	.088	.109	.127	.143	.156	.167	.176	.185
	.025	$.0^211$	.026	.066	.104	.135	.161	.181	.198	.212	.224	.234	.243
	05	$.0^244$	.052	.110	.157	.193	.221	.243	.261	.275	.288	.298	.307
	.10	.018	.108	.187	.243	.284	.314	.338	.356	.371	.384	.394	.403
	.25	.117	.309	.418	.484	.528	.560	.583	.601	.615	.627	.637	.645
	.50	.549	.828	.941	1.00	1.04	1.06	1.08	1.09	1.10	1.11	1.12	1.13
	.75	1.81	2.00	2.05	2.06	2.07	2.08	2.08	2.08	2.08	2.08	2.08	2.08
	.90	4.54	4.32	4.19	4.11	4.05	4.01	3.98	3.95	3.94	3.92	3.91	3.90
	.95	7.71	6.94	6.59	6.39	6.26	6.16	6.09	6.04	6.00	5.96	5.94	5.91
	.975	12.2	10.6	9.98	9.60	9.36	9.20	9.07	8.98	8.90	8.84	8.79	8.75
	.99	21.2	18.0	16.7	16.0	15.5	15.2	15.0	14.8	14.7	14.5	14.4	14.4
	.995	31.3	26.3	24.3	23.2	22.5	22.0	21.6	21.4	21.1	21.0	20.8	20.7
	.999	74.1	61.2	56.2	53.4	51.7	50.5	49.7	49.0	48.5	48.0	47.7	47.4
	.9995	106	87.4	80.1	76.1	73.6	71.9	70.6	69.7	68.9	68.3	67.8	67.4
5	.0005	$.0^643$	$.0^550$	$.0^247$	.014	.025	.038	.050	.061	.070	.081	.089	.096
	.001	$.0^517$	$.0^210$	$.0^275$	.019	.034	.048	.062	.074	.085	.095	.104	.112
	.005	$.0^443$	$.0^250$	.022	.045	.067	.087	.105	.120	.134	.146	.156	.165
	.01	$.0^317$	.010	.035	.064	.091	.114	.134	.151	.165	.177	.188	.197
	.025	$.0^211$	.025	.067	.107	.140	.167	.189	.208	.223	.236	.248	.257
	.05	$.0^243$	.052	.111	.160	.198	.228	.252	.271	.287	.301	.313	.322
	.10	.017	.108	.188	.247	.290	.322	.347	.367	.383	.397	.408	.418
	.25	.113	.305	.415	.483	.528	.560	.584	.604	.618	.631	.641	.650
	.50	.528	.799	.907	.965	1.00	1.02	1.04	1.05	1.06	1.07	1.08	1.09
	.75	1.69	1.85	1.88	1.89	1.89	1.89	1.89	1.89	1.89	1.89	1.89	1.89
	.90	4.06	3.78	3.62	3.52	3.45	3.40	3.37	3.34	3.32	3.30	3.28	3.27
	.95	6.61	5.79	5.41	5.19	5.05	4.95	4.88	4.82	4.77	4.74	4.71	4.68
	.975	10.0	8.43	7.76	7.39	7.15	6.98	6.85	6.76	6.68	6.62	6.57	6.52
	.99	16.3	13.3	12.1	11.4	11.0	10.7	10.5	10.3	10.2	10.1	9.96	9.89
	.995	22.8	18.3	16.5	15.6	14.9	14.5	14.2	14.0	13.8	13.6	13.5	13.4
	.999	47.2	37.1	33.2	31.1	29.7	28.8	28.2	27.6	27.2	26.9	26.6	26.4
	.9995	63.6	49.8	44.4	41.5	39.7	38.5	37.6	36.9	36.4	35.9	35.6	35.2
6	.0005	$.0^643$	$.0^550$	$.0^247$	.014	.026	.039	.052	.064	.075	.085	.094	.103
	.001	$.0^517$	$.0^210$	$.0^275$	.020	.035	.050	.064	.078	.090	.101	.111	.119
	.005	$.0^443$	$.0^250$	.022	.045	.069	.090	.109	.126	.140	.153	.164	.174
	.01	$.0^317$	.010	.036	.066	.094	.118	.139	.157	.172	.186	.197	.207
	.025	$.0^211$	.025	.068	.109	.143	.172	.195	.215	.231	.246	.258	.268
	.05	$.0^243$	.052	.112	.162	.202	.233	.259	.279	.296	.311	.324	.334
	.10	.017	.107	.189	.249	.294	.327	.354	.375	.392	.406	.418	.429
	.25	.111	.302	.413	.481	.524	.561	.586	.606	.622	.635	.645	.654
	.50	.515	.780	.886	.942	.977	1.00	1.02	1.03	1.04	1.05	1.05	1.06
	.75	1.62	1.76	1.78	1.79	1.79	1.78	1.78	1.78	1.77	1.77	1.77	1.77
	.90	3.78	3.46	3.29	3.18	3.11	3.05	3.01	2.98	2.96	2.94	2.92	2.90
	.95	5.99	5.14	4.76	4.53	4.39	4.28	4.21	4.15	4.10	4.06	4.03	4.00
	.975	8.81	7.26	6.60	6.23	5.99	5.82	5.70	5.60	5.52	5.46	5.41	5.37
	.99	13.7	10.9	9.78	9.15	8.75	8.47	8.26	8.10	7.98	7.87	7.79	7.72
	.995	18.6	14.5	12.9	12.0	11.5	11.1	10.8	10.6	10.4	10.2	10.1	10.0
	.999	35.5	27.0	23.7	21.9	20.8	20.0	19.5	19.0	18.7	18.4	18.2	18.0
	.9995	46.1	34.8	30.4	28.1	26.6	25.6	24.9	24.3	23.9	23.5	23.2	23.0

ν_2	p	15	20	24	30	40	50	60	100	120	200	500	∞
4	.0005	.105	.125	.135	.147	.159	.166	.172	.183	.186	.191	.196	.200
	.001	.121	.141	.152	.163	.176	.183	.188	.200	.202	.208	.213	.217
	.005	.172	.193	.204	.216	.229	.237	.242	.253	.255	.260	.266	.269
	.01	.204	.226	.237	.249	.261	.269	.274	.285	.287	.293	.298	.301
	.025	.263	.284	.296	.308	.320	.327	.332	.342	.346	.351	.356	.359
	.05	.327	.349	.360	.372	.384	.391	.396	.407	.409	.413	.418	.422
	.10	.424	.445	.456	.467	.478	.485	.490	.500	.502	.508	.510	.514
	.25	.664	.683	.692	.702	.712	.718	.722	.731	.733	.737	.740	.743
	.50	1.14	1.15	1.16	1.16	1.17	1.18	1.18	1.18	1.18	1.19	1.19	1.19
	.75	2.08	2.08	2.08	2.08	2.08	2.08	2.08	2.08	2.08	2.08	2.08	2.08
	.90	3.87	3.84	3.83	3.82	3.80	3.80	3.79	3.78	3.78	3.77	3.76	3.76
	.95	5.86	5.80	5.77	5.75	5.72	5.70	5.69	5.66	5.66	5.65	5.64	5.63
	.975	8.66	8.56	8.51	8.46	8.41	8.38	8.36	8.32	8.31	8.29	8.27	8.26
	.99	14.2	14.0	13.9	13.8	13.7	13.7	13.7	13.6	13.6	13.5	13.5	13.5
	.995	20.4	20.2	20.0	19.9	19.8	19.7	19.6	19.5	19.5	19.4	19.4	19.3
	.999	46.8	46.1	45.8	45.4	45.1	44.9	44.7	44.5	44.4	44.3	44.1	44.0
	.9995	66.5	65.5	65.1	64.6	64.1	63.8	63.6	63.2	63.1	62.9	62.7	62.6
5	.0005	.115	.137	.150	.163	.177	.186	.192	.205	.209	.216	.222	.226
	.001	.132	.155	.167	.181	.195	.204	.210	.223	.227	.233	.239	.244
	.005	.186	.210	.223	.237	.251	.260	.266	.279	.282	.288	.294	.299
	.01	.219	.244	.257	.270	.285	.293	.299	.312	.315	.322	.328	.331
	.025	.280	.304	.317	.330	.344	.353	.359	.370	.374	.380	.386	.390
	.05	.345	.369	.382	.395	.408	.417	.422	.432	.437	.442	.448	.452
	.10	.440	.463	.476	.488	.501	.508	.514	.524	.527	.532	.538	.541
	.25	.669	.690	.700	.711	.722	.728	.732	.741	.743	.748	.752	.755
	.50	1.10	1.11	1.12	1.12	1.13	1.13	1.14	1.14	1.14	1.15	1.15	1.15
	.75	1.89	1.88	1.88	1.88	1.88	1.88	1.87	1.87	1.87	1.87	1.87	1.87
	.90	3.24	3.21	3.19	3.17	3.16	3.15	3.14	3.13	3.12	3.12	3.11	3.10
	.95	4.62	4.56	4.53	4.50	4.46	4.44	4.43	4.41	4.40	4.39	4.37	4.36
	.975	6.43	6.33	6.28	6.23	6.18	6.14	6.12	6.08	6.07	6.05	6.03	6.02
	.99	9.72	9.55	9.47	9.38	9.29	9.24	9.20	9.13	9.11	9.08	9.04	9.02
	.995	13.1	12.9	12.8	12.7	12.5	12.5	12.4	12.3	12.3	12.2	12.2	12.1
	.999	25.9	25.4	25.1	24.9	24.6	24.4	24.3	24.1	24.1	23.9	23.8	23.8
	.9995	34.6	33.9	33.5	33.1	32.7	32.5	32.3	32.1	32.0	31.8	31.7	31.6
6	.0005	.123	.148	.162	.177	.193	.203	.210	.225	.229	.236	.244	.249
	.001	.141	.166	.180	.195	.211	.222	.229	.243	.247	.255	.262	.267
	.005	.197	.224	.238	.253	.269	.279	.286	.301	.304	.312	.318	.324
	.01	.232	.258	.273	.288	.304	.313	.321	.334	.338	.346	.352	.357
	.025	.293	.320	.334	.349	.364	.375	.381	.394	.398	.405	.412	.415
	.05	.358	.385	.399	.413	.428	.437	.444	.457	.460	.467	.472	.476
	.10	.453	.478	.491	.505	.519	.526	.533	.546	.548	.556	.559	.564
	.25	.675	.696	.707	.718	.729	.736	.741	.751	.753	.758	.762	.765
	.50	1.07	1.08	1.09	1.10	1.10	1.11	1.11	1.11	1.12	1.12	1.12	1.12
	.75	1.76	1.76	1.75	1.75	1.75	1.75	1.74	1.74	1.74	1.74	1.74	1.74
	.90	2.87	2.84	2.82	2.80	2.78	2.77	2.76	2.75	2.74	2.73	2.73	2.72
	.95	3.94	3.87	3.84	3.81	3.77	3.75	3.74	3.71	3.70	3.69	3.68	3.67
	.975	5.27	5.17	5.12	5.07	5.01	4.98	4.96	4.92	4.90	4.88	4.86	4.85
	.99	7.56	7.40	7.31	7.23	7.14	7.09	7.06	6.99	6.97	6.93	6.90	6.88
	.995	9.81	9.59	9.47	9.36	9.24	9.17	9.12	9.03	9.00	8.95	8.91	8.88
	.999	17.6	17.1	16.9	16.7	16.4	16.3	16.2	16.0	16.0	15.9	15.8	15.7
	.9995	22.4	21.9	21.7	21.4	21.1	20.9	20.7	20.5	20.4	20.3	20.2	20.1

ν_2	p	ν_1 1	2	3	4	5	6	7	8	9	10	11	12
7	.0005	$.0^6 42$	$.0^3 50$	$.0^2 48$	.014	.027	.040	.053	.066	.078	.088	.099	.108
	.001	$.0^5 17$	$.0^2 10$	$.0^2 76$	.020	.035	.051	.067	.081	.093	.105	.115	.125
	.005	$.0^4 42$	$.0^2 50$	.023	.046	.070	.093	.113	.130	.145	.159	.171	.181
	.01	$.0^3 17$	.010	.036	.067	.096	.121	.143	.162	.178	.192	.205	.216
	.025	$.0^2 10$	.025	.068	.110	.146	.176	.200	.221	.238	.253	.266	.277
	.05	$.0^2 42$	.052	.113	.164	.205	.238	.264	.286	.304	.319	.332	.343
	.10	.017	.107	.190	.251	.297	.332	.359	.381	.399	.414	.427	.438
	.25	.110	.300	.412	.481	.528	.562	.588	.608	.624	.637	.649	.658
	.50	.506	.767	.871	.926	.960	.983	1.00	1.01	1.02	1.03	1.04	1.04
	.75	1.57	1.70	1.72	1.72	1.71	1.71	1.70	1.70	1.69	1.69	1.69	1.68
	.90	3.59	3.26	3.07	2.96	2.88	2.83	2.78	2.75	2.72	2.70	2.68	2.67
	.95	5.59	4.74	4.35	4.12	3.97	3.87	3.79	3.73	3.68	3.64	3.60	3.57
	.975	8.07	6.54	5.89	5.52	5.29	5.12	4.99	4.90	4.82	4.76	4.71	4.67
	.99	12.2	9.55	8.45	7.85	7.46	7.19	6.99	6.84	6.72	6.62	6.54	6.47
	.995	16.2	12.4	10.9	10.0	9.52	9.16	8.89	8.68	8.51	8.38	8.27	8.18
	.999	29.2	21.7	18.8	17.2	16.2	15.5	15.0	14.6	14.3	14.1	13.9	13.7
	.9995	37.0	27.2	23.5	21.4	20.2	19.3	18.7	18.2	17.8	17.5	17.2	17.0
8	.0005	$.0^6 42$	$.0^3 50$	$.0^2 48$	.014	.027	.041	.055	.068	.081	.092	.102	.112
	.001	$.0^5 17$	$.0^2 10$	$.0^2 76$	.020	.036	.053	.068	.083	.096	.109	.120	.130
	.005	$.0^4 42$	$.0^2 50$	.027	.047	.072	.095	.115	.133	.149	.164	.176	.187
	.01	$.0^3 17$	.010	.036	.068	.097	.123	.146	.166	.183	.198	.211	.222
	.025	$.0^2 10$	.025	.069	.111	.148	.179	.204	.226	.244	.259	.273	.285
	.05	$.0^2 42$	.052	.113	.166	.208	.241	.268	.291	.310	.326	.339	.351
	.10	.017	.107	.190	.253	.299	.335	.363	.386	.405	.421	.435	.445
	.25	.109	.298	.411	.481	.529	.563	.589	.610	.627	.640	.654	.661
	.50	.499	.757	.860	.915	.948	.971	.988	1.00	1.01	1.02	1.02	1.03
	.75	1.54	1.66	1.67	1.66	1.66	1.65	1.64	1.64	1.64	1.63	1.63	1.62
	.90	3.46	3.11	2.92	2.81	2.73	2.67	2.62	2.59	2.56	2.54	2.52	2.50
	.95	5.32	4.46	4.07	3.84	3.69	3.58	3.50	3.44	3.39	3.35	3.31	3.28
	.975	7.57	6.06	5.42	5.05	4.82	4.65	4.53	4.43	4.36	4.30	4.24	4.20
	.99	11.3	8.65	7.59	7.01	6.63	6.37	6.18	6.03	5.91	5.81	5.73	5.67
	.995	14.7	11.0	9.60	8.81	8.30	7.95	7.69	7.50	7.34	7.21	7.10	7.01
	.999	25.4	18.5	15.8	14.4	13.5	12.9	12.4	12.0	11.8	11.5	11.4	11.2
	.9995	31.6	22.8	19.4	17.6	16.4	15.7	15.1	14.6	14.3	14.0	13.8	13.6
9	.0005	$.0^6 41$	$.0^3 50$	$.0^2 48$	.015	.027	.042	.056	.070	.083	.094	.105	.115
	.001	$.0^5 17$	$.0^2 10$	$.0^2 77$	.021	.037	.054	.070	.085	.099	.112	.123	.134
	.005	$.0^4 42$	$.0^2 50$	.023	.047	.073	.096	.117	.136	.153	.168	.181	.192
	.01	$.0^3 17$	.010	.037	.068	.098	.125	.149	.169	.187	.202	.216	.228
	.025	$.0^2 10$	.025	.069	.112	.150	.181	.207	.230	.248	.265	.279	.291
	.05	$.0^2 40$	.052	.113	.167	.210	.244	.272	.296	.315	.331	.345	.358
	.10	.017	.107	.191	.254	.302	.338	.367	.390	.410	.426	.441	.452
	.25	.108	.297	.410	.480	.529	.564	.591	.612	.629	.643	.654	.664
	.50	.494	.749	.852	.906	.939	.962	.978	.990	1.00	1.01	1.01	1.02
	.75	1.51	1.62	1.63	1.63	1.62	1.61	1.60	1.60	1.59	1.59	1.58	1.58
	.90	3.36	3.01	2.81	2.69	2.61	2.55	2.51	2.47	2.44	2.42	2.40	2.38
	.95	5.12	4.26	3.86	3.63	3.48	3.37	3.29	3.23	3.18	3.14	3.10	3.07
	.975	7.21	5.71	5.08	4.72	4.48	4.32	4.20	4.10	4.03	3.96	3.91	3.87
	.99	10.6	8.02	6.99	6.42	6.06	5.80	5.61	5.47	5.35	5.26	5.18	5.11
	.995	13.6	10.1	8.72	7.96	7.47	7.13	6.88	6.69	6.54	6.42	6.31	6.23
	.999	22.9	16.4	13.9	12.6	11.7	11.1	10.7	10.4	10.1	9.89	9.71	9.57
	.9995	28.0	19.9	16.8	15.1	14.1	13.3	12.8	12.4	12.1	11.8	11.6	11.4

ν_2	p	15	20	24	30	40	50	60	100	120	200	500	∞
7	.0005	.130	.157	.172	.188	.206	.217	.225	.242	.246	.255	.263	.268
	.001	.148	.176	.191	.208	.225	.237	.245	.261	.266	.274	.282	.288
	.005	.206	.235	.251	.267	.285	.296	.304	.319	.324	.332	.340	.345
	.01	.241	.270	.286	.303	.320	.331	.339	.355	.358	.366	.373	.379
	.025	.304	.333	.348	.364	.381	.392	.399	.413	.418	.426	.433	.437
	.05	.369	.398	.413	.428	.445	.455	.461	.476	.479	.485	.493	.498
	.10	.463	.491	.504	.519	.534	.543	.550	.562	.566	.571	.578	.582
	.25	.679	.702	.713	.725	.737	.745	.749	.760	.762	.767	.772	.775
	.50	1.05	1.07	1.07	1.08	1.08	1.09	1.09	1.10	1.10	1.10	1.10	1.10
	.75	1.68	1.67	1.67	1.66	1.66	1.66	1.65	1.65	1.65	1.65	1.65	1.65
	.90	2.63	2.59	2.58	2.56	2.54	2.52	2.51	2.50	2.49	2.48	2.48	2.47
	.95	3.51	3.44	3.41	3.38	3.34	3.32	3.30	3.27	3.27	3.25	3.24	3.23
	.975	4.57	4.47	4.42	4.36	4.31	4.28	4.25	4.21	4.20	4.18	4.16	4.14
	.99	6.31	6.16	6.07	5.99	5.91	5.86	5.82	5.75	5.74	5.70	5.67	5.65
	.995	7.97	7.75	7.65	7.53	7.42	7.35	7.31	7.22	7.19	7.15	7.10	7.08
	.999	13.3	12.9	12.7	12.5	12.3	12.2	12.1	11.9	11.9	11.8	11.7	11.7
	.9995	16.5	16.0	15.7	15.5	15.2	15.1	15.0	14.7	14.7	14.6	14.5	14.4
8	.0005	.136	.164	.181	.198	.218	.230	.239	.257	.262	.271	.281	.287
	.001	.155	.184	.200	.218	.238	.250	.259	.277	.282	.292	.300	.306
	.005	.214	.244	.261	.279	.299	.311	.319	.337	.341	.351	.358	.364
	.01	.250	.281	.297	.315	.334	.346	.354	.372	.376	.385	.392	.398
	.025	.313	.343	.360	.377	.395	.407	.415	.431	.435	.442	.450	.456
	.05	.379	.409	.425	.441	.459	.469	.477	.493	.496	.505	.510	.516
	.10	.472	.500	.515	.531	.547	.556	.563	.578	.581	.588	.595	.599
	.25	.684	.707	.718	.730	.743	.751	.756	.767	.769	.775	.780	.783
	.50	1.04	1.05	1.06	1.07	1.07	1.07	1.08	1.08	1.08	1.09	1.09	1.09
	.75	1.62	1.61	1.60	1.60	1.59	1.59	1.59	1.58	1.58	1.58	1.58	1.58
	.90	2.46	2.42	2.40	2.38	2.36	2.35	2.34	2.32	2.32	2.31	2.30	2.29
	.95	3.22	3.15	3.12	3.08	3.04	3.02	3.01	2.97	2.97	2.95	2.94	2.93
	.975	4.10	4.00	3.95	3.89	3.84	3.81	3.78	3.74	3.73	3.70	3.68	3.67
	.99	5.52	5.36	5.28	5.20	5.12	5.07	5.03	4.96	4.95	4.91	4.88	4.86
	.995	6.81	6.61	6.50	6.40	6.29	6.22	6.18	6.09	6.06	6.02	5.98	5.95
	.999	10.8	10.5	10.3	10.1	9.92	9.80	9.73	9.57	9.54	9.46	9.39	9.34
	.9995	13.1	12.7	12.5	12.2	12.0	11.8	11.8	11.6	11.5	11.4	11.4	11.3
9	.0005	.141	.171	.188	.207	.228	.242	.251	.270	.276	.287	.297	.303
	.001	.160	.191	.208	.228	.249	.262	.271	.291	.296	.307	.316	.323
	.005	.220	.253	.271	.290	.310	.324	.332	.351	.356	.366	.376	.382
	.01	.257	.289	.307	.326	.346	.358	.368	.386	.391	.400	.410	.415
	.025	.320	.352	.370	.388	.408	.420	.428	.446	.450	.459	.467	.473
	.05	.386	.418	.435	.452	.471	.483	.490	.508	.510	.518	.526	.532
	.10	.479	.509	.525	.541	.558	.568	.575	.588	.594	.602	.610	.613
	.25	.687	.711	.723	.736	.749	.757	.762	.773	.776	.782	.787	.791
	.50	1.03	1.04	1.05	1.05	1.06	1.06	1.07	1.07	1.07	1.08	1.08	1.08
	.75	1.57	1.56	1.56	1.55	1.55	1.54	1.54	1.53	1.53	1.53	1.53	1.53
	.90	2.34	2.30	2.28	2.25	2.23	2.22	2.21	2.19	2.18	2.17	2.17	2.16
	.95	3.01	2.94	2.90	2.86	2.83	2.80	2.79	2.76	2.75	2.73	2.72	2.71
	.975	3.77	3.67	3.61	3.56	3.51	3.47	3.45	3.40	3.39	3.37	3.35	3.33
	.99	4.96	4.81	4.73	4.65	4.57	4.52	4.48	4.42	4.40	4.36	4.33	4.31
	.995	6.03	5.83	5.73	5.62	5.52	5.45	5.41	5.32	5.30	5.26	5.21	5.19
	.999	9.24	8.90	8.72	8.55	8.37	8.26	8.19	8.04	8.00	7.93	7.86	7.81
	.9995	11.0	10.6	10.4	10.2	9.94	9.80	9.71	9.53	9.49	9.40	9.32	9.26

ν_2	p / ν_1	1	2	3	4	5	6	7	8	9	10	11	12
10	.0005	$.0^641$	$.0^350$	$.0^249$	.015	.028	.043	.057	.071	.085	.097	.108	.119
	.001	$.0^517$	$.0^210$	$.0^277$	.021	.037	.054	.071	.087	.101	.114	.126	.137
	.005	$.0^441$	$.0^250$	.023	.048	.073	.098	.119	.139	.156	.171	.185	.197
	.01	$.0^317$	.010	.037	.069	.100	.127	.151	.172	.190	.206	.220	.233
	.025	$.0^210$	.025	.069	.113	.151	.183	.210	.233	.252	.269	.283	.296
	.05	$.0^241$	.052	.114	.168	.211	.246	.275	.299	.319	.336	.351	.363
	.10	.017	.106	.191	.255	.303	.340	.370	.394	.414	.430	.444	.457
	.25	.107	.296	.409	.480	.529	.565	.592	.613	.631	.645	.657	.667
	.50	.490	.743	.845	.899	.932	.954	.971	.983	.992	1.00	1.01	1.01
	.75	1.49	1.60	1.60	1.59	1.59	1.58	1.57	1.56	1.56	1.55	1.55	1.54
	.90	3.28	2.92	2.73	2.61	2.52	2.46	2.41	2.38	2.35	2.32	2.30	2.28
	.95	4.96	4.10	3.71	3.48	3.33	3.22	3.14	3.07	3.02	2.98	2.94	2.91
	.975	6.94	5.46	4.83	4.47	4.24	4.07	3.95	3.85	3.78	3.72	3.66	3.62
	.99	10.0	7.56	6.55	5.99	5.64	5.39	5.20	5.06	4.94	4.85	4.77	4.71
	.995	12.8	9.43	8.08	7.34	6.87	6.54	6.30	6.12	5.97	5.85	5.75	5.66
	.999	21.0	14.9	12.6	11.3	10.5	9.92	9.52	9.20	8.96	8.75	8.58	8.44
	.9995	25.5	17.9	15.0	13.4	12.4	11.8	11.3	10.9	10.6	10.3	10.1	9.93
11	.0005	$.0^641$	$.0^350$	$.0^249$	.015	.028	.043	.058	.072	.086	.099	.111	.121
	.001	$.0^516$	$.0^210$	$.0^278$	.021	.038	.055	.072	.088	.103	.116	.129	.140
	.005	$.0^440$	$.0^250$	.023	.048	.074	.099	.121	.141	.158	.174	.188	.200
	.01	$.0^316$	.010	.037	.069	.100	.128	.153	.175	.193	.210	.224	.237
	.025	$.0^210$	.025	.069	.114	.152	.185	.212	.236	.256	.273	.288	.301
	.05	$.0^241$	.052	.114	.168	.212	.248	.278	.302	.323	.340	.355	.368
	.10	.017	.106	.192	.256	.305	.342	.373	.397	.417	.435	.448	.461
	.25	.107	.295	.408	.481	.529	.565	.592	.614	.633	.645	.658	.667
	.50	.486	.739	.840	.893	.926	.948	.964	.977	.986	.994	1.00	1.01
	.75	1.47	1.58	1.58	1.57	1.56	1.55	1.54	1.53	1.53	1.52	1.52	1.51
	.90	3.23	2.86	2.66	2.54	2.45	2.39	2.34	2.30	2.27	2.25	2.23	2.21
	.95	4.84	3.98	3.59	3.36	3.20	3.09	3.01	2.95	2.90	2.85	2.82	2.79
	.975	6.72	5.26	4.63	4.28	4.04	3.88	3.76	3.66	3.59	3.53	3.47	3.43
	.99	9.65	7.21	6.22	5.67	5.32	5.07	4.89	4.74	4.63	4.54	4.46	4.40
	.995	12.2	8.91	7.60	6.88	6.42	6.10	5.86	5.68	5.54	5.42	5.32	5.24
	.999	19.7	13.8	11.6	10.3	9.58	9.05	8.66	8.35	8.12	7.92	7.76	7.62
	.9995	23.6	16.4	13.6	12.2	11.2	10.6	10.1	9.76	9.48	9.24	9.04	8.88
12	.0005	$.0^641$	$.0^350$	$.0^249$	.015	.028	.044	.058	.073	.087	.101	.113	.124
	.001	$.0^516$	$.0^210$	$.0^278$	.021	.038	.056	.073	.089	.104	.118	.131	.143
	.005	$.0^439$	$.0^250$	.023	.048	.075	.100	.122	.143	.161	.177	.191	.204
	.01	$.0^316$	.010	.037	.070	.101	.130	.155	.176	.196	.212	.227	.241
	.025	$.0^210$	.025	.070	.114	.153	.186	.214	.238	.259	.276	.292	.305
	.05	$.0^241$	.052	.114	.169	.214	.250	.280	.305	.325	.343	.358	.372
	.10	.016	.106	.192	.257	.306	.344	.375	.400	.420	.438	.452	.466
	.25	.106	.295	.408	.480	.530	.566	.594	.616	.633	.649	.662	.671
	.50	.484	.735	.835	.888	.921	.943	.959	.972	.981	.989	.995	1.00
	.75	1.46	1.56	1.56	1.55	1.54	1.53	1.52	1.51	1.51	1.50	1.50	1.49
	.90	3.18	2.81	2.61	2.48	2.39	2.33	2.28	2.24	2.21	2.19	2.17	2.15
	.95	4.75	3.89	3.49	3.26	3.11	3.00	2.91	2.85	2.80	2.75	2.72	2.69
	.975	6.55	5.10	4.47	4.12	3.89	3.73	3.61	3.51	3.44	3.37	3.32	3.28
	.99	9.33	6.93	5.95	5.41	5.06	4.82	4.64	4.50	4.39	4.30	4.22	4.16
	.995	11.8	8.51	7.23	6.52	6.07	5.76	5.52	5.35	5.20	5.09	4.99	4.91
	.999	18.6	13.0	10.8	9.63	8.89	8.38	8.00	7.71	7.48	7.29	7.14	7.01
	.9995	22.2	15.3	12.7	11.2	10.4	9.74	9.28	8.94	8.66	8.43	8.24	8.08

ν_2	p	15	20	24	30	40	50	60	100	120	200	500	∞
10	.0005	.145	.177	.195	.215	.238	.251	.262	.282	.288	.299	.311	.319
	.001	.164	.197	.216	.236	.258	.272	.282	.303	.309	.321	.331	.338
	.005	.226	.260	.279	.299	.321	.334	.344	.365	.370	.380	.391	.397
	.01	.263	.297	.316	.336	.357	.370	.380	.400	.405	.415	.424	.431
	.025	.327	.360	.379	.398	.419	.431	.441	.459	.464	.474	.483	.488
	.05	.393	.426	.444	.462	.481	.493	.502	.518	.523	.532	.541	.546
	.10	.486	.516	.532	.549	.567	.578	.586	.602	.605	.614	.621	.625
	.25	.691	.714	.727	.740	.754	.762	.767	.779	.782	.788	.793	.797
	.50	1.02	1.03	1.04	1.05	1.05	1.06	1.06	1.06	1.06	1.07	1.07	1.07
	.75	1.53	1.52	1.52	1.51	1.51	1.50	1.50	1.49	1.49	1.49	1.48	1.48
	.90	2.24	2.20	2.18	2.16	2.13	2.12	2.11	2.09	2.08	2.07	2.06	2.06
	.95	2.85	2.77	2.74	2.70	2.66	2.64	2.62	2.59	2.58	2.56	2.55	2.54
	.975	3.52	3.42	3.37	3.31	3.26	3.22	3.20	3.15	3.14	3.12	3.09	3.08
	.99	4.56	4.41	4.33	4.25	4.17	4.12	4.08	4.01	4.00	3.96	3.93	3.91
	.995	5.47	5.27	5.17	5.07	4.97	4.90	4.86	4.77	4.75	4.71	4.67	4.64
	.999	8.13	7.80	7.64	7.47	7.30	7.19	7.12	6.98	6.94	6.87	6.81	6.76
	.9995	9.56	9.16	8.96	8.75	8.54	8.42	8.33	8.16	8.12	8.04	7.96	7.90
11	.0005	.148	.182	.201	.222	.246	.261	.271	.293	.299	.312	.324	.331
	.001	.168	.202	.222	.243	.266	.282	.292	.313	.320	.332	.343	.353
	.005	.231	.266	.286	.308	.330	.345	.355	.376	.382	.394	.403	.412
	.01	.268	.304	.324	.344	366	.380	.391	.412	.417	.427	.439	.444
	.025	.332	.368	.386	.407	.429	.442	.450	.472	.476	.485	.495	.503
	.05	.398	.433	.452	.469	.490	.503	.513	.529	.535	.543	.552	.559
	.10	.490	.524	.541	.559	.578	.588	.595	.614	.617	.625	.633	.637
	.25	.694	.719	.730	.744	.758	.767	.773	.780	.788	.794	.799	.803
	.50	1.02	1.03	1.03	1.04	1.05	1.05	1.05	1.06	1.06	1.06	1.06	1.06
	.75	1.50	1.49	1.49	1.48	1.47	1.47	1.47	1.46	1.46	1.46	1.45	1.45
	.90	2.17	2.12	2.10	2.08	2.05	2.04	2.03	2.00	2.00	1.99	1.98	1.97
	.95	2.72	2.65	2.61	2.57	2.53	2.51	2.49	2.46	2.45	2.43	2.42	2.40
	.975	3.33	3.23	3.17	3.12	3.06	3.03	3.00	2.96	2.94	2.92	2.90	2.88
	.99	4.25	4.10	4.02	3.94	3.86	3.81	3.78	3.71	3.69	3.66	3.62	3.60
	.995	5.05	4.86	4.76	4.65	4.55	4.49	4.45	4.36	4.34	4.29	4.25	4.23
	.999	7.32	7.01	6.85	6.68	6.52	6.41	6.35	6.21	6.17	6.10	6.04	6.00
	.9995	8.52	8.14	7.94	7.75	7.55	7.43	7.35	7.18	7.14	7.06	6.98	6.93
12	.0005	.152	.186	.206	.228	.253	.269	.280	.305	.311	.323	.337	.345
	.001	.172	.207	.228	.250	.275	.291	.302	.326	.332	.344	.357	.365
	.005	.235	.272	.292	.315	.339	.355	.365	.388	.393	.405	.417	.424
	.01	.273	.310	.330	.352	.375	.391	.401	.422	.428	.441	.450	.458
	.025	.337	.374	.394	.416	.437	.450	.461	.481	.487	.498	.508	.514
	.05	.404	.439	.458	.478	.499	.513	.522	.541	.545	.556	.565	.571
	.10	.496	.528	.546	.564	.583	.595	.604	.621	.625	.633	.641	.647
	.25	.695	.721	.734	.748	.762	.771	.777	.789	.792	.799	.804	.808
	.50	1.01	1.02	1.03	1.03	1.04	1.04	1.05	1.05	1.05	1.05	1.06	1.06
	.75	1.48	1.47	1.46	1.45	1.45	1.44	1.44	1.43	1.43	1.43	1.42	1.42
	.90	2.11	2.06	2.04	2.01	1.99	1.97	1.96	1.94	1.93	1.92	1.91	1.90
	.95	2.62	2.54	2.51	2.47	2.43	2.40	2.38	2.35	2.34	2.32	2.31	2.30
	.975	3.18	3.07	3.02	2.96	2.91	2.87	2.85	2.80	2.79	2.76	2.74	2.72
	.99	4.01	3.86	3.78	3.70	3.62	3.57	3.54	3.47	3.45	3.41	3.38	3.36
	.995	4.72	4.53	4.43	4.33	4.23	4.17	4.12	4.04	4.01	3.97	3.93	3.90
	.999	6.71	6.40	6.25	6.09	5.93	5.83	5.76	5.63	5.59	5.52	5.46	5.42
	.9995	7.74	7.37	7.18	7.00	6.80	6.68	6.61	6.45	6.41	6.33	6.25	6.20

ν_2	p / ν_1	1	2	3	4	5	6	7	8	9	10	11	12
15	.0005	$.0^641$	$.0^350$	$.0^249$	.015	.029	.045	.061	.076	.091	.105	.117	.129
	.001	$.0^516$	$.0^210$	$.0^279$	.021	.039	.057	.075	.092	.108	.123	.137	.149
	.005	$.0^439$	$.0^250$	.023	.049	.076	.102	.125	.147	.166	.183	.198	.212
	.01	$.0^316$	.010	.037	.070	.103	.132	.158	.181	.202	.219	.235	.249
	.025	$.0^210$	.025	.070	.116	.156	.190	.219	.244	.265	.284	.300	.315
	.05	$.0^241$	.051	.115	.170	.216	.254	.285	.311	.333	.351	.368	.382
	.10	.016	.106	.192	.258	.309	.348	.380	.406	.427	.446	.461	.475
	.25	.105	.293	.407	.480	.531	.568	.596	.618	.637	.652	.667	.676
	.50	.478	.726	.826	.878	.911	.933	.948	.960	.970	.977	.984	.989
	.75	1.43	1.52	1.52	1.51	1.49	1.48	1.47	1.46	1.46	1.45	1.44	1.44
	.90	3.07	2.70	2.49	2.36	2.27	2.21	2.16	2.12	2.09	2.06	2.04	2.02
	.95	4.54	3.68	3.29	3.06	2.90	2.79	2.71	2.64	2.59	2.54	2.51	2.48
	.975	6.20	4.76	4.15	3.80	3.58	3.41	3.29	3.20	3.12	3.06	3.01	2.96
	.99	8.68	6.36	5.42	4.89	4.56	4.32	4.14	4.00	3.89	3.80	3.73	3.67
	.995	10.8	7.70	6.48	5.80	5.37	5.07	4.85	4.67	4.54	4.42	4.33	4.25
	.999	16.6	11.3	9.34	8.25	7.57	7.09	6.74	6.47	6.26	6.08	5.93	5.81
	.9995	19.5	13.2	10.8	9.48	8.66	8.10	7.68	7.36	7.11	6.91	6.75	6.60
20	.0005	$.0^640$	$.0^350$	$.0^250$	.015	.029	.046	.063	.079	.094	.109	.123	.136
	.001	$.0^516$	$.0^210$	$.0^279$	.022	.039	.058	.077	.095	.112	.128	.143	.156
	.005	$.0^439$	$.0^250$	.023	.050	.077	.104	.129	.151	.171	.190	.206	.221
	.01	$.0^316$	.010	.037	.071	.105	.135	.162	.187	.208	.227	.244	.259
	.025	$.0^210$	.025	.071	.117	.158	.193	.224	.250	.273	.292	.310	.325
	.05	$.0^240$	.051	.115	.172	.219	.258	.290	.318	.340	.360	.377	.393
	.10	.016	.106	.193	.260	.312	.353	.385	.412	.435	.454	.472	.485
	.25	.104	.292	.407	.480	.531	.569	.598	.622	.641	.656	.671	.681
	.50	.472	.718	.816	.868	.900	.922	.938	.950	.959	.966	.972	.977
	.75	1.40	1.49	1.48	1.47	1.45	1.44	1.43	1.42	1.41	1.40	1.39	1.39
	.90	2.97	2.59	2.38	2.25	2.16	2.09	2.04	2.00	1.96	1.94	1.91	1.89
	.95	4.35	3.49	3.10	2.87	2.71	2.60	2.51	2.45	2.39	2.35	2.31	2.28
	.975	5.87	4.46	3.86	3.51	3.29	3.13	3.01	2.91	2.84	2.77	2.72	2.68
	.99	8.10	5.85	4.94	4.43	4.10	3.87	3.70	3.56	3.46	3.37	3.29	3.23
	.995	9.94	6.99	5.82	5.17	4.76	4.47	4.26	4.09	3.96	3.85	3.76	3.68
	.999	14.8	9.95	8.10	7.10	6.46	6.02	5.69	5.44	5.24	5.08	4.94	4.82
	.9995	17.2	11.4	9.20	8.02	7.28	6.76	6.38	6.08	5.85	5.66	5.51	5.38
24	.0005	$.0^640$	$.0^350$	$.0^250$	.015	.030	.046	.064	.080	.096	.112	.126	.139
	.001	$.0^516$	$.0^210$	$.0^279$	.022	.040	.059	.079	.097	.115	.131	.146	.160
	.005	$.0^440$	$.0^250$	.023	.050	.078	.106	.131	.154	.175	.193	.210	.226
	.01	$.0^316$	.010	.038	.072	.106	.137	.165	.189	.211	.231	.249	.264
	.025	$.0^210$	.025	.071	.117	.159	.195	.227	.253	.277	.297	.315	.331
	.05	$.0^240$	.051	.116	.173	.221	.260	.293	.321	.345	.365	.383	.399
	.10	.016	.106	.193	.261	.313	.355	.388	.416	.439	.459	.476	.491
	.25	.104	.291	.406	.480	.532	.570	.600	.623	.643	.659	.671	.684
	.50	.469	.714	.812	.863	.895	.917	.932	.944	.953	.961	.967	.972
	.75	1.39	1.47	1.46	1.44	1.43	1.41	1.40	1.39	1.38	1.38	1.37	1.36
	.90	2.93	2.54	2.33	2.19	2.10	2.04	1.98	1.94	1.91	1.88	1.85	1.83
	.95	4.26	3.40	3.01	2.78	2.62	2.51	2.42	2.36	2.30	2.25	2.21	2.18
	.975	5.72	4.32	3.72	3.38	3.15	2.99	2.87	2.78	2.70	2.64	2.59	2.54
	.99	7.82	5.61	4.72	4.22	3.90	3.67	3.50	3.36	3.26	3.17	3.09	3.03
	.995	9.55	6.66	5.52	4.89	4.49	4.20	3.99	3.83	3.69	3.59	3.50	3.42
	.999	14.0	9.34	7.55	6.59	5.98	5.55	5.23	4.99	4.80	4.64	4.50	4.39
	.9995	16.2	10.6	8.52	7.39	6.68	6.18	5.82	5.54	5.31	5.13	4.98	4.85

ν_2	p \ ν_1	15	20	24	30	40	50	60	100	120	200	500	∞
15	.0005	.159	.197	.220	.244	.272	.290	.303	.330	.339	.353	.368	.377
	.001	.181	.219	.242	.266	.294	.313	.325	.352	.360	.375	.388	.398
	.005	.246	.286	.308	.333	.360	.377	.389	.415	.422	.435	.448	.457
	.01	.284	.324	.346	.370	.397	.413	.425	.450	.456	.469	.483	.490
	.025	.349	.389	.410	.433	.458	.474	.485	.508	.514	.526	.538	.546
	.05	.416	.454	.474	.496	.519	.535	.545	.565	.571	.581	.592	.600
	.10	.507	.542	.561	.581	.602	.614	.624	.641	.647	.658	.667	.672
	.25	.701	.728	.742	.757	.772	.782	.788	.802	.805	.812	.818	.822
	.50	1.00	1.01	1.02	1.02	1.03	1.03	1.03	1.04	1.04	1.04	1.04	1.05
	.75	1.43	1.41	1.41	1.40	1.39	1.39	1.38	1.38	1.37	1.37	1.36	1.36
	.90	1.97	1.92	1.90	1.87	1.85	1.83	1.82	1.79	1.79	1.77	1.76	1.76
	.95	2.40	2.33	2.39	2.25	2.20	2.18	2.16	2.12	2.11	2.10	2.08	2.07
	.975	2.86	2.76	2.70	2.64	2.59	2.55	2.52	2.47	2.46	2.44	2.41	2.40
	.99	3.52	3.37	3.29	3.21	3.13	3.08	3.05	2.98	2.96	2.92	2.89	2.87
	.995	4.07	3.88	3.79	3.69	3.59	3.52	3.48	3.39	3.37	3.33	3.29	3.26
	.999	5.54	5.25	5.10	4.95	4.80	4.70	4.64	4.51	4.47	4.41	4.35	4.31
	.9995	6.27	5.93	5.75	5.58	5.40	5.29	5.21	5.06	5.02	4.94	4.87	4.83
20	.0005	.169	.211	.235	.263	.295	.316	.331	.364	.375	.391	.408	.422
	.001	.191	.233	.258	.286	.318	.339	.354	.386	.395	.413	.429	.441
	.005	.258	.301	.327	.354	.385	.405	.419	.448	.457	.474	.490	.500
	.01	.297	.340	.365	.392	.422	.441	.455	.483	.491	.508	.521	.532
	.025	.363	.406	.430	.456	.484	.503	.514	.541	.548	.562	.575	.585
	.05	.430	.471	.493	.518	.544	.562	.572	.595	.603	.617	.629	.637
	.10	.520	.557	.578	.600	.623	.637	.648	.671	.675	.685	.694	.704
	.25	.708	.736	.751	.767	.784	.794	.801	.816	.820	.827	.835	.840
	.50	.989	1.00	1.01	1.01	1.02	1.02	1.02	1.03	1.03	1.03	1.03	1.03
	.75	1.37	1.36	1.35	1.34	1.33	1.33	1.32	1.31	1.31	1.30	1.30	1.29
	.90	1.84	1.79	1.77	1.74	1.71	1.69	1.68	1.65	1.64	1.63	1.62	1.61
	.95	2.20	2.12	2.08	2.04	1.99	1.97	1.95	1.91	1.90	1.88	1.86	1.84
	.975	2.57	2.46	2.41	2.35	2.29	2.25	2.22	2.17	2.16	2.13	2.10	2.09
	.99	3.09	2.94	2.86	2.78	2.69	2.64	2.61	2.54	2.52	2.48	2.44	2.42
	.995	3.50	3.32	3.22	3.12	3.02	2.96	2.92	2.83	2.81	2.76	2.72	2.69
	.999	4.56	4.29	4.15	4.01	3.86	3.77	3.70	3.58	3.54	3.48	3.42	3.38
	.9995	5.07	4.75	4.58	4.42	4.24	4.15	4.07	3.93	3.90	3.82	3.75	3.70
24	.0005	.174	.218	.244	.274	.309	.331	.349	.384	.395	.416	.434	.449
	.001	.196	.241	.268	.298	.332	.354	.371	.405	.417	.437	.455	.469
	.005	.264	.310	.337	.367	.400	.422	.437	.469	.479	.498	.515	.527
	.01	.304	.350	.376	.405	.437	.459	.473	.505	.513	.529	.546	.558
	.025	.370	.415	.441	.468	.498	.518	.531	.562	.568	.585	.599	.610
	.05	.437	.480	.504	.530	.558	.575	.588	.613	.622	.637	.649	.659
	.10	.527	.566	.588	.611	.635	.651	.662	.685	.691	.704	.715	.723
	.25	.712	.741	.757	.773	.791	.802	.809	.825	.829	.837	.844	.850
	.50	.983	.994	1.00	1.01	1.01	1.02	1.02	1.02	1.02	1.02	1.03	1.03
	.75	1.35	1.33	1.32	1.31	1.30	1.29	1.29	1.28	1.28	1.27	1.27	1.26
	.90	1.78	1.73	1.70	1.67	1.64	1.62	1.61	1.58	1.57	1.56	1.54	1.53
	.95	2.11	2.03	1.98	1.94	1.89	1.86	1.84	1.80	1.79	1.77	1.75	1.73
	.975	2.44	2.33	2.27	2.21	2.15	2.11	2.08	2.02	2.01	1.98	1.95	1.94
	.99	2.89	2.74	2.66	2.58	2.49	2.44	2.40	2.33	2.31	2.27	2.24	2.21
	.995	3.25	3.06	2.97	2.87	2.77	2.70	2.66	2.57	2.55	2.50	2.46	2.43
	.999	4.14	3.87	3.74	3.59	3.45	3.35	3.29	3.16	3.14	3.07	3.01	2.97
	.9995	4.55	4.25	4.09	3.93	3.76	3.66	3.59	3.44	3.41	3.33	3.27	3.22

ν_2	ν_1 / p	1	2	3	4	5	6	7	8	9	10	11	12
30	.0005	$.0^6 40$	$.0^5 50$	$.0^2 50$	.015	.030	.047	.065	.082	.098	.114	.129	.143
	.001	$.0^5 16$	$.0^2 10$	$.0^2 80$	.022	.040	.060	.080	.099	.117	.134	.150	.164
	.005	$.0^4 40$	$.0^2 50$	.024	.050	.079	.107	.133	.156	.178	.197	.215	.231
	.01	$.0^3 16$	.010	.038	.072	.107	.138	.167	.192	.215	.235	.254	.270
	.025	$.0^2 10$	.025	.071	.118	.161	.197	.229	.257	.281	.302	.321	.337
	.05	$.0^2 40$	.051	.116	.174	.222	.263	.296	.325	.349	.370	.389	.406
	.10	.016	.106	.193	.262	.315	.357	.391	.420	.443	.464	.481	.497
	.25	.103	.290	.406	.480	.532	.571	.601	.625	.645	.661	.676	.688
	.50	.466	.709	.807	.858	.890	.912	.927	.939	.948	.955	.961	.966
	.75	1.38	1.45	1.44	1.42	1.41	1.39	1.38	1.37	1.36	1.35	1.35	1.34
	.90	2.88	2.49	2.28	2.14	2.05	1.98	1.93	1.88	1.85	1.82	1.79	1.77
	.95	4.17	3.32	2.92	2.69	2.53	2.42	2.33	2.27	2.21	2.16	2.13	2.09
	.975	5.57	4.18	3.59	3.25	3.03	2.87	2.75	2.65	2.57	2.51	2.46	2.41
	.99	7.56	5.39	4.51	4.02	3.70	3.47	3.30	3.17	3.07	2.98	2.91	2.84
	.995	9.18	6.35	5.24	4.62	4.23	3.95	3.74	3.58	3.45	3.34	3.25	3.18
	.999	13.3	8.77	7.05	6.12	5.53	5.12	4.82	4.58	4.39	4.24	4.11	4.00
	.9995	15.2	9.90	7.90	6.82	6.14	5.66	5.31	5.04	4.82	4.65	4.51	4.38
40	.0005	$.0^6 40$	$.0^5 50$	$.0^2 50$	.016	.030	.048	.066	.084	.100	.117	.132	.147
	.001	$.0^5 16$	$.0^2 10$	$.0^2 80$	.022	.042	.061	.081	.101	.119	.137	.153	.169
	.005	$.0^4 40$	$.0^2 50$	.024	.051	.080	.108	.135	.159	.181	.201	.220	.237
	.01	$.0^3 16$	.010	.038	.073	.108	.140	.169	.195	.219	.240	.259	.276
	.025	$.0^3 99$	.025	.071	.119	.162	.199	.232	.260	.285	.307	.327	.344
	.05	$.0^2 40$	.051	.116	.175	.224	.265	.299	.329	.354	.376	.395	.412
	.10	.016	.106	.194	.263	.317	.360	.394	.424	.448	.469	.488	.504
	.25	.103	.290	.405	.480	.533	.572	.603	.627	.647	.664	.680	.691
	.50	.463	.705	.802	.854	.885	.907	.922	.934	.943	.950	.956	.961
	.75	1.36	1.44	1.42	1.40	1.39	1.37	1.36	1.35	1.34	1.33	1.32	1.31
	.90	2.84	2.44	2.23	2.09	2.00	1.93	1.87	1.83	1.79	1.76	1.73	1.71
	.95	4.08	3.23	2.84	2.61	2.45	2.34	2.25	2.18	2.12	2.08	2.04	2.00
	.975	5.42	4.05	3.46	3.13	2.90	2.74	2.62	2.53	2.45	2.39	2.33	2.29
	.99	7.31	5.18	4.31	3.83	3.51	3.29	3.12	2.99	2.89	2.80	2.73	2.66
	.995	8.83	6.07	4.98	4.37	3.99	3.71	3.51	3.35	3.22	3.12	3.03	2.95
	.999	12.6	8.25	6.60	5.70	5.13	4.73	4.44	4.21	4.02	3.87	3.75	3.64
	.9995	14.4	9.25	7.33	6.30	5.64	5.19	4.85	4.59	4.38	4.21	4.07	3.95
60	.0005	$.0^6 40$	$.0^5 50$	$.0^2 51$	.016	.031	.048	.067	.085	.103	.120	.136	.152
	.001	$.0^5 16$	$.0^2 10$	$.0^2 80$	.022	.041	.062	.083	.103	.122	.140	.157	.174
	.005	$.0^4 40$	$.0^2 50$	.024	.051	.081	.110	.137	.162	.185	.206	.225	.243
	.01	$.0^3 16$	.010	.038	.073	.109	.142	.172	.199	.223	.245	.265	.283
	.025	$.0^3 99$	.025	.071	.120	.163	.202	.235	.264	.290	.313	.333	.351
	.05	$.0^2 40$	.051	.116	.176	.226	.267	.303	.333	.359	.382	.402	.419
	.10	.016	.106	.194	.264	.318	.362	.398	.428	.453	.475	.493	.510
	.25	.102	.289	.405	.480	.534	.573	.604	.629	.650	.667	.680	.695
	.50	.461	.701	.798	.849	.880	.901	.917	.928	.937	.945	.951	.956
	.75	1.35	1.42	1.41	1.38	1.37	1.35	1.33	1.32	1.31	1.30	1.29	1.29
	.90	2.79	2.39	2.18	2.04	1.95	1.87	1.82	1.77	1.74	1.71	1.68	1.66
	.95	4.00	3.15	2.76	2.53	2.37	2.25	2.17	2.10	2.04	1.99	1.95	1.92
	.975	5.29	3.93	3.34	3.01	2.79	2.63	2.51	2.41	2.33	2.27	2.22	2.17
	.99	7.08	4.98	4.13	3.65	3.34	3.12	2.95	2.82	2.72	2.63	2.56	2.50
	.995	8.49	5.80	4.73	4.14	3.76	3.49	3.29	3.13	3.01	2.90	2.82	2.74
	.999	12.0	7.76	6.17	5.31	4.76	4.37	4.09	3.87	3.69	3.54	3.43	3.31
	.9995	13.6	8.65	6.81	5.82	5.20	4.76	4.44	4.18	3.98	3.82	3.69	3.57

ν_2	p	15	20	24	30	40	50	60	100	120	200	500	∞
30	.0005	.179	.226	.254	.287	.325	.350	.369	.410	.420	.444	.467	.483
	.001	.202	.250	.278	.311	.348	.373	.391	.431	.442	.465	.488	.503
	.005	.271	.320	.349	.381	.416	.441	.457	.495	.504	.524	.543	.559
	.01	.311	.360	.388	.419	.454	.476	.493	.529	.538	.559	.575	.590
	.025	.378	.426	.453	.482	.515	.535	.551	.585	.592	.610	.625	.639
	.05	.445	.490	.516	.543	.573	.592	.606	.637	.644	.658	.676	.685
	.10	.534	.575	.598	.623	.649	.667	.678	.704	.710	.725	.735	.746
	.25	.716	.746	.763	.780	.798	.810	.818	.835	.839	.848	.856	.862
	.50	.978	.989	.994	1.00	1.01	1.01	1.01	1.02	1.02	1.02	1.02	1.02
	.75	1.32	1.30	1.29	1.28	1.27	1.26	1.26	1.25	1.24	1.24	1.23	1.23
	.90	1.72	1.67	1.64	1.61	1.57	1.55	1.54	1.51	1.50	1.48	1.47	1.46
	.95	2.01	1.93	1.89	1.84	1.79	1.76	1.74	1.70	1.68	1.66	1.64	1.62
	.975	2.31	2.20	2.14	2.07	2.01	1.97	1.94	1.88	1.87	1.84	1.81	1.79
	.99	2.70	2.55	2.47	2.39	2.30	2.25	2.21	2.13	2.11	2.07	2.03	2.01
	.995	3.01	2.82	2.73	2.63	2.52	2.46	2.42	2.32	2.30	2.25	2.21	2.18
	.999	3.75	3.49	3.36	3.22	3.07	2.98	2.92	2.79	2.76	2.69	2.63	2.59
	.9995	4.10	3.80	3.65	3.48	3.32	3.22	3.15	3.00	2.97	2.89	2.82	2.78
40	.0005	.185	.236	.266	.301	.343	.373	.393	.441	.453	.480	.504	.525
	.001	.209	.259	.290	.326	.367	.396	.415	.461	.473	.500	.524	.545
	.005	.279	.331	.362	.396	.436	.463	.481	.524	.534	.559	.581	.599
	.01	.319	.371	.401	.435	.473	.498	.516	.556	.567	.592	.613	.628
	.025	.387	.437	.466	.498	.533	.556	.573	.610	.620	.641	.662	.674
	.05	.454	.502	.529	.558	.591	.613	.627	.658	.669	.685	.704	.717
	.10	.542	.585	.609	.636	.664	.683	.696	.724	.731	.747	.762	.772
	.25	.720	.752	.769	.787	.806	.819	.828	.846	.851	.861	.870	.877
	.50	.972	.983	.989	.994	1.00	1.00	1.01	1.01	1.01	1.01	1.02	1.02
	.75	1.30	1.28	1.26	1.25	1.24	1.23	1.22	1.21	1.21	1.20	1.19	1.19
	.90	1.66	1.61	1.57	1.54	1.51	1.48	1.47	1.43	1.42	1.41	1.39	1.38
	.95	1.92	1.84	1.79	1.74	1.69	1.66	1.64	1.59	1.58	1.55	1.53	1.51
	.975	2.18	2.07	2.01	1.94	1.88	1.83	1.80	1.74	1.72	1.69	1.66	1.64
	.99	2.52	2.37	2.29	2.20	2.11	2.06	2.02	1.94	1.92	1.87	1.83	1.80
	.995	2.78	2.60	2.50	2.40	2.30	2.23	2.18	2.09	2.06	2.01	1.96	1.93
	.999	3.40	3.15	3.01	2.87	2.73	2.64	2.57	2.44	2.41	2.34	2.28	2.23
	.9995	3.68	3.39	3.24	3.08	2.92	2.82	2.74	2.60	2.57	2.49	2.41	2.37
60	.0005	.192	.246	.278	.318	.365	.398	.421	.478	.493	.527	.561	.585
	.001	.216	.270	.304	.343	.389	.421	.444	.497	.512	.545	.579	.602
	.005	.287	.343	.376	.414	.458	.488	.510	.559	.572	.602	.633	.652
	.01	.328	.383	.416	.453	.495	.524	.545	.592	.604	.633	.658	.679
	.025	.396	.450	.481	.515	.555	.581	.600	.641	.654	.680	.704	.720
	.05	.463	.514	.543	.575	.611	.633	.652	.690	.700	.719	.746	.759
	.10	.550	.596	.622	.650	.682	.703	.717	.750	.758	.776	.793	.806
	.25	.725	.758	.776	.796	.816	.830	.840	.860	.865	.877	.888	.896
	.50	.967	.978	.983	.989	.994	.998	1.00	1.00	1.01	1.01	1.01	1.01
	.75	1.27	1.25	1.24	1.22	1.21	1.20	1.19	1.17	1.17	1.16	1.15	1.15
	.90	1.60	1.54	1.51	1.48	1.44	1.41	1.40	1.36	1.35	1.33	1.31	1.29
	.95	1.84	1.75	1.70	1.65	1.59	1.56	1.53	1.48	1.47	1.44	1.41	1.39
	.975	2.06	1.94	1.88	1.82	1.74	1.70	1.67	1.60	1.58	1.54	1.51	1.48
	.99	2.35	2.20	2.12	2.03	1.94	1.88	1.84	1.75	1.73	1.68	1.63	1.60
	.995	2.57	2.39	2.29	2.19	2.08	2.01	1.96	1.86	1.83	1.78	1.73	1.69
	.999	3.08	2.83	2.69	2.56	2.41	2.31	2.25	2.11	2.09	2.01	1.93	1.89
	.9995	3.30	3.02	2.87	2.71	2.55	2.45	2.38	2.23	2.19	2.11	2.03	1.98

ν_2	p \\ ν_1	1	2	3	4	5	6	7	8	9	10	11	12
120	.0005	$.0^4 40$	$.0^3 50$	$.0^2 51$	.016	.031	.049	.067	.087	.105	.123	.140	.156
	.001	$.0^5 16$	$.0^2 10$	$.0^2 81$	.023	.042	.063	.084	.105	.125	.144	.162	.179
	.005	$.0^4 39$	$.0^2 50$	.024	.051	.081	.111	.139	.165	.189	.211	.230	.249
	.01	$.0^3 16$	.010	.038	.074	.110	.143	.174	.202	.227	.250	.271	.290
	.025	$.0^3 99$	.025	.072	.120	.165	.204	.238	.268	.295	.318	.340	.359
	.05	$.0^2 39$	.051	.117	.177	.227	.270	.306	.337	.364	.388	.408	.427
	.10	.016	.105	.194	.265	.320	.365	.401	.432	.458	.480	.500	.518
	.25	.102	.288	.405	.481	.534	.574	.606	.631	.652	.670	.685	.699
	.50	.458	.697	.793	.844	.875	.896	.912	.923	.932	.939	.945	.950
	.75	1.34	1.40	1.39	1.37	1.35	1.33	1.31	1.30	1.29	1.28	1.27	1.26
	.90	2.75	2.35	2.13	1.99	1.90	1.82	1.77	1.72	1.68	1.65	1.62	1.60
	.95	3.92	3.07	2.68	2.45	2.29	2.18	2.09	2.02	1.96	1.91	1.87	1.83
	.975	5.15	3.80	3.23	2.89	2.67	2.52	2.39	2.30	2.22	2.16	2.10	2.05
	.99	6.85	4.79	3.95	3.48	3.17	2.96	2.79	2.66	2.56	2.47	2.40	2.34
	.995	8.18	5.54	4.50	3.92	3.55	3.28	3.09	2.93	2.81	2.71	2.62	2.54
	.999	11.4	7.32	5.79	4.95	4.42	4.04	3.77	3.55	3.38	3.24	3.12	3.02
	.9995	12.8	8.10	6.34	5.39	4.79	4.37	4.07	3.82	3.63	3.47	3.34	3.22
∞	.0005	$.0^6 39$	$.0^3 50$	$.0^2 51$	.016	.032	.050	.069	.088	.108	.127	.144	.161
	.001	$.0^5 16$	$.0^2 10$	$.0^2 81$	.023	.042	.063	.085	.107	.128	.148	.167	.185
	.005	$.0^4 39$	$.0^2 50$	.024	.052	.082	.113	.141	.168	.193	.216	.236	.256
	.01	$.0^3 16$	.010	.038	.074	.111	.145	.177	.206	.232	.256	.278	.298
	.025	$.0^3 98$	.025	.072	.121	.166	.206	.241	.272	.300	.325	.347	.367
	.05	$.0^2 39$	.051	.117	.178	.229	.273	.310	.342	.369	.394	.417	.436
	.10	.016	.105	.195	.266	.322	.367	.405	.436	.463	.487	.508	.525
	.25	.102	.288	.404	.481	.535	.576	.608	.634	.655	.674	.690	.703
	.50	.455	.693	.789	.839	.870	.891	.907	.918	.927	.934	.939	.945
	.75	1.32	1.39	1.37	1.35	1.33	1.31	1.29	1.28	1.27	1.25	1.24	1.24
	.90	2.71	2.30	2.08	1.94	1.85	1.77	1.72	1.67	1.63	1.60	1.57	1.55
	.95	3.84	3.00	2.60	2.37	2.21	2.10	2.01	1.94	1.88	1.83	1.79	1.75
	.975	5.02	3.69	3.12	2.79	2.57	2.41	2.29	2.19	2.11	2.05	1.99	1.94
	.99	6.63	4.61	3.78	3.32	3.02	2.80	2.64	2.51	2.41	2.32	2.25	2.18
	.995	7.88	5.30	4.28	3.72	3.35	3.09	2.90	2.74	2.62	2.52	2.43	2.36
	.999	10.8	6.91	5.42	4.62	4.10	3.74	3.47	3.27	3.10	2.96	2.84	2.74
	.9995	12.1	7.60	5.91	5.00	4.42	4.02	3.72	3.48	3.30	3.14	3.02	2.90

ν_2	p	ν_1 15	20	24	30	40	50	60	100	120	200	500	∞
120	.0005	.199	.256	.293	.338	.390	.429	.458	.524	.543	.578	.614	.676
	.001	.223	.282	.319	.363	.415	.453	.480	.542	.568	.595	.631	.691
	.005	.297	.356	.393	.434	.484	.520	.545	.605	.623	.661	.702	.733
	.01	.338	.397	.433	.474	.522	.556	.579	.636	.652	.688	.725	.755
	.025	.406	.464	.498	.536	.580	.611	.633	.684	.698	.729	.762	.789
	.05	.473	.527	.559	.594	.634	.661	.682	.727	.740	.767	.785	.819
	.10	.560	.609	.636	.667	.702	.726	.742	.781	.791	.815	.838	.855
	.25	.730	.765	.784	.805	.828	.843	.853	.877	.884	.897	.911	.923
	.50	.961	.972	.978	.983	.989	.992	.994	1.00	1.00	1.00	1.01	1.01
	.75	1.24	1.22	1.21	1.19	1.18	1.17	1.16	1.14	1.13	1.12	1.11	1.10
	.90	1.55	1.48	1.45	1.41	1.37	1.34	1.32	1.27	1.26	1.24	1.21	1.19
	.95	1.75	1.66	1.61	1.55	1.50	1.46	1.43	1.37	1.35	1.32	1.28	1.25
	.975	1.95	1.82	1.76	1.69	1.61	1.56	1.53	1.45	1.43	1.39	1.34	1.31
	.99	2.19	2.03	1.95	1.86	1.76	1.70	1.66	1.56	1.53	1.48	1.42	1.38
	.995	2.37	2.19	2.09	1.98	1.87	1.80	1.75	1.64	1.61	1.54	1.48	1.43
	.999	2.78	2.53	2.40	2.26	2.11	2.02	1.95	1.82	1.76	1.70	1.62	1.54
	.9995	2.96	2.67	2.53	2.38	2.21	2.11	2.01	1.88	1.84	1.75	1.67	1.60
∞	.0005	.207	.270	.311	.360	.422	.469	.505	.599	.624	.704	.804	1.00
	.001	.232	.296	.338	.386	.448	.493	.527	.617	.649	.719	.819	1.00
	.005	.307	.372	.412	.460	.518	.559	.592	.671	.699	.762	.843	1.00
	.01	.349	.413	.452	.499	.554	.595	.625	.699	.724	.782	.858	1.00
	.025	.418	.480	.517	.560	.611	.645	.675	.741	.763	.813	.878	1.00
	.05	.484	.543	.577	.617	.663	.694	.720	.781	.797	.840	.896	1.00
	.10	.570	.622	.652	.687	.726	.752	.774	.826	.838	.877	.919	1.00
	.25	.736	.773	.793	.816	.842	.860	.872	.901	.910	.932	.957	1.00
	.50	.956	.967	.972	.978	.983	.987	.989	.993	.994	.997	.999	1.00
	.75	1.22	1.19	1.18	1.16	1.14	1.13	1.12	1.09	1.08	1.07	1.04	1.00
	.90	1.49	1.42	1.38	1.34	1.30	1.26	1.24	1.18	1.17	1.13	1.08	1.00
	.95	1.67	1.57	1.52	1.46	1.39	1.35	1.32	1.24	1.22	1.17	1.11	1.00
	.975	1.83	1.71	1.64	1.57	1.48	1.43	1.39	1.30	1.27	1.21	1.13	1.00
	.99	2.04	1.88	1.79	1.70	1.59	1.52	1.47	1.36	1.32	1.25	1.15	1.00
	.995	2.19	2.00	1.90	1.79	1.67	1.59	1.53	1.40	1.36	1.28	1.17	1.00
	.999	2.51	2.27	2.13	1.99	1.84	1.73	1.66	1.49	1.45	1.34	1.21	1.00
	.9995	2.65	2.37	2.22	2.07	1.91	1.79	1.71	1.53	1.48	1.36	1.22	1.00

APPENDIX 7

RANDOM NUMBERS

	00 04	05 09	10 14	15 19	20 24	25 29	30 34	35 39	40 44	45 49
00	39591	66082	48626	95780	55228	87189	75717	97042	19696	48613
01	46304	97377	43462	21739	14566	72533	60171	29024	77581	72760
02	99547	60779	22734	23678	44895	89767	18249	41702	35850	40543
03	06743	63537	24553	77225	94743	79448	12753	95986	78088	48019
04	69568	65496	49033	88577	98606	92156	08846	54912	12691	13170
05	68198	69571	34349	73141	42640	44721	30462	35075	33475	47407
06	27974	12609	77428	64441	49008	60489	66780	55499	80842	57706
07	50552	20688	02769	63037	15494	71784	70559	58158	53437	46216
08	74687	02033	98290	62635	88877	28599	63682	35566	03271	05651
09	49303	76629	71897	30990	62923	36686	96167	11492	90333	84501
10	89734	39183	52026	14997	15140	18250	62831	51236	61236	09179
11	74042	40747	02617	11346	01884	82066	55913	72422	13971	64209
12	84706	31375	67053	73367	95349	31074	36908	42782	89690	48002
13	83664	21365	28882	48926	45435	60577	85270	02777	06878	27561
14	47813	74854	73388	11385	99108	97878	32858	17473	07682	20166
15	00371	56525	38880	53702	09517	47281	15995	98350	25233	79718
16	81182	48434	27431	55806	25389	40774	72978	16835	65066	28732
17	75242	35904	73077	24537	81354	48902	03478	42867	04552	66034
18	96239	80246	07000	09555	55051	49596	44629	88225	28195	44598
19	82988	17440	85311	03360	38176	51462	86070	03924	84413	92363
20	77599	29143	89088	57593	60036	17297	30923	36224	46327	96266
21	61433	33118	53488	82981	44709	63655	64388	00498	14135	57514
22	76008	15045	45440	84062	52363	18079	33726	44301	86246	99727
23	26494	76598	85834	10844	56300	02244	72118	96510	98388	80161
24	46570	88558	77533	33359	07830	84752	53260	46755	36881	98535
25	73995	41532	87933	79930	14310	64833	49020	70067	99726	97007
26	93901	38276	75544	19679	82899	11365	22896	42118	77165	08734
27	41925	28215	40966	93501	45446	27913	21708	01788	81404	15119
28	80720	02782	24326	41328	10357	86883	80086	77138	57072	12100
29	92596	39416	50362	04423	04561	58179	54188	44978	14322	97056
30	39693	58559	45839	47278	38548	38885	19875	26829	86711	57005
31	86923	37863	14340	30929	04079	65274	03030	15106	09362	82972
32	99700	79237	18172	58879	56221	65644	33331	87502	32961	40996
33	60248	21953	52321	16984	03252	90433	97304	50181	71026	01946
34	29136	71987	03992	67025	31070	78348	47823	11033	13037	47732
35	57471	42913	85212	42319	92901	97727	04775	94396	38154	25238
36	57424	93847	03269	56096	95028	14039	76128	63747	27301	65529
37	56768	71694	63361	80836	30841	71875	40944	54827	01887	54822
38	70400	81534	02148	41441	26582	27481	84262	14084	42409	62950
39	05454	88418	48646	99565	36635	85496	18894	77271	26894	00889
40	80934	56136	47063	96311	19067	59790	08752	68040	85685	83076
41	06919	46237	50676	11238	75637	43086	95323	52867	06891	32089
42	00152	23997	41751	74756	50975	75365	70158	67663	51431	46375
43	88505	74625	71783	82511	13661	63178	39291	76796	74736	10980
44	64514	80967	33545	09582	86329	58152	05931	35961	70069	12142
45	25280	53007	99651	96366	49378	80971	10419	12981	70572	11575
46	71292	63716	93210	59312	39493	24252	54849	29754	41497	79228
47	49734	50498	08974	05904	68172	02864	10994	22482	12912	17920
48	43075	09754	71880	92614	99928	94424	86353	87549	94499	11459
49	15116	16643	03981	06566	14050	33671	03814	48856	41267	76252

Source: Reproduced from George W. Snedecor, *Everyday Statistics.* Copyright 1950. Published by Wm. C. Brown Company, Dubuque, Iowa. By permission of the author and publishers.

50 54	55 59	60 64	65 69	70 74	75 79	80 84	85 89	90 94	95 99	
25178	77518	41773	39926	09843	29694	43801	69276	44707	23455	00
45803	95106	85816	33366	37383	76832	37024	06581	22587	24827	01
15532	30898	14922	13923	44987	45122	86515	55836	96165	19650	02
99068	35453	42152	12078	04913	06083	06645	93310	40016	85421	03
70983	88359	95583	79848	24101	67502	25692	42496	77732	19278	04
71181	48289	03153	18779	65702	03612	64608	84071	47588	09982	05
44052	59163	74033	86112	27731	46135	63092	59171	44816	12354	06
91555	87708	70964	43346	56811	08725	75139	77674	82467	41899	07
54307	12188	58089	73745	35569	97352	77301	37684	36823	69218	08
63631	23919	06785	13891	89918	76211	09362	34292	17640	65907	09
46832	30801	98898	28954	97793	20825	36775	71974	15574	09184	10
05944	82632	39310	74857	61725	50569	81937	16820	85446	51168	11
28199	90116	59501	49025	73005	84954	11587	97691	90415	84685	12
08391	05600	00624	95068	33776	44985	01505	76911	45539	32181	13
29634	13021	96568	15124	55092	44043	31073	92371	51288	33378	14
61509	18842	79201	46451	68594	98120	68110	91062	42095	61839	15
87888	23033	69837	65661	15130	44649	42515	83861	50721	36110	16
94585	15218	74838	61809	92293	85400	46934	08531	70107	65707	17
82033	93915	34898	79913	70013	27573	39256	35167	35070	47095	18
79131	10022	82199	78976	22702	37936	10445	96846	84927	69745	19
79344	39236	41333	11473	15049	47930	99029	97150	82275	55149	20
15384	44585	18773	89733	40779	59664	83328	25162	58758	17761	21
38802	90957	32910	97485	10358	88588	95310	22252	19143	69011	22
85874	18400	28151	29541	63706	43197	65726	94117	22169	91806	23
26200	72680	12364	46010	92208	59103	60417	45389	56122	85353	24
13772	75282	81418	42188	66529	47981	92548	10079	68179	40915	25
91876	07434	96946	98382	97374	34444	17992	42811	01579	48741	26
31721	21713	83632	40605	24227	53219	05482	86768	53239	24812	27
92570	53242	98133	84706	78048	29645	79336	66091	05793	25922	28
02880	29307	73734	66448	64739	74645	29562	13999	17492	49891	29
80982	14684	31038	85302	98349	57313	86371	33938	10768	60837	30
38000	43364	94825	32413	46781	09685	69058	56644	85531	55173	31
14218	94289	79484	61868	40034	22546	68726	14736	89844	13466	32
74358	21940	40280	22233	09123	49375	55094	46113	54046	51771	33
39049	14986	94000	26649	13037	34609	45186	89515	63214	66886	34
48727	06300	91486	67316	84576	11100	37580	49629	83224	46321	35
22719	29784	40682	96715	40745	57458	70048	48306	50270	87424	36
33980	36769	51977	03689	79071	20279	64787	48877	44063	93733	37
23885	66721	16542	12648	65986	43104	45583	75729	35118	58742	38
85190	44068	78477	69133	58983	96504	44232	74809	25266	73872	39
33453	36333	45814	78128	55914	89829	43251	41634	48488	49153	40
98236	11489	97240	01678	30779	75214	80039	68895	95271	19654	41
21295	53563	43609	48439	87427	88065	09892	58524	43815	31340	42
28335	79849	69842	71669	38770	54445	48736	03242	83181	85403	43
95449	35273	62581	85522	35813	34475	97514	72839	10387	31649	44
88167	03878	89405	55461	73248	48620	31732	47317	06252	54652	45
86131	62596	98785	02360	54271	26242	93735	20752	17146	18315	46
71134	90264	30126	08586	97497	61678	81940	00907	39096	02082	47
02664	53438	76839	52290	77999	05799	93744	16634	84924	31344	48
90664	96876	16663	25608	67140	84619	67167	13192	81774	58619	49

	00 04	05 09	10 14	15 19	20 24	25 29	30 34	35 39	40 44	45 49
50	93873	86558	72524	02542	73184	37905	05882	15596	73646	50798
51	08761	47547	02216	48086	56490	89959	69975	04500	23779	76697
52	61270	98773	40298	26077	80396	08166	35723	61933	13985	19102
53	73758	15578	95748	02967	35122	36539	72822	68241	34803	42457
54	17132	32196	60523	00544	73700	70122	27962	85597	36011	79971
55	26175	29794	44838	84414	82748	22246	70694	57953	39780	17791
56	06004	04516	06210	03536	84451	30767	37928	26986	07396	64611
57	34687	73753	36327	73704	61564	99434	90938	03967	97420	19913
58	27865	08255	57859	04746	79700	68823	16002	58115	07589	12675
59	89423	51114	90820	26786	77404	05795	49036	34686	98767	32284
60	99030	80312	69745	87636	10058	84834	89485·	08775	19041	61375
61	02852	54339	45496	20587	85921	06763	68873	35367	42627	54973
62	10850	42788	94737	74549	74296	13053	46816	32141	02533	25648
63	38301	18507	33151	69434	80103	02603	61110	89395	67621	67025
64	48181	95478	62739	90148	00156	09338	44558	53271	87549	45974
65	23098	23720	76508	69083	56584	90423	21634	35999	09234	95116
66	25104	82019	21120	06165	44324	77577	15774	44091	69687	67576
67	22205	40198	86884	28103	57306	54915	03426	66700	45993	36668
68	64975	05064	29617	40622	20330	18518	45312	57921	23188	82361
69	58710	75278	47730	26093	16436	38868	76861	85914	14162	21984
70	12140	72905	26022	07675	16362	34504	47740	39923	04081	03162
71	73226	39840	47958	97249	14146	34543	76162	74158	59739	67447
72	12320	86217	66162	70941	58940	58006	80731	66680	02183	94678
73	41364	64156	23000	23188	64945	33815	32884	76955	56574	61666
74	97881	80867	70117	72041	03554	29087	19767	71838	80545	61402
75	88295	87271	82812	97588	09960	06312	03050	77332	25977	18385
76	95321	89836	78230	46037	72483	87533	74571	88859	26908	55626
77	24337	14264	30185	36753	22343	81737	62926	76494	93536	75502
78	00718	66303	75009	91431	64245	61863	16738	23127	89435	45109
79	38093	10328	96998	91386	34967	40407	48380	09115	59367	49596
80	87661	31701	29974	56777	66751	35181	63887	95094	20056	84990
81	87142	91818	51857	85061	17890	39057	44506	00969	32942	54794
82	60634	27142	21199	50437	04685	70252	91453	75952	66753	50664
83	73356	64431	05068	56334	34487	78253	67684	69916	63885	88491
84	29889	11378	65915	66776	95034	81447	98035	16815	68432	63020
85	48257	36438	48479	72173	31418	14035	84239	02032	40409	11715
86	38425	29462	79880	45713	90049	01136	72426	25077	64361	94284
87	48226	31868	38620	12135	28346	17552	03293	42618	44151	78438
88	80189	30031	15435	76730	58565	29817	36775	64007	47912	16754
89	33208	33475	95219	29832	74569	50667	90569	66717	46958	04820
90	19750	48564	49690	43352	53884	80125	47795	99701	06800	22794
91	62820	23174	71124	36040	34873	95650	79059	23894	58534	78296
92	95737	34362	81520	79481	26442	37826	76866	01580	83713	94272
93	64642	62961	37566	41064	69372	84369	92823	91391	61056	44495
94	77636	60163	14915	50744	95611	99346	39741	04407	72940	87936
95	43633	52102	93561	31010	11299	52661	79014	17910	88492	60753
96	93686	41960	61280	96529	52924	87371	34855	67125	40279	10186
97	23775	33402	28647	42314	51213	29116	26243	40243	32137	25177
98	91325	64698	58868	63107	08993	96000	66854	11567	80604	72299
99	58129	44367	31924	73586	24422	92799	28963	36444	01315	10226

50 54	55 59	60 64	65 69	70 74	75 79	80 84	85 89	90 94	95 99	
37686	78520	31209	83677	99115	94024	09286	58927	24078	16770	50
58108	29344	11825	51955	50618	99753	02200	50503	32466	50055	51
71545	42326	66429	93607	55276	85482	24449	41764	19884	46443	52
93303	90557	79166	90097	01627	96690	77434	06402	05379	59549	53
36731	37929	13079	83036	31525	35811	59131	65257	03731	86703	54
49781	31581	80391	84608	23390	30433	08240	85136	80060	43651	55
65995	94208	68785	04370	44192	91852	01129	28739	08705	54538	56
19663	09309	02836	10223	90814	92786	96747	46014	54765	76001	57
88479	24307	63812	47615	17220	27942	11785	49933	03923	35432	58
95407	95006	95421	20811	76761	47475	58865	06204	36543	81002	59
22789	87011	61926	97996	10604	80855	48714	52754	98279	96467	60
96783	18403	36729	18760	30810	73087	94565	68682	15792	60020	61
68933	05665	12264	23954	01583	75411	04460	83939	66528	22576	62
68794	13000	20066	98963	93483	51165	63358	12373	13877	37580	63
40537	31604	60323	51235	65546	85117	15647	09617	73520	48525	64
41249	42504	91773	81579	02882	74657	73765	10932	74607	83825	65
08813	84525	30329	33144	76884	89996	07834	67266	96820	15128	66
46609	30917	29996	10848	39555	09233	58988	82131	69232	76762	67
68543	69424	92072	57937	05563	80727	67053	35431	00881	56541	68
09926	84219	30089	08843	24998	27105	18397	79071	40738	73876	69
30515	76316	49597	37900	98604	05857	51729	19006	15239	27129	70
21611	26346	04877	71584	55724	39616	64648	36811	60915	34108	71
47410	83767	56454	96768	27001	83712	01245	27256	57991	75758	72
18572	31214	41015	64110	61807	72472	78059	69701	78681	17356	73
28078	02819	02459	33308	96540	15817	78694	81476	87856	99737	74
56644	50430	34562	75842	67724	02918	55603	55195	88219	39676	75
27331	48055	18928	47763	61966	64507	06559	81329	29481	03660	76
32080	21524	32929	07739	00836	39497	94476	27433	96857	52987	77
27027	69762	65362	90214	89572	52054	43067	73017	87664	03293	78
56471	68839	09969	45853	72627	71793	49920	64544	71874	74053	79
22689	19799	18870	49272	74783	38777	76176	40961	18089	32499	80
71263	82247	66684	90239	67686	48963	30842	59354	33551	87966	81
64084	57386	89278	27187	52142	96305	87393	80164	95518	82742	82
23121	10194	09911	37062	43446	09107	47156	70179	00858	92326	83
78906	48080	76745	65814	51167	87755	66884	12718	14951	47937	84
87257	26005	21544	37223	53288	72056	96396	67099	49416	91891	85
39529	98126	33694	29025	94308	24426	63072	51444	04718	49891	86
89632	11606	87159	89408	06295	31055	15530	46432	49871	37982	87
23708	98919	14407	53722	58779	92849	04176	24870	56688	25405	88
51445	46758	42024	27940	64237	10086	95601	53923	85209	79385	89
23849	65272	24743	39960	27313	99925	29743	87270	05773	21797	90
78613	15441	34568	57398	25872	61792	94599	60944	90908	38948	91
90694	27996	94181	87428	41135	29461	72716	68956	67871	72459	92
96772	86829	36403	40087	67456	21071	39039	91937	45280	00066	93
24527	40701	56894	73327	00789	97573	09303	41704	05772	95372	94
31596	70876	46807	06741	29352	23829	52465	00336	24155	61871	95
31613	99249	17260	05242	19535	52702	64761	66694	06150	13820	96
02911	09514	50864	80622	20017	59019	43450	75942	08567	40547	97
02484	74068	04671	19646	41951	05111	34013	57443	87481	48994	98
69259	75535	73007	15236	01572	44870	53280	25132	70276	87334	99

APPENDIX 8

CONTROL CHART CONSTANTS

Source: Reprinted from ASTM Manual on Quality Control of Materials, Special Technical Publication 15-C, Table B2, American Society for Testing Materials, Philadelphia, 1951. By permission of the authors and publishers.

Number of Observations in Sample, n	Chart for Averages — Factors for control limits			Chart for Standard Deviations — Factors for central line		Chart for Standard Deviations — Factors for control limits				Chart for Ranges — Factors for central line			Chart for Ranges — Factors for control limits			
	A	A_1	A_2	c_2	$1/c_2$	B_1	B_2	B_3	B_4	d_2	$1/d_2$	d_3	D_1	D_2	D_3	D_4
2	2.121	3.760	1.880	0.5642	1.7725	0	1.843	0	3.267	1.128	0.8865	0.853	0	3.686	0	3.267
3	1.732	2.394	1.023	0.7236	1.3820	0	1.858	0	2.568	1.693	0.5907	0.888	0	4.358	0	2.575
4	1.500	1.880	0.729	0.7979	1.2533	0	1.808	0	2.266	2.059	0.4857	0.880	0	4.698	0	2.282
5	1.342	1.596	0.577	0.8407	1.1894	0	1.756	0	2.089	2.326	0.4299	0.864	0	4.918	0	2.115
6	1.225	1.410	0.483	0.8686	1.1512	0.026	1.711	0.030	1.970	2.534	0.3946	0.848	0	5.078	0	2.004
7	1.134	1.277	0.419	0.8882	1.1259	0.105	1.672	0.118	1.882	2.704	0.3698	0.833	0.205	5.203	0.076	1.924
8	1.061	1.175	0.373	0.9027	1.1078	0.167	1.638	0.185	1.815	2.847	0.3512	0.820	0.387	5.307	0.136	1.864
9	1.000	1.094	0.337	0.9139	1.0942	0.219	1.609	0.239	1.761	2.970	0.3367	0.808	0.546	5.394	0.184	1.816
10	0.949	1.028	0.308	0.9227	1.0837	0.262	1.584	0.284	1.716	3.078	0.3249	0.797	0.687	5.469	0.223	1.777
11	0.905	0.973	0.285	0.9300	1.0753	0.299	1.561	0.321	1.679	3.173	0.3152	0.787	0.812	5.534	0.256	1.744
12	0.866	0.925	0.266	0.9359	1.0684	0.331	1.541	0.354	1.646	3.258	0.3069	0.778	0.924	5.592	0.284	1.716
13	0.832	0.884	0.249	0.9410	1.0627	0.359	1.523	0.382	1.618	3.336	0.2998	0.770	1.026	5.646	0.308	1.692
14	0.802	0.848	0.235	0.9453	1.0579	0.384	1.507	0.406	1.594	3.407	0.2935	0.762	1.121	5.693	0.329	1.671
15	0.775	0.816	0.223	0.9490	1.0537	0.406	1.492	0.428	1.572	3.472	0.2880	0.755	1.207	5.737	0.348	1.652
16	0.750	0.788	0.212	0.9523	1.0501	0.427	1.478	0.448	1.552	3.532	0.2831	0.749	1.285	5.779	0.364	1.636
17	0.728	0.762	0.203	0.9551	1.0470	0.445	1.465	0.466	1.534	3.588	0.2787	0.743	1.359	5.817	0.379	1.621
18	0.707	0.738	0.194	0.9576	1.0442	0.461	1.454	0.482	1.518	3.640	0.2747	0.738	1.426	5.854	0.392	1.608
19	0.688	0.717	0.187	0.9599	1.0418	0.477	1.443	0.497	1.503	3.689	0.2711	0.733	1.490	5.888	0.404	1.596
20	0.671	0.697	0.180	0.9619	1.0396	0.491	1.433	0.510	1.490	3.735	0.2677	0.729	1.548	5.922	0.414	1.586
21	0.655	0.679	0.173	0.9638	1.0376	0.504	1.424	0.523	1.477	3.778	0.2647	0.724	1.606	5.950	0.425	1.575
22	0.640	0.662	0.167	0.9655	1.0358	0.516	1.415	0.534	1.466	3.819	0.2618	0.720	1.659	5.979	0.434	1.566
23	0.626	0.647	0.162	0.9670	1.0342	0.527	1.407	0.545	1.455	3.858	0.2592	0.716	1.710	6.006	0.443	1.557
24	0.612	0.632	0.157	0.9684	1.0327	0.538	1.399	0.555	1.445	3.895	0.2567	0.712	1.759	6.031	0.452	1.548
25	0.600	0.619	0.153	0.9696	1.0313	0.548	1.392	0.565	1.435	3.931	0.2544	0.709	1.804	6.058	0.459	1.541
Over 25	$\dfrac{3}{\sqrt{n}}$	$\dfrac{3}{\sqrt{n}}$	......	......	......	*	**	*	**	......	......	......	......	......	......	......

$$* \; 1 - \frac{3}{\sqrt{2n}} \qquad\qquad ** \; 1 + \frac{3}{\sqrt{2n}}$$

APPENDIX 9

NUMBER OF OBSERVATIONS FOR
t TEST OF MEAN

Source: Reproduced from Table E of Owen L. Davies, *The Design and Analysis of Industrial Experiments,* 2nd ed., Oliver and Boyd, Edinburgh, 1956. By permission of the author and publishers.

Note: The entries in this table show the number of observations needed in a *t* test of the significance of a mean in order to control the probabilities of errors of the first and second kinds at α and β respectively.

Level of t Test

	0.01					0.02					0.05					0.1					
Single-Sided Test Double-Sided Test	α=0.005 α=0.01					α=0.01 α=0.02					α=0.025 α=0.05					α=0.05 α=0.1					
β=	0.01	0.05	0.1	0.2	0.5	0.01	0.05	0.1	0.2	0.5	0.01	0.05	0.1	0.2	0.5	0.01	0.05	0.1	0.2	0.5	$D=\dfrac{\delta}{\sigma}$
0.05																					0.05
0.10																					0.10
0.15																				122	0.15
0.20										139					99					70	0.20
0.25					110					90				128	64			139	101	45	0.25
0.30				134	78				115	63			119	90	45		122	97	71	32	0.30
0.35			125	99	58			109	85	47		109	88	67	34		90	72	52	24	0.35
0.40		115	97	77	45		101	85	66	37	117	84	68	51	26	101	70	55	40	19	0.40
0.45		92	77	62	37	110	81	68	53	30	93	67	54	41	21	80	55	44	33	15	0.45
0.50	100	75	63	51	30	90	66	55	43	25	76	54	44	34	18	65	45	36	27	13	0.50
0.55	83	63	53	42	26	75	55	46	36	21	63	45	37	28	15	54	38	30	22	11	0.55
0.60	71	53	45	36	22	63	47	39	31	18	53	38	32	24	13	46	32	26	19	9	0.60
0.65	61	46	39	31	20	55	41	34	27	16	46	33	27	21	12	39	28	22	17	8	0.65
0.70	53	40	34	28	17	47	35	30	24	14	40	29	24	19	10	34	24	19	15	8	0.70
0.75	47	36	30	25	16	42	31	27	21	13	35	26	21	16	9	30	21	17	13	7	0.75
0.80	41	32	27	22	14	37	28	24	19	12	31	22	19	15	9	27	19	15	12	6	0.80
0.85	37	29	24	20	13	33	25	21	17	11	28	21	17	13	8	24	17	14	11	6	0.85
0.90	34	26	22	18	12	29	23	19	16	10	25	19	16	12	7	21	15	14	11	6	0.90
0.95	31	24	20	17	11	27	21	18	14	9	23	17	14	11	7	19	14	11	9	5	0.95
1.00	28	22	19	16	10	25	19	16	13	9	21	16	13	10	6	18	13	11	8	5	1.00
1.1	24	19	16	14	9	21	16	14	12	8	18	13	11	9	6	15	11	9	7		1.1
1.2	21	16	14	12	8	18	14	12	10	7	15	12	10	8	5	13	10	8	6		1.2
1.3	18	15	13	11	8	16	13	11	9	6	14	10	9	7		11	8	7	6		1.3
1.4	16	13	12	10	7	14	11	10	9	6	12	9	8	7		10	8	7	5		1.4
1.5	15	12	11	9	7	13	10	9	8	6	11	8	7	6		9	7	6			1.5
1.6	13	11	10	8	6	12	10	9	7	5	10	8	7	6		8	6	6			1.6
1.7	12	10	9	8	6	11	9	8	7		9	7	6	5		8	6	5			1.7
1.8	12	10	9	8	6	10	8	7	7		8	7	6			7	6				1.8
1.9	11	9	8	7	6	10	8	7	6		8	6	6			7	5				1.9
2.0	10	8	8	7	5	9	7	7	6		7	6	5			6					2.0
2.1	10	8	7	7		8	7	6	6		7	6				6					2.1
2.2	9	8	7	6		8	7	6	5		7	6				6					2.2
2.3	9	7	7	6		8	6	6			6	5				5					2.3
2.4	8	7	7	6		7	6	6			6										2.4
2.5	8	7	6	6		7	6	6			6										2.5
3.0	7	6	6	5		6	5	5			5										3.0
3.5	6	5	5			5															3.5
4.0	6																				4.0

NUMBER OF OBSERVATIONS FOR *t* TEST OF DIFFERENCE BETWEEN TWO MEANS

Source: Reproduced from Table E.1 of Owen L. Davies, *The Design and Analysis of Industrial Experiments,* 2nd ed., Oliver and Boyd, Edinburgh, 1956. By permission of the author and publishers.

Note: The entries in this table show the number of observations needed in a *t* test of the significance of the difference between two means in order to control the probabilities of the errors of the first and second kinds at α and β respectively. "It should be noted that the entries in the table show the number of observations needed in *each* of two samples of equal size."

Level of t Test

Level of t Test	Single-Sided Test	Double-Sided Test
0.01	$\alpha = 0.005$	$\alpha = 0.01$
0.02	$\alpha = 0.01$	$\alpha = 0.02$
0.05	$\alpha = 0.025$	$\alpha = 0.05$
0.1	$\alpha = 0.05$	$\alpha = 0.1$

Value of $D = \dfrac{\delta}{\sigma}$

D	0.01	0.05	0.1	0.2	0.5	0.01	0.05	0.1	0.2	0.5	0.01	0.05	0.1	0.2	0.5	0.01	0.05	0.1	0.2	0.5	D
0.05																					0.05
0.10																					0.10
0.15																					0.15
0.20																					0.20
0.25															124					137	0.25
0.30															87					88	0.30
0.35					110					123					64					61	0.35
0.40					85					90				100	50				102	45	0.40
0.45				118	68				101	70			105	79	39			108	78	35	0.45
0.50				96	55			106	82	55		106	86	64	32		88	70	51	23	0.50
0.55			101	79	46		106	88	68	38		87	71	53	27	112	73	58	42	19	0.55
0.60		101	85	67	39		90	74	58	32	104	74	60	45	23	89	61	49	36	16	0.60
0.65		87	73	57	34	104	77	64	49	27	88	63	51	39	20	76	52	42	30	14	0.65
0.70	100	75	63	50	29	90	66	55	43	24	76	55	44	34	17	66	45	36	26	12	0.70
0.75	88	66	55	44	26	79	58	48	38	21	67	48	39	29	15	57	40	32	23	11	0.75
0.80	77	58	49	39	23	70	51	43	33	19	59	42	34	26	14	50	35	28	21	10	0.80
0.85	69	51	43	35	21	62	46	38	30	17	52	37	31	23	12	45	31	25	18	9	0.85
0.90	62	46	39	31	19	55	41	34	27	15	47	34	27	21	11	40	28	22	16	8	0.90
0.95	55	42	35	28	17	50	37	31	24	14	42	30	25	19	10	36	25	20	15	7	0.95
1.00	50	38	32	26	15	45	33	28	22	13	38	27	23	17	9	33	23	18	14	7	1.00
1.1	42	32	27	22	13	38	28	23	19	11	32	23	19	14	8	27	19	15	12	6	1.1
1.2	36	27	23	18	11	32	24	20	16	9	27	20	16	12	7	23	16	13	10	5	1.2
1.3	31	23	20	16	10	28	21	17	14	8	23	17	14	11	6	20	14	11	9	5	1.3
1.4	27	20	17	14	9	24	18	15	12	8	20	15	12	10	6	17	12	10	8	6	1.4
1.5	24	18	15	13	8	21	16	14	11	7	18	13	11	9	5	15	11	9	7	4	1.5
1.6	21	16	14	11	7	19	14	12	10	6	16	12	10	8	5	14	10	8	6	4	1.6
1.7	19	15	13	10	7	17	13	11	9	6	14	11	9	7	4	12	9	7	6	3	1.7
1.8	17	13	11	10	6	15	12	10	8	5	13	10	8	6	4	11	8	7	5		1.8
1.9	16	12	11	9	6	14	11	9	8	5	12	9	7	6	4	10	7	6	5		1.9
2.0	14	11	10	8	6	13	10	9	7	5	11	8	7	6	4	9	7	6	4		2.0
2.1	13	10	9	8	5	12	9	8	7	5	10	8	6	5	3	8	6	5	4		2.1
2.2	12	10	8	7	5	11	9	7	6	4	9	7	6	5		8	6	5	4		2.2
2.3	11	9	8	7	5	10	8	7	6	4	9	7	6	5		7	5	5	4		2.3
2.4	11	9	8	6	5	10	8	7	6	4	8	6	5	4		7	5	4	4		2.4
2.5	10	8	7	6	4	9	7	6	5	4	8	6	5	4		6	5	4	3		2.5
3.0	8	6	6	5	4	7	6	5	4	3	6	5	4	4		5	4	3			3.0
3.5	6	5	5	4	3	6	5	4	4		5	4	4	3		4	3				3.5
4.0	6	5	4	4		5	4	4	3		4	4	3			4					4.0

NUMBER OF OBSERVATIONS REQUIRED FOR THE COMPARISON OF A POPULATION VARIANCE WITH A STANDARD VALUE USING THE χ^2 TEST

ν	$\alpha=0.01$				$\alpha=0.05$			
	$\beta=0.01$	$\beta=0.05$	$\beta=0.1$	$\beta=0.5$	$\beta=0.01$	$\beta=0.05$	$\beta=0.1$	$\beta=0.5$
1	42,240	1,687	420.2	14.58	24,450	977.0	243.3	8.444
2	458.2	89.78	43.71	6.644	298.1	58.40	28.43	4.322
3	98.79	32.24	19.41	4.795	68.05	22.21	13.37	3.303
4	44.69	18.68	12.48	3.955	31.93	13.35	8.920	2.826
5	27.22	13.17	9.369	3.467	19.97	9.665	6.875	2.544
6	19.28	10.28	7.628	3.144	14.44	7.699	5.713	2.354
7	14.91	8.524	6.521	2.911	11.35	6.491	4.965	2.217
8	12.20	7.352	5.757	2.736	9.418	5.675	4.444	2.112
9	10.38	6.516	5.198	2.597	8.103	5.088	4.059	2.028
10	9.072	5.890	4.770	2.484	7.156	4.646	3.763	1.960
12	7.343	5.017	4.159	2.312	5.889	4.023	3.335	1.854
15	5.847	4.211	3.578	2.132	4.780	3.442	2.925	1.743
20	4.548	3.462	3.019	1.943	3.802	2.895	2.524	1.624
24	3.959	3.104	2.745	1.842	3.354	2.630	2.326	1.560
30	3.403	2.752	2.471	1.735	2.927	2.367	2.125	1.492
40	2.874	2.403	2.192	1.619	2.516	2.103	1.919	1.418
60	2.358	2.046	1.902	1.490	2.110	1.831	1.702	1.333
120	1.829	1.661	1.580	1.332	1.686	1.532	1.457	1.228
∞	1.000	1.000	1.000	1.000	1.000	1.000	1.000	1.000

Source: Reproduced from Table G of Owen L. Davies, *The Design and Analysis of Industrial Experiments,* 2nd ed., Oliver and Boyd, Edinburgh, 1956. By permission of the author and publishers.

Note: The entries in this table show the value of the ratio R of the population variance σ^2 to a standard variance σ_0^2 that will be undetected with probability β in a χ^2 test at the 100α percent significance level of an estimate s^2 of σ^2 based on ν degrees of freedom.

APPENDIX 12

NUMBER OF OBSERVATIONS REQUIRED FOR THE COMPARISON OF TWO POPULATION VARIANCES USING THE F TEST

ν	$\alpha=0.01$				$\alpha=0.05$				$\alpha=0.5$			
	$\beta=0.01$	$\beta=0.05$	$\beta=0.1$	$\beta=0.5$	$\beta=0.01$	$\beta=0.05$	$\beta=0.1$	$\beta=0.5$	$\beta=0.01$	$\beta=0.05$	$\beta=0.1$	$\beta=0.5$
1	16,420,000	654,200	161,500	4,052	654,200	26,070	6,436	161.5	4,052	161.5	39.85	1.000
2	9,801	1,881	891.0	99.00	1,881	361.0	171.0	19.00	99.00	19.00	9.000	1.000
3	867.7	273.3	158.8	29.46	273.3	86.06	50.01	9.277	29.46	9.277	5.391	1.000
4	255.3	102.1	65.62	15.98	102.1	40.81	26.24	6.388	15.98	6.388	4.108	1.000
5	120.3	55.39	37.87	10.97	55.39	25.51	17.44	5.050	10.97	5.050	3.453	1.000
6	71.67	36.27	25.86	8.466	36.27	18.35	13.09	4.284	8.466	4.284	3.056	1.000
7	48.90	26.48	19.47	6.993	26.48	14.34	10.55	3.787	6.993	3.787	2.786	1.000
8	36.35	20.73	15.61	6.029	20.73	11.82	8.902	3.438	6.029	3.438	2.589	1.000
9	28.63	17.01	13.06	5.351	17.01	10.11	7.757	3.179	5.351	3.179	2.440	1.000
10	23.51	14.44	11.26	4.849	14.44	8.870	6.917	2.978	4.849	2.978	2.323	1.000
12	17.27	11.16	8.923	4.155	11.16	7.218	5.769	2.687	4.155	2.687	2.147	1.000
15	12.41	8.466	6.946	3.522	8.466	5.777	4.740	2.404	3.522	2.404	1.972	1.000
20	8.630	6.240	5.270	2.938	6.240	4.512	3.810	2.124	2.938	2.124	1.794	1.000
24	7.071	5.275	4.526	2.659	5.275	3.935	3.376	1.984	2.659	1.984	1.702	1.000
30	5.693	4.392	3.833	2.386	4.392	3.389	2.957	1.841	2.386	1.841	1.606	1.000
40	4.470	3.579	3.183	2.114	3.579	2.866	2.549	1.693	2.114	1.693	1.506	1.000
60	3.372	2.817	2.562	1.836	2.817	2.354	2.141	1.534	1.836	1.534	1.396	1.000
120	2.350	2.072	1.939	1.533	2.072	1.828	1.710	1.352	1.533	1.352	1.265	1.000
∞	1.000	1.000	1.000	1.000	1.000	1.000	1.000	1.000	1.000	1.000	1.000	1.000

Source: Reproduced from Table H of Owen L. Davies, *The Design and Analysis of Industrial Experiments,* 2nd ed., Oliver and Boyd, Edinburgh, 1956. By permission of the author and publishers.

Note: The entries in this table show the value of the ratio R of two population variances σ_2^2/σ_1^2 that will be undetected with probability β in a variance ratio test at the 100α percent significance level of the ratio s_2^2/s_1^2 of estimates of the two variances, both being based on ν degrees of freedom.

APPENDIX 13

CRITICAL VALUES OF *r* FOR THE SIGN TEST

n	1%	5%	10%	25%	n	1%	5%	10%	25%
1					46	13	15	16	18
2					47	14	16	17	19
3				0	48	14	16	17	19
4				0	49	15	17	18	19
5			0	0	50	15	17	18	20
6		0	0	1	51	15	18	19	20
7		0	0	1	52	16	18	19	21
8	0	0	1	1	53	16	18	20	21
9	0	1	1	2	54	17	19	20	22
10	0	1	1	2	55	17	19	20	22
11	0	1	2	3	56	17	20	21	23
12	1	2	2	3	57	18	20	21	23
13	1	2	3	3	58	18	21	22	24
14	1	2	3	4	59	19	21	22	24
15	2	3	3	4	60	19	21	23	25
16	2	3	4	5	61	20	22	23	25
17	2	4	4	5	62	20	22	24	25
18	3	4	5	6	63	20	23	24	26
19	3	4	5	6	64	21	23	24	26
20	3	5	5	6	65	21	24	25	27
21	4	5	6	7	66	22	24	25	27
22	4	5	6	7	67	22	25	26	28
23	4	6	7	8	68	22	25	26	28
24	5	6	7	8	69	23	25	27	29
25	5	7	7	9	70	23	26	27	29
26	6	7	8	9	71	24	26	28	30
27	6	7	8	10	72	24	27	28	30
28	6	8	9	10	73	25	27	28	31
29	7	8	9	10	74	25	28	29	31
30	7	9	10	11	75	25	28	29	32
31	7	9	10	11	76	26	28	30	32
32	8	9	10	12	77	26	29	30	32
33	8	10	11	12	78	27	29	31	33
34	9	10	11	13	79	27	30	31	33
35	9	11	12	13	80	28	30	32	34
36	9	11	12	14	81	28	31	32	34
37	10	12	13	14	82	28	31	33	35
38	10	12	13	14	83	29	32	33	35
39	11	12	13	15	84	29	32	33	36
40	11	13	14	15	85	30	32	34	36
41	11	13	14	16	86	30	33	34	37
42	12	14	15	16	87	31	33	35	37
43	12	14	15	17	88	31	34	35	38
44	13	15	16	17	89	31	34	36	38
45	13	15	16	18	90	32	35	36	39

Source: Reproduced from W. J. Dixon and F. J. Massey, Jr., *An Introduction to Statistical Analysis*, McGraw-Hill, New York, 1951, p. 324. By permission of the authors and publishers.

Notes: The values of *r* given in the table are the two-tailed percentage points for the binomial for $p = 0.5$.

For values of *n* larger than 90, approximate values of *r* may be found by taking the nearest integer less than $(n - 1)/2 - k\sqrt{n + 1}$, where *k* is 1.2879, 0.9800, 0.8224, 0.5752 for the 1, 5, 10, 25% values respectively.

CRITICAL VALUES OF *T* IN
THE WILCOXON SIGNED RANK TEST

	Level of Significance for One-Tailed Test		
	.025	.01	.005
	Level of Significance for Two-Tailed Test		
n	.05	.02	.01
6	0	—	—
7	2	0	—
8	4	2	0
9	6	3	2
10	8	5	3
11	11	7	5
12	14	10	7
13	17	13	10
14	21	16	13
15	25	20	16
16	30	24	20
17	35	28	23
18	40	33	28
19	46	38	32
20	52	43	38
21	59	49	43
22	66	56	49
23	73	62	55
24	81	69	61
25	89	77	68

Source: Adapted from Table I of F. Wilcoxon, *Some Rapid Approximate Statistical Procedures,* American Cyanamid Co., Stamford, Conn., 1949, p. 13. By permission of the author and publishers.

Notes: The values of *T* given in the table are critical values associated with selected values of *n*. Any value of *T* that is less than or equal to the tabulated value is significant at the indicated level of significance.

For $n > 25$, *T* is approximately normally distributed with mean $n(n + 1)/4$ and variance $n(n + 1)(2n + 1)/24$.

APPENDIX 15

CRITICAL VALUES OF r IN THE RUN TEST

TABLE 1

n_1 \ n_2	2	3	4	5	6	7	8	9	10	11	12	13	14	15	16	17	18	19	20
2											2	2	2	2	2	2	2	2	2
3					2	2	2	2	2	2	2	2	2	3	3	3	3	3	3
4				2	2	3	3	3	3	3	3	3	3	3	4	4	4	4	4
5			2	2	3	3	3	3	3	4	4	4	4	4	4	4	5	5	5
6		2	2	3	3	3	3	4	4	4	4	5	5	5	5	5	5	6	6
7		2	2	3	3	3	4	4	5	5	5	5	5	6	6	6	6	6	6
8		2	3	3	3	4	4	5	5	5	6	6	6	6	6	7	7	7	7
9		2	3	3	4	4	5	5	5	6	6	6	7	7	7	7	8	8	8
10		2	3	3	4	5	5	5	6	6	7	7	7	7	8	8	8	8	9
11		2	3	4	4	5	5	6	6	7	7	7	8	8	8	9	9	9	9
12	2	2	3	4	4	5	6	6	7	7	7	8	8	8	9	9	9	10	10
13	2	2	3	4	5	5	6	6	7	7	8	8	9	9	9	10	10	10	10
14	2	2	3	4	5	5	6	7	7	8	8	9	9	9	10	10	10	11	11
15	2	3	3	4	5	6	6	7	7	8	8	9	9	10	10	11	11	11	12
16	2	3	4	4	5	6	6	7	8	8	9	9	10	10	11	11	11	12	12
17	2	3	4	4	5	6	7	7	8	9	9	10	10	11	11	11	12	12	13
18	2	3	4	5	5	6	7	8	8	9	9	10	10	11	11	12	12	13	13
19	2	3	4	5	6	6	7	8	8	9	10	10	11	11	12	12	13	13	13
20	2	3	4	5	6	6	7	8	9	9	10	10	11	12	12	13	13	13	14

Source: Adapted from Frieda S. Swed and C. Eisenhart, Tables for testing randomness of grouping in a sequence of alternatives, *Ann. Math. Stat.* 14:83–86, 1943. By permission of the authors and publishers.

Note: The values of r given in Tables 1 and 2 are various critical values of r associated with selected values of n_1 and n_2. For the one-sample run test, any value of r that is equal to or less than the value shown in Table 1 or equal to or greater than the value shown in Table 2 is significant at the 5 percent level. For the two-sample run test, any value of r that is equal to or less than the value shown in Table 1 is significant at the 5 percent level.

TABLE 2

n_1 \ n_2	2	3	4	5	6	7	8	9	10	11	12	13	14	15	16	17	18	19	20
2																			
3																			
4				9	9														
5			9	10	10	11	11												
6			9	10	11	12	12	13	13	13	13								
7				11	12	13	13	14	14	14	14	15	15	15					
8				11	12	13	14	14	15	15	16	16	16	16	17	17	17	17	17
9					13	14	14	15	16	16	16	17	17	18	18	18	18	18	18
10					13	14	15	16	16	17	17	18	18	18	19	19	19	20	20
11					13	14	15	16	17	17	18	19	19	19	20	20	20	21	21
12					13	14	16	16	17	18	19	19	20	20	21	21	21	22	22
13						15	16	17	18	19	19	20	20	21	21	22	22	23	23
14						15	16	17	18	19	20	20	21	22	22	23	23	23	24
15						15	16	18	18	19	20	21	22	22	23	23	24	24	25
16							17	18	19	20	21	21	22	23	23	24	25	25	25
17							17	18	19	20	21	22	23	23	24	25	25	26	26
18							17	18	19	20	21	22	23	24	25	25	26	26	27
19							17	18	20	21	22	23	23	24	25	26	26	27	27
20							17	18	20	21	22	23	24	25	25	26	27	27	28

APPENDIX 16

CRITICAL VALUES OF D IN THE KOLMOGOROV-SMIRNOV GOODNESS OF FIT TEST

Sample Size (n)	Level of Significance for $D = $ Maximum $\lvert F(x) - S_n(x) \rvert$				
	.20	.15	.10	.05	.01
1	.900	.925	.950	.975	.995
2	.684	.726	.776	.842	.929
3	.565	.597	.642	.708	.828
4	.494	.525	.564	.624	.733
5	.446	.474	.510	.565	.669
6	.410	.436	.470	.521	.618
7	.381	.405	.438	.486	.577
8	.358	.381	.411	.457	.543
9	.339	.360	.388	.432	.514
10	.322	.342	.368	.410	.490
11	.307	.326	.352	.391	.468
12	.295	.313	.338	.375	.450
13	.284	.302	.325	.361	.433
14	.274	.292	.314	.349	.418
15	.266	.283	.304	.338	.404
16	.258	.274	.295	.328	.392
17	.250	.266	.286	.318	.381
18	.244	.259	.278	.309	.371
19	.237	.252	.272	.301	.363
20	.231	.246	.264	.294	.356
25	.21	.22	.24	.27	.32
30	.19	.20	.22	.24	.29
35	.18	.19	.21	.23	.27
Over 35	$\dfrac{1.07}{\sqrt{n}}$	$\dfrac{1.14}{\sqrt{n}}$	$\dfrac{1.22}{\sqrt{n}}$	$\dfrac{1.36}{\sqrt{n}}$	$\dfrac{1.63}{\sqrt{n}}$

Source: Adapted from F. J. Massey, Jr., The Kolmogorov-Smirnov test for goodness of fit, *J. Am. Stat. Assoc.* 46:68–78, 1951. By permission of the author and publishers.

Note: The values of D given in the table are critical values associated with selected values of n. Any value of D that is greater than or equal to the tabulated value is significant at the indicated level of significance.

PERCENTAGE POINTS OF PSEUDO *t* AND *F* STATISTICS

TABLE 1–Table for Testing the Significance of the Deviation of the Mean of a Small Sample (of Size *n*) from Some Preassigned Value

n	*p* 0.95	0.975	0.99	0.995	0.999	0.9995
2	3.196	6.353	15.910	31.828	159.16	318.31
3	0.885−	1.304	2.111	3.008	6.77	9.58
4	.529	0.717	1.023	1.316	2.29	2.85+
5	.388	.507	0.685+	0.843	1.32	1.58
6	0.312	0.399	0.523	0.628	0.92	1.07
7	.263	.333	.429	.507	.71	0.82
8	.230	.288	.366	.429	.59	.67
9	.205−	.255+	.322	.374	.50	.57
10	.186	.230	.288	.333	.44	.50
11	0.170	0.210	0.262	0.302	0.40	0.44
12	.158	.194	.241	.277	.36	.40
13	.147	.181	.224	.256	.33	.37
14	.138	.170	.209	.239	.31	.34
15	.131	.160	.197	.224	.29	.32
16	0.124	0.151	0.186	0.212	0.27	0.30
17	.118	.144	.177	.201	.26	.28
18	.113	.137	.168	.191	.24	.26
19	.108	.131	.161	.182	.23	.25+
20	.104	.126	.154	.175−	.22	.24

$$P\{\tau_1 < \text{value in table}\} = p$$

Sources: Tables 1 and 2 are reproduced with the permission of Professor E. S. Pearson from E. Lord, The use of the range in place of the standard deviation in the *t* test, *Biometrika* 34:66, 1947.

Table 3 is reproduced from J. E. Walsh, On the range-midrange test and some tests with bounded significance levels, *Ann. Math. Stat.* 20:257, 1949. By permission of the author and publishers.

Table 4 is reproduced from R. F. Link, On the ratio of two ranges, *Ann. Math. Stat.* 21:112, 1950. By permission of the author and publishers.

TABLE 2–Table for Testing the Significance of the Difference between the
Means of Two Small Samples of Equal Size n

n	p					
	0.95	0.975	0.99	0.995	0.999	0.9995
2	1.161	1.713	2.776	3.958	8.90	12.62
3	.487	.636	.857	1.046	1.63	2.09
4	.322	.406	.523	.618	.87	.99
5	.246	.306	.386	.448	.60	.67
6	.202	.249	.310	.357	.47	.51
7	.173	.213	.262	.300	.38	.42
8	.153	.186	.229	.260	.33	.36
9	.137	.167	.204	.232	.29	.32
10	.125	.152	.185	.209	.26	.29
11	.116	.140	.170	.192	.24	.26
12	.107	.130	.157	.177	.22	.24
13	.100	.121	.147	.165	.20	.22
14	.094	.114	.138	.155	.19	.21
15	.089	.108	.130	.146	.18	.19
16	.085	.102	.123	.139	.17	.18
17	.081	.097	.118	.132	.16	.17
18	.077	.093	.112	.126	.15	.17
19	.074	.089	.108	.121	.15	.16
20	.071	.086	.103	.116	.14	.15

$$P\{\tau_d < \text{value in table}\} = p$$

TABLE 3–Cumulative Percentage Points for
$$\tau_2 = [(X_{min}+X_{max})/2-\mu]/R$$

Sample Size	$p=.95$	$p=.975$	$p=.99$	$p=.995$
2	3.16	6.35	15.91	31.83
3	.90	1.30	2.11	3.02
4	.55	.74	1.04	1.37
5	.42	.52	.71	.85
6	.35	.43	.56	.66
7	.30	.37	.47	.55
8	.26	.33	.42	.47
9	.24	.30	.38	.42
10	.22	.27	.35	.39

$p=P\{\tau_2 < \text{value in the table}\}$

TABLE 4–Substitute F Ratio, Cumulative Percentage Points

$p=.95$

Sample Size for Denominator	Sample Size for Numerator								
	2	3	4	5	6	7	8	9	10
2	12.7	19.1	25	28	29	31	32	34	36
3	3.19	4.4	5.0	5.7	6.2	6.6	6.9	7.2	7.4
4	2.02	2.7	3.1	3.4	3.6	3.8	4.0	4.2	4.4
5	1.61	2.1	2.4	2.6	2.8	2.9	3.0	3.1	3.2
6	1.36	1.8	2.0	2.2	2.3	2.4	2.5	2.6	2.7
7	1.26	1.6	1.8	1.9	2.0	2.1	2.2	2.3	2.4
8	1.17	1.4	1.6	1.8	1.9	1.9	2.0	2.1	2.1
9	1.10	1.3	1.5	1.6	1.7	1.8	1.9	1.9	2.0
10	1.05	1.3	1.4	1.5	1.6	1.7	1.8	1.8	1.9

$p=.975$

Sample Size for Denominator	Sample Size for Numerator								
	2	3	4	5	6	7	8	9	10
2	25.5	38.2	52	57	60	62	64	67	68
3	4.61	6.3	7.3	8.0	8.7	9.3	9.8	10.2	10.5
4	2.72	3.5	4.0	4.4	4.7	5.0	5.2	5.4	5.6
5	2.01	2.6	2.9	3.2	3.4	3.6	3.7	3.8	3.9
6	1.67	2.1	2.4	2.6	2.8	2.9	3.0	3.1	3.2
7	1.48	1.9	2.1	2.3	2.4	2.5	2.6	2.7	2.8
8	1.36	1.7	1.9	2.0	2.2	2.3	2.3	2.4	2.5
9	1.27	1.6	1.8	1.9	2.0	2.1	2.1	2.2	2.3
10	1.21	1.5	1.6	1.8	1.9	1.9	2.0	2.0	2.1

$$p = .99$$

Sample Size for Denominator	Sample Size for Numerator								
	2	3	4	5	6	7	8	9	10
2	63.7	95	125	140	150	150	160	160	160
3	7.37	10	12	13	14	15	15	16	17
4	3.83	5.0	5.5	6.0	6.4	6.7	7.0	7.2	7.5
5	2.64	3.4	3.8	4.1	4.3	4.6	4.7	4.9	5.0
6	2.16	2.7	3.0	3.2	3.4	3.6	3.7	3.8	3.9
7	1.87	2.3	2.6	2.8	2.9	3.0	3.1	3.2	3.3
8	1.69	2.1	2.3	2.4	2.6	2.7	2.8	2.8	2.9
9	1.56	1.9	2.1	2.2	2.3	2.4	2.5	2.6	2.6
10	1.47	1.8	1.9	2.1	2.2	2.2	2.3	2.4	2.4

$$p = .995$$

Sample Size for Denominator	Sample Size for Numerator								
	2	3	4	5	6	7	8	9	10
2	127	191	230	250	260	270	280	290	290
3	10.4	14	17	18	20	21	22	23	25
4	4.85	6.1	7.0	7.6	8.1	8.5	8.8	9.3	9.6
5	3.36	4.1	4.6	4.9	5.2	5.5	5.7	5.9	6.1
6	2.67	3.1	3.5	3.8	4.0	4.1	4.3	4.5	4.6
7	2.28	2.7	2.9	3.1	3.3	3.5	3.6	3.7	3.8
8	2.03	2.3	2.6	2.7	2.9	3.0	3.1	3.2	3.3
9	1.87	2.1	2.4	2.5	2.6	2.7	2.8	2.9	3.0
10	1.75	2.0	2.2	2.3	2.4	2.5	2.6	2.6	2.7

$$p = P\{R_1/R_2 < \text{value in the table}\}$$

GREEK ALPHABET

A	α	alpha	N	ν	nu
B	β	beta	Ξ	ξ	xi
Γ	γ	gamma	O	o	omicron
Δ	δ	delta	Π	π	pi
E	ϵ	epsilon	P	ρ	rho
Z	ζ	zeta	Σ	σ	sigma
H	η	eta	T	τ	tau
Θ	θ	theta	Υ	υ	upsilon
I	ι	iota	Φ	ϕ	phi
K	κ	kappa	X	χ	chi
Λ	λ	lambda	Ψ	ψ	psi
M	μ	mu	Ω	ω	omega

INDEX